Essentials of Environmental Health:
A Public Health Perspective

Essentials of Environmental Health: A Public Health Perspective

Editor: Olivia Cunningham

Callisto Reference
www.callistoreference.com

Callisto Reference,
118-35 Queens Blvd., Suite 400,
Forest Hills, NY 11375, USA

Visit us on the World Wide Web at:
www.callistoreference.com

ISBN: 978-1-64116-610-2 (Hardback)

Cataloging-in-Publication Data

Essentials of environmental health : a public health perspective / edited by Olivia Cunningham.
p. cm.
Includes bibliographical references and index.
ISBN 978-1-64116-610-2
1. Environmental health. 2. Public health. 3. Health risk assessment.
4. Environmental engineering. I. Cunningham, Olivia.

RA565 .E77 2022
616.98--dc23

Table of Contents

Preface

The branch of public health that delves into an analysis of the different aspects of the natural and built environment to determine how they affect human health is termed as environmental health. The assessment and control of all environmental factors which have negative impact on human health, as well as the creation of health-conducive environments and prevention of disease, are under the purview of this field. Some of the health concerns addressed within this field are air quality, climate change, food safety, waste disposal, biosafety, hazardous materials management, etc. Such studies are collectively approached from the sub-domains of environmental science, epidemiology, environmental and occupational medicine, and toxicology. The book studies, analyzes and upholds the pillars of environmental health and its utmost significance in modern times. It picks up individual branches and explains their need and contribution in the context of this area of study. This book will prove to be immensely beneficial to students and researchers in this field.

This book is a comprehensive compilation of works of different researchers from varied parts of the world. It includes valuable experiences of the researchers with the sole objective of providing the readers (learners) with a proper knowledge of the concerned field. This book will be beneficial in evoking inspiration and enhancing the knowledge of the interested readers.

In the end, I would like to extend my heartiest thanks to the authors who worked with great determination on their chapters. I also appreciate the publisher's support in the course of the book. I would also like to deeply acknowledge my family who stood by me as a source of inspiration during the project.

Editor

Study on the adsorption isotherms of chromium (VI) by means of carbon nano tubes from aqueous solutions

Ali Asghar Najafpoor[1], Ghorban Soleimani[2], Mohammad Hasan Ehrampoush[3], Mohammad Taghi Ghanelan[4], Elham Rahmanpour Salmani[5*], Maryam Dolatabadi Takabi[5]

[1]Associate Professor, Health Sciences Research Center, Department of Environmental Health Engineering, School of Health, Mashhad University of Medical Sciences, Mashhad, Iran

[2]MS of Environmental Health Engineering, Shahid Sadoughi University of Medical Sciences, Yazd, Iran

[3]Professor, Environmental Health Engineering Department, Shahid Sadoughi University of Medical Sciences, Yazd, Iran

[4]Associate Professor, Environmental Health Engineering Department, Shahid Sadoughi University of Medical Sciences, Yazd, Iran

[5]MSc student of Environmental Health Engineering, Student Research Committee, Department of Environmental Health, School of Health, Mashhad University of Medical Sciences, Mashhad, Iran

Abstract

Background: The presence of heavy metals in the environment especially in water supplies have caused many concerns because of their toxicity and non-degradability. Hexavalent Chromium (Cr) is one of the most toxic metals which is used in many industries, so it is essential to remove it from industrial wastewater. In this study, we made a comparison between different adsorption isotherms in the chromium (VI) removal process using carbon nano tubes from aqueous solutions.

Methods: This experimental study used atomic absorption spectrophotometry. To determine the adsorption isotherms, a synthetic sample with defined concentration of Cr (VI) was prepared and different doses of adsorbent were added to it. The effect of initial Cr concentration, pH, adsorbent dose, and reaction time on removal of Cr was investigated. Temperature and mix rate were steady during a defined time. At the end, Cr (VI) concentration measured and adsorbents equality capacities were calculated via formulas and graphs. Data analysis were performed using descriptive statistics.

Results: Adsorption capacities (q_e) increased with increasing of initial Cr concentration, and reaction time decreased with increasing adsorbent dose and pH. Correlation coefficients for Langmuir, and Freundlich isotherms in oxidized Multi-Walled Carbon Nano Tubes (MWCNTs) were 0.93, 0.874, and 0.714 and in oxidized Single-Walled Carbon Nano Tubes (SWCNTs) were 0.904, 0.868, and 0.711 respectively.

Conclusion: Chromium ions adsorption in carbon nano tubes is accordant to Langmuir isotherm model, and MWCNTs have more cc than SWCNTs. Carbon nano tubes are effective adsorbents in removal of Cr (VI) from aqueous solutions.

Keywords: Adsorption isotherm, Carbon nano tubes, Hexavalent chromium

***Correspondence to:**
Elham Rahmanpour Salmani
Email: Rahmanpoure921@mums.ac.ir

Introduction

Waters have been polluted with heavy metals due to rapid growth of industries. Discharging wastewater from various industries like electro plating, leather tanning, glass, acid mine drainage, and ceramic enters hexavalent Chromium (Cr) which is a very toxic and harmful metal in to water supplies (1). Cr does not exist in water naturally (2). There are multiple states of Cr including (III) and (VI) forms as the most common (3,4). Cr (VI) is a toxic metal, whereas Cr (III) is almost non-toxic (5). Researches on human and animals revealed the carcinogenic effects of Cr (VI) (6). United States Environmental Protection Agency (USEPA) listed Cr (VI) as one of the most critical pollutants. Based on World Health Organization (WHO) recommendations, 0.05 ppm of Cr in water was assumed as the maximum permissible level (7). The acceptable amount of Cr based on Institute of Standards and Industrial Research of Iran (ISIRI) directions is 0.05 (8). Cr (VI) in concentrations more than 50 has irreversible effects on kidney and liver, can cause skin and lung cancer, and also can bring allergic problems (9). As Cr (VI) is a rare and expensive metal, its recovery is important (10). There are several methods for chromium removal, such as chemical precipitation, ion exchange resin, reverse osmosis and electro dialysis, although a favorable removal does not occur in any of the (11). Common removal processes, such as chemical precipitation has low performance, and advanced methods are costly and need expert operators (12). Another removal method is adsorption defined as an effective way. It does not need special equipment, ions recovery can be done easily, and adsorbents reuse is possible (13). One of the most effective adsorbents that can be used in this process is active carbon, but its preparation and revival has a high expense (14). In recent years, the use of nano technology especially carbon nano tubes for environmental pollutants removal has been developed (15). Nano filtration is one of the most important applications in nano technology scope. Studies in China showed reduction in cardiovascular diseases and cancer after long-term consumption of water treated by nano filters (16). In this article, we focused on the adsorption isotherms of Cr as one of the secondary purposes in our study on the Cr removal process by use of carbon nano tubes from aqueous solutions. So we do not report our study results about the effect of initial Cr concentration, pH, adsorbent dose and reaction time on removal of chromium by carbon nano tubes here.

Methods

In this experimental study, we used chemical materials, produced by German Merck Company. Oxidized carbon nano tubes were purchased from Research Institute of Petroleum Industry (RIPI). The unknown Cr (VI) concentration measured by atomic absorption T80+ UV/VIS spectrophotometer. To have an adequate mixing, rotomix shaker made by Iranian Behdad Company was applied. Three different concentrations of Cr (0.5, 1, and 2) were prepared and the effect of three different pH (3, 5, and 7) and three different masses of carbon nano tubes (0.4, 0.8, and 1) during the adsorption process were investigated. Because of the low removal percentage in pH=7, tests continued in pH= 3 and 5. To find the effect of contact time, three different times (30, 60, and 120 minutes) were selected. A discontinuous method was used to determine the pollutant adsorption isotherm. First, a synthetic sample with a defined Cr (VI) concentration was prepared. We solved $K_2Cr_2O_7$ in deionized water to prepare Cr (VI) solution. After that, different doses of adsorbent were added to sample. This container mixed for 24 hours in a condition that temperature and mix rate were steady at 20 and 120 rpm, respectively. A vacuum pump and a 0.45 μ filter were used to separate the sample from adsorbent. Then Cr concentration in liquid phase was measured, and adsorbents equality capacities were calculated by use of formulas and graphs.

Langmuir equation:

$Qe = \theta.b.Ce/1+(b.Ce)$

Ce: It shows the concentration in equality conditions (mg/l)

Qe: It introduces the amount of adsorbate on one unit of adsorbent weight in equality conditions (mg/g)

θ & b: These are Langmuir constants related to maximum adsorption and adsorption energy, respectively.

Freundlich equation:

$Qe=K.Ce.^{1/n}$

K: It is an index for adsorption capacity

1/n: It demonstrates the adsorption intensity.

Data analysis were performed using descriptive statistics.

Results

Figures 1A and 1B show Langmuir adsorption isotherm for Cr (VI) ion. As seen in the figures, Cr (VI) adsorption by carbon nano tubes is adjusted to Langmuir isotherm. With respect to other figures, Cr (VI) adsorption does not follow Freundlich isotherm (Figures 1C and 1D). Constants of isotherms are placed in Table 1. As shown in

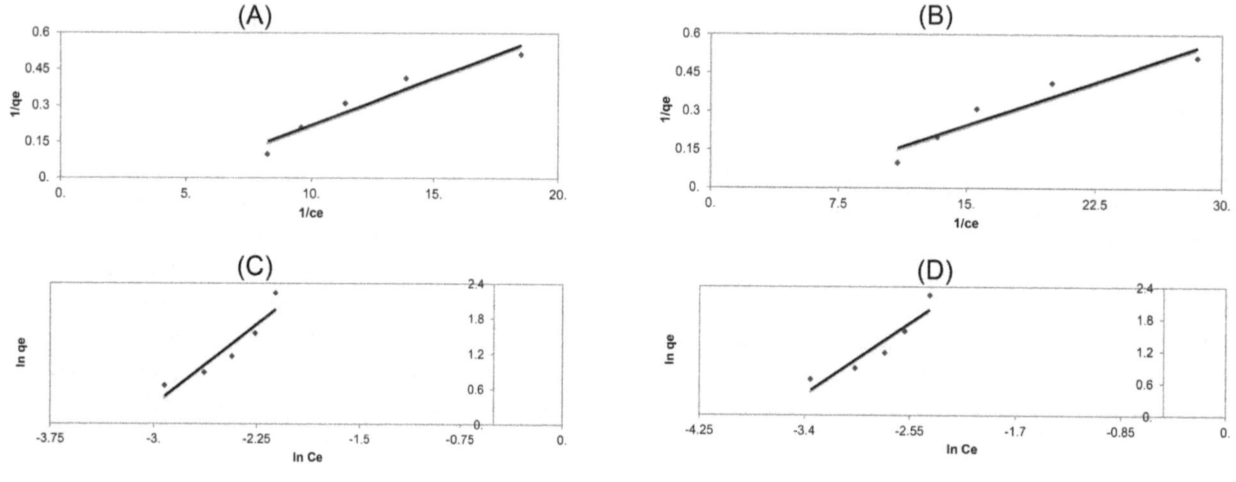

Figure 1. A) Langmuir isotherm for Cr adsorption by different masses of MWCNTs; pH= 3; pollutant= 2 mg/l. B) Langmuir isotherm for Cr adsorption by different masses of SWCNTs; pH= 3; pollutant= 2 mg/l. C) Freundlich isotherm for Cr adsorption by different masses of MWCNTs; pH= 3; pollutant= 2 mg/l. D) Freundlich isotherm for Cr adsorption by different masses of SWCNTs; pH=3; pollutant= 2 mg/l

Table 1. Constants of isotherms

Type of adsorbent	Langmuir isotherm			Freundlich isotherm		
	q_m	b	R_2	n	k_f	R_2
SWCNTs	44.64	0.251	0.9045	0.174	4.77	0.8686
MWCNTs	25.64	0.228	0.9309	0.171	6.27	0.8747

SWCNTs= Single-Walled Carbon Nano Tubes; MWCNTs= Multi-Walled Carbon Nano Tubes

Table 2. The effect of adsorbent dose on Cr (VI) remained concentration (C_e), adsorption capacity (q_e), and removal percentage (r_e) by use of carbon nano tubes in time= 120 minutes, pH= 3, and initial Cr concentration= 2

Adsorbent dose (g/l)	Adsorbent types					
	MWCNTs			SWCNTs		
	C_e	R_e	q_e	C_e	R_e	q_e
0.4	0.501	74.95	3.74	0.422	78.9	3.94
0.8	0.412	79.4	1.98	0.334	83.3	2.08
1	0.351	82.45	1.65	0.222	88.9	1.778

SWCNTs= Single-Walled Carbon Nano Tubes; MWCNTs= Multi-Walled Carbon Nano Tubes

Table 2, in the solution of 2 Cr with an increase in adsorption mass, removal percentage (r_e) increased, and adsorption capacity (q_e) decreased. As you can see, adsorption capacities for 0.4 and 1 of adsorption masses, for Single-Walled Carbon Nano Tubes (SWCNTs) in time= 120 minutes and in pH=3 are 3.94 and 1.78 respectively and in the same condition for Multi-Walled Carbon Nano Tubes (MWCNTs) are 3.74, and 1.65 respectively. Table 3 shows the effect of initial Cr concentration on adsorption by carbon nano tubes. According to Table 3, when Cr concentration increased from 0.5 to 2, removal efficiency decreased and adsorption capacity increased. According to the results on Table 4, there is a direct relation between contact time and adsorption capacity and between contact time and removal percentage. The amount of q_e and r_e were calculated to gain a better understanding of the role of pH data earned from analyzed experiments in a defined time and initial Cr concentration. These findings are presented in Table 5.

Discussion

The relationship between adsorbed Cr (VI) and its concentration in solution under equilibrium conditions is described by isotherm models. In one study, bone char used as an adsorbent for Cr (VI) adsorption. The researchers found that adsorption process is done better in acidic pH and follows Langmuir (=0.98) and Freundlich (=0.99) isotherm models (17). Cr (VI) removal by using sulfonated lignite showed maximum adsorption in low pH, and data were well analyzed with Langmuir isotherm (18). Various adsorbents for Cr removal were applied in the acidic conditions which showed the predominant

Table 3. The effect of initial Cr concentration on q_e, r_e, and C_e in adsorbent dose= 1 g/l, time= 120 minutes, pH= 3

Initial Cr concentration mg/l	Adsorbent types					
	MWCNTs			SWCNTs		
	C_e	R_e	q_e	C_e	R_e	q_e
0.5	0.023	95.4	0.477	0.009	98.2	0.49
1	0.113	88.7	0.887	0.071	92.9	0.93
2	0.351	82.5	1.65	0.222	88.9	1.778

SWCNTs= Single-Walled Carbon Nano Tubes; MWCNTs= Multi-Walled Carbon Nano Tubes

Table 4. The effect of contact time on q_e, r_e, and C_e in adsorbent dose= 0.4 g/l, pH= 3, and initial Cr concentration= 2 mg/l

Time (min)	Adsorbent types					
	MWCNTs			SWCNTs		
	C_e	R_e	q_e	C_e	R_e	q_e
0	2	0	0	2	0	0
30	0.961	51.95	2.6	0.796	60.2	3.01
60	0.797	60.15	3.01	0.591	70.45	3.52
120	0.501	74.95	3.74	0.422	78.9	3.94

SWCNTs= Single-Walled Carbon Nano Tubes; MWCNTs= Multi-Walled Carbon Nano Tubes

Table 5. The effect of pH on q_e, r_e, and C_e in time= 30 minutes and Cr concentration = 2 mg/l

Adsorbent dose (g/l)	pH	Adsorbent types					
		MWCNTs			SWCNTs		
		C_e	R_e	q_e	C_e	R_e	q_e
0.4	5	1.176	41.2	2.06	0.982	50.9	2.54
	3	0.961	51.95	2.6	0.796	60.2	3.01
0.8	5	1.057	47.15	1.18	0.843	57.85	1.44
	3	0.776	61.2	1.53	0.591	70.45	1.76
1	5	0.883	55.85	1.12	0.664	66.8	1.33
	3	0.625	68.75	1.37	0.503	74.85	1.49

SWCNTs= Single-Walled Carbon Nano Tubes; MWCNTs= Multi-Walled Carbon Nano Tubes

sorption for Cr (VI) species. Isotherm data fitted well for both Langmuir and Freundlich models at the lower initial Cr concentration, although the Langmuir isotherm was determined to be the most suitable model over the entire range (19). In our study for Freundlich isotherm, Correlation coefficient with adsorption data is less than 0.9, so adsorption process is not similar to this isotherm. x^2= 0.796 and 0.0025 were gained for Freundlich model and Langmuir model respectively. Because value of x^2 is very low for Langmuir model, it adjusts to adsorption process. In this study, the efficiency of carbon nano tubes as a new adsorbent in removal of Cr (VI) was examined. The results from the study of Hu *et al* (20), and Moradi *et al* (21) on Cr and other heavy metals removal by means of carbon nano tubes have shown that these adsorbents are effective for the removal of heavy metals especially Cr (VI). In the study of Hu *et al* (20), the effect of initial Cr concentration, pH, and temperature on Cr removal were investigated. Their study results showed decreasing removal percentage after pH increase and increasing removal percentage after temperature increase. In our study, removal percentage decreased after Cr concentration increase from 0.5 to 2 mg/l. To cite the cause, we can say that adsorbents have a limited number of active areas, so active areas become saturated quickly by pollutant and the efficiency of removal process goes down in high concentrations. In Ghanizadeh *et al* (17) study, the efficiency of removal process also came down by increasing initial pollutant concentration. In Hu *et al* study (20), you can see a diverse relationship between pH and removal percentage like our study. A good reason for that can be $HCrO_4$ and CrO_4 ions production under these conditions which facilitate adsorption process.

Conclusion

Based on findings adsorption capacities (qe) increased with increasing of initial Cr (VI) concentrations and reaction time and decreased when adsorbent dose and pH increased. According to the results Langmuir isotherm model can better define this adsorption process. Carbon nano tubes are proper adsorbents for the Cr (VI) ion removal from aqueous solutions and MWCNTs have better efficiency than SWCNTs.

Competing interests

The authors have declared that no competing interest exists.

Authors' contributions

All authors contributed equally and all authors participated in the data acquisition, analysis and interpretation. All authors critically reviewed, refined and approved the manuscript.

References

1. Poonkuzhali K, Rajeswari V, Saravanakumar T, Viswanathamurthi P, Park SM, Govarthanan M, et al. Reduction of hexavalent chromium using aerva lanata L.: elucidation of reduction mechanism and identification of active principles. J Hazard Mater 2014; 272: 89-95.
2. Samani MR, Borghai SM, Olad A, Chaichi MJ. Adsorption of chromium from aqueous solution using polyaniline. Water & Wastewater 2011; 22(3): 2-9. [In Persian]
3. Tsopmo A, Gao Q, Baakdah MM. Reduction of hexavalent chromium by digested oat bran proteins. Food Chem 2014; 153: 171-6.
4. Kamakura N, Inui T, Kitano M, Nakamura T. Determination of Chromium(III), Chromium(VI), and Chromium(III) acetylacetonate in water by ion-exchange disk extraction/metal furnace atomic absorption spectrometry. Spectrochimica Acta Part B: Atomic Spectroscopy 2014; 93: 28-33.
5. Şahan S, Saçmacı Ş, Kartal Ş, Saçmacı M, Şahin U, Ülgen A. Development of a new on-line system for the sequential speciation and determination of chromium species in various samples using a

combination of chelating and ion exchange resins. Talanta 2014; 120: 391-7.

6. Haney JT Jr, Erraguntla N, Sielken RL Jr, Valdez-Flores C. Development of an inhalation unit risk factor for hexavalent chromium. Regul Toxicol Pharmacol 2014; 68(2): 201-11.
7. Bertagnolli C, da Silva MGC, Guibal E. Chromium biosorption using the residue of alginate extraction from Sargassum filipendula. Chemical Engineering Journal 2014; 237: 362-71.
8. Institute of Standards and Industrial Research of Iran (ISIRI). Quality standards of drinking water. Institute of Standards & Industrial Research of Iran; 1997.
9. American Public Health Association (APHA), American Water Works Association (AWWA), Water Environment Federation (WEF). Standard Methods For Examination of Water And Wastewater. Washington, DC: American Public Health Association; 2005.
10. Gupta VK, Rastogi A. Sorption and desorption studies of chromium(VI) from nonviable cyanobacterium Nostoc muscorum biomass. J Hazard Mater 2008; 154(1-3): 347-54.
11. Bhattacharya M, Dutta SK, Sikder J, Mandal MK. Computational and experimental study of chromium (VI) removal in direct contact membrane distillation. Journal of Membrane Science 2014; 450: 447-56.
12. Shirzad Siboni M, Azizian S, Maleki A, Zarrabi M. Removal of chromium by using of adsorption onto strong base anion resin: study of equilibrium and kinetic. Water and Wastewater 2011; 22(3): 10-18. [In Persian]
13. Sun X, Yang L, Li Q, Zhao J, Li X, Wang X, *et al.* Amino-functionalized magnetic cellulose nanocomposite as adsorbent for removal of Cr (VI): synthesis and adsorption studies. Chemical Engineering Journal 2014; 241: 175-83.
14. Selvi K, Pattabhi S, Kadirvelu K. Removal of Cr (VI) from aqueous solution by adsorption onto activated carbon. Bioresour Technol 2001; 80(1): 87-9.
15. Gupta VK, Agarwal S, Saleh TA. Chromium removal by combining the magnetic properties of iron oxide with adsorption properties of carbon nanotubes. Water Res 2011; 45(6): 2207-12.
16. Delnavaz M, Fartous S, Aiati B. Study on the new technologies of nano particles in environmental engineering, water and wastewater. Future studies, technology and development prospect convention, Tehran: Amir Kabir Industrial University; 2006. [In Persian]
17. Ghanizadeh GH, Ghaneian MT, Asgari GH, Shahbazi F, Taheri Z. Removal of excess Cr (VI) from water by adsorption on to bone char. 13th National Conference on Environmental Health. 2010 Nov 2-4; Kerman, Iran; 2010. [In Persian]
18. Zhang R, Wang B, Ma H. Studies on Chromium (VI) adsorption on solfonated lignite. desalination 2010; 255 (1-3): 61-6.
19. Jung C, Heo J, Han J, Her N, Lee SJ, Oh J, *et al.* Hexavalent chromium removal by various adsorbents: powdered activated carbon, chitosan, and single/multi-walled carbon nanotubes. Separation and Purification Technology 2013; 106, 63-71.
20. Hu J, Chen C, Zhu X, Wang X. Removal of chromium from aqueous solution by using oxidized multiwalled carbon nanotubes. J Hazard Mater 2009; 162(2-3): 1542-50.
21. Moradi O, Zare K, Yari M. Interaction of some heavy metal ions with single walled carbon nanotube. Int J Nano Dim 2011; 1(3): 203-20.

2

Kinetic study of heavy metal ions removal from aqueous solutions using activated pumice stone

Naseh Babakhani[1], Mahdi Reyahi-Khoram[2*], Soheil Sobhanardakani[3]

[1]MSc Student, Department of Environment, Faculty of the Basic Knowledge, Hamedan Branch, Islamic Azad University, Hamedan, Iran
[2]Assistant Professor, Department of Environment, Faculty of the Basic Knowledge, Hamedan Branch, Islamic Azad University, Hamedan, Iran
[3]Associate Professor, Department of Environment, Faculty of the Basic Knowledge, Hamedan Branch, Islamic Azad University, Hamedan, Iran

Abstract
Background: The presence of heavy metals in aquatic and terrestrial ecosystems causes harmful effects to living organisms in the environment. This research aimed to determine the potential of activated pumice stone (APS) as a sorbent for the removal of Cd, Cu, and Zn from wastewater.
Methods: This research was conducted during 2013 on a laboratory scale. The study was performed using batch experiments with synthetic wastewater having Cd, Cu, and Zn concentrations of 10, 20, 40, and 80 mg/L. Various isotherm models, including Langmuir and Freundlich, were used to evaluate the sorption data. The influence of contact time and amount of sorbent on the removal of Cd, Cu, and Zn from wastewater was studied. All experiments were done at pH = 7 and at room temperature (20 ± 1°C). The solution pH was adjusted using 1N NaOH or 1N HCl solutions, and the pH value was determined by a pH meter.
Results: The results showed that the adsorption of Cd, Cu, and Zn approaches equilibrium after about 2 hours, while the rates of removal efficiency for Cd, Cu, and Zn at equilibrium were 91.6%, 91.8%, and 82.9%, respectively. Kinetic studies showed that the sorption of Cd, Cu, and Zn onto APS were fitted to the pseudo-second order kinetic models.
Conclusion: The results indicate that the APS is a good choice, because it is a low-cost and effective sorbent. The sorption capacity of APS as a sorbent was considerably affected by the initial concentration of metal ions in the solution and by contact time.
Keywords: Adsorption, Aqueous solution, Environment, Heavy metals, Pumice.

***Correspondence to:**
Mahdi Reyahi-Khoram
Email: phdmrk@gmail.com

Introduction

Heavy metal ions such as zinc, lead, mercury, chromium, nickel, copper, and cadmium are dangerous materials, non-biodegradable, and can be toxic and carcinogenic, even at very low levels. Hence, they usually cause a serious threat to the environment (1). Heavy metals are generated from various industries, including but not exclusive to metal plastering, paint manufacturing, textile, steel fabrication, storage batteries alloy industries, fertilizers, dyeing, and other chemical industries (2). Cadmium is extensively used and highly toxic in relatively low doses. It is one of the most toxic heavy metals causing renal disorder, kidney damage, high blood pressure, destruction of red blood cells, and bone fracture. Human beings have reported nausea and vomiting at a cadmium level of 15 mg/L (3). Copper is known to be one of the heavy metals most toxic to living organisms, and it is one of the more widespread heavy metal contaminants of the environment. The main sources of copper in industry include plating baths, metal cleaning, pulp, paper board mills, wood pulp production, fertilizer industry, paints and pigment (4).

Zinc is one of the most commonly occurring elements in the earth's crust and an essential micronutrient for all living things. Zn has a multitude of industrial uses; the major industrial use of Zn is as a corrosion resistant coating for Fe and steel. It is used in dry cell batteries, mixed with other metals to form alloys, and used in the manufacture of products such as ceramics, paints, rubber, wood preservatives, dyes and fertilizers, dietary foods, drugs, and other medicinal preparations. Exposure to Zn in large concentrations is highly toxic to living organisms. In humans, it can cause a range of serious ailments including damage to the pancreas, decreased immune functions, growth retardation, anemia, metal fume fever, birth defects, and cancer (5).

Common methods for the removal of heavy metals from water environments include chemical precipitation coagulation, oxidation, reduction, solvent extraction, electrolytic processes, membrane separation, ion exchange, reverse osmosis, membrane filtration, ultra-filtration, biological systems, and adsorption. Among these methods, adsorption is most highly recommended because it is effective and economical (1,2,4). The adsorption process has been identified as an efficient, cheap, and promising technology in wastewater treatment for the removal of metal ions (6). Several authors have reported on various adsorbents. A number of adsorbent materials, such as zeolites, activated carbon, tea factory waste, amorphous iron oxide, kaolinite, phenolated wood resin, modified cellulose, mesoporous silica, egg shell, hydrogle, protein granules produced, cationic resin Purolite, nano-materials, bentonite, and modified jute, have been used to remove heavy metals from wastewaters (7-13).

Pumice is a porous igneous rock; it is formed during massive volcanic eruptions when fluid lava is emitted into the air as a set containing mass of gas bubbles (14). Pumice has been used in ceramic, cement, brick, and concrete industries as an additive and coating material (15).

In the last two decades, many researchers have used pumice for the removal of fluoride (16), nitrates and nitrites (17), disinfection by-products (18), phosphorous ions (19), organic matter (20), pathogens (21), arsenic (22), azo dye (22), and p-chloronitrobenzene (23). In Iran, many studies have focused on the use of pumice for the adsorption of metal ions and other solutes (14,16,19,20), but not much research related to activated pumice stone (APS) absorption has been done. Access to sources of pumice, its lower costs, its availability in Iran, and innovation are the motivations behind this research, which aimed to determine the adsorption capacity of APS for removing (Zn), (Cu), and (Zn)ions from aqueous solutions through a batch process.

Methods

This study was carried out in the laboratory of Kermanshah Provincial Directorate of Environmental Protection during 2013. First, a pre-test was performed to identify possible issues and problems for this study and to determine optimum pH. Based on the pre-test results, the optimum pH was found to be pH 7.

Preparation of sorbent

The sorbent used in this study was made in the lab from natural sources. Pumice stone was obtained from Ghorveh township of Kurdistan province in the western part of Iran. To remove contaminants, the pumice was washed thoroughly with double-distilled water until the wash water ran free from color and turbidity. Subsequently, the prepared sorbent was kept in an oven at 110 ± 1°C for 6 hours and was then transferred to a furnace and held at a temperature of 550°C for one hour. In the next stage, the activated sample was crushed with laboratory-scale crushers and granulated using standard ASTM sieves with a mesh size of 20 (0.85 mm) and an effective size of 2 mm. Finally, the sorbent was rinsed with distilled water and kept at 110 ± 1°C for 14 hours until it was completely dried.

Preparation of chemical agents

All employed chemicals were of the purest grade. Cd, Cu, and Zn in granular form were provided by Merck Co. (Germany).The stock solutions (1000 mg/L) of each element were prepared by dissolving an adequate quantity of Cd, Cu, and Zn powder in double-distilled water. Then, different working concentrations (10, 20, 40, and 80 mg/L) for each metal were prepared by sufficiently diluting stock solutions.

Sorption experiments

This research was performed on a laboratory scale. Erlenmeyer flasks with 250 mL were used as a simple batch reactor; initial concentrations of Cd, Cu, and Zn solutions were selected at four different values (10, 20, 40, and 80 mg/L), and 2, 4, 6, and 8 g of APS sorbent were added to the batch reactor.

Erlenmeyer flasks were filled with 200 mL of solution. All experiments were done at pH = 7 and at room temperature (20 ± 1°C). The solution pH was adjusted using 1N NaOH or 1N HCl solutions; pH values were determined by a pH meter. The mixture was agitated on a shaker at a speed of 200 rpm. In order to ensure reliability and reproducibility, all experiments were performed in duplicate, and the results are reported as averages. Sorption studies were also carried out at different contact times (10, 20, 40, 80, 120, and 180 minutes) to determine the effect of contact time on the removal of heavy metals (Cd, Zn, and Cu) from aqueous solutions.

After each sorption experiment, the mixture was centrifuged at 200 rpm for 5 minutes and the supernatant fluid was decanted. Then, the final concentrations of metal ions in the samples were determined by ICP (Varian ES-710) in ppb.

All analyses were performed according to the procedures outlined in standard methods (24). Excel software was used to input, analyze, and graphically display the results. The effects of contact time, initial metal ion concentration, and sorbent dosage on sorption efficiency were investigated. Pseudo-first order and pseudo-second order kinetic models were used to represent the kinetics or rate of the process and obtain the main kinetic parameters. Common linear isotherms, such as the Langmuir and Freundlich isotherms, are usually used by researchers to study the adsorption mechanism (7-10,22); here, they were assayed for their ability to fit the sorption data. The quantity of metal ions sorbed per gram of sorbent was fulfilled according to the equation below (25):

$$q_t = \frac{(c_1 - c_t)}{m}.v \qquad (1)$$

where, q_t is the mass of metal ion sorbed as mg metal/g of the sorbent at time t; c_1 is the initial concentration of metal in the solution in mg/L; c_t is the equilibrium concentration of the metal in the solution; m is the amount of the sorbent sample on a dry basis in g, and v, the liquid volume in L. The proportion of metal ions decreased from the aqueous solution was determined using the following equation (25):

$$R = \frac{(c_1 - c_t)}{C_1}.100 \qquad (2)$$

Results

Effect of contact time

Contact time is an important factor in evaluating sorption efficiency. The removal rate of Cd, Cu, and Zn from solution is shown in Figure 1, from which it can be seen that the adsorption of Cd, Cu, and Zn approached equilibrium after about two hours, while removal efficiency rates for Cd, Cu, and Zn at equilibrium were 91.6%, 91.8%, and 82.9%, respectively. Initially, the uptake rate was higher, because all sites on the APS were vacant and the concentration was high; however, the decrease in sorption sites reduced the uptake rate. As shown in Figure 1, considering technical and economic aspects, a contact time of 120 minutes was chosen for Cd, Cu, and Zn removal from aqueous solutions by sorption onto APS.

Effect of the sorbent dosage

Adsorption capacity is highly dependent on the amount or concentration of the sorbent. The effect of sorbent dosage

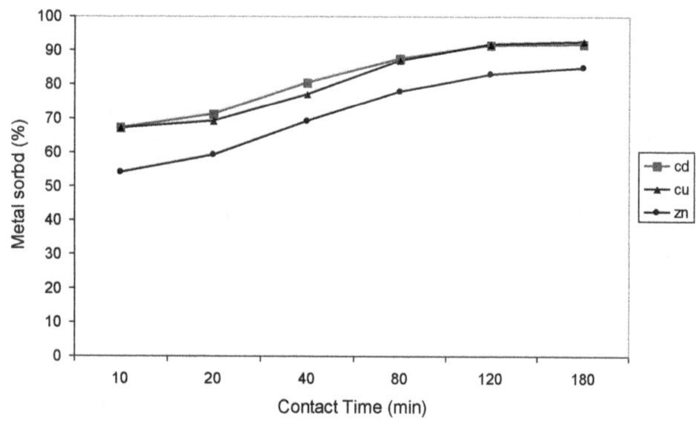

Figure 1. Effect of contact time on the sorption of Cd, Cu, and Zn onto APS. Conditions: 10 g/L APS, pH 7, 200 mL 10 mg/L metal ions solution, temperature = 20 ± 1°C, agitation speed = 200 rpm.

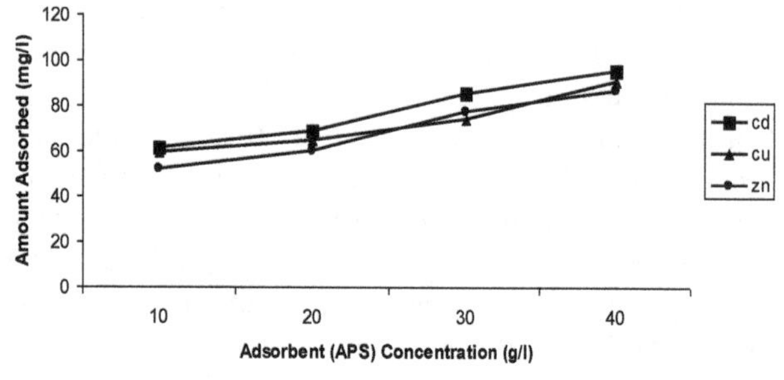

Figure 2. Effect of sorbent concentration on the sorption of Cd, Cu, and Zn onto APS. Conditions: pH = 7, 200 mL 40 mg/L metal ions solution, contact time = 120 min, temperature = 20 ± 1°C, agitation speed = 200 rpm.

on the removal of Cd, Cu, and Zn from aqueous solutions was studied using batch sorption experiments. Experiments were conducted by contacting known amounts of the adsorbent with sorbent concentrations varying from 10 to 40 g/L. The results are presented in Figure 2. It is apparent that the amount of metal ions sorbed increased with increasing concentrations of sorbent. This is an expected result, because as the amount of sorbent increased, the available surface area increased, thereby exposing more active sites for the binding of metal ions. A similar trend was observed for the effect of sorbent concentration and is discussed in the literature (26).

Sorption isotherm studies

Several mathematical relationships have been developed to describe the equilibrium distribution of solute between the solid and liquid phases and thus aid in the interpretation of adsorption data. These relationships apply when sorption tests are conducted at a constant temperature and are referred to as sorption isotherms (27). Analytical isotherm equations, such as Langmuir and Freundlich isotherms, are widely used for modeling sorption data (28). In this research, sorption isotherms were obtained from the previous data using four different concentrations of APS and under constant pH conditions for the solution (pH=7).

Langmuir equation

The Langmuir equation is generally written as (27):

$$\frac{1}{q_e} = \frac{1}{q_m} + \frac{1}{b.q_m}(\frac{1}{c_e}) \qquad (3)$$

where, C_e is the equilibrium concentration of sorbent (mg/L), q_m is the maximum sorption capacity of the adsorbent (mg/g), b is the Langmuir constant, and q_e is the equilibrium sorption capacity of the sorbent in mg metal/g sorbent. The plots of specific sorption ($1/q_e$) against equilibrium concentration ($1/C_e$) gave the linear isotherm parameters of q_m and b. These coefficients are presented in Table 1. On this basis, the monolayer (maximum) adsorption capacities (q_m) were found to be 3.01, 2.62, and 2.53 mg/g for Cd, Cu, and Zn, respectively. The basic factor of the Langmuir adsorption isotherm can be obtained from the analysis of a distinctive feature commonly known as R_L and described as (22):

$$R_L = \frac{1}{1 + b.c_0} \qquad (4)$$

where, C_0 is the initial concentration and R_L indicates the confirmation of the Freundlich adsorption isotherm ($R_L>1$, unfavorable, $R_L=1$ linear, $0<R_L<1$, favorable, $R_L<0$

Table 1. Langmuir isotherm parameters for adsorption of Cd, Cu, and Zn from aqueous solution on APS adsorbent

Element	q_m(mg/g)	b (L/mg)	R^2
Cu	2.616	0.739	0.987
Zn	2.526	0.347	0.993
Cd	3.008	1.141	0.996

irreversible). The Langmuir isotherm model is based on the assumption of homogenous surface energy distribution (22).

For the present study, the values of R_L are shown in Table 2 and indicate that sorption intensity for Cd, Cu, and Zn is most favorable.

It can be observed from Figure 3 that the isotherm data fits the Langmuir equation well as it is a straight line. The values of q_m and b were determined from the slope and the intercept of the plot of $1/q_e$ versus $1/C_e$ (Figure 3) and are reported in Table 1.

The Langmuir isotherm yielded a better fit to the experimental data for sorbent, probably because of the homogenous natures of their surface sites involved in the uptake of cadmium, copper, and zinc.

Freundlich equation

Freundlich developed an empirical equation that is used mainly to describe the results of adsorption isotherm experiments. One form of the equation is:

$$q_e = K_f C^{1/n} \qquad (5)$$

In this equation, *qe* is the mass of solute adsorbed per mass of absorbent, mg/g; C is the equilibrium concentration of solute concentration after sorption is complete (mg/L); *Kf* and *n* are Freundlich adsorption parameters that must be determined for each temperature and solute (27).

The log form of the Freundlich equation is:

$$\log q_e = \log K_f + \frac{1}{n}\log C \qquad (6)$$

The Freundlich model is an empirical and suitable isotherm model for nonhomogenous systems (22). A plot of log q_e against log C_e yielding a straight line was made to confirm the model. The values of K_f and *1/n* were obtained from the intercept and slope of the straight line.

In this research, the experimental data from the sorption study of the three metal ions on APS were plotted logarithmically (Figure 4) using the linear Freundlich isotherm equation.

The linear Freundlich isotherm constants obtained for the sorption of Cd, Cu, and Zn onto APS are presented in Table 3. The Freundlich isotherm parameter *1/n* measures the adsorption intensity of metal ions on sorbent. In this research, the values of the constant *1/n* were found to be 0.291, 0.259, and 0.258 mg/g for Cd, Cu, and Zn, respectively. The value of *1/n* is less than 1, indicating the favorable adsorption of Cd, Cu, and Zn, probably due to the ability of the APS to remove these three metal ions from aqueous solutions (3).

Table 2. Dimensionless Langmuir parameter R_L for Cd, Cu, and Zn sorption onto APS

Concentration mg/L	Cd	Cu	Zn
10	0.08	0.12	0.22
20	0.04	0.06	0.12
40	0.02	0.03	0.07
80	0.01	0.02	0.03

Table 3. Freundlich isotherm parameters

Element	K_f	N	1/n	R^2
Cu	1.108	3.862	0.259	0.984
Zn	0.912	3.866	0.258	0.855
Cd	1.101	3.433	0.291	0.941

Sorption kinetic studies

Kinetic studies are important in determining the sorption capacity of the sorbent. Most sorption (or desorption) processes of the solid phases are time dependent (29). To study the control and operation of sorption processes such as chemical reaction and mass transfer, various kinetic models were applied to the experimental data obtained from sorption processes.

Pseudo-first order kinetic model:

The pseudo-first order is one of the most commonly used kinetic models for describing the adsorption mechanism (22). It is expressed as:

$$\log(q_e - q_t) = \log q_e - \frac{k_1}{2.303}t \qquad (7)$$

In this equation, q_e and q_t represent the sorption capacity of sorbent material (mg/g) at equilibrium and at time (min), respectively, and k_1 is the rate constant of the model (1/min).

The straight line plot of log (q_e –q_t) against time (t) should give a linear relationship, and in this situation, k_1 and q_e

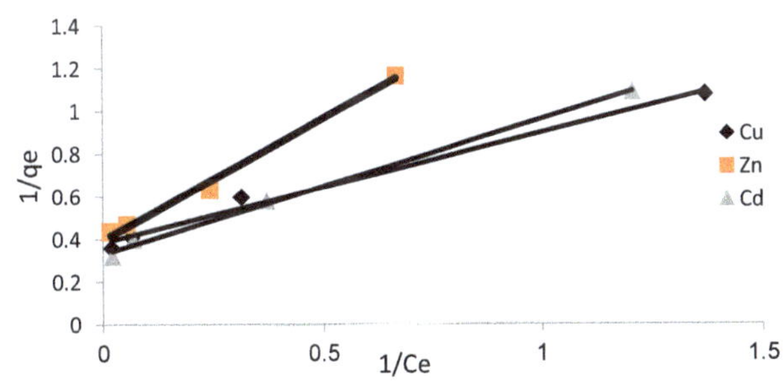

Figure 3. Fit of experimental data to Langmuir isotherm model.

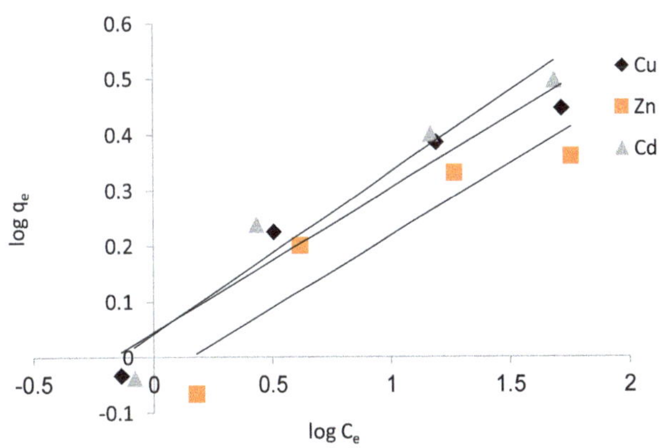

Figure 4. Fit of experimental data to Freundlich isotherm model.

can be calculated from the slope and intercept, respectively. The pseudo-first order kinetic constant k_1 and the correlation coefficients for Cd, Cu, and Zn adsorption were calculated from the linear plots of log $(q_e - q_t)$ versus t (Figure 5) and are listed in Table 4. The correlation coefficients for the pseudo-first order kinetic model are at below 0.9. As observed in Figure 5, the pseudo-first order model gives a good description of the sorption. The experimental kinetic data for Cd, Cu, and Zn onto APS were fitted to the pseudo-first order kinetic model, and its expressions are shown in Table 4.

Pseudo-second order kinetic model

The pseudo-second order equation has many advantages and is one of the best models to describe adsorption behavior (30). The pseudo-second order equation is represented as shown in Equation 8:

$$\frac{t}{q_t} = \frac{1}{k_2 q_e^2} + \frac{1}{q_e} t \qquad (8)$$

where k_2 is the constant rate of pseudo-second order kinetic (g/(mg.min)), and q_e and q_t are defined in the preceding paragraph.

Also:

$$h = k_2 q_e^2 \qquad (9)$$

where h can be regarded as the initial sorption rate since q_t/t, when t approaches 0 (30).

In the present study, the pseudo-second order model showed the best fit to the experimental data related to the adsorption of Cd, Cu, and Zn onto APS with the highest squared correlation coefficient (Figure 6).

The pseudo-second order kinetic rate constant (k_2) and the theoretical q_e can be calculated from the plots of t/qt versus t as shown in Figure 6.

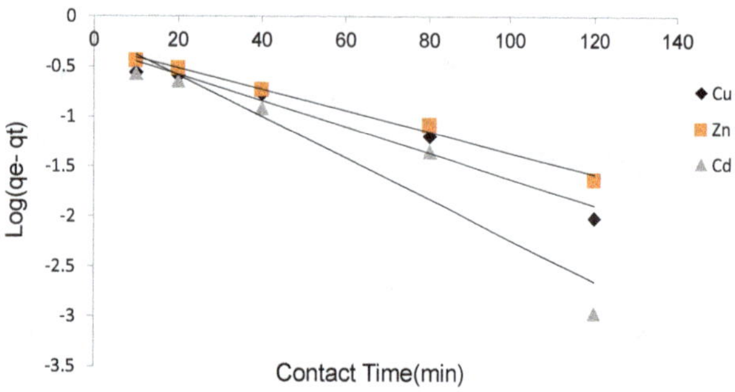

Figure 5. Pseudo-first order model for Cd, Cu, and Zn sorbtion onto APS.

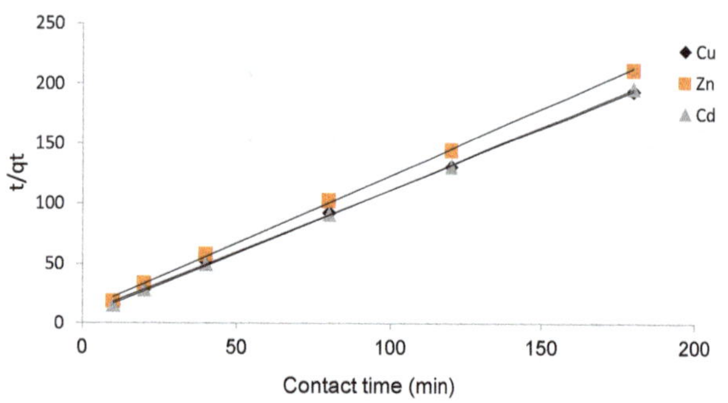

Figure 6. Pseudo-second order model for Cd, Cu, and Zn sorption onto APS.

The ways to obtain the kinetic rate constant k_2 (g/mg min), initial sorption rate h (mg/g min) and the amount of metals adsorbed at equilibrium q_e (mg/g) are presented in Table 5. The correlation coefficients for the pseudo-second order kinetic model are all extremely high (greater than 0.99), indicating that the adsorption system belongs to the pseudo-second order kinetic model (31).

This study has developed a new medium, pumice stone, for the removal of Cd, Cu, and Zn from aqueous solutions. Based on the results and on the correlation coefficients calculated, the sorption kinetics were found to follow a pseudo-second order kinetic model. In general, the isotherm gives the R^2 value closest to unity, but the Langmuir isotherm is better fitted for describing the equilibrium adsorption data for Cd, Zn, and Cu than the Freundlich isotherm.

Discussion

Comparable results were obtained by Nasseri and Heidari (14). Their study investigated and compared arsenic removal from water resources using aluminum-coated pumice (ACP) and aluminum-coated zeolite (ACZ) as new adsorbents. The effects of parameters including adsorbent doses, pH, contact time, and initial arsenic concentration were studied. They concluded that more than 96% of arsenic with an initial concentration of 250 μg/L was removed by 10 g/L aluminum-coated zeolite at pH = 7 and in 60 minutes to achieve an arsenic concentration of 10 μg/L, while only 71% of arsenic could be removed by aluminum-coated pumice. In a recent study done in Iran, the removal of acid red 14 and acid red 18 azo dyes by acid-treated pumice stone as an efficient adsorbent was investigated under various experimental conditions. It was revealed that maximum sorption capacities were 3.1 and 29.7 mg/g for acid red 14 and acid red 18, respectively. These results were significantly higher than those reported in the literature, even for activated carbon (22). Another study performed by Kaplan Bekaroglu et al (18) investigated the effectiveness of iron oxide-coated pumice and volcanic slag particles in removing disinfection by-product precursors from a raw drinking water source with high specific UV absorbance. It was found the iron oxide-coated pumice/slag particles are effective adsorbents for removing natural organic matter and

Table 4. Pseudo-first order rate equations; parameters related to the kinetic models for APS sorbent

R^2	q_e (mg/g)	k_1 (1/min)	Element	Initial concentration (mg/L)
0.958	0.725	0.013	Cu	10
0.990	0.738	0.010	Zn	10
0.904	0.835	0.020	Cd	10

Table 5. Pseudo-second order rate equations

Initial concentration (mg/L)	Element	k_2 (g/mg-min)	qe (mg/g)	R2	h (mg/g. min)
10	Cu	0.139	0.964	0.999	0.129
10	Zn	0.116	0.891	0.999	0.092
10	Cd	0.178	0.949	0.999	0.161

control disinfection by-product formation in waters; but, they may not be effective for waters with alkalinity levels above 250 mg $CaCO_3$/L. Also, Asgari et al (16) reported that surfactant-modified pumice was shown to be an affordable and promising option for the removal of fluoride from drinking water.

Conclusion

The removal of cadmium, copper, and zinc from aqueous solutions using APS as a sorbent is possible. The results of the current study indicate that APS is a good choice for a low-cost and effective sorbent. The sorption capacity of APS was considerably affected by the initial concentration of metal ions in the solution and by contact time. The Langmuir isotherm provided the best correlation for the sorption of cadmium, copper, and zinc onto APS. At a constant pH of 7, the sorption capacities as calculated from the Langmuir isotherm were 3.01, 2.62, and 2.53 mg/g for Cd, Cu, and Zn, respectively. The kinetic data was well fitted by a pseudo-second order model. The authors believe that the sorbent APS can be recommended as a suitable alternative for the control and treatment of industrial wastewater where heavy metals or toxic chemicals threaten the quality of ground or surface waters. Because of the excellent qualitative and quantitative capabilities of new instruments, it is recommended that future researchers apply the latest technology such as x-ray diffraction (XRD), x-ray fluorescence (XRF), and scanning electron microscopy (SEM) to evaluate the capability of APS to absorb heavy metals and characterize their mechanisms of action.

Acknowledgements

This research was carried out as a Master's Thesis at the Department of Environment, financially supported by Hamedan Branch, Islamic Azad University, and technically supported by Kermanshah Provincial Directorate of Environmental Protection, to which the authors' thanks are due.

Competing interests

The authors declare that they have no competing interests.

Authors' contributions

All authors participated in the study design, literature search, writing of the manuscript, and data acquisition, analysis, and interpretation. All authors critically reviewed, refined, and approved the manuscript.

References

1. Zavvar Mousavi H, Hosseinifar A, Jahed V. Studies of the adsorption thermodynamics and kinetics of Cr (III) and Ni (II) removal by polyacrylamide. J Serb Chem Soc 2012; 77(3): 393-405.
2. Abuh MA, Akpomie GK, Nwagbara NK, Abia-Bassey N, Ape DI, Ayabie BU. kinetic rate equations application on the removal of copper (II) and zinc (II) by unmodified lignocellulosic fibrous layer of palm tree trunk- single component system studies. International Journal of Basic and Applied Science 2013; l(4): 800-9.
3. Kumar PS, Ramakrishnan K, Kirupha SD, Sivanesan S. Thermodynamic and kinetic studies of cadmium adsorption from aqueous solution onto rice husk. Braz J Chem Eng 2010; 27(2): 347-55.
4. Ackacha MA, Meftah SA. Elimination of cu (II) cations from water media using clean adsorbent as application of green chemistry: acacia tortilis seeds. Journal of Clean Energy Technologies 2013; 1(1): 57-61.
5. Tiwari RP, Bala Ramudu P, Srivastava RK, Gupta MK. Sorption and desorption studies of metallic zinc on an alluvial soil. Iranian J Environ Health Sci Eng 2007; 4(3): 139-46.
6. Radina H, Ghoreyshi AA, Younesi H. Isotherm and kinetics of Fe(II) adsorption onto Chitosan in a batch process. Iranian Journal of Energy & Environment 2011; 2(3): 250-7.
7. Moradi M, Dehpahlavan A, Rezaei Kalantary R, Ameri A, Farzadkia M, Izanloo H. Application of modified bentonite using sulfuric acid for the removal of hexavalent chromium from aqueous solutions. Environmental Health Engineering and Management Journal 2015; 2(3): 99-106.
8. Najafpoor AA, Soleimani G, Ehrampoush MH, Ghaneian MT, Rahmanpour Salmani E, Dolatabadi Takabi M. Study on the adsorption isotherms of chromium (VI) by means of carbon nano tubes from aqueous solutions. Environmental Health Engineering and Management Journal 2014; 1(1): 1-5.
9. Dowlatshahi S, Haratinezhad Torbati AZ, Loloei M. Adsorption of copper, lead and cadmium from aqueous solutions by activated carbon prepared from saffron leaves. Environmental Health Engineering and Management Journal 2014; 1(1): 37-44.
10. Murkani M, Nasrollahi M, Ravanbakhsh M, Bahrami P, Jaafarzadeh Haghighi Fard N. Evaluation of natural zeolite clinoptilolite efficiency for the removal of ammonium and nitrate from aquatic solutions. Environmental Health Engineering and Management Journal 2015; 2(1): 17-22.
11. Merganpour AM, Nekuonam G, Alipour Tomaj O, Kor Y, Safari H, Karimi K, et al. Efficiency of lead removal from drinking water using cationic resin Purolite. Environmental Health Engineering and Management Journal 2015; 2(1): 41-5.
12. Salmani Nodoushan MH, Ehrampoush MH, Rafati L, Farzi M, Barzegar K. Study of Co (II) adsorption from aqueous solution using protein granules produced from chicken feather. Environmental Health Engineering and Management Journal 2015; 2(4): 193-7.
13. Moradi O, Mirza B, Norouzi M, Fakhri A. Removal of Co(II), Cu(II) and Pb(II) ions by polymer based 2-hydroxyethyl methacrylate: thermodynamics and desorption studies. Iranian J Environ Health Sci Eng 2012; 9(1): 31.
14. Nasseri S, Heidari M. Evaluation and comparison of aluminum-coated pumice and zeolite in arsenic removal from water resources. Iranian J Environ Health 2012; 9(1): 38.
15. Tapan M, Yalçın Z, İçelli O, Kara H, Orak S, Özvan A, et al. Effect of physical, chemical and electro-kinetic properties

of pumice samples on radiation shielding properties of pumice material. Ann Nucl Energy 2014; 65: 290-98.
16. Asgari G, Roshani B, Ghanizadeh G. The investigation of kinetic and isotherm of fluoride adsorption onto functionalize pumice stone. J Hazard Mater 2012; 217-218: 123-32.
17. Deganello F, Liotta LF, Macaluso A, Venezia AM, Deganello G. Catalytic Reduction of nitrates and nitrites in water solution on pumice-supported Pd–Cu catalysts. Appl Catal B 2000; 24(3-4): 265-73.
18. Kaplan Bekaroglu SS, Yigit NO, Karanfil T, Kitis M. The adsorptive removal of disinfection by-product precursors in a high-SUVA water using iron oxide-coated pumice and volcanic slag particles. J Hazard Mater 2010; 183(1-3): 389-94.
19. Karimaian KA, Amrane A, Kazemian H, Panahi R, Zarrabi M. Retention of phosphorus ions on natural and engineered waste pumice: characterization, equilibrium, competing ions, regeneration, kinetic, equilibrium and thermodynamic study. Appl Surf Sci 2013; 284: 419-31.
20. Kitis M, Kaplan SS, Karakaya E, Yigit NO, Civelekoglu G. Adsorption of natural organic matter from waters by iron coated pumice. Chemosphere 2007; 66(1): 130-8.
21. Subrahmanyam M, Boule P, DurgaKumari V, Naveen Kumar D, Sancelme M, Rachel A. Pumice stone supported titanium dioxide for removal of pathogen in drinking water and recalcitrant in wastewater. Solar Energy 2008; 82(12): 1099-1106.
22. Samarghandi MR, Zarrabi M, Noorisepehr M, Amrane A, Safari GH, Bashiri S. Application of acidic treated pumice as an adsorbent for the removal of azo dye from aqueous solutions: kinetic, equilibrium and thermodynamic studies. Iranian J Environ Health Sci Eng 2012; 9(1): 9.
23. Yuan L, Shen J, Chen Z, Liu Y. Pumice-catalyzed ozonation degradation of p-chloronitrobenzene in aqueous solution. Appl Catal B Environ 2012; 117–118: 414-9.
24. American Public Health Association (APHA). Standard Methods for the Examination of Water and Wastewater. 20th ed. Washington DC: APHA; 2005.
25. Babatunde AO, Zhao YQ. Equilibrium and kinetic analysis of phosphorus adsorption from aqueous solution using waste alum sludge. J Hazard Mater 2010; 184(1-3): 746-52.
26. Nemr AE, Abdelwahab O, Khaled A, Sikaily AE. Removal of chrysophenine dye (dy-12) from aqueous solution using dried ulvalactuca. Egyptian Journal of Aquatic Research 2005; 31(1): 86-98.
27. Benefield LD, Judkins JF, Weand BL. Process Chemistry for Water and Wastewater Treatment. 1st ed. New Jersey: Prentice-Hall; 1981.
28. Jeppua GP, Clementa TP. A modified Langmuir-Freundlich isotherm model for simulating pH-dependent adsorption effects. J Contam Hydrol 2012; 129-130: 46-53.
29. Shahmohammadi Kalalagh SH, Babazadeh H, Nazemi AH, Manshouri M. Isotherm and Kinetic Studies on Adsorption of Pb, Zn and Cu by Kaolinite. Caspian Journal Of Environmental Sciences 2011; 9(2): 243-55.
30. Ho YS, Ofomaja AE. Pseudo-second-order model for lead ion sorption from aqueous solutions onto palm kernel fiber. J Hazard Mater 2006; 129(1-3): 137-42.
31. Ho YS, McKay G. Pseudo-second order model for sorption processes. Process Biochem 1999; 34(5): 451-65.

3

Humic acid removal from aqueous solutions by peroxi-electrocoagulation process

Ahmad Reza Yazdanbakhsh[1], Majid Kermani[2], Sanaz Komasi[3], Ehsan Aghayani[4], Amir Sheikhmohammadi[5*]

[1]Associate Professor, Department of Environmental Health Engineering, Shahid Beheshti University of Medical Sciences, Tehran, Iran
[2]Assistant Professor, Department of Environmental Health Engineering, Iran University of Medical Sciences, Tehran, Iran
[3]MSc of Environmental Engineering, Department of Environmental Engineering, Islamic Azad University, West Tehran Branch, Tehran, Iran
[4]PhD Student, Department of Environmental Health Engineering, Tarbiat Modares University of Medical Sciences, Tehran, Iran
[5]PhD Student, Department of Environmental Health Engineering, Shahid Beheshti University of Medical Sciences, Tehran, Iran

Abstract

Background: Natural organic matter is the cause of many problems associated with water treatment such as the presence of disinfection by-products (DBPs) and membrane fouling during water filtration. In this study, the performance of the peroxi-electrocoagulation process (PEP) was investigated for the removal of humic acids (HAs) from aqueous solutions.

Methods: PEP was carried out for the removal of HA using a plexiglas reactor with a volume of 2 L and fitted with iron electrodes and a direct current supply (DC). Samples were taken at various amounts of pH (2-4), current density (1 and 2A/cm^2), hydrogen peroxide (50-150 mg/L) and reaction time (5-20 minutes) and then filtered to remove sludge formed during reaction. Finally, the HA concentration was measured by UV absorbance at 254 nm (UV_{254}).

Results: Results indicated that increasing the concentration of H_2O_2 from 50 to 150 mg/L increased HA removal efficiency from 83% to 94.5%. The highest removal efficiency was observed at pH 3.0; by increasing the pH to the alkaline range, the efficiency of the process was reduced. It was found that HA removal efficiency was high in current density 1A/cm^2. Increasing current density up to 1 A cm^{-2} caused a decrease in removal efficiency. Results of this study showed that under the optimum operating range for the process ([current density] = 1A/cm^2, [hydrogen peroxide concentration] = 150 mg/L, [reaction time]= 20 minutes and [pH]= 3.0), HA removal efficiency reached 98%.

Conclusion: It can be concluded that PEP has the potential to be utilized for cost-effective removal of HA from aqueous solutions.

Keywords: Humic acid, Peroxi-electrocoagulation, Iron electrode, UV_{254}, Water solutions

***Correspondence to:**
Amir Sheikhmohammadi
Email: asheikh1359@gmail.com

Introduction

Humic acid (HA) is one of the major components of humic substances which arise from the microbial degradation of biomolecules. Natural organic materials (NOM) such as HA enter into water from natural and artificial sources. They pose a variety of problems in treatment operations and distribution systems because of their specific attributes including an unpleasant smell and taste, a yellow to brown color, reaction with chlorine, and the formation disinfection by-products (DBPs), eg, trihalomethanes (THMs). These exhibit mutagenic properties during the chlorination step in drinking water production or water treatment.

The guideline for DBPs in drinking water announced by the World Health Organization state that DBPs should not exceed 100 µg/L (1-3). Consequently, the removal of HA from surface water or wastewater is important because of health and environmental concerns. HA may account for up to 90% of NOM. It is not possible to completely remove NOM in conventional water treatment plants (4,5). Different treatment technologies have been used in practice to improve NOM removal. These methods included ion exchange, sorption, membrane processes (such as reverse osmosis), and ultra-filtration, but they either provide only a partial solution or are expensive. Coagulation is a key process which may be carried out in the conventional mode (chemical) or by peroxi-electrocoagulation process (PEP) (6). Thus, PEP has been suggested as an alternative to conventional coagulation because it provides better removal capabilities, produces less sludge, and has a more readily filterable, minimal startup time (7-9).

Some limitations of the PEP are the periodical replacement of sacrificial anodes, minimum solution conductivity, the formation of toxic chlorinated organic compounds from effluent containing chlorides, the formation of an impermeable oxide film on the cathode, and resistance to the flow of electric current, which increases the cost of electricity thus increasing operational costs (10).

Stages in the PEP process include anode dissolution, formation of OH^- ions and H_2 at the cathode, electrolytic reactions at electrode surfaces, adsorption of coagulant on colloidal pollutants, and removal of colloids by sedimentation or flotation. Reactions carried out in the cathode and anode for the production of metal hydroxides include:

Mechanism 1: Anode:

$$4Fe_{(s)} \rightarrow 4Fe^{2+}_{(aq)} + 8e^- \quad (1)$$

$$4Fe^{2+}_{(aq)} + 10H_2O_{(l)} + O_{2(aq)} \rightarrow 4Fe(OH)_{3(s)} + 8H^+_{(aq)} \quad (2)$$

Cathode:

$$8H^+_{(aq)} + 8e^- \rightarrow 4H_{2(g)} \quad (3)$$

Overall:

$$4Fe_{(s)} + 10\,H_2O_{(l)} + O_{2(aq)} \rightarrow 4Fe(OH)_{3(s)} + 4H_{2(g)} \quad (4)$$

In recent years, it has been recommended to combine electrocoagulation with hydrogen peroxide to increase the power of the electrocoagulation process. This method is based on the generation of a powerful oxidant (the hydroxyl radical [$^{\bullet}OH$]), because the OH radical attacks the organic substrates present in the water and wastewater and then degrades them (11-15). During the electrocoagulation process, Fe^{2+} is the common ion generated at the anode; in contrast, OH^- ions are produced at the cathode (16-19). By adding hydrogen peroxide into the PEP reactor, Fenton reactions are carried out in accordance with the following:

$$Fe^{2+} + H_2O_2 \rightarrow Fe^{3+} + OH^- + {}^{\bullet}OH \quad (5)$$

$$Fe^{3+} + H_2O_2 \leftrightarrow Fe\text{-}OOH^{2+} + H^+ \leftrightarrow Fe^{2+} + HO_2^{\bullet} + H^+ \quad (6)$$

Vasudevan (20) studied the removal of phenol from water by the PEP. Their results showed that phenol was removed by PEP at 92% efficiency. Evrim Yüksel et al (21) achieved a removal efficiency of 81.6% in the removal of sodium dodecyl sulfate from synthetic wastewater by the peroxi-electrocoagulation method. Vasudevan and Lakshmi (22) also studied the removal of boron from water by electrochemical processes and investigated various operating parameters such as initial boron ion concentration, initial pH, current density, and temperature. Their results showed a removal efficiency of 93.2% under optimum operating conditions. Kamaraj et al (23) studied the adsorption of 2,4-dichlorophenoxyacetic acid (2,4-D) from water by in situ generated metal hydroxides using sacrificial anodes. The results showed that the maximum removal efficiency of 91.0% was achieved with iron as the anode. The current study examined the possibility of HA removal from aqueous solutions by PEP and evaluated the influence of operation parameters such as initial pH, current density, amount of hydrogen peroxide, and reaction time on HA removal efficiency.

Methods

Material

All chemical reagents used in bench-scale tests were analytical grade. The materials used in this study included sulfuric acid (96%) and NaOH (98%) for pH adjustment and technical grade hydrogen peroxide (30% w/w and density of 1.13 kg/L), purchased from the Merck company.

Set-up and Procedure

A stock solution of HA was prepared by dissolving HA powder in 100 mL NaOH (0.1N). The prepared solution was kept for a day, and then its volume was increased to 1 L with distilled water. The solution was filtered using a Whatman filter, which was then placed in an oven at a temperature of 103°C-105°C. The difference between the primary and secondary weight of the filter was determined to be 314.5 mg/L. Working solutions (4, 8, and 10 mg/L) were prepared by diluting the stock solution in deionized water.

The batch experimental setup is schematically shown in Figure 1. Water was electrochemically oxidized in an undivided cell reactor equipped with a magnetic stirrer using a constant speed of 200 rpm in order to get a correct homogenization of the wastewater-floc mixture. The reactor consisted of an EC cell, a DC power supply, and monopolar electrodes in parallel connections. All experiments were run under isoperibolic conditions and at a temperature of 20°C. The reaction was timed starting when the DC power supply was switched on. Two flat plates with smooth surfaces were connected as cathodes and 2 as anodes. Each electrode measured 4 cm × 3 cm, and the distance between the iron electrodes was 1 cm. To remove the influence mass transfer rate of pollutants to the electrodes (iron salts produce electrode passivation and cause a 50% increase in treatment time and power requirements), high mixing was ensured by a magnetic stirrer. To maintain

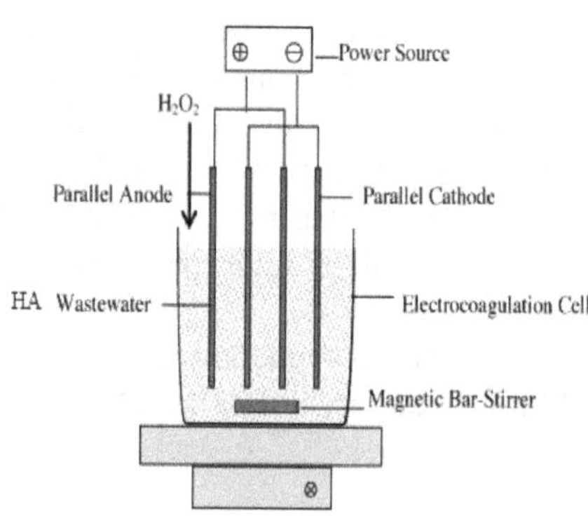

Figure 1. A schematic diagram of the electrochemical reactor.

the same experimental conditions, the Fe electrodes were polished with grinding papers and then rinsed with soapy water and deionized water after each experiment.

The effects of the main operational variables (including solution pH [2-4], current density [1 and 2 A/cm^2], reaction time [5–20 minutes], and H_2O_2 [50-150 mg/L]) on the electrochemical process in treating HA were investigated in various runs. At the beginning of a run, the HA solution was fed into the reactor; then the electrodes were connected to the DC power supply and the reaction was started while the cell content was gently stirred (200 rpm). H_2O_2 was added at desired rates to the reactor for enhanced pollutant removal.

Analytic Procedure

The HA samples were measured by HACH spectrophotometer (DR 5000) at a wavelength of 254 nm after filtration through a Millipore membrane filter with a 0.45 µm pore size. Experiments with 3 replications were conducted according to the standard methods for examination of water and wastewater (24). Data was analyzed using Excel software.

Results

To determine the optimal amount of H_2O_2, concentrations in the range of 50 to 150 mg/L were investigated, and the results are presented in Figure 2. It was observed that increasing the concentration of H_2O_2 increased the removal of HA. For example, in the initial HA concentration of 4 mg/L, increasing the H_2O_2 concentration from 50 to 150 mg/L increased HA removal efficiency from 83% to 94.5% at pH 3.0 and current density 1A after 20 minutes.

pH plays a significant role in the performance of the Fenton process. Various studies have shown that an acidic pH of 2-4 is the best range for the Fenton reaction; therefore, the performance of PEP for the removal of HA was investigated in a pH range of 2-4 (Figure 3). The highest removal efficiency was observed at pH 3.0; increasing pH to the alkaline range reduced the efficiency of the process.

The results of the influence of the current density on the HA removal are presented in Figure 4. It was found that the HA removal efficiency was high in current density 1A/cm^2. Increasing current density up to 1A/cm^2 caused a decrease in removal efficiency. The minimum energy consumption, was 10 kWh (kg HA)$^{-1}$ at current density 1 1A/cm^2 for a 20 minutes reaction time.

As shown in Figure 5, early on, the pollutant decomposition rate was very high, and increasing reactive time did not significantly affect the removal of HA. The effects of initial HA on removal efficiency are shown in Figure 6. Results of this study confirm that increasing the HA concentration increases the efficiency of the electrochemical process.

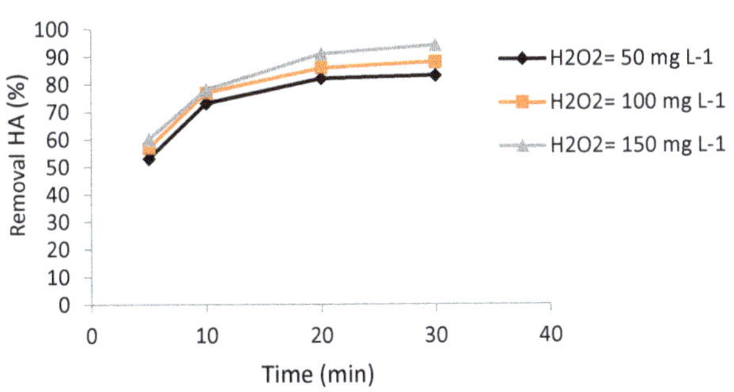

Figure 2. Effect of hydrogen peroxide on HA removal efficiency (peroxi-electrocoagulation process [PEP]; initial HA, 4 mg/L; pH, 3; current density, 1A/cm^2; distance between plates, 1 cm).

Discussion

Results of the current study indicated that increasing the H_2O_2 concentration increased HA removal efficiency. Determining the optimal amount of hydrogen peroxide in PEP is very important and necessary because of: (*a*) the high cost of hydrogen peroxide in high concentrations; (*b*) reduced production of hydroxyl radical exceeding H_2O_2

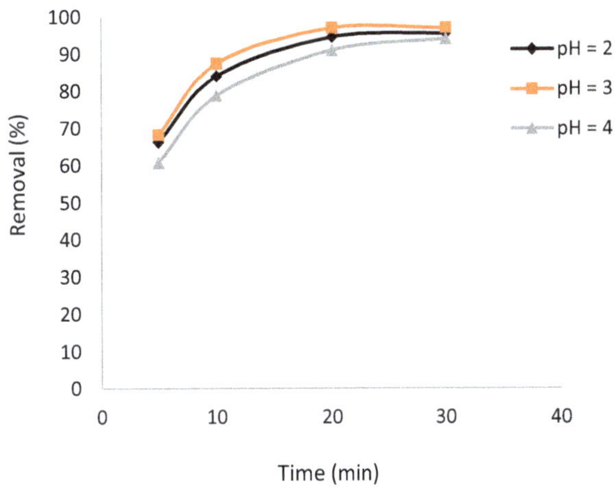

Figure 3. Effect of initial pH on HA removal efficiency (peroxi-electrocoagulation process [PEP]; initial HA, 4 mg/L; hydrogen peroxide, 150 mg/L; current density, 1A/cm^2; distance between plates, 1 cm).

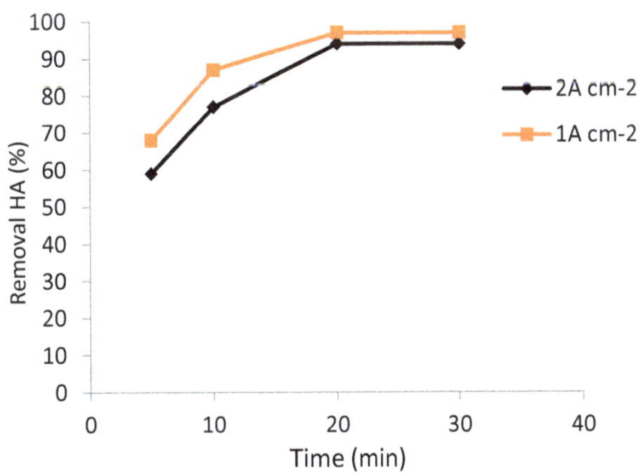

Figure 4. Effect of current densityon HA removal (initial HA, 4 mg/L; pH, 3; H_2O_2, 150 mg/L; distance between plates, 1 cm).

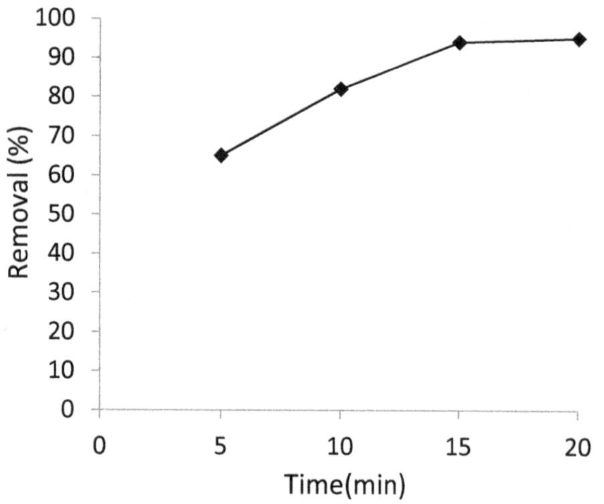

Figure 5. Effect of reaction time on HA removal efficiency (peroxi-electrocoagulation process [PEP]; initial HA, 4 mg/L; hydrogen peroxide, 150 mg/L; pH, 3; current density, 1A/cm^2; distance between plates, 1 cm).

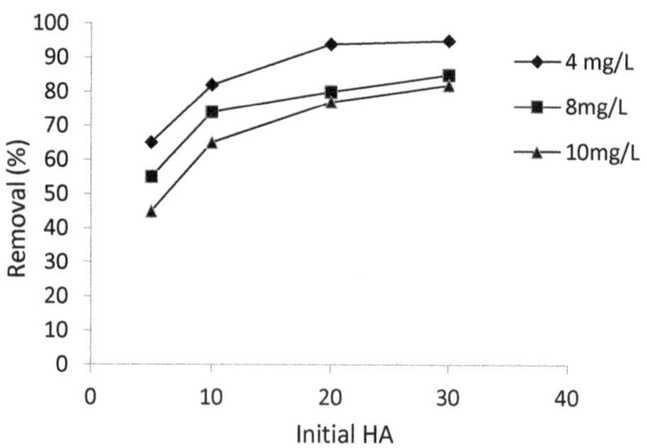

Figure 6. Effect of initial HA (peroxi-electrocoagulation process [PEP]; hydrogen peroxide,150 mg/L; Initial HA, 4-10 mg/L; pH, 3; current density, 1A/cm^2; distance between plates, 1 cm).

(H_2O_2 acts as a scavenger of the OH according to reaction 7); and (*c*) increased chemical oxygen demand by residual H_2O_2 (15,16,25,26).

$$H_2O_2 + {}^{\bullet}OH \rightarrow H_2O + HO_2^{\bullet} \quad (7)$$

It can be concluded that the removal of HA increases when the concentration of hydrogen peroxide is increased. Mahesh et al (26) examined the effects of different concentrations of hydrogen peroxide on the reduction of humic substances from leachate using the Fenton oxidation process and chemical coagulation. They concluded that the removal of humic substances increases when the concentration of hydrogen peroxide is increased to a certain concentration due to the production of hydroxyl radicals. Increasing the hydrogen peroxide concentration over this amount caused a reduction in removal efficiency. They also determined that the decreased removal efficiency in excessive amounts of hydrogen peroxide can be caused by the production of OOH radicals that consume the OH. Farhadi et al (27) used PEP for COD removal from pharmaceutical wastewater. Their results showed that the maximum removal rate was nearly 42.2% for the optimum level of 300 mg/L H_2O_2 after 20 minutes of electrolysis time for the voltage of 40 V.

The decrease in removal efficiency in alkaline pH is probably due to the formation of ferric species and decreased production of ferrous ions, both of which reduce the production of hydroxyl radicals and so decrease the efficiency of removal. A decrease in the efficiency of HA removal at pH <3 could be explained by the following:

1. $Fe(OH)^{2+}$, $Fe(OH)_2^{+}$, and $Fe_2(OH)_2^{4+}$ were produced, which react more slowly with H_2O_2.
2. The scavenging effect of OH by H^+ becomes significant at very low ranges of pH (at lower pH the protons in the solution are reduced to H_2 at the cathode, and the ions of H^+ may inhibit the generation of hydroxyl and perhydroxyl radicals) (28).

$${}^{\bullet}OH + H^+ + e^- \rightarrow H_2O \quad (8)$$

Farhadi et al (27) used PEP for COD removal from pharmaceutical wastewater. The optimum pH for this process was determined to be 3.0. Su et al (29) used electro-Fenton processes for the degradation of acetaminophen (ACTP). The experimental results showed that the electro-Fenton processes had maximum ACTP degradation efficiencies of 100%. ACTP degradation efficiency increased 74% at pH 4 in the electro-Fenton process. Cheng et al (30) used the electro-Fenton process with a Ce/SnO_2–Sb-coated titanium anode for the mineralization of metronidazole. A decrease in pH from 7.0 to 2.0 increased removal efficiency from 10% to 37%.The optimum pH was found to be about 2.0.

Current density is an important factor in the electrochemical processes because of its effects on the coagulant dosage rate, bubble production rate, size and growth of the flocs. Increasing the current density increased HA removal. This is in accordance with Faraday's law; with an increase in current density and reaction time, the anodic dissolution rate increases that will give rise to Fe^{2+} being released from the anode as in equation 9 (31,32).

$$C_{Fe} = \frac{M_w I t_{EC}}{ZFV} \quad (9)$$

where C_{Fe}, Z, F, V, and M_w are the theoretical concentration of Fe^{2+} (g/m^3), the chemical equivalence, Faraday's constant (96,487 C/mol), reactor volume (m^3), and molecular weight of iron (g/mol), respectively. The increase of removal efficiency upon increasing current density is the result of the extent of anodic dissolution of iron, increased hydroxyl radical generation, increased bubble generation rate, and decreased bubble size. These trends are beneficial in terms of high pollutant removal efficiency by H_2 flotation (33). Increasing current density up to 1A/cm^2 caused a decrease in the efficiency of removal. This decrease could be due to the decomposition of hydrogen peroxide in higher current densities. Wang (34) studied COD removal from real dyeing wastewater by electro-Fenton technology using an activated carbon fiber cathode. In this study, 75% COD removal was achieved under the optimum conditions of current density= 32 mA/cm^2, pH = 3.0, Fe^{2+}= 2 mM/L, and T= 20°C (34). Vasudevan et al (35) studied the removal of copper by electrochemically-assisted coagulation. Their results showed that the optimum removal efficiency of 96.6% was achieved at a current density of 0.05 A/dm^2 and a pH of 7.0. Babuponnusami and Muthukumar (36) used the electro-Fenton process for the advanced oxidation of phenol. Their results showed 62.5% phenol degradation for the EF process at 16 mA/cm^2 in 60 minutes. Increasing current density from 0 to 16 mA/cm^2 offered faster phenol degradation and COD removal. The important economical parameter in PEP is electrical energy consumption which is the major operating cost of the electrochemical process. The minimum amount of energy consumption was 10 kWh (kg HA)$^{-1}$ at a current density of 1A/cm^2 for a 20 minutes reaction time that was calculated using the following equation:

$$E = UIt_{EC} \tag{10}$$

where E is the electrical energy in Wh, U= the cell voltage in volt (V), I= the current in ampere (A), and t_{EC} is the time of EC process per hour.

Reactive time influences the treatment efficiency of PEP. To explore the effect of operating time, the optimal conditions of these processes were used, and all processes were started simultaneously. Early on, the pollutant decomposition rate was very high, possibly due to the appropriate amounts of Fe (II) and hydrogen peroxide used to produce hydroxyl radicals. It was observed the percentage of HA removal was higher for PEP at 20 minutes, pH= 3.0, and current density 1A/cm^2. The decrease of HA as a function of elapsed time might be caused by the production of Fe^{2+} or Fe^{3+} ions from iron and the formation of OH, because by adding H_2O_2 into the system, hydrogen peroxide is catalyzed by ferrous ion to produce hydroxyl radicals, and a reaction is propagated from ferrous ion regeneration mainly by the reduction of the produced ferric species with hydrogen peroxide. Therefore, the dosage of H_2O_2 could enhance the performance of these processes. Ruppert et al (37) investigated the mineralization of several organic matters by hydroxyl radicals. They reported that organic material only affected the behavior of ferrous ions. Ferrous ions compete with organic material for the use of hydroxyl radicals. At low concentrations of organic material, ferrous ions react with hydroxyl radicals and reduce the amount of them in the environment. In the absence of organic compounds, hydrogen peroxide is decomposed slowly by ferric ions. Excessive amounts of organic matter prevent the reaction between ferrous ions and hydroxyl radicals with the increase in reactivity between ferrous ions and hydrogen peroxide.

Conclusion

The PEP was tested on HA synthetic wastewater, and the effects of operational parameters such as pH, amount of hydrogen peroxide, current density, and reaction time on HA removal efficiency were investigated. The HA removal efficiency that was achieved using PEP was 98% for the initial HA concentration of 4 mg/L. The ferric hydroxide and hydroxyl radical generated in the cell can decrease the HA present in the wastewater. The optimal conditions for the process were determined to be current density = 1A/cm^2, hydrogen peroxide concentration= 150 mg/L, reaction time= 20 minutes, mean energy consumption= 10 kWh (kg HA)$^{-1}$, and pH=3.0. Finally, given the results ultimately obtained in this research, it became clear that the process is technologically feasible.

Competing interests

The authors have declared that no competing interests exist.

Authors' contributions

All authors contributed equally, and all authors participated in the data acquisition, analysis, and interpretation. All authors critically reviewed, refined, and approved the manuscript.

References

1. Jacangelo JG, Demarco J, Owen DM, Randtke SJ. Selected processes for removing NOM: an overview. J Am Water Works Assoc 1995; 87(1): 64-77.
2. Katsumata H, Sada M, Kaneco S, Suzuki T, Ohta K, Yobiko Y. Humic acid degradation in aqueous solution by the photo-Fenton process. Chem Eng J 2008; 137(2): 225-30.
3. Magdaleno GB, Coichev N. Chemiluminescent determination of humic substances based on the oxidation by peroxymonosulfate. Anal Chim Acta 2005; 552(1-2): 141–46.
4. Panyapinyopol B, Marhaba TF, Kanokkantapong V, Pavasant P. Characterization of precursors to trihalomerthanes formation in Bangkok source water. J Hazard Mater 2005; 120(1-3): 229-36.
5. Yoon Y, Lueptow RM. Removal of organic contaminants by RO and NF membranes. J Memb Sci 2005; 261(1-2): 76-86.
6. Abdelwahab O, Amin NK, El-Ashtoukhy ES. Electrochemical removal of phenol from oil refinery wastewater. J Hazard Mater 2009; 163(2-3): 711–6.
7. Apaydin O, Kurt U, Gonullu MT. An investigation on the treatment of tannery wastewater by electrocoagulation. Global Nest Journal 2009; 11(4): 546-55.
8. Enric B, Ignasi S, Mehmet A. Electro-Fenton process

and related electrochemical technologies based on Fenton's reaction chemistry. Chem Rev 2009; 109(12): 6570-631.

9. Cho JH, Lee JE, Ra CS. Effects of electric voltage and sodium chloride level on electrolysis of swine wastewater. J Hazard Mater 2010; 180(1-3): 535-41.
10. Chou W-L, Wang C-T, Huang K-Y. Investigation of process parameters for the removal of polyvinyl alcohol from aqueous solution by iron electrocoagulation. Desalination 2010; 251(1-3): 12-19.
11. Kabdasl I, Vardar B, Arslan-Alaton I, Tunay O. Effect of dye auxiliaries on color and COD removal from simulated reactive dye bath effluent by electrocoagulation. Chem Eng J 2009; 148(1): 89–98.
12. Kurt U, Apaydin O, Gonullu MT. Reduction of COD in wastewater from an organized tannery industrial region by Electro-Fenton process. J Hazard Mater 2007; 143(1-2): 33-40.
13. Nidheesh PV, Gandhimathi R. Trends in electro-Fenton process for water and wastewater treatment: an overview. Desalination 2012; 229: 1-15.
14. Bagga A, Chellam S, Clifford DA. Evaluation of iron chemical coagulation and electrocoagulation pretreatment for surface water microfiltration. J Memb Sci 2008; 309(1-2): 82-93.
15. Ben Sasson M, Adin A. Fouling mitigation by iron-based electroflocculation in microfiltration: mechanisms and energy minimization. Water Res 2010; 44(13): 3973-81.
16. Gao P, Chen X, Shen F, Chen V. Removal of chromium(VI) from wastewater by combined electrocoagulation - electroflotation without a filter. Sep Purif Technol 2005; 43(2): 117-23.
17. Harif T, Hai M, Adin A. Electroflocculation as potential pretreatment in colloid ultrafiltration. Water Supply 2006; 6(1): 69-78.
18. Harif T, Adin A. Characteristics of aggregates formed by electroflocculation of a colloidal suspension. Water Res. 2007; 41(13): 2951-61.
19. Holt PK, Barton GW, Wark M, Mitchell CA. A quantitative comparison between chemical dosing and electrocoagulation. Colloids Surf A Physicochem Eng Asp 2002; 211(2–3): 233-48.
20. Vasudevan S. An efficient removal of phenol from water by peroxi-electrocoagulation processes. Journal of Water Process Engineering 2014; 2: 53-7.
21. Yüksel E, Şengil İA, Özacar M. The removal of sodium dodecyl sulfate in synthetic wastewater by peroxi-electrocoagulation method. Chem Eng J 2009; 152(2-3): 347-53.
22. Vasudevan S, Lakshmi J. Electrochemical removal of boron from water: adsorption and thermodynamic studies. Can J Chem Eng 2012; 90(4): 1017-26.
23. Kamaraj R, Davidson DJ, Sozhan G, Vasudevan S. Adsorption of 2,4-dichlorophenoxyacetic acid (2,4-D) from water by in situ generated metal hydroxides using sacrificial anodes. Journal of the Taiwan Institute of Chemical Engineers 2014; 45(6): 2943-9.
24. APHA, AWWA, WPCF. Standard Methods for the Examination of Water and Wastewater. 21st ed. Washington, DC: American Public Health Association; 2005.
25. Jiang J-Q, Graham N, André C, Kelsall GH, Brandon N. Laboratory study of electro-coagulation–flotation for water treatment. Water Res 2002; 36(16): 4064-78.
26. Mahesh S, Prasad B, Mall ID, Mishra IM. Electrochemical degradation of pulp and paper mill wastewater. Part 1. COD and color removal. Ind Eng Chem Res 2006; 45(8): 2830-9.
27. Farhadi S, Aminzadeh B, Torabian A, Khatibikamal V, Fard MA. Comparison of COD removal from pharmaceutical wastewater by electrocoagulation, photoelectrocoagulation, peroxi-electrocoagulation and peroxi-photoelectrocoagulation processes. J Hazard Mater 2012; 219-220: 35-42.
28. Vasudevan S, Lakshmi J, Jayaraj J, Ganapathy S. Remediation of phosphate-contaminated water by electrocoagulation with aluminum, aluminum alloy and mild steel anodes. J Hazard Mater 2009; 164(2-3): 1480-86.
29. Su C-C, Chang A-T, Bellotindos LM, Lu M-C. Degradation of acetaminophen by Fenton and electro-Fenton processes in aerator reactor. Sep Purif Technol 2012; 99: 8-13.
30. Cheng W, Yang M, Xie Y, Liang B, Fang Z, Tsang EP. Enhancement of mineralization of metronidazole by the electro-Fenton process with a Ce/SnO2–Sb coated titanium anode. Chem Eng J 2013; 220: 214-20.
31. Wu Y, Zhou S, Qin F, Peng H, Lai Y, Lin Y. Removal of humic substances from landfill leachate by Fenton oxidation and coagulation. Process Saf Environ Prot 2010; 88(4): 276-84.
32. Meunier N, Drogui P, Montané C, Hausler R, Mercier G, Blais JF. Comparison between electrocoagulation and chemical precipitation for metals removal from acidic soil leachate. J Hazard Mater 2006; 137(1): 581-90.
33. Saleem M, Bukhari A, Akram MN. Electrocoagulation for the treatment of wastewater for reuse in irrigation and plantation. Journal of Basic and Applied Sciences 2011; 7(1): 11-20
34. Wang S. A comparative study of Fenton and Fenton-like reaction kinetics in decolourisation of wastewater. Dyes Pigments 2008; 76(3): 714-20.
35. Vasudevan S, Lakshmi J, Kamaraj R, Sozhan G. A critical study on the removal of copper by an electrochemically assisted coagulation: equilibrium, kinetics, and thermodynamics. Asia-Pacific Journal of Chemical Engineering 2012; 8(1): 162-71.
36. Babuponnusami A, Muthukumar K. Advanced oxidation of phenol: a comparison between Fenton, electro-Fenton, sono-electro-Fenton and photo-electro-Fenton processes. Chem Eng J 2012; 183: 1-9.
37. Ruppert G, Bauer R, Heisler G. The photo-Fenton reaction — an effective photochemical wastewater treatment process. J Photochem Photobiol A Chem 1993; 73(1): 75-8.

Optimization of culture media for lipid production by *Nannochloropsis oculata* for Biodiesel production

Mohammad Malakootian[1*], Behnam Hatami[2], Shidwash Dowlatshahi[3], Ahmad Rajabizadeh[3]

[1]Professor, Environmental Health Engineering Research Center, Department of Environmental Health, Kerman University of Medical Sciences, Kerman, Iran
[2]MSc, Department of Environmental Health, Kerman University of Medical Sciences, Kerman, Iran
[3]Instructure, Environmental Health Engineering Research Center, Department of Environmental Health, Kerman University of Medical Sciences, Kerman, Iran

Abstract
Background: This study quantified the effect of four popular culture media in a biodiesel production study on the qualitative and quantitative lipid content, dry biomass, and lipid productivity of *Nannochloropsis oculata*.
Methods: Culture of microalgae was done separately in Walne, F/2, Sato, and TMRL media. In the logarithmic and stationary growth phases, biomass production and lipid accumulation of microalgae were measured and the constituents were identified by gas chromatography.
Results: *N. oculata* exhibited the highest rate of cell growth and biomass productivity of 0.2616 day^{-1} and 2.652 gl^{-1} in the Walne medium. The highest level of biomass conversion into lipids in TMRL medium revealed a cell dry weight of 37.22%. Walne medium proved to have the most efficient lipid productivity which was 0.1057 gl^{-1} day^{-1}. The highest amount of triacylglycerol (TAG) was obtained in Sato medium in the stationary growth phase and was 75.25% of the fatty acids.
Conclusion: The present study provides a practical benchmark, which allows the introduction of Walne as a suitable culture medium for *N. oculata* in biodiesel studies.
Keywords: *Nannochloropsis oculata*, Culture, Medium, Biofuel, Biodiesel, Lipid productivity

***Correspondence to:**
Mohammad Malakootian
Email: m.malakotian@gmail.com

Introduction

Energy crisis is one of the most society' s daunting challenge, as a result of the quick development of human activities and rapid depletion of fossil fuels (1,2). The use of fossil fuels as an energy source is intimately linked with the ever-increasing emission of carbon dioxide (3), climatic changes and global warming effects (4). In the past decade, attempts have been made to control these effects by reducing the level of CO_2 in the atmosphere, implementing the microalgae and producing renewable energy (5-8). Biodiesel as a sustainable and environmentally friendly alternative is a renewable, non-toxic, biodegradable and CO_2 neutral energy source (1,9-11). Microalgae are photosynthetic organisms and are a promising source for biodiesel production (12,13). Reacting microalgal oil in the form of triacylglycerol (TAG) with simple alcohol (known as "transesterification"), results in the formation of a chemical composition known as alkyl ester or biodiesel (14-16). In addition to fuel production, microalgae are commercially important for aquaculture and the food industry because they possess valuable products such as fatty acids, steroids, carotenoids and polysaccharides (17). *Nannochloropsis oculata* is a marine unicellular microalgae belonging to the Eustigmatophyceae class (18-21), alongside their ability to synthesize polyunsaturated fatty acids (PUFAs) and carotenoids for human and marine aquaculture consumption, they can also accumulate large amounts of neutral lipids in the form of TAG (13,22-25). In order to exploit these microalgae for biodiesel production and reduce the total cost of it, it is necessary to optimize biomass and lipid accumulation by obtaining a better understanding of the essential parameters contributing to the microalgae culture media (1,26,27). The composition of the culture media affects the specific growth rate, the maximum level of biomass production and change the biochemical composition of the biomass and lipids (5). For example, nutrient stress conditions is one of the most efficient ways of increasing lipid accumulation in cells and storage in the form of TAG with change in fatty acid (FA) composition in single cell microalgae (1,11). Studies have shown that the quality and quantity of microalgae lipid content can be changed as a result of changes in growth conditions (temperature, light intensity) or medium composition (nitrogen, phosphate and iron concentration) (15,28,29). Furthermore, in order to achieve mass cultivation of microalgal biomass in an industrial scale, optimization of the appropriate culture medium is one of the most important factors (7).

This study quantified the effect of 4 popular culture media in a biodiesel production study on the qualitative and quantitative lipid content, dry biomass, and lipid productivity of *N. oculata*.

Methods

Microalgae

N. oculata was obtained from the research institute for Aquaculture in the south of the country in the form of stock culture with high density (25×10^6 cell/ml). This microalga is a eukaryotic photosynthetic microorganism and given its simple structure, it has a fast growth rate (28). According to the study of Chen et al (13), which determined the effects of cell density on microalgae growth and lipid composition, this study used microalgae with high density.

Culture conditions of *N. oculata*

In order to determine the most appropriate medium, microalgae was cultured separately in four media namely Walne, F/2, Sato, and TMRL. Table 1 shows the composition of each medium including every elemental nutrient and concentration. In all the media, the Gillard vitamins were used. Growth experiments were repeated three times using a 2L-Erlenmeyer flask and in refrigerated incubators equipped with temperature and light intensity. The optimum value of temperature, 20°C, was chosen on the basis of data reported in the literature (28) and according to the study of Banerjee et al (30) and Sen et al (31), the light intensity would be 70 $\mu E\ m^{-2}\ s^{-1}$. In order to achieve higher efficiency, on the basis of Chiu et al study (14), air flow containing 2% of carbon dioxide was used for aeration, after being saturated in water and passed through a 0.45 μm filter. To avoid any kind of pollution during the different steps, the media and containers were sterilized in an autoclave at a temperature of 121°C.

Microalgae cell counting and dry weight

Cell density (cells mL^{-1}) was measured using a ultraviolet-visible spectrophotometer(UV-Vis) spectrophotometer (Shimadzu Corporation) at an absorbance wavelength of 680 nm. Each sample was diluted to give an absorbance in the range of 0.1–1.0, if optical density was greater than 1.0 (14).

Microalgae dry weight per liter (g L^{-1}) was measured according to the method previously reported (15). Microalgae cells were collected by centrifugation of wet biomass for 30 minutes in 15°C with 3000 rpm. The dry weight of marine microalgae samples was affected by salt absorbed on the cell surface and its presence in the intercellular water ensured error in estimating the amount of biomass. This explains the differentiations in the amount of dry cell weight in various papers. Hence, before gravimetric analysis, to remove salts, the centrifuged cells were again solved in 200 ml Ammonium format (0.5M, pH 8.0, adjusted with 1M NaOH) and centrifuged under the mentioned circumstances (30,32). The microalgae pellet was dried at 100°C for 4 hours for dry weight measurement (13,30,32).

Measurement of growth rate

Specific growth rate of microalgae in logarithmic phase was calculated as follows:

$$\mu = \frac{\ln N_f - \ln N_i}{t_2 - t_1} \quad \text{(Eq. 1)}$$

Where; μ (day^{-1}) is the specific growth rate, $\ln N_i$ and $\ln N_f$ are the cell densities (cell/ml) at the beginning and end of the logarithmic growth phase, respectively and t is the time (day) (1,14,5,30).

Extraction and measurement of lipid content and triacylglycerol

The methanol-chloroform (1/1, V/V) extraction method was used to extract total lipids from the dried cells (13,33,34). To remove residual microalgae, the extraction

Table 1. The media composition used in this study (m_{mol}/L)

Element's nutrients	Medium			
	Walne	F/2	Sato	TMRL
$MnCl_{2.}4H_2o$	7.6717×10^{-4}	0.3452	5.1784×10^{-4}	-
$CoCl_{2.}6H_2o$	0.084	0.042	3.3623×10^{-4}	-
$CuSo_4.5H_2o$	0.0801	0.04	1.602×10^{-5}	-
$FeCl_3$	4.932×10^{-3}	0.0194	1.8934×10^{-5}	10^{-5}
$NaHCo_3$	-	-	1.9998	-
$(NH_4)_6Mo_7O_{24.}4H_2O$	7.2814×10^{-4}	4.8543×10^{-6}	-	-
$ZnCl_2$	0.154	0.0765	2.2012×10^{-4}	-
$NaNO_3$	1.1765	1.1765	0.8824	0.8824
H_3BO_3	0.5434	-	0.0556	-
$Na_2EDTA.2H_2O$	0.1208	0.0117	8.0593×10^{-3}	-
$NaH_2PO_4.2H_2O$	0.0755	0.0387	0.0193	0.0193

lipids were filtered on membranes with 0.45 μm mean pore diameter. After washing twice with methanol and its complete evaporation, gravimetric analysis was done and part of the lipid fraction expressed as the percent of dry cell weight (28). Lipid productivity was calculated by (9,35):

$$P_{Lipid}\left(gl^{-1}day^{-1}\right) = \frac{(C_f \times DCW_f)-(C_i \times DCW_i)}{T} \quad \text{(Eq. 2)}$$

Where; P_{Lipid} is the lipid productivity, C_f and D_{CWf} are the lipid content (g/g) and biomass (g/l) of the microalgae in the final stationary growth phase, respectively; C_i and D_{CWi} are the lipid content (g/g) and biomass (g/l) of the microalgae in the initial stationary growth phase, respectively; and T is the cultivation time (day). After the measurement of total lipid, the dried lipid was solved in 0.4 ml of Isopropyl alcohol and the TAG was estimated by an enzymatic colorimetric method using a commercial kit (36).

Determination of fatty acid profiles

The direct esterification method was used to measure the fatty acid property. A mixture of 100 mg of lyophilized microalgae and 8 ml of KOH was sonicated for 3 minutes. For saponification, the mixture was heated to 100°C for 15 minutes and cooled to room temperature. For esterification, 8 ml of 0.7 N HCl in methanol and BF_3/CH_3OH was added to the mixture (14% V/V) and again was heated to 100°C for 15 minutes. After cooling to room temperature, to avoid emulsification, 2 ml of a saturated solution of sodium chloride was added. The FAMEs were extracted by adding aliquots of n-hexane. The FAMEs in the hexane layer were analyzed using standard gas chromatography (Agilent technologies 7890A-5975c) with a capillary column and a flame ionization detector. Nitrogen was used as the carrier gas and delivered at a rate of 1.5 mL min^{-1}. The temperature was programmed to increase from 130°C to 180°C at a rate of 10°C min^{-1} and thereafter ramped to 210°C at a rate of 2°C min^{-1}. The injector and detector were kept at 220°C and 250°C, respectively (1,35).

Statistics

All values were expressed as mean ± standard deviation (SD). Data were analyzed using one-way analysis of variance (ANOVA) and in order to determine the statistical difference between media, the Tukey test was used. A value of $P<0.05$ was considered statistically significant (21,24).

Results

Effects of medium on cell growth rate of N. oculata

Figure 1 shows the cell growth rate of *N. oculata* in four different media. *N. oculata* exhibited the highest growth rate in the Walne medium. Except for the TMRL medium, in which there was no logarithmic phase, in other media, the microalgae after a lag phase of 24 hours, entered the logarithmic growth phase. The results of the one-way ANOVA showed that the observed differences between media types were significant ($P<0.05$). Tukey test also confirmed that they were not put in homogenous groups and the Walne medium had considerable difference with F/2 ($P=0.008$), Sato ($P=0.002$) and TMRL ($P=0.000$) media. There was no significant statistical difference between F/2, Sato and TMRL media ($P>0.05$).

Effects of medium on biomass and lipid production

Figure 2 shows the results of medium effects on biomass production in two logarithmic and stationary growth phases. Walne medium recorded the highest biomass production. The results of one-way ANOVA showed a large variation in biomass production between the culture media used. Tukey test also confirmed that they were not in homogenous groups and the Walne medium had a significant difference with the F/2 ($P=0.005$), Sato ($P=0.001$) and TMRL ($P=0.000$) media. Biomass production did not vary greatly between the F/2, Sato and TMRL media ($P>0.05$).

The results of medium effects on lipid production in 2 logarithmic and stationary growth phases are shown in Figure 3. The TMRL medium recorded the highest percentage of biomass conversion to lipid. The results of one-way ANOVA showed that the difference between lipid conversion percentages in the used media was significant. Also, the Tukey test showed that the differences between all media were significant ($P<0.001$) but between Walne and F/2 media, no significant differences were observed ($P=0.065$).

Effect of medium on fatty acids composition

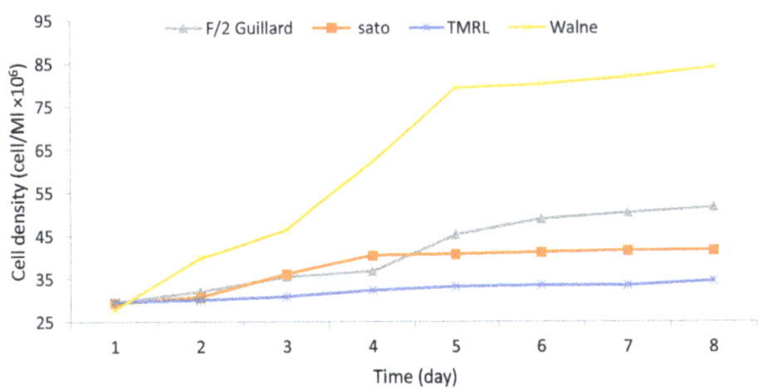

Figure 1. Cell growth rate of *Nannochloropsis oculata* in Walne, F/2, Sato and TMRL media.

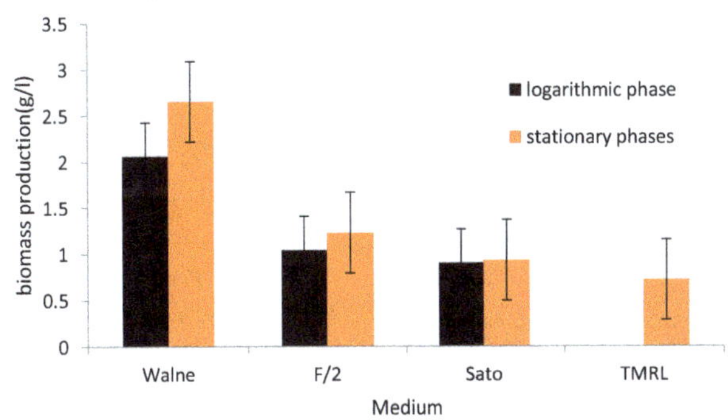

Figure 2. Biomass production by *Nannochloropsis oculata* in logarithmic and stationary growth phases in various media (in TMRL medium, there was no logarithmic growth phase).

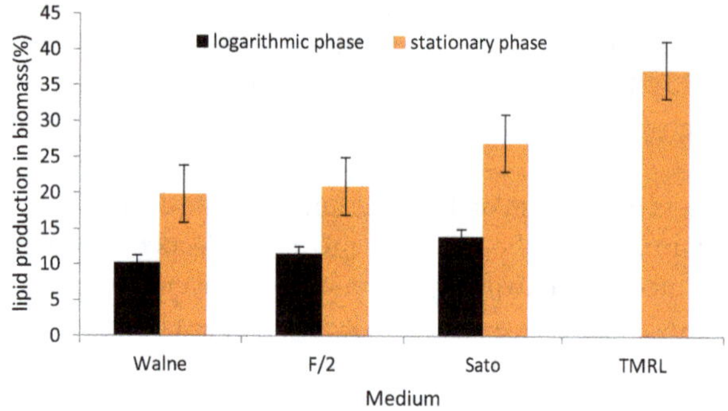

Figure 3. Lipid production by *Nannochloropsis oculata* in logarithmic and stationary growth phases in various mediums (in TMRL medium, there was no logarithmic growth phase).

The results of fatty acids composition of *N. oculata* grown in various media during logarithmic and stationary growth phases are shown in Table 2.

Discussion

Maximal cell densities, specific growth rates and biomass production

N, P, K, Mg, Ca, S, Fe, Cu, Mn, and Zn are essential elements for the growth of green algae added to culture media in the form of salts (37). Due to the existence of enough nutrients in the Walne and F/2 media, the logarithmic growth phase continued until the fifth day but it reached the end of the logarithmic phase on the fourth day, in the Sato medium. In the TMRL medium, the logarithmic phase was so short and could be ignored.

N. oculata reached the highest cell density (84×10^6 cell/ml) (Figure 1) and highest biomass production (2.652 g/l) (Figure 2) in the Walne medium at the end of the stationary growth phase and conformed to the values reported by Olofsson et al (11), Solovchenko et al (25) and Chiu et al (14). According to Table 1, unlike other media, the Walne medium used two nitrogen sources (sodium nitrate and ammonium molybdate). Nitrogen as an important constituent of cellular protein and chlorophyll molecules are required for microalgae cell growth (21). Another reason for higher growth rate and biomass production in the Walne medium was the existence of ammonium which is a necessary element for the microalgae and its concentration in the Walne medium was 150 times more than the F/2 medium. This was not observed in other media. It takes more energy to assimilate NO3-N than to assimilate $NH4^+$-N, hence microalgae prefer $NH4^+$-N in the medium (21). On the other hand, in consideration of the role of phosphate in producing ATP required for photosynthesis and rapid microalgae growth (38), its concentration in Walne medium was double that of the F/2 medium and 4 times the other media. Copper is one of the required elements for microalgae and is an important part of the plastocyanin protein in the electron transport chain. The concentration of copper in the Walne medium is considerably higher than other media.

The fastest specific growth rate among the media was recorded in the Walne medium (0.2616 day^{-1}), while the F/2 and Sato media were 0.1066 day^{-1} and 0.08 day^{-1}, respectively. The specific growth rates found in the present study were not in the median range of other published studies. In the study of Chiu et al (14) and Banerjee et al (30) *on N. oculata* cultivation, the specific growth rate was 0.571 day^{-1} and 0.004 day^{-1}. Differences in growth rates compared to the other studies could be due to differences in culturing methods, reactor geometry, prior culture history, light intensity and temperature (15,20).

Lipid accumulation and lipid productivity

Algal biomass is composed of three main components namely carbohydrates, proteins and lipids (natural oils) (16). To increase the proportion of the biomass that contains a useful lipid, different strategies including nutrient starvation, bioprocess optimization and genetic modification (39,40) were used. Under stress conditions, photosynthetic activity in microalgae decreases and the excess energy might be stored in the form of valuable compounds such as lipids (17,24). Nutrient starvation is one

Table 2. Fatty acid composition in dry weight, percentage of *Nannochloropsis oculata* grown in Walne, F/2, Sato and TMRL medium during logarithmic and stationary phases

Fatty acid	Name	Medium						
		Walne		F/2		Sato		TMRL
		Logarithmic	Stationary	Logarithmic	Stationary	Logarithmic	Stationary	Stationary
C14:0	Myristic acid	3.5[a]	5.59	3.1	6.45	3.1	4.79	3.3
C15:0	-	0.2	0.34	0.1	0	0.2	0.12	0.2
C16:0	Palmitic acid	22.76	28.57	29.97	34.36	30.5	35.17	34.3
C16:1n-7	Palmitoleic acid	18.45	25.04	19.3	23.19	20.11	24.16	22.56
C18:0	Stearic acid	1.01	0.73	0.8	0.8	1	0.89	0.7
C18:1n-9	Oleic acid	6.3	5.24	7.4	7.85	9.6	11.01	7.5
C18:2n-6	Linoleic acid	3.15	1.12	1.7	0.23	2.4	1.15	1.2
C20:0	Eicosanoic acid	0.1	0	0.1	0.03	0.1	0	0.1
C20:4n-6	Arachidonic acid	3.73	2.46	3.25	1.1	2.98	0.76	3.21
C20:5n-3	Eicosapentaenoic acid (EPA)	36.11	22.19	27.6	19.98	24.2	16.17	17.3

[a] Values represent mean of two replicate samples.

of the stress conditions that can be applied to microalgae by culture medium modifications and therefore change the biochemical composition of the biomass. Nitrogen and phosphorus are the most common restrictive factors in the media that can lead to lipid accumulation. In nitrogen limiting media, the lipid content usually increases in the algae due to less susceptibility of lipid-synthesizing enzymes for disorganization than carbohydrate synthesizing enzymes due to nitrogen deprivation (2). Also, decreasing the nitrate concentration of the medium causes a decrease in the amount of chlorophyll II and limits biochemical protein synthesis (28,37). According to the results in Figure 3, the maximum percentage of converting biomass to lipid was attained in the TMRL medium (37.22% of dry cell weight). Also the results of lipid analysis in two logarithmic and stationary growth phases showed that lipid accumulation in microalgae cells had a direct relation to their growth phases and growth from the logarithmic to the stationary growth phase was accompanied by increase in lipid percentage. These results matched the results of Nigam et al (6), Hu and Gao (19) and Gouveia and Oliveira (4) studies.

The lipid content or biomass productions are not appropriate scales for microalgae lipid yields alone in biodiesel production and the most crucial comparative measure is lipid productivity that must be calculated using Equation 2 (15). The maximum lipid productivity was related to the Walne medium 0.1057 and for the F/2, Sato and TMRL was 0.0462, 0.0417, 0.023 gl^{-1} day^{-1} sequence, respectively. In Gouveia and Oliveira study (4), the lipid productivity was 0.09 gl^{-1} day^{-1} and in Griffiths and Harrison study (10), it was 0.082 gl^{-1} day^{-1} and matched the results obtained in this study.

Fatty acid compositions

Vegetable oils currently used for biodiesel productions are mainly C16 and C18. Olofsson et al (11) proposed myristic acid, palmitic acid, palmitoleic acid, stearic acid and oleic acid as important fatty acids for biodiesel production and consisted more than 45%-78% of all fatty acid compositions. According to the results of this study, these fatty acids consist of 65.7%, 72.65%, 76.02% and 68.38% of fatty acid composition in the dry weight of *N. oculata* grown in the Walne, F/2, Sato and TMRL media during the stationary phase.

Also, the microalgae oil in the form of TAG can be converted to biodiesel. The TAG of *N. oculata* consists of saturated and monounsaturated fatty acids and is mainly stored in vacuoles within the cell (11,24). In this study, TAG consisted of 64.78%, 71.88%, 75.25% and 67.96% of fatty acid composition in dry weight of *N. oculata* grown in the Walne, F/2, Sato and TMRL media during the stationary phase. Therefore, the Sato medium quantitatively provided the maximum amount of TAG for biodiesel production.

Conclusion

Lab scale experiments have an important role in developing biodiesel studies and finally scale-up production. In this study, a microalgae culture medium was surveyed as a major aspect of biodiesel production. The results showed that the best medium for *N. oculata* cultivation is Walne and the microalgae had maximum efficiency in the stationary growth phase. One advantage of this study is the use of *N. oculata* for aquaculture and human consumptions.

Acknowledgements

The work was financially supported by the Research and Technology Assistance Division of Kerman University of Medical Sciences, Kerman, Iran. This work was also partially supported by the Environmental Health Engineering Research Center of Kerman University of Medical Sciences, Kerman, Iran.

Competing interests

Authors declare that they have no competing interests.

Authors' contributions

SD and BH conceived and designed the study. MM and AR performed the literature search and wrote the manuscript. All authors participated in the data acquisition, analysis and interpretation. All authors critically reviewed, refined and approved the manuscript.

References

1. Xin L, Hong-ying H, Yu-ping Z. Growth and lipid accumulation properties of a freshwater microalga Scenedesmus sp. under different cultivation temperature. Bioresour Technol 2011; 102(3): 3098-102.
2. Rajasri Y, Ramgopal RS, Rao CS. Lipid accumulation studies in chlorella pyrenoidosa using customized photobioreactor - effect of nitrogen source, light intensity and mode of operation. International Journal of Engineering Research and Applications 2012; 2(3): 2446-53.
3. Xiaodong D, Yajun L, Xiaowen F. Effects of selective medium on lipid accumulation of chlorellas and screening of high lipid mutants through ultraviolet mutagenesis. African Journal of Agricultural Research 2011; 6(16): 3768-74.
4. Gouveia L, Oliveira A. Microalgae as a raw material for biofuels production. J Ind Microbiol Biotechnol 2009; 36(2): 269-74.
5. Sánchez S, Martínez M, Espinola F. Biomass production and biochemical variability of the marine microalga Isochrysis galbana in relation to culture medium. Biochem Eng J 2000; 6(1): 13-8.
6. Nigam S, Rai MP, Sharma R. Effect of nitrogen on growth and lipid content of chlorella pyrenoidosa.

Am J Biochem Biotechnol 2011; 7(3): 124-9.
7. Sankar M, Ramasubramanian V. Biomass production of commercial algae Chlorella vulgaris on different culture media. E-Journal of Life Science 2012; 1(1): 56-60.
8. Wang G, Wang T. Characterization of lipid components in two microalgae for biofuel application. J Am Oil Chem Soc 2012; 89(1): 135-43.
9. Li Y, Horsman M, Wang B, Wu N, Lan C. Effects of nitrogen sources on cell growth and lipid accumulation of green alga Neochloris oleoabundans. Appl Microbiol Biotechnol 2008; 81(4): 629-36.
10. Griffiths M, Harrison SL. Lipid productivity as a key characteristic for choosing algal species for biodiesel production. J Appl Phycol 2009; 21(5): 493-507.
11. Olofsson M, Lamela T, Nilsson E, Bergé JP, del Pino V, Uronen P, et al. Seasonal variation of lipids and fatty acids of the microalgae Nannochloropsis oculata grown in outdoor large-scale photobioreactors. Energies 2012; 5(5): 1577-92.
12. Banskota A, Stefanova R, Gallant P, McGinn P. Mono- and digalactosyldiacylglycerols: potent nitric oxide inhibitors from the marine microalga Nannochloropsis granulata. J Appl Phycol 2013; 25(2): 349-57.
13. Chen Y, Wang J, Liu T, Gao L. Effects of initial population density (IPD) on growth and lipid composition of Nannochloropsis sp. J Appl Phycol 2012; 24(6): 1623-7.
14. Chiu SY, Kao CY, Tsai MT, Ong SC, Chen CH, Lin CS. Lipid accumulation and CO2 utilization of Nannochloropsis oculata in response to CO2 aeration. Bioresour Technol 2009; 100(2): 833-8.
15. Huerlimann R, de Nys R, Heimann K. Growth, lipid content, productivity, and fatty acid composition of tropical microalgae for scale-up production. Biotechnol Bioeng 2010; 107(2): 245-56.
16. Karampudi S, Chowdhury K. Effect of media on algae growth for bio-fuel production. Notulae Scientia Biologicae 2011; 3(3): 33-41.
17. Recht L, Zarka A, Boussiba S. Patterns of carbohydrate and fatty acid changes under nitrogen starvation in the microalgae Haematococcus pluvialis and Nannochloropsis sp. Appl Microbiol Biotechnol 2012; 94(6): 1495-503.
18. Sandnes JM, Källqvist T, Wenner D, Gislerød HR. Combined influence of light and temperature on growth rates of Nannochloropsis oceanica: linking cellular responses to large-scale biomass production. J Appl Phycol 2005; 17(6): 515-25.
19. Hu H, Gao K. Response of growth and fatty acid compositions of Nannochloropsis sp. to environmental factors under elevated CO2 concentration. Biotechnol Lett 2006; 28(13): 987-92.
20. Spolaore P, Joannis-Cassan C, Duran E, Isambert A. Optimization of Nannochloropsis oculata growth using the response surface method. J Chem Technol Biotechnol 2006; 81(6): 1049-56.
21. Durmaz Y. Vitamin E (α-tocopherol) production by the marine microalgae Nannochloropsis oculata (Eustigmatophyceae) in nitrogen limitation. Aquaculture 2007; 272(1–4): 717-22.
22. Zhang J, Liu S, Sun X, Yang G, Zhang X, Gao Z. Fatty acid composition analyses of the DCMU resistant mutants of Nannochloropsis oculata (Eustigmatophyceae). Journal of Ocean University of Qingdao 2003; 2(1): 65-8.
23. Solovchenko AE, Khozin-Goldberg I, Didi-Cohen S, Cohen Z, Merzlyak MN. Effects of light intensity and nitrogen starvation on growth, total fatty acids and arachidonic acid in the green microalga Parietochloris incisa. J Appl Phycol 2008; 20(3): 245-51.
24. Pal D, Khozin-Goldberg I, Cohen Z, Boussiba S. The effect of light, salinity, and nitrogen availability on lipid production by Nannochloropsis sp. Appl Microbiol Biotechnol 2011; 90(4): 1429-41.
25. Solovchenko A, Khozin-Goldberg I, Recht L, Boussiba S. Stress-induced changes in optical properties, pigment and fatty acid content of Nannochloropsis sp.: implications for non-destructive assay of total fatty acids. Mar Biotechnol 2011; 13(3): 527-35.
26. Simionato D, Sforza E, Corteggiani Carpinelli E, Bertucco A, Giacometti GM, Morosinotto T. Acclimation of Nannochloropsis gaditana to different illumination regimes: effects on lipids accumulation. Bioresour Technol 2011; 102(10): 6026-32.
27. Huang X, Huang Z, Wen W, Yan J. Effects of nitrogen supplementation of the culture medium on the growth, total lipid content and fatty acid profiles of three microalgae (Tetraselmis subcordiformis, Nannochloropsis oculata and Pavlova viridis). J Appl Phycol 2013; 25(1): 129-37.
28. Converti A, Casazza AA, Ortiz EY, Perego P, Del Borghi M. Effect of temperature and nitrogen concentration on the growth and lipid content of Nannochloropsis oculata and Chlorella vulgaris for biodiesel production. Chemical Engineering and Processing: Process Intensification 2009; 48(6): 1146-51.
29. Srinivas R, Ochs C. Effect of UV-A irradiance on lipid accumulation in Nannochloropsis oculata. Photochem Photobiol 2012; 88(3): 684-9.
30. Banerjee S, Hew WE, Khatoon H, Shariff M, Yusoff FM. Growth and proximate composition of tropical marine Chaetoceros calcitrans and Nannochloropsis oculata cultured outdoors and under laboratory conditions. Afr J Biotechnol 2011; 10(8): 1375-83.
31. Sen B, Koser M, Alp M, Erbas H. Studies on growth of marine microalgae in batch cultures:III. Nannochloropsis Oculata (Eustigmatophyceae). Asian Journal of Plant Science 2005; 4(6): 642-4.
32. Zhu CJ, Lee YK. Determination of biomass dry weight of marine microalgae. J Appl Phycol 1997; 9(2): 189-94.
33. Xin L, Hu HY, Ke G, Sun YX. Effects of different nitrogen and phosphorus concentrations on the

growth, nutrient uptake, and lipid accumulation of a freshwater microalga Scenedesmus sp. Bioresour Technol 2010; 101(14): 5494-500.
34. Wan M, Liu P, Xia J, Rosenberg J, Oyler G, Betenbaugh M, et al. The effect of mixotrophy on microalgal growth, lipid content, and expression levels of three pathway genes in Chlorella sorokiniana. Appl Microbiol Biotechnol 2011; 91(3): 835-44.
35. Su CH, Chien LJ, Gomes J, Lin YS, Yu YK, Liou JS, et al. Factors affecting lipid accumulation by Nannochloropsis oculata in a two-stage cultivation process. J Appl Phycol 2011; 23(5): 903-8.
36. Xin L, Hong-Ying H, Jia Y, Yin-Hu W. Enhancement effect of ethyl-2-methyl acetoacetate on triacylglycerols production by a freshwater microalga, Scenedesmus sp. LX1. Bioresour Technol 2010; 101(24): 9819-21.
37. Mandalam RK, Palsson BO. Elemental balancing of biomass and medium composition enhances growth capacity in high-density Chlorella vulgaris cultures. Biotechnol Bioeng 1998; 59(5): 605-11.
38. Hu H, Zhou Q. Regulation of inorganic carbon acquisition by nitrogen and phosphorus levels in the Nannochloropsis sp. World J Microbiol Biotechnol 2010; 26(5): 957-61.
39. Van Wagenen J, Miller TW, Hobbs S, Hook P, Crowe B, Huesemann M. Effects of light and temperature on fatty acid production in Nannochloropsis salina. Energies 2012; 5(3): 731-40.
40. Malakootian M, Hatami B, Dowlatshahi S, Rajabizadeh A. Evaluation of factors affecting on lipid extraction for recovery of fatty acids from Nannochloropsis oculata micro-algae to biodiesel production. Environ Health Eng Manag J 2014; 1(1): 19-24.

5

Temporal and spatial variation of hardness and total dissolved solids concentration in drinking water resources of Ilam City using Geographic Information System

Zabihollah Yousefi[1], Reza Ali Mohammadpour Tahmtan[2], Farzad Kazemi[3*]

[1]Professor, Department of Environmental Health Engineering, School of Public Health, Mazandaran University of Medical Sciences, Sari, Iran
[2]Associate Professor, Department of Biostatistics, School of Public Health, Mazandaran University of Medical Sciences, Sari, Iran
[3]MSc Student, Department of Environmental Health Engineering, School of Public Health, Mazandaran University of Medical Sciences, Sari, Iran

Abstract
Background: In recent times, the decreasing groundwater reserves due to over-consumption of water resources and the unprecedented reduction of precipitation, during the past 1 decades, have resulted in a change in the volume and quality of water with time. The aim of this study was to determine the spatial and temporal variations of hardness and total dissolved solids in drinking water resources of Ilam city, using the GIS system.
Methods: This cross-sectional study was carried out on 20 sources of drinking water in Ilam and the results of 5 years archived by the Water and Sewage Co were analyzed using geographic information system (GIS) software version 9.3, SPSS version 16 and one-way analysis of variance (ANOVA). The sampling and measurement were also performed in this study based on the *Standard Method* book.
Results: The ordinary kriging method and spherical model are the best interpolation methods for hardness and total dissolved solid, due to the normal distribution of data. The highest values of parameters in most cases are related to the western parts based on maps. The one-way ANOVA test showed that the average amount of total hardness ($P=0.68$) and total dissolved solids ($P=0.6$) in different seasons of the year are the same.
Conclusion: Overuse of groundwater due to illegal digging and permanent easy access to water, increased the salinity of water in the central sections of the studied area. Proper planning that allows the withdrawal of water from authorized underground aquifers or water supply from surface water or dams should be done to overcome this problem.
Keywords: Geographic information system (GIS), Temporal and spatial variation, Hardness, Total dissolved solid, Drinking water resources, Ilam city.

***Correspondence to:**
Farzad Kazemi
Email: farzadkazemi1368@yahoo.com

Introduction

Water electrical conductivity is a total parameter for soluble materials and ion separation, its ability have a determined ratio with the total dissolved solid (TDS) of water, for this reason, it is important for simple control of water quality. The amount of TDSs is calculated by multiplying the electrical conductivity ability by an experimental coefficient (0.7 to 0.55) (1). The extreme amounts of TDS can have different negative effects such as increase of soap usage, create speckle on dishes, blasting vapor pots, etc (2). Hard water is defined as water that requires a considerable amount of soap for spume and creates germ in warm water pipelines, heaters, boilers and other units, when warm water comes in contact with them (3). Water with hardness greater than 200 mg/L can create germ deposit in the distribution and consumption system of soap; on the other hand, water with hardness of less than 100 mg/L is corrosive to pipes (4).
Water can cause sedimentation problems when associated with the digestive system. Studies have shown that the corrosion products at the surfaces of pipes can be deposited or accumulated in distribution networks and may protect microorganisms from being disinfected, resulting in the multiplication of these microorganisms and problems such as taste and odor, mass biological and increase in corrosion may occur (5,6).
The corrosive waters rise to secondary pollutants, such as iron, zinc, copper and manganese, in concentrations higher than the standard drinking water, problems, such as taste, smell, color and stains on the toilet may occur

(7). A review by Avazpour et al (8) on Ilam city's drinking water revealed that it is corrosive.

Due to the significant expansion of carbonate rocks in the province and several appropriate conditions, the aquifer water potential fit with the rock formations. The major factors in the formation of limestone aquifers in the province are "open striking limestone, performance of tectonic forces that result in the collapse of the compact and close together, average annual temperatures in mountainous areas and good rainfalls.

TDS is one of the most important factors in water quality and has impact on transport, chemical transformation and ionization of material. The quality of irrigation water, TDS, salinity risk can be divided into four namely low risk (<500 mg/l), medium risk (500-1000 mg/l), high risk (1000-2000 mg/l) and very high risk (>2000 mg/l) (9).

Hardness is a concept by which the amount of calcium and magnesium ions is expressed and is a measure of water quality. Despite the importance of hard water to the body, its disadvantages are numerous. Hard water causes loss of taste to drink. The late maturation of crops as a result of water hardness is another complication (10).

Studies that stressed on the hydrological and hydro dynamical aspects of water resources have been carried out, however, it was reported that a qualitative research on underground resources is a fundamental and applied movement for obtaining qualitative information on water resources (3).

In a study by Ehsani et al (11), the total amount of dissolve solids throughout the plateau was investigated and informational layers were designed. In a study by Rahmani and Shokohi (3) on the quality of Hamadan underground water, the results obtained showed that, according to aquifer quantity drop off, the quality changes of underground water resources are yet to attain an intensive state and one third of the stations have TDS that are more than the standard level (3). Therefore, the aim of this study was to determine the spatial and temporal variations of hardness and TDSs in the drinking water of Ilam city using geographical information system (GIS).

Methods

According to the "geographical situation and political division of Ilam city" in Ilam province, 19086 km^2 forms 1.4% of the total area of the country. It is placed in the west edge of Zagrous mountains between 31°58' and 34°15' N in width from the equator, and 45°24' to 48°10' E longitude from the Greenwich meridian, in the west of Iran. Ilam shares a common border with Khozestan from the south, Lorestan from the east, Kermanshah from the north (inside provinces) and Iraq with 425 km from the west (has common border). According to the recent political divisions in 1996, Ilam has 7 city, 17 town, 15 state, 36 rural district and 753 villages.

In determining the region in this descriptive research,, the maps of management and programming organization were used. Wells sampled according to latitude and longitude on a map were found. The obtained results of the chemical parameters in 2009-2013, which are being used once in a season by water and wastewater organization, were entered into the GIS software and were reserved as an information bank. Data were collected from all water sources in the city of Ilam (20 wells), respectively. The research results after the parameters' process were colored by GIS that can describe the process of parameters' changes in spatial and temporal situations as shown in Figure 1. This figure shows provision of field maps of different parameters in a thematic model.

Statistical population: All sources of drinking water (20 wells).

Sampling method: Experiments on chemical parameters were conducted once in a season by water and wastewater organization.

Data analysis and comparison were done using SPSS® 16 software and analysis of variance (ANOVA) statistical test, respectively. Thereafter, the data were entered into data base GIS program (ArcGIS® 9.3). After the entering of data, the preparation of informational layers parameters' maps was extended (Geostatistic). Data ordering, processing and analyzing were carried out in order to provide suitable diagrams for spatial modeling in ArcGIS® 9.3 software.

Results

Figure 2 shows the spatial changes of total resources of drinking hard water in Ilam in different seasons of 2009.

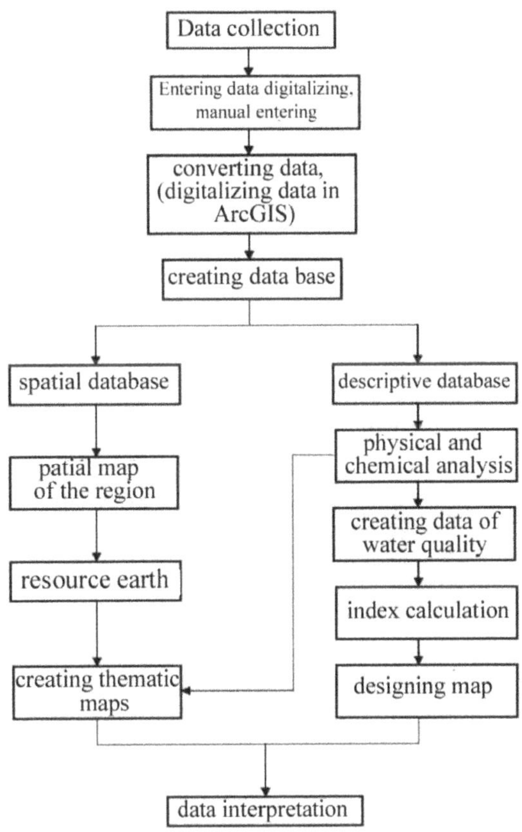

Figure 1. The process for mapping.

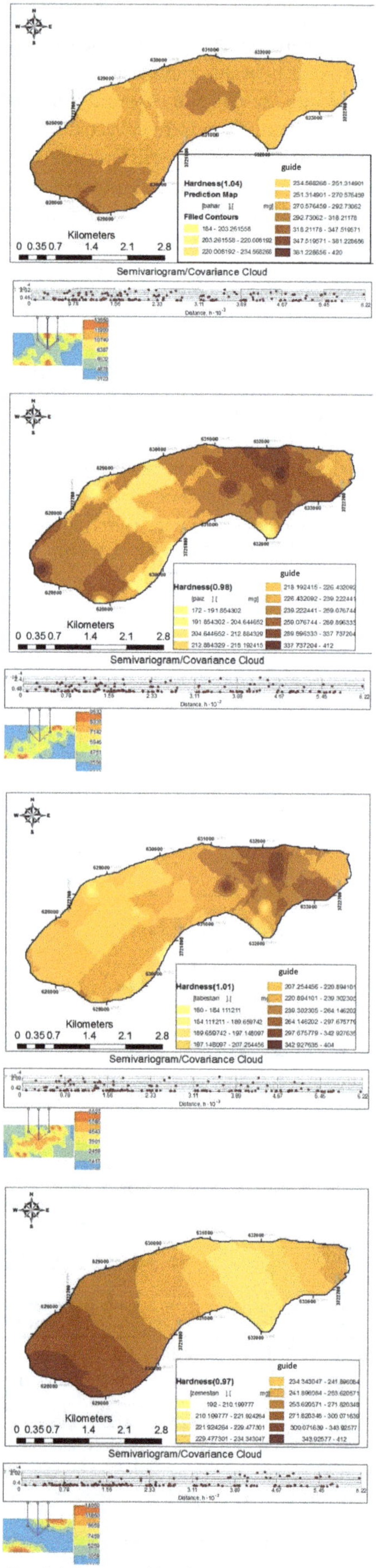

Figure 2. Spatial changes of hardness in drinking water resources of Ilam city in different seasons in 2009.

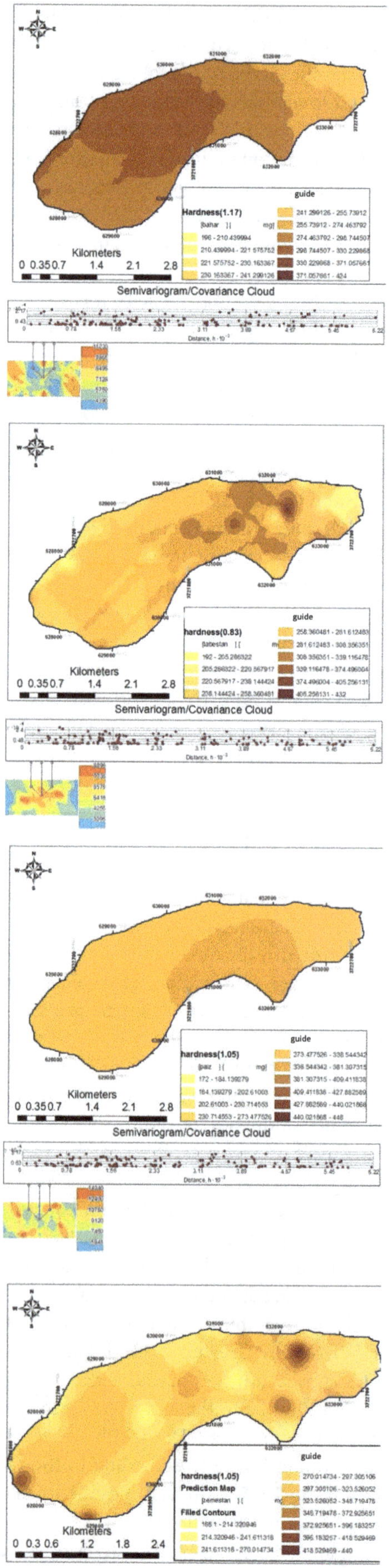

Figure 3. Spatial changes of hardness in drinking water resources of Ilam City in different seasons in 2013.

The best interpolation methods for hardness according to normal data distribution were the normal kriging and spherical models. The root mean square errors (RMSE) for spring, summer, fall and winter were 1.04, 1.01, 0.98 and 0.97, respectively. According to the interpolation map, the amount of water hardness was varied in different seasons, so as to have high concentration during spring and winter in the western areas, but in the other two seasons, the concentration was, low. In general, the concentration of hardness was high in all areas.

Figure 3 shows the spatial changes of hardness in drinking water resources of Ilam in different seasons in 2013. The RMSE for spring, summer, fall and winter were 1.17, 0.83, 1.05 and 1.05 respectively. According to the interpolation map, the amount of water hardness was varied in different seasons, in order to have high concentration in the western areas during winter and spring and low concentration in the other two seasons in the west. In general hardness concentration is high in all areas. The ANOVA test shows that the total mean hardness in different seasons of the year is similar ($P=0.68$), while the mean of hardness in wells and different resources have significant differences ($P<0.001$).

The total hardness mean of all experiments on water resources under examination was 316.4±78.5 mg/l and the variance analysis test shows that the mean results of the verification tests and measurements in different seasons in different years were significant ($P=0.1$).

Figure 4 shows the spatial changes of TDSs in Ilam city's drinking water resources in all seasons of year 2009. The best interpolation method for dissolved solids according to data distribution were normal kriging and spherical model. The RMSE for spring, summer, fall and winter were observed as 1.08, 0.88, 1.02 and 1.04 respectively. According to the interpolation maps, high amount of dissolve solids was observed in the western areas, except for summer in which high concentration was seen in eastern areas. The high amounts were seen during spring and fall.

Figure 5 shows the spatial changes of TDSs of Ilam city's drinking water resources in different seasons in 2013. The semivariogram drawn on each map represents spatial correlation between samples. The RMSE for spring, summer, fall and winter are 1.14, 0.99, 1.08 and 1.12, respectively. The high amounts of dissolved solids based on the interpolation map were observed in the western areas. However, high concentration was observed during spring. The one-way ANOVA test shows that the means of the TDSs in different seasons were similar ($P=0.6$), while the mean of wells and other resources had significant difference ($P<0.001$).

Furthermore, the results of the verification test show that the total means of the TDSs on water resources was 414.2±99.14 mg/l, and the ANOVA test shows that the mean of verification test results had significant differ-

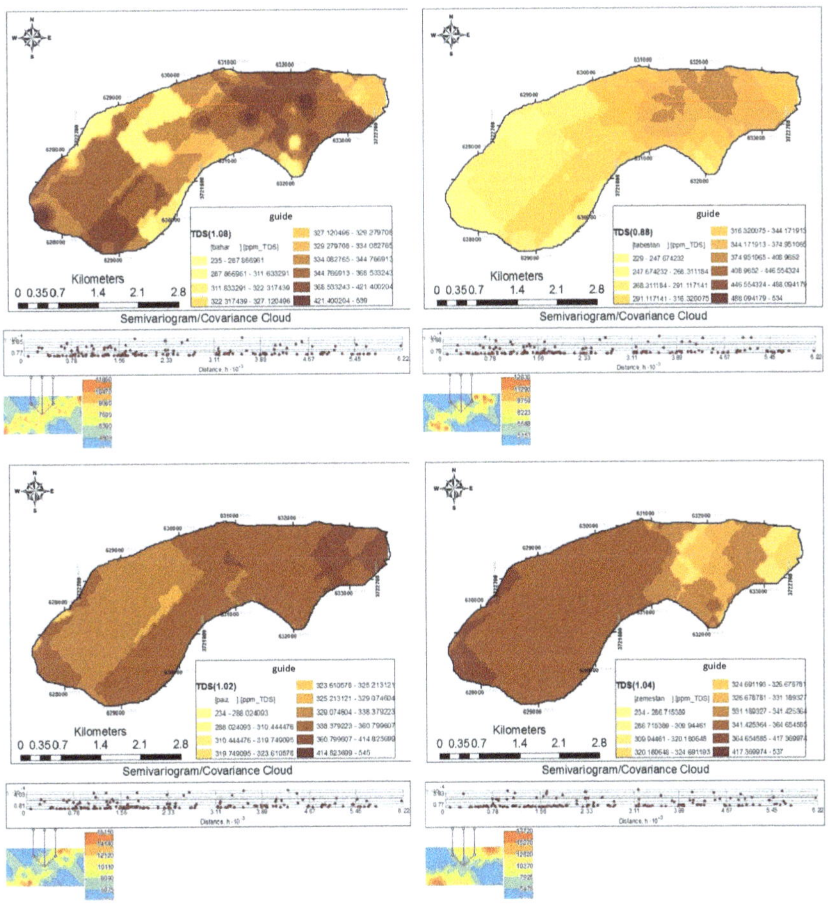

Figure 4. Spatial changes of TDSs in drinking water resources of Ilam city in different seasons in 2009.

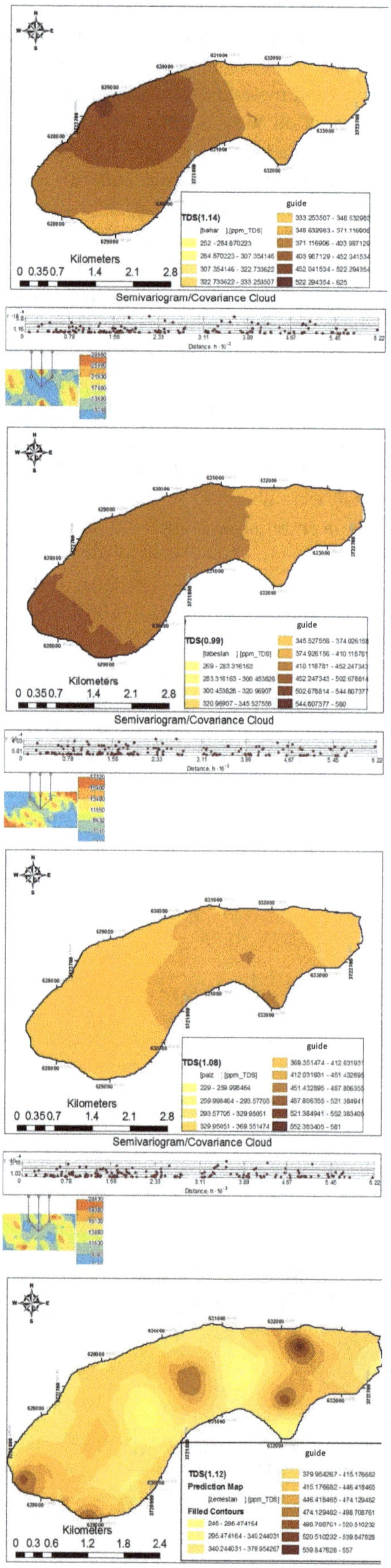

Figure 5. Spatial changes of TDSs in drinking water resources of Ilam in different seasons in 2013.

ence with different seasons measured in different years except for during spring ($P=0.012$), summer ($P=0.04$), fall ($P=0.072$) and winter ($P=0.05$), based on the process in different years as well as increase and decrease in the amount of rainfalls, these differences were predictable. Table 1 shows the comparison between delivered data and analyzed samples. As the table shows, the trend of these data is on the increase.

Also, the increasing trends of TDS and hardness are shown in Figures 6 and 7.

Discussion

In the optimal management of water resources, reversing and promoting the water resources require data in relation to situation, quantity and water chemical reactions in a special geographical area. The aim of this study, therefore, was to determine spatial and temporal changes of

Table 1. Comparison of the results of water, wastewater and verification

Chemical parameters	Average data for 20 wells in the years 2009-2013	The average results of verification for 20 wells in 2015
TDS (mg/l)	364.4761	414.2
Total Hardness (mg/l $CaCO_3$)	273.7018	316.4

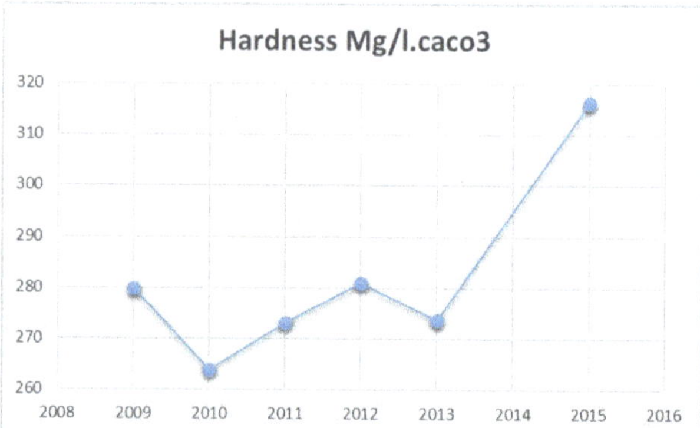

Figure 6. Changes in the value of total hardness in years 2009-2013 and verification in year 2015

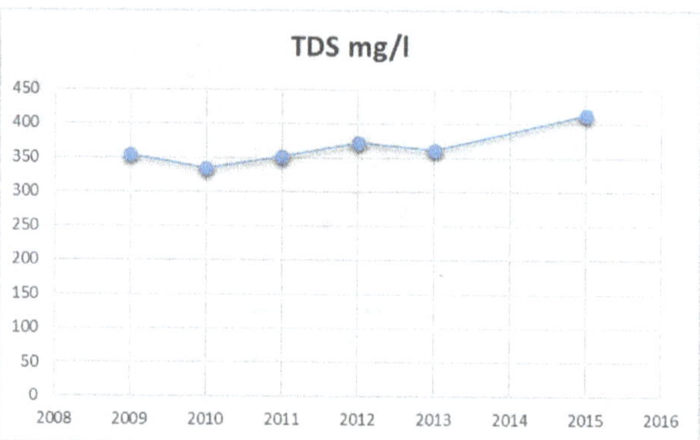

Figure 7. Changes in the value of TDS in years 2009-2013 and verification in year 2015..

hardness and TDSs of Ilam's drinking water resources, the most suitable interpolation method was used to spatially investigate and analyze the parameter amount in Ilam.

Determining the best method of interpolation for water chemical quality:

According to all obtained maps in this study, the best method of interpolation was kriging. The study of spatial changes of underground salty water based on statistical earth (case study: Rafsanjan) by Taghizadeh Mehrjerdy et al (12) showed that in determining the salinity of water kriging method has priority over the inverse distance weighted (IDW) method (12).

A study was carried out to determine the best methods of statistics maps changes in TDS and TH groundwater city of Fasa, it was shown that among the deterministic methods for mapping changes in TDS in the area, the ordinary kriging, using Gaussian model that is lower than other methods as result of the small RMSE and ME value, was more suitable (13).

Gholami et al (14) researched on the modeling of groundwater salinity on the Caspian southern coasts and reported that there is a relation between water EC and TDS (TDS = 0.6 EC).

In order to better understand the spatial constructions of the adopted samples and select the best interpolation method, spatial correlation between the samples were studied. This investigation was obtained by drawing a distance between samples and samples' value variance or by drawing semivariograms. Based on the theory that the characteristics of different phenomena have more similarities in near places than far places, the degree of correlation and dependency was measured (15).

Also, in order to investigate the interpolation methods, the RMSE statistical parameter was used, so that the lesser the amount, the better the method. In this study, the least amount of root mean square error in all two cases showed that except for kriging method, the spherical model was the most used for the interpolation of hardness variables and TDSs.

In a study by Yousefi and Naeej (16) on Amol nitrate in rural wells and deep wells, it was concluded that nitrate was far lesser than the shallow wells (16). This result is consistent with the results of the present study.

In this study, hardness and dissolved solids in the year 2015 has increased when compared with previous years, and this could be attributed to pollution as a result of the industrial and agricultural activities in the study area. Dianati et al (17) in their study on the comparison of zoning maps, pointed out that GIS represents an increase of over 1380-1390 average of these parameters.

Dianati and Falah (18) conducted a study on the change of electrical conductivity and hardness of groundwater in the coastal city of Sari. The results showed that the electrical conductivity and hardness of groundwater of Sari city dropped from north to south (the sea coast) and in places where the electrical conductivity was high, but was hardly significant, however, lake or near to river does not have significant impact on the hardness. The average level of electrical conductivity in the dry season was higher than in the wet season and the average level of hardness of groundwater in the wet season was higher than in the dry season (18).

Samadi et al (19) in their study on the survey of water quality in Moradbeik river based on the water quality index (WQI) index by GIS, showed that the distribution of pollution, river pollution, and low rainfall for several months is more rain. Pollution of rivers in the rainy months cannot generally be reversed.

The increasing trend of total dissolved solids (TDS) and hardness (displayed in Table 1 as well as Figures 6 and 7) has become an alarm to the authorities of water and wastewater organizers to design proper program that will help prevent the increase.

Conclusion

Normal kriging, which involves the use of Gaussian model due to its smaller ME & RMSE value, is more suitable for the preparation of total hardness (TH) changes maps among other methods.

In the examination of the zoning maps, salinity problem was seen in the western parts of the Central Highlands. The reasons for the development of salt can be attributed to excessive use of underground water due to illegal digging of wells, making permanent access to water easy. Detailed plan for irrigation and agricultural justification, as well as the monitoring of the numbers of drilled wells every day, can be effective in improving these conditions. The obtained results, based on the software used in this study, on the relationship between variables, total hardness and TDSs in groundwater, can be used as a guide in order to harvest this water with less salt or stop the harvest of water in areas that are very salty. Therefore, further extensive studies are needed in order to determine other qualitative parameters of water.

Acknowledgements

The authors gratefully acknowledge the assistance and cooperation of the research committee of the Mazandaran University of Medical Science. Also, the efforts of Abdolazim Alinejad and Rouhollah Khodadadi are appreciated. This research was the subject of MSc thesis of Farzad Kazemi as a student of Mazandaran University of Medical Sciences.

Competing interests

We affirm that this article is the original work of the authors and have no conflict of interest to declare.

Authors' contributions

All authors were involved in all stages of the article. On

behalf of the co-authors, the corresponding author bears full responsibility for this submission.

References

1. Samaei MR, Ebrahimy MH, Ehrampoosh MH, Talebi P, Khalili MH, Morovati R. A survey on chemical quality of Yazd drinking water. Toloo-e-Behdasht 2007; 6(2): 51-7. [In Persian].
2. Shariat Panahi M. Principals of Quality and Water and Wastes' Purification. 5th ed. Tehran: Tehran University Publication; 1992. [In Persian].
3. Rahmani A, Shokohi R. Investigating the Hamadan underground water quality - spring. The 10th Hamadan Health Environment National Congress; 2007; 67-77. [In Persian].
4. Imandel K. Physical and Chemical Qualities of Drinking Water. Standard Handbook, number 52. 5th ed. Iran: Industrial Researches and Standard Institution; 1997. [In Persian].
5. NasehiNia H, Naghizadeh A, Ravankhah M. Determining of corrosion or sedimentation of drinking water in Dameghan city with corrision indexes. The 4th conference & exhibition on environmental engineering; Tehran University; 2010.
6. Mazloomi S, Babaie AA, Fazlzadeh Davil M, Abouee E, Badiee Nejad A, Hajpoor Sough KH. Corrosion and Scaling Potentiality of Shiraz Drinking Water. Journal of Toloo-e-behdasht 2008; 7(1-2): 64-71. [In Persian].
7. Mokhtari S, Aliqadri M, Hazrati S, Sadeghi H, Gharari N, Ghorbani L. Evaluation of corrosion and precipitation potential in Ardebil drinking water distribution system by using langelier & ryznar indexes. J Health Ardabil Univ Med Sci 2010; 1(1): 14-23. [In Persian].
8. Avazpour M, Gholami M, Aeli R. Evaluation of corrosion and sedimentation potential of drinking water in Ilam city. Eleventh National Conference on Environmental Health; Zahedan University; 2007. p. 10. [In Persian].
9. Ahmadpour H, Khaledian M, Ashrafzadeh A, Rezaei M. Spatial and Temporal Zoning of Groundwater Electrical Conductivity and Total Dissolved Solids in Guilan Plain. Journal of Water Research in Agricultural 2014; 28(3): 667-76.
10. Parast N, Behnia M, Sodmand M, Bahrami H. Survey the water hardness Bojnoord in 1390. Proceedings of the First National Conference on applied research in public health and sustainable development; 2012.
11. Ehsani HR, Javid AH, Hasani AH, Shariat SM. Investigating the process of Nitrat's and total solids changes in Hamadan underground drinking water in winter by use of geographical information system (GIS). Tenth National Congress of Environment Health; Hamadan; 2007. [In Persian].
12. Taghizadeh Mehrjerdi RA, Mahmoudi SH, Khazaei S, Heidari A. Spatial variability of groundwater salinity using geostatistical study (case study: Rafsanjan). Second Symposium of Environmental Engineering; Tehran University; 2008. [In Persian].
13. Zahedifar M, Moosavi AA, Rajabi M. Zoning the groundwater chemical quality attributes of Fasa plain using geostatistical approaches. ournal OF Water and Soil (Agricultural Sciences and Technology) 2013; 27(4): 812-2. [In Persian].
14. Gholami V, Yousefi Z, Zabardast Rostami H. Modeling of ground water salinity on the Caspian southern coasts. Water Resour Manag 2010; 24(7): 1415-24.
15. Ghaderpoori M, Dehghani MH, Fazlzadeh M , Zarei A. Survey of microbial quality of drinking water in rural areas of Saqqez, Iran. American-Eurasian J Agric & Environ Sci 2009; 5(5): 627-32.
16. Yousefi Z, Naeej O. Study on nitrate value in rural area in Amol city. Journal of Mazandaran University of Medical Sciences 2007; 17(61): 161-5. [In Persian].
17. Dianati R, Alam Ghalilo M, Barafrashteh Pour M. Study of temporal and spatial changes in electrical conductivity and hardness of water resources underground city of Ghaem Shahr-Joibar using geographic information system (GIS). 6th National Conference on Environmental Health; Tabriz University of Medical Sciences; 2013.
18. Dianati RA, Falah F. Change of electrical conductivity and hardness of groundwater in coastal areas of the city of Sari. 12th National Conference of Environmental Health; 2009.
19. Samadi MT, Sadeghi S, Rahmani A, Saghi MH. Survey of water quality in Moradbeik river basis on WQI index by GIS. Environmental Health Engineering and Management Journal 2015; 2(1): 7–11.

6

Water desalination by membrane technology (RO) in southern Iran (Jask city)

Zabihollah Yousefi[1*], Ramazan Motallebi[2], Afshin Takdastan[3]

[1]Professor, Department of Environmental Health, Mazandaran University of Medical Sciences, Sari, Iran
[2]MSc student, Department of Water Engineering, Islamic Azad University, Bandar Abbas Branch, Hormozgan, Iran
[3]Associate Professor, Environmental Technologies Research Center, Ahvaz Jundishapur University of Medical Sciences, Ahvaz, Iran

Abstract

Background: Reverse Osmosis (RO) is an increasingly common method of desalination. A full scale water desalination system by membrane technology (RO) evaluated in a southern city (Jask) in Iran.

Methods: First, data collection on water supply and network were performed. Analysis on most of the water quality parameters (Turbidity, pH, EC, Cl^-, Na^+, Alkalinity, Ca, Na, K, No_3, No_2, Fe, Mg, Mn, NH_4, Po_4, HCo_3^-, So_4^{2-} etc.) was performed as standard methods. The membranes of the RO in the desalination system were Poly-Amid (CSM type).

Results: The efficiency of the RO water desalination system was 94.16, 84.12, 92.00, and 96.17% respectively for Turbidity, Na^+, Mg^{2+}, So_4^{2-}. The result shows a significant difference between influent and effluent water of the RO system. The produced water is in agreement with national standard of drinking water. Furthermore, water exited from the RO system for TDS, Ca^{+2}, and Mg^{2+} was less than minimum limit of the guideline.

Conclusion: The quality parameters of the water resource (EC, TDS, Cl^-, Na^+ etc.) were higher than Iranian drinking water standards. The RO technology modified the quality of the water parameters.

Keywords: Membrane technology, Reverse osmosis, Water desalination, Southern Iran

***Correspondence to:**
Zabihollah Yousefi
Email: zyousefi2004@gmail.com

Introduction

Dissolved salts removal from brackish and saline water was difficult and also expensive in the past, so saline water was not as a drinking water source. From 1950s, desalination process was considered as an economic option for traditional usage. Isolation of dissolved salts from a saline or brackish water, is called desalination. Every water that contains Total Dissolved Salts (TDS) less than 1000 ppm, is called fresh water (1). Range of TDS for injection to desalination process is vary from 1000 to 60,000 ppm. Usually seawater contain TDS between 30,000-45,000 ppm, which can be removed by Reverse Osmosis (RO) membrane. RO membrane can be applied for TDS variation from 10,000 to 60,000 ppm. TDS range from 1000-10,000 ppm (usually related to groundwater resources), can be removed by brackish water RO membrane (2,3). Use of membrane technology in desalination process, increased in recent years (4,5). Recently, different industrial applications, use desalination process by RO membranes. This new technology, increased our potential for improvement of environmental protection and sustainable growth (6,7).

Currently, RO method is considered as the best technology for brackish and seawater treatment and also consumes less energy than the other desalination processes (1,8). Requirement of RO system to energy is less than the other desalination processes. It well ensures the global marketing of RO system (1,9). RO membrane is a basic treatment process for groundwater contaminated with different pollutants (10,11).

Method of treatment for contaminated aquifer by RO process is the same as brackish water RO (11-14).

Methods

This is an experimental and intermediate study on a full-scale water desalination system in Jask city of southern Iran (Figures 1-3). In this study, a full-scale water desalination system by membrane technology (RO) in a Southern city (Jask) in Iran, has been evaluated. First, data collection on water supply and network was performed. Then, weekly, water samples were taken from inlet and outlet of the system and transferred to the water laboratory of the University and Jask water organization. Analysis on most of the water quality parameters (Turbidity, PH,

EC, Cl^-, Na^+, Alkalinity, Ca, Na, K, No_3, No_2, Fe, Mg, Mn, NH_4, Po_4, HCo_3^-, So_4^{2-} etc.) was performed as standard methods. The membranes of the RO in the desalination system were Poly-Amid (CSM type). All experiments and preparation of the solutions were carried out based on the guidelines of a reference book titled "standard methods for water and wastewater experiments" (15).

Results

The Tables 1 and 2 show, the water analysis for the inlet and outlet samples of the RO system. As Tables 1 and 2

Figure 1. Satellite image of Jask city

Figure 2. Full scale MBR system of Jask city - side a

Figure 3. Full scale MBR system of Jask city - side b

Table 1. Water analysis for the inlet samples of the RO system

Parameters	Stage											
	1	2	3	4	5	6	7	8	9	10	11	12
Turbidity	0.53	2.240	1.70	1.64	19.100	6.57	1.78	8.32	0.507	0.54	0.47	1.77
pH	7.76	8.33	7.72	7.86	7.50	7.78	7.72	8.130	7.100	7.60	7.47	7.69
TDS	1441	1289	1379	1308	1484	1271.4	1540	1218	1603	1669	1326.5	1340.5
EC	2484	2223	2377	2255	2558	2192	2656	2100	2762.50	2878	2287	2311
T, °C	25	24	23	23	22.80	22.70	22.70	18.50	19	19	15	14.50
Total hardness	546	458	500	544	562	716	740	442	664	643	489	489
Flouride	0.78	0.42	0.48	0.61	0.610	0.570	0.62	0.31	0.92	0.80	0.59	0.46
Chloride	493	369	369	524.60	493	641	404	330	571.60	571.60	385	385
SO_4	515	498	616	240	508	685	500	450	500	520	514	490
CO_3	0	0	0	0	0	0	0	0	0	0	0	0
HCO_3	117.4	113.20	110	127	120.60	119.60	121	126.80	116.20	116.80	130	131
NO_2^-	0.001	_	0	0.012	0	0	0.001	0.001	0.004	0.005	0	0.001
PO_4	0.018	_	0.028	0.025	0.016	0.0068	0.012	0.0135	0.023	0.020	0.0098	0.0079
NO_3	0.485	_	1.260	0.094	0.08	0.58	0.123	0.213	0.394	0.43	0.43	0.711
Ca^{+2}	151.5	111.60	124	137.70	153.100	185.200	190	95	164.30	158.30	115	117.50
Mg^{+2}	40.82	43.64	46.41	48.69	43.74	61.70	66.100	49.81	61.73	60.260	49.100	47.50
Na^+	320	275	285	285	340	385	300	250	320	330	300	300
K^+	10	10	9.810	10.180	10.100	11.50	10	12	10.43	10.120	10.200	10
Fe^{+2}	0.039	0.075	0.669	0.146	0.043	0.001	0	0.015	0.354	0.029	0.01	0
Mn^{+2}	0	0.003	0.005	0.015	0.130	0	0.111	0	0.001	0.005	0	0
NH_3	0	0	0.308	0.183	0	0	0	0	0	0		0

Table 2. Water analysis for the outlet samples of the RO system

Parameters	Stage											
	1	2	3	4	5	6	7	8	9	10	11	12
Turbidity	0.278	0.157	0.139	0.493	0.492	0.04	0.140	0.143	0.23	0.145	0.201	0.18
PH	7.61	7.83	7.210	6.56	6.670	7.53	7.050	7.240	7.62	6.95	6.150	7.45
TDS	165	130	144.7	138.60	176	213	232	163	130	110.90	248	236
EC	300	235.70	263	252	319.1	386.60	422.50	297	236	201.6	450.30	429
T, °C	25	14.500	14.500	18.60	18.60	18.50	22.70	22.80	22	23	23	24
Total Hardness	33.200	15.400	39	25	24.20	64.40	46	41	17.40	28.40	45	64.40
Flouride	0.220	0.120	0.23	0.17	0.32	0.18	0.17	0.20	0.21	0.14	0.15	0.12
Chloride	80	59	60	75	58.100	63.130	46	49.54	59.100	42.74	107.8	49.53
SO_4	11	20.400	29.80	14.34	15.84	17.21	15.90	41.040	10.62	9.300	29.070	16.50
CO_3	0	0	0	0	0	0	0	0	0	0	0	0
HCO_3	26.80	16.60	19.60	20	17	25.20	19.80	24.40	18.80	23	20.20	16.50
NO_2^-	0	0.001	0.001	0.008	0.009	0.005	0.001	0.0008	0.013	0.0004	0	_
PO_4	0.014	0.007	0.0063	0.016	0.013	0.007	0.006	0.005	0.013	0.024	0.029	_
NO_3	0.331	0	0.42	0.286	0.394	0.2	0.4	0.3	0.394	0.22	4.918	_
Ca^{+2}	7.300	3.85	11.50	6.090	6.65	43.50	12.80	9.78	4.240	6.250	12.42	16.67
Mg^{+2}	3.64	1.40	2.50	2.38	1.80	10.72	10.40	4.030	1.65	3.010	3.40	5.46
Na^+	50	46	42	50	40	50	46	48	41	34	74	65
K^+	1.620	0.001	1.40	1.120	1.43	2.80	1.200	1.68	2.70	1.190	2.31	1.81
Fe^{+2}	0.231	0	0.02	0.041	0.028	0.03	0	0	0.006	0.147	0.147	0.111
Mn^{+2}	0	0	0	0.001	0.001	0	0	0	0	0.001	0.001	0
NH_3	0	0	0	0	0	0	0	0	0	0.210	0.035	0

and Figures 4-6 indicate, this system can be getting the government standards of drinking water. RO system, inlet and outlet parameters statistical results were indicated in Tables 3 and 4. Efficiency of the RO system that applied in this study, represented in Figure 4.

Range of operating pressures is different for brackish and seawater, 250 to 400 and 800 to 1000 psi, respectively.

The quality of feed water is a determining factor to decide on the type of membrane process to use. Surface water (as compared to groundwater) represents the most variable water quality, particularly in terms of particle loadings and turbidity.

Some problems associated with using membranes may include short design life; membrane cleaning (backwashing or chemical treatment); high membrane replacement costs; low resistance to chlorine and lack of resistance to fouling.

As Figure 5 showed, except for nitrogen, phosphorus, iron and sulphate, removal efficiency for other parameters was between 83 to 99.6%. Removal efficiency for turbidity was about 65%. The most efficiency removal related to Mg^{+2} and the minimum was related to Po_4.

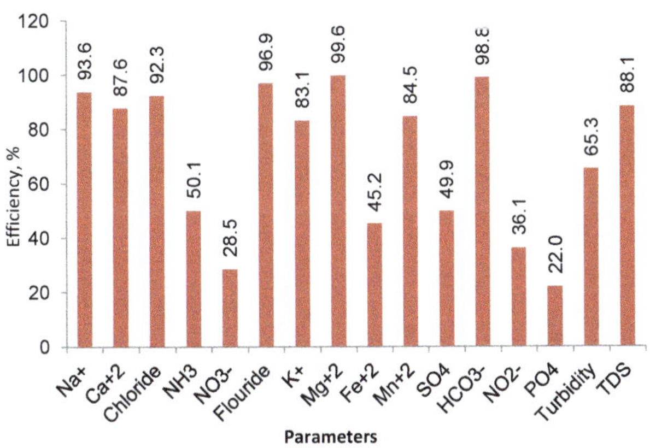

Figure 4. Removal efficiency of different parameters by RO system

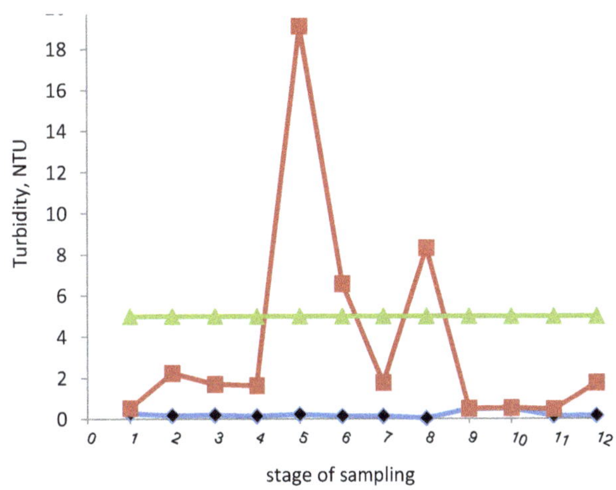

Figure 5. Turbidity status of the RO system in Jask city

Table 3. RO system, inlet parameters statistical results

	Turbidity	pH	TDS	EC	T	TH	F	CL	SO_4	CO_3	HCO_3	NO_2	PO_4	NO_3	Ca	Mg	Na	K	Fe	Mn	NH_3
Number of values	12	12	12	12	12	12	12	12	12	12	12	11	11	11	12	12	12	12	12	12	12
Minimum	0.4700	7.100	1218	2100	14.50	442.0	0.3100	330.0	240.0	0.0	110.0	0.0	0.0068	0.0800	95.00	40.82	250.0	9.810	0.0	0.0	0.0
25% Percentile	0.5325	7.525	1294	2231	18.63	489.0	0.4650	373.0	492.0	0.0	116.4	0.0	0.0098	0.1230	115.6	44.41	285.0	10.00	0.00325	0.0	0.0
Median	1.735	7.720	1360	2344	22.70	545.0	0.6000	448.5	504.0	0.0	120.1	0.0010	0.0160	0.4300	144.6	48.90	300.0	10.11	0.0340	0.0020	0.0
75% Percentile	5.488	7.840	1526	2632	23.00	658.8	0.7400	559.9	518.8	0.0	127.0	0.0040	0.0230	0.5800	162.8	61.34	327.5	10.37	0.1283	0.0125	0.0
Maximum	19.10	8.330	1669	2878	25.00	740.0	0.9200	641.0	685.0	0.0	131.0	0.0120	0.0280	1.260	190.0	66.10	385.0	12.00	0.6690	0.1300	0.3080
Mean	3.764	7.722	1406	2424	20.77	566.1	0.5975	461.4	503.0	0.0	120.8	0.002273	0.01636	0.4364	141.9	51.63	307.5	10.36	0.1151	0.0225	0.04092
Std. Deviation	5.437	0.3127	141.9	244.5	3.489	101.1	0.1724	100.8	103.9	0.0	6.659	0.003636	0.007096	0.3422	30.08	8.490	34.94	0.6741	0.2012	0.04615	0.09921
Std. Error	1.569	0.09027	40.95	70.58	1.007	29.19	0.04976	29.11	29.98	0.0	1.922	0.001096	0.002139	0.1032	8.683	2.451	10.08	0.1946	0.05809	0.01332	0.02864
Lower 95% CI of mean	0.3096	7.523	1316	2268	18.55	501.8	0.4880	397.3	437.0	0.0	116.6	-0.0001697	0.01160	0.2065	122.8	46.23	285.3	9.933	-0.01277	-0.006824	-0.02212
Upper 95% CI of mean	7.218	7.920	1496	2579	22.98	630.3	0.7070	525.5	569.0	0.0	125.0	0.004715	0.02113	0.6663	161.0	57.02	329.7	10.79	0.2429	0.05182	0.1039
Coefficient of variation	144.44%	4.05%	10.09%	10.09%	16.80%	17.86%	28.85%	21.85%	20.65%		5.51%	159.97%	43.36%	78.42%	21.19%	16.45%	11.36%	6.51%	174.86%	205.12%	242.46%
Standard, WHO	5	6.5-8.5	1000	2000	-	500	1.5	250	250	-	-	0.2	-	50	100	60	200	5	0.3	0.05	-
Standard, IRAN	5	6.5-9.0	1500			500	1.5	400	400	-	-	3	-	50	200	-	200	-	-	0.4	

Table 4. RO system, outlet parameters statistical results

	Turbidity	pH	TDS	EC	T	TH	F	CL	SO_4	CO_3	HCO_3	NO_2	PO_4	NO_3	Ca	Mg	Na	K	Fe	Mn	NH_3
Number of values	12	12	12	12	12	12	12	12	12	12	12	11	11	11	12	12	12	12	12	12	12
Minimum	0.0400	6.150	110.9	201.6	14.50	15.40	0.1200	42.74	9.300	0.0	16.50	0.0	0.0050	0.0	3.850	1.400	34.00	0.0010	0.0	0.0	0.0
25% Percentile	0.1408	6.740	132.2	240.0	18.53	24.40	0.1425	49.53	11.84	0.0	17.45	0.0004	0.0063	0.2200	6.130	1.945	41.25	1.193	0.0015	0.0	0.0
Median	0.1685	7.225	164.0	298.5	22.35	36.10	0.1750	59.05	16.20	0.0	19.90	0.0010	0.0130	0.3310	8.540	3.205	47.00	1.525	0.0290	0.0	0.0
75% Percentile	0.2660	7.590	227.3	413.5	23.00	45.75	0.2175	72.03	26.90	0.0	24.05	0.0080	0.0160	0.4000	12.71	5.103	50.00	2.185	0.1380	0.0010	0.0
Maximum	0.4930	7.830	248.0	450.3	25.00	64.40	0.3200	107.8	41.04	0.0	26.80	0.0130	0.0290	4.918	43.50	10.72	74.00	2.800	0.2310	0.0010	0.2100
Mean	0.2198	7.156	173.9	316.1	20.60	36.95	0.1858	62.50	19.25	0.0	20.66	0.003564	0.01275	0.7148	11.75	4.199	48.83	1.605	0.06342	0.0003333	0.02042
Std. Deviation	0.1397	0.5017	47.07	85.60	3.574	16.27	0.05600	18.00	9.458	0.0	3.450	0.004507	0.007850	1.399	10.72	3.180	10.96	0.7621	0.07654	0.0004924	0.06055
Std. Error	0.04034	0.1448	13.59	24.71	1.032	4.698	0.01616	5.196	2.730	0.0	0.9959	0.001359	0.002367	0.4219	3.095	0.9180	3.164	0.2200	0.02209	0.0001421	0.01748
Lower 95% CI of mean	0.1310	6.837	144.0	261.7	18.33	26.61	0.1503	51.06	13.24	0.0	18.47	0.0005360	0.007481	-0.2253	4.942	2.179	41.87	1.121	0.01479	2.050e-005	-0.01805
Upper 95% CI of mean	0.3086	7.475	203.8	370.5	22.87	47.29	0.2214	73.93	25.26	0.0	22.85	0.006591	0.01803	1.655	18.57	6.220	55.80	2.089	0.1120	0.0006462	0.05888
Standard, WHO	5	6.5-8.5	1000	2000	-	500	1.5	250	250	-	-	0.2	-	50	100	60	200	5	0.3	0.05	-
Standard, IRAN	5	6.5-9.0	1500			500	1.5	400	400	-	-	3	-	50	200	-	200	-	-	0.4	

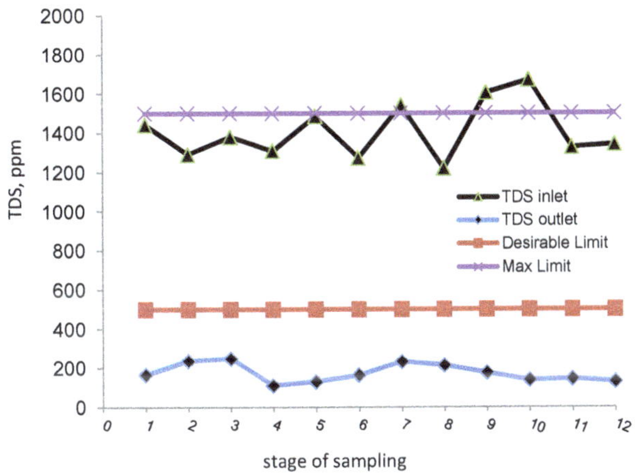

Figure 6. TDS status of the RO system in Jask city

Discussion

RO membrane systems have applied increasingly for seawater and brackish water. Although, permeate flux for brackish water is higher than seawater RO system, operational pressure and salt rejection are lower.

Saudi Arabia and the United States, with 26% and 17% have the first and second ranks in the global desalination capacity, respectively (16).

As the previous researches indicated, rejection of monovalent ions (such as Cl^-, Na^+) and salt rejections by RO membranes, can be higher than 99% (16).

According to WHO and IR standards, the maximum allowable value for the parameters were illustrated in the Tables 3 and 4. As shown in Figure 6, hardness in the feeds of Jask city plant was higher than the allowable limit recommended by WHO and IR standards. Treated water by RO system in the Jask city plant was agreeable with both WHO and IR standards. Moreover, efficiency of RO system for hardness reduction was found 98.5%. As presented in Tables 3 and 4, the influent and effluent of Jask city plant have lower turbidity, fluoride, magnesium and nitrate concentration levels than WHO and IR standards (17,18).

Applied RO desalination plant in Jask city is for brackish water. As Figure 4 indicates, removal efficiency of differr ent parameters by RO system has an important role to deliver drinking water to the Jask city people.

Applied pressure is the basic agent for driving the RO system. The level of energy requirement for RO system severance is directly dependent on level of the salinity of the solution. Increasing of salinity can cause high pressure requirement for drinking water production and high energy and cost (19,20). Jask city desalination plant is designed based on brackish water than seawater.

Conclusion

This research indicated that the efficiency of MBR system for removing the pollutants is effectively high and applicable in all seaside cities and other places which encounter with anions and cations so it is suggested that with respect to the deficiency of the water treatment system of this plant, MBR process can practically be used for the water treatment.

Application of this system in water systems with high coloured and turbidity is relatively low. Finally, we can conclude that this system is very suitable and ideal for water treatment with low concentration of color and turbidity such as groundwater.

Acknowledgements

Our thanks go to the Islamic Azad University of Bandar Abbas for the financial support of this study. We would also like to thank Dr. Mortazavi Dr. Dehghani in the University for Co-operation to do this research.

Competing interests

The authors declare that they have no competing interests.

Authors' contributions

All authors participated in the design study, data acquisition, and interpretation. RM involved in the analyse and ZY wrote the manuscript.

References

1. Ahuja S, Larsen MC, Eimers JE, Patterson CL, Sengupta S. Comprehensive Water Quality and Purification. Waltham, MA: Elsevier; 2014.
2. Abdessemed D, Hamouni S, Nezzal G. State of the reverse osmosis membrane of sea water corso plant desalination (Algiers). Physics Procedia 2009; 2(3): 1469-74.
3. Kim YM, Lee YS, Lee YG, Kim SJ, Yang DR, Kim IS, *et al.* Development of a package model for process simulation and cost estimation of seawater reverse osmosis desalination plant. Desalination 2009; 247(1-3): 326-35.
4. Zhao L, Ho WS. Novel reverse osmosis membranes incorporated with a hydrophilic additive for seawater desalination. Journal of Membrane Science 2014; 455: 44-54.
5. Zhang Z, Zhu M, Li Z. Applications of reverse osmosis desalination membranes for afforestation. Procedia Engineering 2012; 27: 457-64.
6. Zheng X, Chen D, Wang Q, Zhang Z. Seawater desalination in China: retrospect and prospect. Chem Eng J 2014; 242: 404-13.
7. Zhu B, Hong Z, Milne N, Doherty CM, Zou L, Lin YS, *et al.* Desalination of seawater ion complexes by

MFI-type zeolite membranes: temperature and long term stability. Journal of Membrane Science 2014; 453: 126-35.
8. Noshita M. Reverse osmosis seawater desalination for power plant. Desalination 1994; 96(1–3): 359-68.
9. Murrer J, Rosberg R. Desalting of seawater using UF and RO — results of a pilot study. Desalination 1998; 118(1–3): 1-4.
10. Murakami M. Hydro-powered reverse osmosis (RO) desalination for co-generation: a Middle East case study. Desalination 1994; 97(1–3): 301-11.
11. Marquardt K, Nagel R, Seeger H. Seawater desalination by reverse osmosis. Desalination 1983; 46(1–3): 135-43.
12. Morillo J, Usero J, Rosado D, El Bakouri H, Riaza A, Bernaola FJ. Comparative study of brine management technologies for desalination plants. Desalination 2014; 336: 32-49.
13. Molina VG, Casañas A. Reverse osmosis, a key technology in combating water scarcity in Spain. Desalination 2010; 250(3): 950-5.
14. Voros NG, Kiranoudis CT, Maroulis ZB. Solar energy exploitation for reverse osmosis desalination plants. Desalination 1998; 115(1): 83-101.
15. American Public Health Association, American Water Works Association, Water Pollution Control Federation. Standard methods for the examination of water and wastewater. 14th edition. Washington D.C. : American Public Health Association; 1975.
16. Mogheir Y, Foul AA, Abuhabib A, Mohammad A. Assessment of large scale brackish water desalination plants in the Gaza Strip. Desalination 2013; 314: 96-100.
17. World Health Organization. Guidelines for drinking-water quality. Geneva: WHO; 2011.
18. Samadi M, Rahmani A, Sedehi M, Sonboli N. Evaluation of chemical quality in 17 brands of Iranian bottled drinking waters. J Res Health Sci 2009; 9(2): 25-31.
19. Karagiannis IC, Soldatos PG. Water desalination cost literature: review and assessment. Desalination 2008; 223(1-3): 448-56.
20. Gholami V, Yousefi Z, Zabardast Rostami H. Modeling of ground water salinity on the caspian southern coasts. Water Resour Manage 2010; 24: 1415-24.

7

Evaluation of heavy metal concentrations in Rutilus frisii kutum on the southern coast of the Caspian Sea (northern Iran)

Dariush Naghipour[1], Zeinab Shaabaninezhad[2], Abdoliman Amouei[3*]

[1]Assistant Professor of Environmental Health Engineering, Department of Environmental Health Engineering, School of Health, Guilan University of Medical Sciences, Rasht, Iran
[2]MSc of Environmental Health Engineering, Student Research Committee, School of Health, Guilan University of Medical Sciences, Rasht, Iran
[3]Associate Professor of Environmental Health Engineering, Department of Environmental Health Engineering, School of Health, Babol University of Medical Sciences, Babol, Iran

Abstract

Background: Marine products, especially fish, play an increasingly significant role in human diets and health. This research evaluated the concentrations of cadmium, lead, chromium, and nickel in liver and muscle tissues of Rutilus frisii kutum along the Caspian Sea coasts of northern Iran.

Methods: Thirty samples of fishes were taken from five stations (Astara, Rezvanshahr, Anazali, Ramsar, and Tonekabon) during fishing season. An ICP-OES spectrophotometer was used to determine concentrations of the studied metals. Data was analyzed by SPSS 19 using descriptive statistical index, *t* test, and analysis of variance (ANOVA) test; $P < 0.05$ was considered significant.

Results: Mean concentrations of cadmium, lead, chromium and nickel in the livers were 0.06, 1.1, 2.1 and 1.5 µg/g respectively and in the muscles were 0.05, 0.8, 1.7 and 1.1 µg/g, respectively. In liver: cadmium= 0.01-0.13; lead= 0.3-2.7; chromium = 0.9-4.5; nickel = 0.1-2.8. In muscle: cadmium = 0.01-0.08; lead = 0.2-1.5; chromium = 0.7-2.7; nickel = 0.5-1.5. Mean contents of these heavy metals were higher in the liver than in muscle. The order of heavy metal concentrations in the liver and muscle tissues were: Cr>Ni>Pb>Cd. The results showed statistically significant differences in the concentrations of these metals in liver and muscle tissues ($P < 0.05$), but no significant difference was observed between males and females.

Conclusion: Based on calculations, the estimated permissible tolerable weekly intakes (PTWIs) of cadmium (0.042 and 0.567 mg) and lead (0.0021 and 0.027 mg) in this study are lower than the levels established by the World Health Organization (WHO) (0.455 mg for cadmium and 1.625 mg for lead).

Keywords: Heavy metals, Rutilus frisii kutum, Caspian Sea

***Correspondence to:**
Abdoliman Amouei
Email: iamouei1966@gmail.com

Introduction

The Caspian Sea is located in the northern part of Iran and has an area of 378 400 km² and a basin volume of 78 170 km³. Its average and maximum depths are 210 m and 1025 m, respectively (1).

Heavy metals in organic and inorganic compounds that are released into aquatic ecosystems are very dangerous because of their various toxicity levels and potential bio-accumulation in aquatic species (2,3). Unlike other pollutants, heavy metals are not degraded and eliminated by biological processes in the environment (2). Heavy metals can accumulate in different organs of fish through food and water (4,5), and their contamination of aquacultures can be proven in water, sediments, and organisms (6). Heavy metal contents in different tissues of fishes are determined by their concentrations in water and food, and therefore can serve as a pollution bioindicator of aquatic ecosystems (7,8). Fish have been widely used in assessing the quality of aquatic environments. These organisms are often at the top of the aquatic food chain and may accumulate large amounts of heavy metals from the surrounding waters (9,10).

Cadmium, one of the most toxic heavy metals, is carcinogenic and has a long half-life (10 15 years), which leads to its gradual accumulation in the human body (6). Lead usually disrupts blood-forming and the nervous and reproductive systems (7,8). Chromium causes dermatitis and irritates peripheral membranes (1,4). Nickel is widely used in industry and is a common aquatic pollutant (5). In aquatic systems, nickel interacts with numerous inorganic and organic compounds and occurs as a soluble salt adsorbed into substances of different chemical origins (4). Other studies have detected the levels of some heavy metals in water, sediments, and several organs of fish.

Rajkowska and Protosowicki assessed heavy metals accumulation in aquacultures consumed by humans and its effect on human health (3). Hassanpour et al determined the maximum concentrations of lead in water and sediments were 0.154 and 2.04 mg/kg and in the liver and kidney of wild birds (the common coot) were 1.2 and 2.1 mg/kg, respectively (11). Other investigations measured the accumulation of heavy metals in different tissues of fish, birds, and other aquacultures in the Caspian Sea. Najm et al found the concentrations of lead and chromium in the tissues of *Clupeonella cultriventris* and *Gasterosteus aculeatus* from the southern coast of the Caspian Sea to be less; however, the concentration of cadmium was higher than the recommended consumption limit for human health (12). Rajkowska and Protosowicki studied heavy metals accumulation (Fe, Mn, Zn, and Cu) in different tissues of fish in two lakes of northern Poland (3). They suggested that the lowest levels of the studied metals were always detectable in muscle. The highest accumulations were Mn in the skin, gills, and gonads; Zn in the digestive tract and gills; and Cu in the liver. Nasrollahzadeh et al studied heavy metal (Ni, Cd, Pb, and Hg) contents in the edible tissue of *Cyprinus carpio* (13). They showed that Ni, Cd, and Pb contents in the muscle of this fish were lower than standard levels; however, the concentration of Hg was higher than the standard levels recommended by the World Health Organization (WHO) and Food and Agriculture Organization (FAO). Khoshnoud et al investigated the concentrations of metals (Zn, Cu, Pb, Cd, and Hg) in two commercial fish species of the Persian Gulf (14). They stated that the concentrations of all metals had mean values lower than permissible limits set by WHO and FAO. Malakootian et al determined the concentrations of Pb, Cd, Ni, and Zn in canned fish in southern Iran. Their results indicated that the concentrations of these toxic metals in tuna fish produced and marketed in Iran were much lower than the standard levels set by FAO and WHO (15). Considering the economic and nutritional values of white fish (Rutilus frisii kutum) from the Caspian Sea and the adverse effects of heavy metals on water quality and the health of fish, this study aimed to determine the concentrations of cadmium, lead, chromium, and nickel in the muscle and liver of fish in five fishing center stations on the southern coast of the Caspian Sea.

Methods

Description of the study area

The studied area included the southern coast of the Caspian Sea between Astara and Tonekabon located in Guilan and Mazandaran provinces. This area is located between longitudes 48°52′ and 50°57′ and latitudes 38°26′ and 36°52′ (Figure 1). The major rivers of this area, Sefid Rud, Cheshmeh Khileh, Pol Rud, Astara Chai, Chelvand, Garakan Rud, Kia Rud, Shalman Rud, Langrud Khan, and Masouleh Rud Khan, flow into the Caspian Sea.

Sample preparation

Thirty fish samples were collected from five fishery stations from November 2013 to April 2014 (Figure 1). In the laboratory, the mass and total lengths of the fish were recorded. The fish were dissected and gender was determined by inspecting the gonads after the body cavity was opened. The muscles and liver were removed from each fish and preserved in the freezer (-20°C) in clean, dry polyethylene bags. To prepare analytical samples, 1 g of the selected organ was weighed with an accuracy of 0.001 g. Hotplate digestion was performed using a mixture (7:3 v/v) of nitric acid (65%, Merck, Germany) and perchloric acid (70%, Merck, Germany). Then, the samples were dissolved in 5 mL of 15% nitric acid (9). At the end of the digestion stage, the samples were cooled and passed through Whatman filter paper 42. After that, pH values were reduced to below 2 using nitric acid, and volumes were increased to 25 ml (16,17).

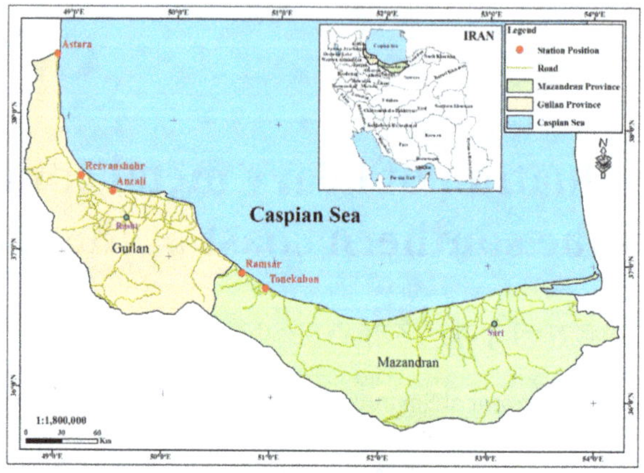

Figure 1. Location of the studied stations

Determination of Cd, Pb, Cr, and Ni contents

Concentrations of Cd, Pb, Cr, and Ni were determined using inductively coupled plasma (ICP) optic emission spectrometry. Metal concentrations in tissues were expressed in micrograms per gram as wet weight . The wave lengths used for determining Cd, Pb, Cr, and Ni were 228.8, 283.3, 267.7, and 231.6 nanometers, and detection limits were 0.05, 2.1, 0.1, and 0.3 μgrm/l, respectively. After optimizing the ICP-OES device (Germany), its calibration curve was drawn using standard heavy metal levels to measure their concentrations in the prepared solutions.

Statistical analysis

All statistical analyses in this study were performed using SPSS version 19. The independent *t* test was used to evaluate the statistical significance of differences in concentrations of heavy metals in fish based on gender. Analysis of variance (ANOVA) was used to identify significant differences in length and weight of fish, heavy metal contents in organs and gender by *t* test. The level of significance was considered to be 95% and an error rate of less than 5% was acceptable ($P < 0.05$).

Results

Biometry study of the fish samples

In Table 1, the biometric characteristics of the fish samples are presented. The average length and weight of fish were higher at Anzali station than any other stations, but the difference was not statistically significant ($P > 0.05$).

Heavy metal contents in fish organs

The average concentrations of Cd, Pb, Cr, and Ni in fish livers were 0.06 ± 0.01, 1.1 ± 0.1, 2.1 ± 0.2, and 1.5 ± 0.1 µg/g, respectively and in muscle were 0.05 ± 0.01, 0.8 ± 0.1, 1.7 ± 0.2, and 1.1 ± 0.1 µg/g, respectively (Figure 2). The mean content of all metals except cadmium was higher in the liver than in the muscle, and these differences were statistically significant ($P < 0.05$). The order of heavy metal concentrations in the muscle and liver of fish was Cr>Ni>Pb>Cd.

In Figure 3, the average concentrations of Cd, Pb, Cr, and Ni in male and female fish are presented. According to this figure, the mean contents of all metals were higher in the tissue of female fish than male fish, but these differences were not statistically significant ($P > 0.05$).

Maximum and minimum Cd, Pb, Cr, and Ni concentrations in the muscle of Caspian kutum were 0.08 and 0.01

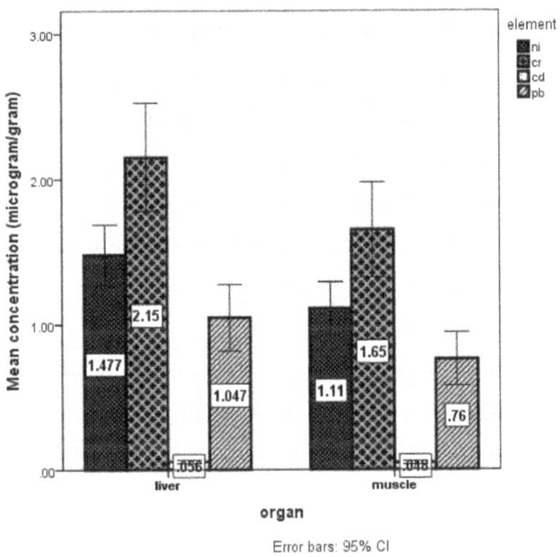

Figure 2. Mean heavy metal contents in liver and muscle tissues of Caspian kutum.

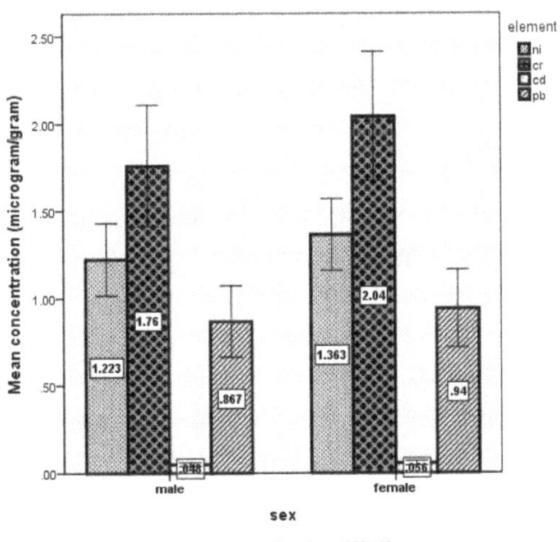

Figure 3. Mean heavy metal contents in male and female Caspian kutum.

µg/g, 1.5 and 0.2 µg/g, 2.7 and 0.7 µg/g, 1.5 and 0.1 µg/g, and in the livers, 0.13 and 0.01 µg/g, 2.7 and 0.3 µg/g, 4.5 and 0.9 µg/g, and 2.8 and 0.1 µg/g as dry weight of the sample, respectively (Table 2). The *t* test results showed that the differences in heavy metal concentrations in the muscle and liver of fish in the studied stations were statistically significant ($P < 0.05$).

Discussion

In the current study, the average length and weight of the fish was highest in Anzali station, but this difference was not statistically significant ($P > 0.05$). Moreover, there were no statistically significant differences between the heavy metals accumulation in liver and muscle and weight and length of the analyzed fishes ($P > 0.05$). Chakeri et al showed that there is no correlation between lead and cadmium contents in the muscle and liver of *Rastrelliger kanagurta* in the Persian Gulf and the size of the analyzed fish (18).

This study investigated the liver and muscle of fish because of their high nutritional value and importance in metabolism and the storage of chemicals (17,19,20). The accumulation of lead, chromium, and nickel was higher in the liver of all fishes than in the muscle, and these differences were statistically significant ($P < 0.05$). The high contents of metal accumulation in the liver compared to the muscle of fish has been confirmed by other studies (14,21). Elsagh investigated the bioaccumulation of heavy metals in the muscles of Rutilus kutum (17) and showed that the average concentrations of Cd, Pb, Cr, and Ni in fish muscles were 2.6 ± 1.3, 1.2 ± 0.3, 0.65 ± 0.12 and 1.7 ± 0.3 µg/g, respectively. Furthermore, it was found that the mean accumulations of nickel and chromium in the muscle and liver of white fish along the coasts of the Caspian Sea were higher than those found in the Persian Gulf and along the coast of Bushehr. Yaghobzadeh et al determined that the mean accumulated concentrations of lead in kutum from the Bandar Anzali and Roodsar stations were 2.55 ± 0.54 and 1.66 ± 0.25 µg/g dry weight, respectively (22). Results of the current study confirmed these investigations. In this study, the mean contents of all metals were higher in the tissue of female fish than male fish, but these differences were not statistically significant ($P > 0.05$). Turkmen et al found that in fish of the Aegean Sea, the accumulated contents of heavy metals were higher in the tissue of male fish than female fish (23). This result is not consistent with the results of this study. In the current study, the concentrations of all metals were higher in the analyzed fish from Bandar Anzali (station 3) than the other stations. The differences in mean contents of all metals in the liver and muscle between stations were statistically significant ($P < 0.05$). Studies have shown changes in the accumulation of metals in fish tissue may be caused by various factors such as type of metal, the target tissue or organ, gender, weight and age of fish, their feeding habits, duration of exposure to the metal, ecological features, environmental conditions, and physical and chemical properties of the environment (17,23,24).

Table 1. Weight and length (mean ± SD) of Caspian kutum collected from the studied stations

Station number	Station name	Sample number (n)	Weight (g)	Length (cm)
1	Astara (S_1)	6	996.5 ± 158.6	40.7 ± 6.3
2	Rezvanshahr (S_2)	6	1084.8 ± 176.2	43.4 ± 7.8
3	Anzali (S_3)	6	1138.7 ± 168.6	48.2 ± 6.8
4	Ramsar (S_4)	6	1112.2 ± 165.5	44.6 ± 6.5
5	Tonekabon (S_5)	6	1025.4 ± 182.5	42.8 ± 7.4

Table 2. Concentrations of metals as μg/g (dry weight) in selected organs of Caspian kutum from different stations

Station	Nickel				Chromium				Cadmium				Lead			
	Liver		Muscle		Liver		Muscle		Liver		Muscle		Liver		Muscle	
	Min	Max	Min	Max	Min	Max	Min	Max	Min	Max	Min	Max	Min	Max	Min	Max
S_1 (n = 6)	0.1	2.3	0.7	1.4	1.5	3.2	0.9	1.9	0.05	0.1	0.03	0.07	0.8	1.9	0.5	1.1
S_2 (n = 6)	0.7	1.6	0.5	0.9	0.9	2.2	0.7	1.4	0.01	0.03	0.01	0.02	0.3	1.1	0.2	0.6
S_3(n = 6)	1.4	2.8	0.9	1.5	2.3	4.5	1.4	2.7	0.06	0.13	0.04	0.08	1.3	2.7	0.8	1.5
S_4 (n = 6)	0.9	1.8	0.6	1.2	1.2	2	0.8	1.5	0.04	0.09	0.03	0.05	0.5	1.3	0.3	0.8
S_5 (n = 6)	0.8	1.4	0.5	0.8	1	2.4	0.7	1.4	0.03	0.05	0.02	0.04	0.3	1.2	0.2	0.7
Mean ± SD	1.5 ± 0.1		1.1 ± 0.1		2.1 ± 0.2		1.7 ± 0.2		0.06 ± 0.01		0.05 ± 0.01		1.1 ± 0.1		0.8 ± 0.1	
P value	*P* < 0.05		*P* < 0.05		*P* < 0.05		P<0.05		*P* < 0.05		*P* < 0.05		*P* < 0.05		*P* < 0.05	

The minimum and maximum cadmium and lead concentrations in analyzed fish were 0.01 and 0.13 $\mu g.g^{-1}$ and 0.2 and 2.7 $\mu g.g^{-1}$, respectively. The joint FAO/WHO expert committee on food additives established the permissible tolerable weekly intake (PTWI) for cadmium as 0.007 and for lead as 0.025 mg per kg^{-1} body weight per week (22), which are equivalent to 0.455 and 1.625 mg per week, respectively, for a 65 kg adult. Using the average weekly fish consumption in Iran of 210 g per person (16) and the minimum and maximum cadmium and lead levels measured in fish by this study, the weekly intake was calculated to range from 0.0021 to 0.027 mg of cadmium per person and 0.252 to 0.504 mg of lead per person in the muscle and liver of fish. As can be seen, the estimated PTWI of cadmium (0.042 and 0.567 mg) and lead (0.0021 and 0.027 mg) in this study is below the established PTWI (0.455 mg for cadmium and 1.625 mg for lead). The minimum and maximum chromium and nickel concentrations were 0.7 and 4.5 $\mu g.g^{-1}$ and 0.1 and 2.8 $\mu g.g^{-1}$, respectively. Although there was no information about maximum permissible chromium and nickel levels in fish tissue in Iranian standards, the maximum permissible chromium and nickel reported by FAO for fish is 1 $\mu g.g^{-1}$ (22).

The high accumulation of heavy metals in the tissue of the analyzed fish clearly shows the considerable amount of pollution with these elements in the Caspian Sea waters at the studied stations. The regions covered in this study are considered centers of agricultural production. Because of the density of cultivation of various crops, large quantities of chemical fertilizers, pesticides, fungicides, and herbicides are used. The increase in Cd, Pb, Cr, and Ni concentrations in the tissue of the fish at Bandar Anzali (station 3) is related to the hydraulic and hydrodynamic flows of the Caspian Sea, heavy marine traffic, and industrial wastewaters discharge, especially wastewater from the electroplating industry (21).

Conclusion

This research evaluated the contents of Cd, Pb, Cr, and Ni in the tissue of Caspian kutum in northern Iran. Mean concentrations of these heavy metals were higher in liver tissue than in muscle tissue. The order of heavy metal concentrations in the muscle and liver tissues of fish were Cr>Ni>Pb>Cd. The results indicated that there are statistically significant differences among the accumulation of these metals in the liver and muscle tissues, but there was no significant difference in heavy metal concentrations between male and female fish. Based on calculations, the estimated PTWIs of cadmium (0.042 and 0.567 mg) and lead (0.0021 and 0.027 mg) in this study are lower than the PTWI established by WHO (0.455 mg for cadmium and 1.625 mg for lead).

Acknowledgments

The authors would like to thank Reyhaneh Barari for her English editing of the manuscript and the Chancellor of Research and Technology in Guilan University of Medical Sciences for financial support of this work.

Competing interests

The authors declare that they have no competing interests.

Authors' contribution

All authors contributed equally and participated in the

collection, analysis, and interpretation of the data. All authors critically reviewed, refined, and approved the manuscript.

References

1. Dumont HJ. The Caspian lake: history, biota, structure and function. Limnol Oceanogr 1998; 43(1): 44-52.
2. Birungi Z, Masola B, Zaranyika MF, Naigaga I, Marsall B. Active biomonitoring of trace heavy metals using fish (Oreochromis niloticus) as bioindicator species. The case of Nakivubo wetland along Lake Victoria. Physics and Chemistry of the Earth 2007; 32(15-18): 1350-8.
3. Rajkowska M, Protosowicki M. Distribution of metals (Fe, Mn, Zn, Cu) in fish tissues in two lakes of different trophy in northern Poland. Environ Monit Assess 2013; 185(4): 3493-502.
4. Golovanova IL. Effect of heavy metals on the physiological and biochemical status of fishes and aquatic invertebrates. Inland Water Biology 2008; 1: 93.
5. Kraal MH, Kraak MH, De Groot CJ, Davids C. Uptak e and tissues distribution of dietary and aqueous cadmium by Carp (Cyprinus carpio). Ecotoxicol Environ Saf 1995;31(2): 179-83.
6. Moiseenko TI, Kurdyavtseva LP. Trace metal accumulation and fish pathologies in areas affected by mining and metallurgical enterprises in the Kola Region, Russia. Environ Pollut 2001; 114(2): 285-97.
7. Fan W, Wang WX, Chen J. Geochemistry of Cd, Cr and Zn in highly contaminated sediments and its influences on assimilation by marine bivalves. Environ Sci Technol 2002; 36(23): 5164-71.
8. Farkas A, Salanki J, Varanka I. Heavy metal concentrations in fish of Lake Balaton. Lakes Reservoirs: research and management 2000; 5: 271-9.
9. Protasowicki M. Sex effects on Cd, Pb, Cu and Zn contents in selected fish organs. Baltic Sea Environment Proceedings. 1986; 19: 433-41.
10. Ptashynski MD, Pedlar RM, Evans RE, Baron CL, Klaverkamp JF. Toxicology of dietary nickel in Lake Whitefish (Coregonus clupeaformis). Aquatic toxicology, 58 (3-4): 229-47.
11. Hassanpour M, Pourkhabbaz A, Ghorbani R. The measurement of heavy metals in water, sediments and wild bird (Comoon coot) from the southeast Caspian Sea. J Mazand Univ Med Sci 2012; 22 (suppl 1): 184- 94. [In Persian].
12. Najm M, Shokrzadeh M, Fakhar M, Sharif M, Hosseini SM, Rahimi Esboei B, et al. Concentration of heavy metals (Cd, Cr, and Pb) in the tissues of Clipeounella cultriventis and Gasterosteus aculeatus from Babolsar coastal waters of Mazandaran province, Caspian Sea. J Mazand Univ Med Sci 2014; 24(113): 185- 92. [In Persian].
13. Nasrollahzadeh Saravi H, Pourgholam R, Pourang N, Rezaei M, Makhlough A, Unesipour H. Heavy metal concentrations in edible tissues of Cyprinus carpio and its Target Hazard Quotients in the southern Iranian Caspian Sea coast (2010). J Mazand Univ Med Sci 2013; 23(103): 33-44. [In Persian]
14. Khoshnoud MJ, Mobini K, Javidnia K, Hoosseinkhezri P, Aeen Jamshid Kh. Heavy metals (Zn, Cu, Pb, Cd, and Hg) contents and fatty acid ratio in two fish species (Scomberomorus commerson and Otolithes ruber) of the Persian Gulf. Iranian Journal of Pharmaceutical Sciences 2011; 7(3): 191-6.
15. Malakootian M, Tahergorabi M, Daneshpajooh M, Amirtaheri K. Determination of Pb, Cd, Ni and Zn concentrations in canned fish in southern Iran. Sacha Journal of Environmental Studies 2011; 1(1): 94-100.
16. Agah H, Leermakers M, Elskens M, Fatemi SM, Baeyens W. Accumulation of trace metals in the muscle and liver tissues of five species from the Persian Gulf. Environ Monit Assess 2009; 157(1-4): 499-514.
17. Elsagh A. Bioaccumulation of heavy metals levels in muscles of Rutilus frisii kutum and Cyprinus carpio Fishes of coastal waters of the Mazandaran Province, Caspian Sea. Quarterly Veterinary Journal (Pajouhesh-va-Sazandegi) 2012; 95(2): 41-8. [In Persian].
18. Chakeri R, Sajadi MM, Kamrani E, Aghajari N. Determination of heavy metal (lead and cadmium) concentrations in liver and muscle tissue of Indian mackerel (Rastrelliiger kanagurta) in Persian gulf. Iranian Scientific Fisheries Journal 2015; 24(2): 115-25.
19. Filazi A, Baskaya R, Kum C, Hismiogullari SE. Metal concentration in tissues of the Black Sea fish (Mugil auratus) from Sinop-Icliman, Turkey. Hum Exp Toxicol 2003; 22(2): 85-7.
20. Amini Ranjbar G, Sotodehnia F. Investigation of heavy metals accumulation in muscle tissue of mugil auratus in relation to standard length, weight, age and sex). Iranian Fisheries Scientific Journal. 2005; 16(3): 1-19.
21. Uluozlu OD, Tuzen M, Mendil D, Soylak M. Trace metal content in nine species of fish from the Black and Agean seas, Turkey. Food Chem 2007; 104(2): 835-40.
22. Yaghobzadeh Y, Hossein-Nezhad M, Asadi-Shiran G, Pourali M. An investigation of Lead concentration in Rutilus frisii kutum from Caspian Sea: Case study of Bandar Anzali and Roodsar, Iran. J Mazand Univ Med Sci 2014; 23(110): 102-8. [In Persian].
23. Turkmen M, Turkmen A, Tepe Y. Metal concentrations in five fish species from Black, Marmara, Agean and Mediterranean seas, Turkey. J Chil Chem Soc 2008; 53(1): 1435-39.
24. Tepe Y, Turkmen M, Turkmen A. Assessment of heavy metals in two commercial fish species of four Turkish seas. Environ Monit Assess. 2008; 146(1): 277-84.

8

Feasibility study of green wastes composting with digested and dewatering sludge from municipal wastewater treatment plant in Iran

Neamat Jaafarzadeh Haghighi Fard[1,2], Behnam Moradi[3,4], Mokhtar Abbasi[3], Rahman Alivar Babadi[3], Hossein Bahrani[3], Azadeh Mirzaie[3], Ahmad Zare Javid[5], Maryam Ravanbakhsh[4,6*]

[1]Professor, Environmental Technologies Research Center, Ahvaz Jundishapur University of Medical Sciences, Ahvaz, Iran
[2]Professor, Department of Environmental Health Engineering, School of Public Health, Ahvaz Jundishapur University of Medical Sciences, Ahvaz, Iran
[3]BSc, Department of Environmental Health Engineering, School of Public Health, Ahvaz Jundishapur University of Medical Sciences, Ahvaz, Iran
[4]Student Research Committee, Ahvaz Jundishapur University of Medical Sciences, Ahvaz, Iran
[5]PhD, Nutrition and Metabolic Diseases Research Center, Ahvaz Jundishapur University of Medical Sciences, Ahvaz, Iran.
[6]MSc, Department of Environmental Health Engineering, School of Public Health, Ahvaz Jundishapur University of Medical Sciences, Ahvaz, Iran

Abstract

Background: Composting as a waste management technology is becoming more widespread. The purpose of this study was to assess the feasibility and to find the most effective composting process for the ratio of green waste, digested and dewatered sludge from Chonibieh wastewater treatment plant in the west region of Ahvaz.

Methods: The composting time was 23 days and the evaluated parameters in this period of the study were organic carbon, total nitrogen, phosphorus, carbon to nitrogen ratio (C/N), moisture content and pH. The C/N ratio was maintained at 30 with weight:weight ratio of 1:1, 1:2, 1:3 (digested and dewatered sludge to green waste).

Results: It was observed that vessel R3 produced higher quality of compost with final total nitrogen (1.28%), final total phosphorus (0.71%), final total organic carbon (TOC) (25.78%) and C/N (20.65%) within the 23 days of composting. While vessel R1 produced higher final total nitrogen and total phosphorus with lower amount of total coliform indicating suitable quality of composting. Therefore, the results showed that the characteristics of dewatered sludge mixed with green waste proportion of green waste significantly influenced the compost quality and process dynamics. The results also showed that the quality of final products in all the conditions was in agreement with Global Organic Textile Standard (GOTS) and World Health Organization (WHO) guidelines. However, the moisture content ratios were lower than the mentioned guidelines. With regards to microbial quality, all three ratios were in agreement with US Environmental Protection Agency (EPA) and Iranian guidelines.

Conclusion: It is suggested that the final product of composting can be safely used in farmland and green space.

Keywords: Composting, Septic sludge, Dewatered sludge, Green waste

***Correspondence to:**
Maryam Ravanbakhsh
Email: mapseh@gmail.com

Introduction

In Iran, more than 80% of agricultural land is arid and semi-arid with poor organic material in their soil. Adding organic matter is essential to improve the fertility of agricultural soils. The composting process is a practical method for achieving this goal. Composting of garden waste and sewage sludge is an effective method for the production of stable material that can be used as a source of nutrients in the soil (1).

Composting is a controlled process of microbial decomposition, oxidation of heterogeneous mass of organic matter. This process of degradation of organic material is intermediated by microorganisms decomposers, therefore, the process is influenced by optimal conditions of moisture, aeration, temperature, carbon to nitrogen (C/N) ratio, and pH (2).

The treatment process and disposal of sewage sludge are considered as a complicated issue for wastewater treatment experts (3). Although the quantity of producing sludge is near to 1% of the total treated sewage in wastewater treatment plants; its treatment costs meet 40% to 50% of the total treatment costs. Moreover, its treatment needs complex equipment and the long time usage from one week to several weeks (4).

Green wastes may result from gardening, landscaping and agriculture activities, such as planting, pruning, lawn mowers and cutting off the extra stem and foliage. Leaves, grasses, herb fines and foliage cuts are usually considered as the most common raw materials for composting of gardening and green wastes (5).

Composting process is a type of post-treatment of waste and may be an economical process of post-treatment. In comparison with other processes of post-treatment, composting process is a more economical process (6). Municipal sludge usually has 18% to 35% solids depending on the type of dewatering method (65% to 82% moisture). These sludge cakes have higher moisture content, so its composting in order to reduce the moisture content would be difficult. However, it can be mixed with a dry bulking agent (7). Using bulking agents have some advantages, such as reducing the moisture content, NH_3 and greenhouse gas emissions, control of initial C/N ratio and aeration. So, adding bulking agents would be an acceptable practice (8). Bulking agents, such as sawdust, wood chips, and leaves have higher carbon content which can be added to compost in order to create C/N ratio in balance, increase mixture porosity and increase the resulted aeration. Carbon and nitrogen are considered as 2 essential energy sources for microorganisms. In sewage sludge, C/N ratio ranges usually from 20:1 to 40:1; so, microbial growth requires additional source of biodegradable carbon with ratios lower than 25:1. Bulking agent and added carbon source can supply additional carbon and improve C/N ratio and energy balance (9). Moreover, composting of bulking agent and sludge would be considered as a safe method to produce humic substances during composting process and reduce odor emission (1).

Several studies (5,10-24) were conducted to evaluate the effects of type and amount of bulking agents in sludge composting to find the most suitable ratio of sludge/bulking agent. However, further studies are needed for better understanding of composting processes. Therefore, this study aimed to evaluate the feasibility of producing usable compost from digested and dewatered sludge and green waste as bulking agents in Ahvaz, Iran.

Methods

In this study, digested and dewatered sewage sludge, passing through filter press, digestion and sludge drying beds, were obtained from wastewater treatment plant in the West of Ahvaz. The composting was carried out in Ahvaz Jundishapur University of Medical Sciences in an uncovered composting pilot scale unit. The harvested wastes from green space were used as a bulking agent. Table 1 shows the main characteristics of composted wastes. Three series of composting vessels were prepared by mixing sludge with green wastes in different weight ratios of bulking agent to sludge (1:1 in vessel R1; 1:2 in vessel R2 and 1:3 in vessel R3). Table 2 shows the features of vessels, sludge and bulking agent. The baseline for composting was defined as the day when all the vessels were completely formed. Vessels with cylinder shape were placed in pilot scale units, which were not sheltered from the rain and sunlight. The sludge was dried under direct sunlight for 7 days and entered into the composting process. Green waste (a mixture of grasses, plants, leaves, branches and roots from the university green space area) was dried and crushed in 7 to 10 cm and then used as a bulking agent. Then, sludge and green wastes were mixed together and poured into vessels. The composting procedure was carried out in 23 days. All vessels were weekly turned over with no other forced aeration provided. To maintain sufficient and adequate water content in compost process, the vessels were periodically monitored and water content was determined by laboratory analysis. Then more water was added when its content was insufficient.

The basic principle for the quantitation of total organic carbon (TOC) relies on the destruction of organic matter present in the compost. The destruction of the organic matter can be performed chemically or via heat at elevated temperatures (25). In this study, combustion method was used to determine the TOC (26). After the weighted 2 g of the sample was dewatered in oven at 105°C for 24 hours. Then, it was weighted again (weight A). Dried sample was burned in 550°C and weighted (weight B). TOC was quantified using this equation (Equation 1).

Table 1. The main characteristics of composted wastes

Material	Moisture content (%)	Total coliform[b]	Fecal coliform[b]	Organic C (db, %)	Total phosphorus (db,%)	Total N (db, %)	C/N	pH
Sludge	15.3	28×10^{16}	28×10^{16}	27.85	0.48	1.084	N/M	N/M
Bulking agent	1.84	9×10^{7}	9×10^{7}	40.95	0.31	1.22	N/M	N/M
R1[a]	20.2	N/M	N/M	27.17	0.67	0.91	30	6
R2[a]	18.2	N/M	N/M	26.6	0.31	0.87	30	6
R3[a]	21	N/M	N/M	26.43	0.71	0.88	30	6

Abbreviations: N/M, not measured; C/N, carbon to nitrogen; MPN, most probable number.

[a]Properties determined once the vessel was built.

[b]MPN/g of total dry solids.

Table 2. Features of vessels, sludge and bulking agent

Material	Bulking agent (w/w[a], kg)	Sludge (w/w, kg)
Vessel R1	1.850	1.850
Vessel R2	1.100	2.200
Vessel R3	0.850	2.550

[a] w/w: wet weight.

TOC = ((A-B)/A)×100

To determine the total Kjeldahl nitrogen, Kjeltec Analyser unit 2300 was used. A test portion of the dried and grinded sample of about 0.5 g was placed in the digestion flask. 10 ml sulfuric acid was added and swirl until the acid was thoroughly mixed with the sample. The mixture was allowed to stand for cooling. Then, 2.5 g of catalyst mixture was added and heated until the digestion mixture became clear. The mixture was gently boiled for up to 5 hours so that the sulfuric acid would condensed to about 1/3 of the way up to the neck of the flask. It must be ensured that the temperature of the solution does not exceed 400°C. After completion of the digestion step, the flask was allowed to cool and 20 ml of water was slowly added while shaking. Then the flask was swirled to bring any insoluble material into suspension and transfer the contents to the distillation apparatus. The flask was rinsed three times with water to complete the transfer (27).The percentage of total nitrogen of the samples was shown by apparatus.

Total phosphorus was determined by stannous chloride method once a week (27). The acid-dried soil was dried to a fine powder and 0.5 g of it was weighed in round-bottomed flask. 2 ml each of concentrated nitric and perchloric acid were added to the sample. The flask was heated till the content became dry. Then 2 ml of sulphuric acid was added and heated for 15 minutes. The digested content was filtered through Whatman No. 44 filter paper and the filtrate was made up to 250 ml with distilled water. The phosphorus content of the filtrate was determined by adding 4 ml of ammonium molybdate solution.

In order to determine the pH, 5 g of the sample was stirred in 50 ml distilled water and pH was measured using a digital pH meter with a glass electrode, previously calibrated and corrected for temperature (28).

The moisture content (4) was measured once every 3 days and was determined by drying a moist sample (10 g/wet weight) in an oven that was set at 105°C until constant sample weight and expressed on a wet weight basis.

Total coliform and fecal coliform were determined once a week using the multiple-tube fermentation direct test (29). One gram of the compost material sample was mixed with 100 ml of sterile water and was blended for 40 seconds in a sterile stainless-steel blender at low speed. Dilution of the slurry was made and 10 ml of the dilutions was added to lactose broth medium. Tube contents were incubated at 37°C for 3 hours, small bubbles were removed, and the materials were again incubated at 44.5°C for 21 hours before results were recorded. Gas in the inverted Durham tube and turbidity of the medium indicated positive results, and the most probable number (MPN) was estimated using an MPN table (30).

Statistical analysis

The statistical analysis was done by SPSS version 16 using variance analysis and correlation coefficients. $P < 0.05$ was considered as significant.

Results

Initial and final features of the composting mixtures are shown in Tables 1 and 3, respectively. Figure 1 shows the evolution of moisture based on wet weight during composting process. The water content was determined once in 3 days and was checked and controlled to maintain vessels moisture in suitable level according to standard limits. The dewatered sludge was very low in water content, so more water was added to raise the vessels manually up to World Health Organization (WHO) limits for composting.

Figure 2 shows the trends for pH in vessels. With regards to R1, there was an increasing trend with slight reduction on day 19. Maximum pH value of 7.5, 7 and 7 was ob-

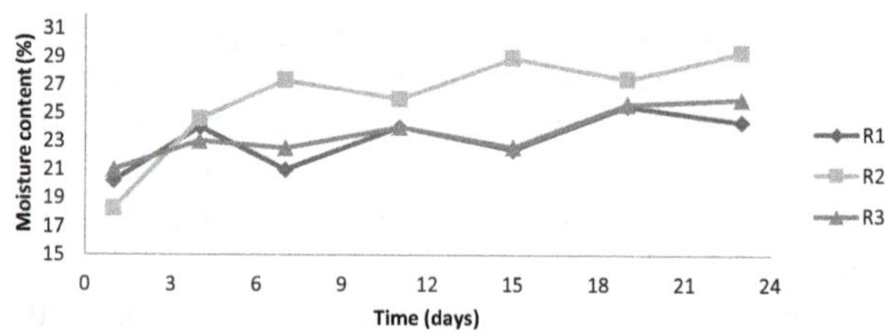

Figure 1. Evolution of moisture based on wet weight during composting process for vessels R1, R2 and R3.

Table 3. Properties of the final products obtained from the three composting experiments (Mean values ± standard deviation).

Material	Moisture content (%)	Total coliform	Fecal coliform	Organic C (db, %)	Total phosphorus (db, %)	Kjeldal N (db, %)	C/N	pH
R1	21.22 ± 1.42	15×10^{17}[a] 120[b]	20×10^{17} 168[b]	24.01 ± 5.03	0.67 ± 0.01	1.96 ± 0.032	13.8	7 ± 0.48
R2	24.03 ± 2.01	21×10^{17}[a] 157[b]	21×10^{17} 190[b]	25.02 ± 2.31	0.31 ± 0.05	1.49 ± 0.06	17.9	6 ± 0.39
R3	25.1 ± 3.47	150×10^{17}[a] 978[b]	93×10^{17} 569[b]	25.78 ± 0.93	0.71 ± 0.02	1.28 ± 0.07	20.65	6.5 ± 0.48

Abbreviations: db, Dry basis; C/N, carbon to nitrogen.

[a] MPN/100 ml.

[b] MPN/gTotaldry solid.

served on the 15th, and 11th day, which steadily decreased to 7, 7 and 6.5 after the 23 days of composting in R1, R2 and R3, respectively.

For organic carbon (Figure 3), a similar pattern was seen in all the vessels with a peak rise on day 8 and thereafter a sharp fall until day 15, followed by gradual decrease until the end of composting process (Figure 3). In this study, the overall decreasing trend for organic carbon is as follows: R1: from 27.17 to 24.01% (17.25% of reduction rate); R2: from 26.6% to 25.02% (2.5% of reduction rate) and R3: from 26.43% to 25.78% (11.63% of reduction rate).

For nitrogen content, an increasing trend was observed in all the vessels. Figure 3 shows the time course of the total nitrogen (NT) in the vessels. The overall increasing trend in this study for total nitrogen is as follows: R1: from 0.91 to 1.96% (115% rising); R2: from 0.87% to 1.49% (71.26% rising) and R3: from 0.88 to 1.28% (45.45% rising). The highest increase was observed in vessel R1.

With regards to total phosphorous, a decrease was observed in all the vessels in the pre and post composting process (Figure 4). Vessel R3 had the highest initial content of total phosphorous (0.71%) when compared with the other vessels (0.67% and 0.31% in vessels R1 and R2, respectively). No changes was found at the final phosphorous for the vessels.

The initial C/N ratio of composting materials was adjusted to about 30 and was decreased to 13.8%, 17.9% and 20.65% in vessels R1, R2 and R3, respectively at the end of the process (Figure 5), which are below the optimal C/N ratio of 25.

The MPN of the total and fecal coliforms in different stages of the composting process was in day 15 which was the highest in vessel R2 and day 8 in vessels R1 and R3 (Figures 6 and 7). The number of both total and fecal coliforms bacteria declined with time except in vessel R2. As shown in Figures 6 and 7, the highest decrease in the number of coliform occurred after the 23 days of composting in vessel R3. In vessel R2, the number of coliforms increased slightly from day 8 to day 15 of composting and thereafter it decreased until day 23.

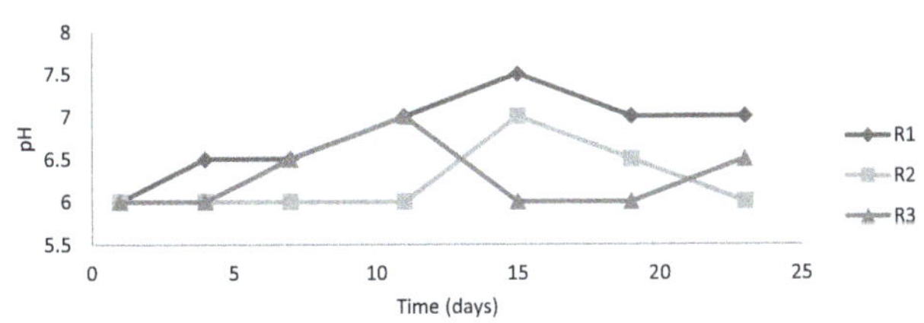

Figure 2. Evolution of pH during the composting process for vessels R1, R2 and R3.

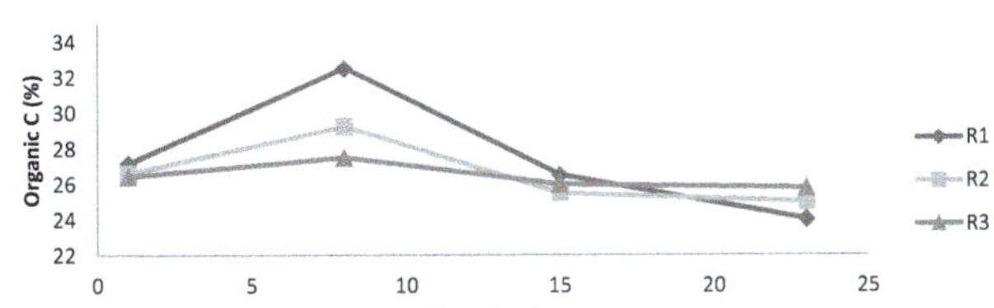

Figure 3. Changes in organic carbon during composting process in vessels R1, R2 and R3.

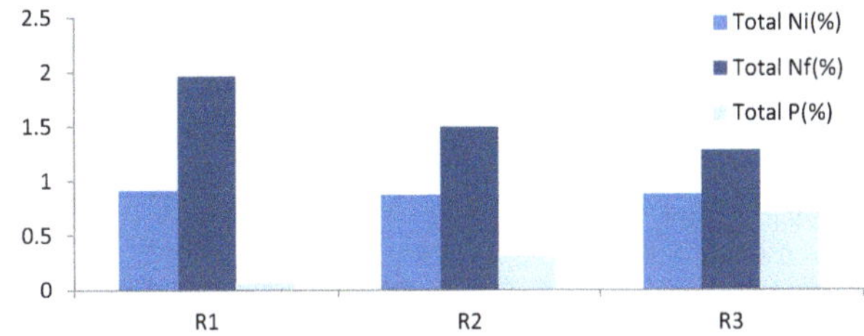

Figure 4. Changes in total nitrogen and total phosphorous during composting process in vessels R1, R2 and R3. Abbreviations: Ni, Initial in total nitrogen content; Nf, Final in total nitrogen content; TP, Total phosphorous.

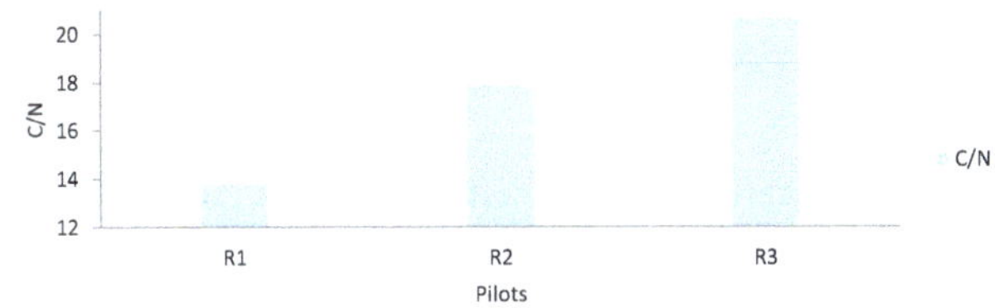

Figure 5. Changes of C/N ratio in the final product in vessels R1, R2 and R3.

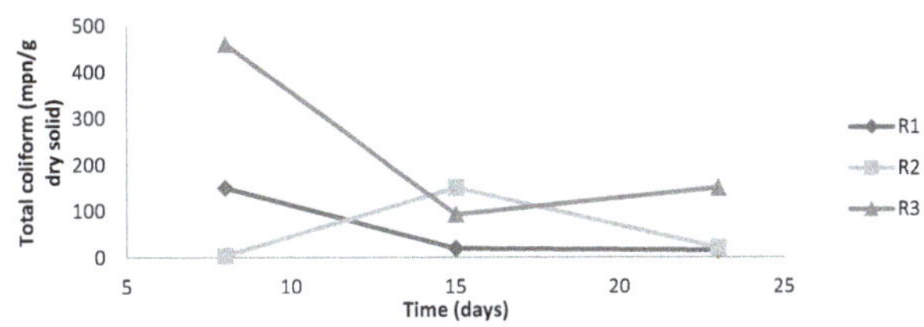

Figure 6. Total coliform number during compost process (expressed as MPN/g dry solid) in vessels R1, R2 and R3.

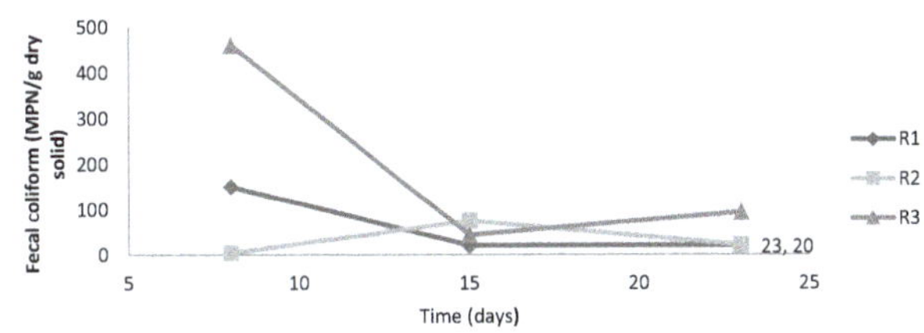

Figure 7. The fecal coliform number during compost process (expressed as MPN/g dry soild) in vessels R1, R2 and R3.

Discussion

Changes of moisture content during composting is shown in Figure 1. As shown in Figure 1, the moisture content is below the standard limitation (Table 4). Since the initial mixture of digested and dewatered sludge and bulking agent was not water-saturated, some water was added to all the mixtures. The final moisture content in all the vessels was under 40% during the composting process which is considered as the least value for the optimal composting moisture (40% to 60%). Thus, it is suggested that composting may be limited by moisture content and need further addition of water and control. It is suggested that the fluctuations observed in the overall trend were due to increase

Table 4. Physical and chemical properties of compost (WHO Standard, GOTS and Institute of Standard and Industrial Research, Iran, 2007) (1,2,22)

Parameters- unit	Acceptable limit (level 1)	Acceptable limit (level 2)
Organic matter (based on dry weight)	35%	25%
Organic carbon (based on dry weight)	Minimum 35%	Minimum 15%
Moisture %	Maximum 15%	Maximum 35%
pH	6-8	6-8
Phosphorus (Based on dry weight)	1-3.8%	3- 3.8%
Total nitrogen (N) (g/100 g)	1.25-1.66%	0.1-1.5%
C/N	20-15	15-10

in the evaporation rate that resulted from microbial activity and the aeration intervals by turning over the vessels. The decrease in moisture content may be as a result of biological degradation of organic compound in the vessels. At the end of the study, vessel R2 had the best moisture content. These results indicated that limitation in sewage composting is expected owing to the dewatering process on sludge moisture. So it is suggested that with regards to sewage sludge natural moisture content (50% to 70%), dewatered sludge should not be used for sewage composting. Digested sludge may be better option to choose. This could reduce the sludge treatment costs remarkably. In this study, higher moisture loss was observed in vessel R1 with the lowest portion of bulking agent (green waste). Higher moisture loss observed in R1 would be due to higher maximum microbial activities in this vessel. Varma and Kalamdhad (18) studied the composting of municipal solid waste mixed with cattle manure and observed higher moisture loss in trial with more microbial degradations.
Figure 2 shows the of the monitored pH of the composting material for all the vessels. Composting proceeds most efficiently at the thermophilic temperature when the pH is approximately 8 (31). Maximum pH value of 7.5 for vessel R1 was observed on day 15 (pH=7.5). The increase in pH level at the beginning of composting is as a result of increase in the volume of ammonia released due to protein degradation and the decrease in pH level at the later stage of composting was caused by the volatilization of ammoniacal nitrogen and H+ released due to microbial nitrification process by nitrifying bacteria (32). The values of pH followed similar pattern as reported by Ponsá et al (33) and Cabañas-Vargas et al (34), rising progressively over the first few days; further decreased steadily after that point and remained fairly stable with values between 6 and 7. The pH in all the vessels was in range with WHO, GOTS and Iranian National Standards (Table 4).
Change in the TOC content during the composting period is shown in Figure 3. The content of organic carbon decreased as the decomposition progressed. Initially, the amount of the TOC were 27.17%, 26.6% and 26.43%, which then reduced to 24.01%, 25.02% and 25.78%, respectively in R1, R2 and R3. Around 17.25% of the available carbon in R1 and 2.5% in R2 and 11.36% in R3 were utilized by microorganism as a source of energy.
This finding corresponds with the findings of Wong et al (5) who suggested that such a decrease in organic carbon at the end of composting process may be due to active microbial decomposition of organic substrates. The highest reduction rate was found in vessel R1 (17.25%) suggesting higher metabolic activity of microorganisms. A possible reason for this is that in vessel R1, the ratio of sludge to bulking agent was higher when compared with the other vessels. The low organic carbon contents of vessel R1 indicates the maturity of compost in this vessel which corresponds with the findings of Garcia et al (35) and Hernández et al(13), who studied the evolution of organic, chemistry and biological parameters in composting process and concluded that the percentage of carbon fraction declined due to mineralization. The findings of this study showed that the final organic carbon content in all the vessels was in the range of GOTS, WHO and Iranian National Standards.
Total nitrogen contents in all the vessels increased within 23 days of composting period due to the net loss of dry mass in terms of carbon dioxide, as well as the water loss by evaporation due to heat evolution during oxidization of organic matter (36). Nitrogen fixing bacteria might also contribute to the increase in total nitrogen in the later stage of composting. Nitrogen may be lost due to escape of ammonia (37) or other volatile nitrogenous gases from the compost material to the atmosphere. Vessel R1 contained significantly higher total nitrogen than vessels R2 and R3 (1.98%) throughout the composting process. Total nitrogen behaved in a similar manner in all vessels as observed by Sánchez-Monedero et al (37). The findings of the current study concur with the findings of Wong et al (5) and Bishop and Godfrey (38). Therefore, all the vessels were primarily stabilized after 23 days, but vessel R3 showed better stabilization, which contains the lowest nitrogen content.
Phosphorus in organic material is released by a mineralization process consisting of microorganisms. Inorganic phosphorus negatively charged, reacts readily with positively charged iron (Fe), aluminum (Al), and calcium (Ca) ions to form relatively insoluble substances. When this occurs, phosphorus is considered fixed or tied up. In this regard, phosphorus does not behave like nitrate (NO_3-N), which also has a negative charge but does not form insoluble complexes (18). The final total phosphorus of vessel R3 (0.71%) was higher than that of R1 (0.67%) and R3 (0.31%), respectively. With regards to the total phosphorous, no changes was observed in all the vessels in the pre- and post-composting process. Total phosphorous in vessels R2 and R3 (0.31% and 0.71%) are within the WHO range (0.3% to 3.5%).
The change in C/N ratios shows the achieved organic matter decomposition and stabilization during composting. The decomposition of organic matter was conducted by living organisms, which utilize the carbon as energy source and nitrogen for building cell structures. The C/N

ratio decreased rapidly from initial values of 30 to 13.8, 17.99 and 20.65 in R1, R2 and R3, respectively after 23 days of composting (Figure 5). The excess carbon tends to utilize nitrogen in the soil to build cell protoplasm, if the C/N ratio of compost is more. This results in the loss of nitrogen in the soil and is known as robbing of nitrogen in the soil. On the other hand, if C/N ratio is too low, the resultant product does not help improve the structure of the soil. It is therefore desirable to control the process so that the final C/N ratio is less than or equal to 20 (39). Thus, it can be concluded that compost from all the vessels were primary stabilized after 23 days of composting, similarly as Huang et al (36) and Varma and Kalamdhad (18), but are below the optimal C/N ratio of 25. It is suggested that this amount of bulking agent does not require further nitrogen to obtain a compost mixture of C/N ratio between 20 and 25. Although, the C/N ratio cannot be used as an absolute indicator of compost maturation due to the large differences that is dependent on the initial compounds (5); a value around or below 20 could be considered satisfactory (39). It is only the values of C/N ratio that are not sufficient to say that organic compost is mature, or that, the composting process has reached the maturation phase, because the maturation phase is related to the synthesis of humic substances.

The MPN of the total and fecal coliforms in different stages of composting process decreased. It is suggested that reduction in MPN of all the vessels may result from the occurring thermophile phase during composting process. However, this phase occurred with delay in vessel R2 when compared with the other 2 vessels, in accordance with pH changes (Figure 2). The final compost product is classified into classes A and B according to EPA standard (16). In class A, there should be <1000 MPN and in class B $<2 \times 10^6$ MPN fecal coliforms/g of compost. In this study, all the vessels with MPN less than 1000 fecal coliforms per g are were considered to be in class A, therefore, the final product of composting can be used safely as a soil amendment in agriculture.

Conclusion

Composting of green waste with dewatered sludge in R3 and R1 produced higher final total phosphorus and total nitrogen when compared with other vessels after 23 days of composting. Lower pH and nitrogen, indicated its less phytoxicity effects on the growth of plant if applied to the soil. In addition, composts with lower microbial activity indicated that the compost was ready for usage as a soil conditioner. Therefore, it can be invented that composting of green waste with dewatered sludge in ratio 1.5:1 is more efficient in rotary drum composter. It is concluded that the final product of composting of dewatered and digested sludge and green waste in Ahvaz, Iran have good microbial quality of level A according to EPA classification and is acceptable according to WHO, GOTS and Iran National Standards with regards to the most physical and chemical parameters, therefore, it can be used in farmland and agriculture as a soil fertilizer and conditioner.

Acknowledgements

This research was financially supported by the Student Research Committee, Deputy of Research and Technology, Ahvaz Jundishapur University of Medical Sciences; project number 90s125. The authors hereby express their gratitude.

Competing interests

The authors declare that they have no competing interests.

Authors' contributions

NJHF conceived and designed the study. BM, MA, RAB, HB and AM did the lab work and MR and AZJ performed literature search and wrote the manuscript. All authors participated in the data acquisition, analysis and interpretation. All authors critically reviewed, refined and approved the manuscript.

References

1. Bazrafshan E, Parvaresh A, Kord Mostafa Pur M, Kamani H. Feasibility study of application compost produced from sewage sludge. Third Agriculture Congress of Recycling and Using Renewable Organic Resources in Agriculture; Sanandaj, Iran; 2009.
2. Zazouli MA, Bagheri Ardebilian M, Ghahramani E, Ghorbanian Alah Abad M. Principles of Compost Production Technology. Tehran: Khaniran; 2009. [In Persian]
3. Zazouli MA, Yousefi Z, Eslami A, Bagheri Ardebilian M. Municipal solid waste landfill leachate treatment by fenton, photo-fenton and fenton-like processes: Effect of some variables. Iranian J Environ Health Sci Eng 2012; 9(1): 3.
4. Tchobanoglous G, Theisen H, Vigil S. Integrated Solid Waste Management: Engineering Principles and Management Issues. New York: McGraw-Hill; 1993.
5. Wong J, Mak K, Chan N, Lam A, Fang M, Zhou L, et al. Co-composting of soybean residues and leaves in Hong Kong. Bioresour Technol 2001; 76(2): 99-106.
6. Agency USEP. Standards for the use or disposal of sewage sludge. Part 503. Washington DC; 2007.
7. Turovskiy IS, Mathai PK. Wastewater Sludge Processing. John Wiley & Sons; 2006.
8. Paredes C, Roig A, Bernal M, Sánchez-Monedero M, Cegarra J. Evolution of organic matter and nitrogen during co-composting of olive mill wastewater with solid organic wastes. Biol Fertil Soils 2000; 32(3): 222-7.
9. Sanin FD, Clarkson WW, Vesilind PA. Sludge Engineering: The Treatment and Disposal of Wastewater Sludges. Lancaster, PA: DEStech Publications; 2011.
10. Brinton WF. Compost Quality Standards and

Guidelines: An International View. New York: Wood End Research Laboratory; 2000. p. 15-72.
11. Gea T, Artola A, Sánchez A. Composting of de-inking sludge from the recycled paper manufacturing industry. Bioresour Technol 2005; 96(10): 1161-7.
12. Haug RT. The Practical Handbook of Compost Engineering. Boca Raton: Lewis Publishers; 1993.
13. Hernández T, Masciandaro G, Moreno JI, García C. Changes in organic matter composition during composting of two digested sewage sludges. Waste Manag 2006; 26(12): 1370-6.
14. Ponsá S, Gea T, Sánchez A. The effect of storage and mechanical pretreatment on the biological stability of municipal solid wastes. Waste Manag 2010; 30(3): 441-5.
15. Abid N, Sayadi S. Detrimental effects of olive mill wastewater on the composting process of agricultural wastes. Waste Manag 2006; 26(10): 1099-107.
16. USEPA. Plain English guide to the EPA Part 503 biosolids rule US EPA/832/r-93/003. Washington, DC: United States Environmental Protection Agency Office of Wastewater Management; 1994.
17. Ali dadi H, Parvaresh AAR, Pour Moghadas H, Shah Mansouri MR. Usage of recycle compost as bulking agents in sewage sludge composting. Journal of Research in Agricultural Science 2007; 3(1): 109-16. [In persian].
18. Varma VS, Kalamdhad AS. Composting of municipal solid waste (MSW) mixed with cattle manure. Int J Environ Sci 2013; 3(6): 2068-79.
19. Mara DD, Cairncross S. Guidelines for the safe use of wastewater and excreta in agriculture and aquaculture: measures for public health protection. Geneva: WHO; 1989. p. 187.
20. World Health Organization (WHO). WHO guidelines on good agricultural and collection practices (GACP) for medicinal plants. Geneva: WHO; 2003.
21. Farzadkia M, Salehi S, Aameri A, Joneidy Jafari A, Nabizadeh R. Study on the quality and comparing of the compost produced by Khomain and Tehran compost factories. Iran J Health Environ 2009; 2(3): 160-9.
22. Sefidkar E Kazemi MA, Mohebbrad B, Sadeghi A. Chemical analysis of the compost produced in Mashhad city and comparison with standards. Journal of North Khorasan University of Medical Sciences 2013; 5(4): 775-82. [In Persian]
23. Karimi B, Ehrampoush MH, Jabary H. Indicator pathogens, organic matter and LAS detergent removal from wastewater by constructed subsurface wetlands. J Environ Health Sci Eng 2014; 12: 52.
24. Azadi N, Falahzadeh R, Sadeghi S. Dairy wastewater treatment plant in removal of organic pollution: a case study in Sanandaj, Iran. Environ Health Eng Manag J 2015; 2(2): 73-7.
25. Schumacher BA. Methods for the determination of total organic carbon (TOC) in soils and sediments. Ecological Risk Assessment Support Center; 2002.
26. Ogunwande G, Osunade J, Adekalu K, Ogunjimi L. Nitrogen loss in chicken litter compost as affected by carbon to nitrogen ratio and turning frequency. Bioresour Technol 2008; 99(16): 7495-503.
27. WI TC. Determination of Kjeldahl Nitrogen in soil, biowaste and sewage sludge. https://www.ecn.nl/docs/society/horizontal/STD6161_Kj-N.pdf. 2005.
28. WPC SIfWSSP. Methods of Sampling and Analysis of Solid Wastes Swiss Federal Institute for Water Supply. Switzerland : Swiss Institute for Water Supply Sewage Purification & WPC; 1970. p. 72.
29. Redlinger T, Graham J, Corella-Barud V, Avitia R. Survival of fecal coliforms in dry-composting toilets. Appl Environ Microbiol 2001; 67(9): 4036-40.
30. Clesceri LS, Greenberg AE, Eaton A. Standard Methods for the Examination of Water and Wastewater. Washington DC: American Public Health Association; 1998.
31. Liao P. Composting of fish wastes in a full-scale in-vessel system using different amendments. Journal of Environmental Science & Health Part A 1997; 32(7): 2011-25.
32. Eklind Y, Kirchmann H. Composting and storage of organic household waste with different litter amendments. II: nitrogen turnover and losses. Bioresour Technol 2000; 74(2): 125-33.
33. Ponsá S, Pagans E, Sánchez A. Composting of dewatered wastewater sludge with various ratios of pruning waste used as a bulking agent and monitored by respirometer. Biosystems Engineering 2009; 102(4): 433-43.
34. Cabañas Vargas DD, Sánchez-Monedero MA, Urpilainen ST, Kamilaki A, Stentiford EI. Assessing the stability and maturity of compost at large-scale plants. Ingeniería 2005; 9(2): 25-30.
35. Garcia CH, Hernandez T, Costa F. Changes in carbon fractions during composting and maturation of organic wastes. Environ Manag 1991; 15(3): 433-9.
36. Huang GF, Wong JW, Wu QT, Nagar BB. Effect of C/N on composting of pig manure with sawdust. Waste Manag 2004; 24(8): 805-13.
37. Sánchez-Monedero MA, Roig A, Paredes C, Bernal MP. Nitrogen transformation during organic waste composting by the Rutgers system and its effects on pH, EC and maturity of the composting mixtures. Bioresour Technol 2001; 78(3): 301-8.
38. Bishop P, Godfrey C. Nitrogen transformations during sludge composting. Biocycle 1983; 24(5): 34-9.
39. Hirai MF, Chamyasak V, Kubota H. A standard measurement for compost maturity. Biocycle 1983; 24(6): 54-6.

9

Heavy metals bioaccumulation in fish of southern Iran and risk assessment of fish consumption

Mohammad Malakootian[1*], Mohammad Seddiq Mortazavi[2], Abdolkarim Ahmadi[3]

[1]Professor, Environmental Health Engineering Research Center, Department of Environmental Health, School of Public Health, Kerman University of Medical Sciences, Kerman, Iran
[2]Associate Professor, Persian Gulf and Oman Sea Ecological Research Institute, IFSRI, AREEO, Bandar Abbas, Hormozgan, Iran
[3]MSc, Department of Environmental Health, School of Public Health, Kerman University of Medical Sciences, Kerman, Iran

Abstract
Background: This study purposed to assess the health risks of heavy metals including nickel (Ni), chromium (Cr), mercury (Hg), and lead (Pb) in three widely-consumed fish from the coast of Bandar Abbas (Persian Gulf) and to determine the maximum amounts allowed for consumption according to the degree of contamination.
Methods: This cross-sectional study was conducted in the summer and fall of 2013. *Thunnus tonggol* and *Liza klunzingeri* were selected from the Bandar Abbas market, and Pleuronectiformes were caught from the sea. Heavy metal concentrations were determined by atomic absorption. Risk assessment and allowable consumption levels were determined according to methods provided by the US Environmental Protection Agency (USEPA).
Results: The highest average levels of Ni and Cr in *Thunnus tonggol* were 0.059 and 0.234 (μg/g dry weight), respectively. The highest average level of Hg in Pleuronectiformes was 0.095 (μg/g dry weight), and the highest average level of Pb in *Liza klunzingeri* was 0.006 (μg/g dry weight). The target hazard quotient (THQ) for all analyses of metals in the studied species and the hazard index (HI) were less than 1 for all three species.
Conclusion: The results showed that despite Ni, Cr, Hg, and Pb pollution, the consumption of up to 4 (kg/d) of *Thunnus tonggol* by adults and up to 2.2 (kg/d) by children does not cause health problems in terms of the tested heavy metals. For *Liza klunzingeri*, these levels are 7.32 and 4.02 (kg/d), respectively.
Keywords: Nickel; Chromium; Mercury; fish; Risk assessment.

***Correspondence to:**
Mohammad Malakootian
Email: m.malakootian@yahoo.com

Introduction

The presence of heavy metals in the environment is a cause for major concern. Even at low concentration levels, they can cause problems as a result of toxicity and their tendency to permeate the food chain (1). The term 'heavy metals' refers to metals whose specific gravity is greater than 5 g/cm^3 in their standard state (2). These metals penetrate aquatic ecosystems through agricultural effluents and discharged industrial wastewater which have different chemical compounds, including heavy metals (3). Fish accumulate metals by ingesting particulate material suspended in water, through food, through the ion-exchange of dissolved metals across lipophilic membranes (e.g., the gills), and through adsorption onto tissue and membrane surfaces (4). Fish are considered bio-indicators of the quality of ecosystems because of their ability to bioconcentrate and integrate a contaminant load (5).

According to statistics compiled by Food and Agriculture Organization (FAO), the consumption of fish accounts for 16% of animal protein consumed and 6% of total protein intake by people of the world (6). Having high levels of unsaturated fats and low cholesterol and being high in protein makes fish one of the most important sources of protein. It is considered an important food ingredient in many countries, including Iran, and its production is increasing (7).

Heavy metals may enter the human body through the consumption of fish and thereby create a serious health hazard (8). The measurement of heavy metals in fish can be useful to assessing potential health risks to humans associated with the consumption of fish (9).

Health risk assessment is defined by the US Environmental Protection Agency (USEPA) as the description of the potential adverse health effects of humans as a result of exposure to contaminants (10).

Because of the importance of measuring heavy metal concentrations in fish and assessing the health risk associated with exposure to heavy metals through the consumption of fish, the accumulation of heavy metals in fish has been widely studied throughout the world (3,11-13). Furthermore, numerous studies have evaluated potential health risks associated with exposure to heavy metals through fish consumption (7,9,14-17).

The present study was undertaken to determine the levels of heavy metals in the three most frequently used fish caught from Bandar Abbas waters. As a result of this study, consumers and authorities can become better aware of the possible risks of contamination through daily intake of fish.

Methods

This cross-sectional study was conducted in the summer and fall of 2013. Sampling was conducted on three of the most widely consumed fish caught from the coast of Bandar Abbas, i.e. *Thunnus tonggol*, Pleuronectiformes (*Pseudorhombus elevatus* and *Euryglossa orientalis*), and Klunzinger mullet that were being sold at the town market or were caught directly from the sea. Five samples from each fish species were tested three times at 2-month intervals (overall 15 samples for each fish species). A total of 180 tests were conducted on the samples. Concentrations of the four metals nickel (Ni), chromium (Cr), mercury (Hg), and lead (Pb) were measured.

Sampling

Thunnus tonggol and *Liza klunzingeri* samples were selected from the town market, and Pleuronectiforme samples were caught from the sea and transferred to the Persian Gulf and Oman Sea Ecological Research Laboratory. The standard length of the fish after cleaning was measured by biometric boards, and their weight was measured by a laboratory scale with an accuracy of 0.1 g. Fish muscle is not an active site for heavy metal accumulation. In other words, fish muscle compared with other tissue has the lowest concentration of heavy metals (6). Therefore, because of its role in the transfer of heavy metals to the human body, muscle tissue was chosen as the target organ in risk assessment. Samples of 15-20 g of fish muscle tissue were selected and washed once with both municipal and distilled water. They were stored in plastic bags and, after coding, were frozen at -19°C.

Sample preparation and device analysis

Samples were prepared according to the Manual of oceanographic observations and pollutant analyses methods (MOOPAM), 1999 method. Heavy metal concentrations (Cr, Ni, and Pb) were assessed using graphite furnace atomic absorption spectrometry. Hg concentrations were measured using an atomic absorption spectrometer equipped with a hydride generator. Data quality was checked by analyzing standard reference materials (DORM-2 National Research Council, Canada). Recovery values of collaborative recovery model (CRM) samples were in the range of 95.2%-97%.

Statistical analysis

Data was tested for normality using the Shapiro-Wilk test. Since the data did not show a normal distribution, the Mann-Whitney and Kruskal-Wallis tests were used to compare average heavy metal concentrations in different fish species. The Pearson correlation coefficient was used for normal data to test the correlation between fish length and concentration of heavy metals; for abnormal data, the Spearman correlation coefficient was used. A level of 0.05 was considered significant. All tests were performed by SPSS version 18, and Excel software was used to draw diagrams.

Comparison with guidelines and standards

Since there is no standard for heavy metals involving fish in Iran, amounts of Ni, Cr, Hg, and Pb in the tested fish were compared with guidelines from the World Health Organization (WHO), FAO, The Joint FAO/WHO Expert Committee Report on Food Additives (JECFA), the European Commission, standards formulated by the American Food and Drug Administration (USFDA), the Ministry of Agriculture, Fisheries, and Food of the United Kingdom (MAFF), and China's national regulations regarding (Table 1).

Estimated daily intake and estimated weekly intake of heavy metals

Based on the method proposed by Fathi et al (23), daily and weekly absorption rates were calculated using equation 1:

$$EWI = C \times IR/BW \quad (1)$$

where estimated weekly intake (EWI) (μg/kg-week) is the EWI of pollutant; C (μg/g) is the average concentration of pollutant; IR (g/week) is the amount of food intake; and BW (kg) is the body weight of the consumer.

Table 1. Guidelines of WHO, FAO, JECFA, MAFF, the European Commission, standards formulated by the USFDA, and China's national regulations regarding heavy metals (mg/kg)

Codify by	Year	Ni	Cr	Hg	Pb	References
JECFA	1989	-	-	0.5	0.5	(18)
Europe Commission(EC)	2006	-	-	0.5-1	0.3	(18)
WHO	1996	-	-	-	2	(18)
FAO	1983	-	-	-	0.5	(19)
USFDA	1993	70-80	12-13	-	-	(20)
MAFF	2000	-	-	0.3	2	(21)
China national regulations	-	-	0.5	0.3	1	(22)

Risk assessment

Risk assessment is a process to calculate or estimate the risk for a given population exposed to certain pollutants (24). This risk assessment study investigated the target hazard quotient (THQ) for three groups: children, normal adults (non-fishermen), and fishermen. THQ is defined as the daily intake of pollutants higher than the reference dose. Estimates were based on the method proposed by Yi et al (25) and equation 2:

$$THQ = (EFr \times EDtot \times FIR \times C)/(RFDo \times BWa \times ATn) \times 10^{-3} \quad (2)$$

where THQ (unitless) is the THQ; EFr (day/year) is the exposure frequency; EDtot (year) is the exposure duration; FIR (g/d) is the food ingestion rate; C (μg/g) is the concentration of heavy metals in fish; BWa (kg) is the average body weight of consumers; RFDo (μg g^{-1} day^{-1}) or (mg kg^{-1} day^{-1}) is the reference dose; and ATn (day) is the average time for exposure to non-carcinogens.

A THQ of THQ < 1 indicates that, based on current consumption rates and current concentrations of pollutants in fish, the consumption of fish has no adverse effects on the health of consumers over a lifetime (26).

EFr: In this study, an exposure frequency of 128 days was considered (7).

EDtot: Exposure time was considered to be 70 years based on life expectancy in Hormozgan province (27).

FIR: Based on the General Department of Fisheries of Hormozghan province, this study used a dose of 50 mg per day (28). This amount was considered to be 20 g per day for children and 2.4 times more than the normal range for fishermen (29).

The parameter of heavy metal concentrations in fish was extracted from the mean heavy metal concentrations in fish muscle in the present study.

BW: In this study, the average adult weight was considered to be 60 kg, based on a 2009 research in Iran by Hajizadeh et al (30), and child weight was considered to be 33 kg (29).

RFDo: The USEPA in 2008 determined the reference doses for Cr and Ni to be 0.003 and 0.02 (mg kg^{-1} day^{-1}), respectively (31), in 2009 determined the reference dose of 0.004 (mg kg^{-1} day^{-1}) for Pb (32), and in 2011 set 0.0003 (mg kg^{-1} day^{-1}) as the reference dose of Hg. These values were used in calculations in this study (16).

ATn: This time comes from multiplying the number of days during a year and number of years of exposure.

Total THQ of heavy metals in fish or a hazard index (HI) was calculated for each species (16).

Calculating permissible limits of consumption

Since man uses several fish species in his diet, he risks possible contamination from multiple pollutants. Equation 3, given by the USEPA (33), was used to calculate the allowable amount of exposure to multiple contaminants in many fish species:

$$CR_{lim} = \Sigma\,(RfD.\,Pm/\,(Cmj \times Pj)) \times BW \quad (3)$$

where, CR_{lim} is the allowable amount of fish consumption (kg/d); RFD is the reference dose (mg/kg-d); Pj is the portion of fish species in the food diet; Pm is the portion of contaminants in the food diet; BW is the body weight of consumers (kg); and Cm is the concentrations of chemical matter (mg/kg).

Results

The amount of heavy metals in fish

The concentrations of Ni, Cr, Hg, and Pb were tested in the muscles of three species of fish (*Thunnus tonggol*, *Liza klunzingeri*, and Pleuronectiformes) collected from the Bandar Abbas coast during the summer and autumn of 2013. Results are listed in Table 2 as mean ± standard deviation (SD), maximum and minimum amounts according to μg/g dry weight.

In order of concentration amounts of Cr and Ni, the fish species are ranked as *Thunnus tonggol* > Pleuronectiformes > *Liza klunzingeri*; in terms of the amount of Hg they are ranked as Pleuronectiformes > *Thunnus tonggol* > *Liza klunzingeri*. In terms of contamination levels of Pb they are ranked as *Liza klunzingeri*> Pleuronectiformes > *Thunnus tonggol*.

Statistical analyses indicated a significant difference between the concentrations of Ni, Cr, and Pb in the tested fish species ($P<0.05$).

Correlation between heavy metal concentrations and fish length

The correlation between Ni, Cr, Hg, and Pb concentrations fish length in tested species is shown in Table 3. Table 3 shows that there is no significant correlation between Ni, Cr, or Hg and the length of the tested fish species, but there is a strong, significant correlation between Pb and fish length.

Estimated daily intake and estimated weekly intake of heavy metals

To describe the safe levels of absorption of various pollutants, tolerable daily intake (TDI) and tolerable weekly intake (TWI) values have been determined by the JEC-

Table 2. Mean concentrations and standard deviation of heavy metals (μg/g dry weight) in three tested fish species

Heavy metals	Fish species	Mean± SD	Min.	Max.
Ni	*Thunnus tonggol*	0.234 ± 0.06	0.179	0.36
Cr	*Thunnus tonggol*	0.059 ± 0.03	0.018	0.097
Hg	*Thunnus tonggol*	0.083 ± 0.03	0.065	0.151
Pb	*Thunnus tonggol*	0.003 ± 0.001	0.0018	0.005
Ni	*Pleuronectiformes*	0.184 ± 0.07	0.118	0.36
Cr	*Pleuronectiformes*	0.033 ± 0.01	0.02	0.05
Hg	*Pleuronectiformes*	0.095 ± 0.003	0.036	0.153
Pb	*Pleuronectiformes*	0.004 ± 0.002	0.002	0.01
Ni	*Liza klunzingeri*	0.116 ± 0.04	0.081	0.232
Cr	*Liza klunzingeri*	0.011 ± 0.01	0.004	0.052
Hg	*Liza klunzingeri*	0.078 ± 0.003	0.054	0.175
Pb	*Liza klunzingeri*	0.006 ± 0.003	0.018	0.013

Table 3. Correlation between Ni, Cr, Hg, and Pb concentrations and fish length in tested species

Fish type	Ni		Cr		Hg		Pb	
	R	P	R	P	R	P	R	P
Thunnus tonggol	0.29	0.29	0.07	0.48	0.34	0.22	0.06	0.83
Pleuronectiformes	0.3	0.27	0.45	0.21	0.11	0.7	0.6	0.02
Liza klunzingeri	0.2	0.46	0.49	0.19	0.1	0. 7	0.14	0.62

Table 4. EDI (µg/kg-day) and EWI (µg/kg-week) for heavy metals

Fish type	Ni		Cr		Hg		Pb	
	EDI	EWI	EDI	EWI	EDI	EWI	EDI	EWI
Thunnus tonggol	0.195	1.36	0.049	0.344	0.07	0.487	0.003	0.021
Pleuronectiformes	0.15	1.07	0.027	0.19	0.079	0.55	0.004	0.028
Liza klunzingeri	0.096	0.676	0.009	0.065	0.065	0.457	0.005	0.036

Abbreviations: EDI, Estimated daily intake; EWI, estimated weekly intake.

FA. They are shown as Provisional tolerable daily intake (PTDI) and Provisional tolerable weekly intake (PTWI) because of their temporary nature. PTWI is the maximum amount of a pollutant that a man can ingest during a lifetime without any acceptable risk to his health (16). Weekly and daily intake rates of heavy metals including Ni, Cr, Hg, and Pb through the consumption of *Thunnus tonggol*, *Liza klunzingeri*, and *Pleuronectiformes* were calculated in micrograms per day and micrograms per week, respectively. The results are given in Table 4.

Result of THQ values

In the present study, THQ was calculated for each of the heavy metals (Ni, Cr, Hg, and Pb) in three fish species (*Thunnus tonggol, Liza klunzingeri*, and Pleuronectiformes) and for three different groups of people (children, normal adults, and fishermen). The results are given in Table 5. It is to be noted that results are unitless.

Because of the possibility of interference and intensification of the effects of heavy metals, the HI was calculated for Ni, Cr, Hg, and Pb in three species (*Thunnus tonggol, Liza klunzingeri*, and Pleuronectiformes), and the results are shown in Figure 1.

The lowest and highest HIs were associated with the *Liza klunzingeri* and Pleuronectiformes species, respectively. The maximum HI was seen in the fishermen group, and the lowest was in the category of children.

Calculating the permissible limit of consumption

The consumers of three different types of fish were divided into seven groups (*Thunnus tonggol* consumers, *Liza klunzingeri* consumers, Pleuronectiformes consumers, *Thunnus tonggol* and Liza *klunzingeri* consumers, *Thunnus tonggol* and Pleuronectiformes consumers, *Liza klunzingeri* and Pleuronectiformes consumers, and consumers of all three species). Permissible intake limits were calculated for the various consumers groups considering the degree of heavy metal contamination from fish consumption for children and adult groups. The results are given in Table 5. It was necessary to determine the proportion of each species in the diet of each consumer group in which members consume two or three species. The ratio of each fish species in the diet was determined based on information regarding consumption. These results are given in Table 6.

Maximum allowable levels of fish consumption belonged to the *Liza klunzingeri* fish and for adults and children were 7.32 and 4.02 kg/d, respectively. Minimum allowable levels of fish consumption belonged to the *Thunnus tonggol* fish and were 4 and 2.2 kg/d, respectively.

Discussion

Heavy metal concentrations

Various concentrations of heavy metals ranging widely from 0.0018 to 0.36 were observed in fish muscle tissue. In order of Cr and Ni concentrations, the fish species are classified as: *Thunnus tonggol* > Pleuronectiformes > *Liza klunzingeri*; in order of Hg concentrations, they are ranked as Pleuronectiformes > *Thunnus tonggol* > *Liza*

Table 5. THQ values for heavy metals in three fish species for children, normal adults, and fishermen

Heavy metals	Fish species	THQ		
		Children	Adults	Fishermen
Ni	*Thunnus tonggol*	0.0025	0.0034	0.0082
Cr	*Thunnus tonggol*	0.0042	0.0057	0.014
Hg	*Thunnus tonggol*	0.0592	0.0814	0.19
Pb	*Thunnus tonggol*	0.0002	0.0002	0.0005
Ni	Pleuronectiformes	0.002	0.0027	0.006
Cr	Pleuronectiformes	0.0023	0.0032	0.008
Hg	Pleuronectiformes	0.067	0.092	0.21
Pb	Pleuronectiformes	0.0002	0.0003	0.0008
Ni	*Liza klunzingeri*	0.0012	0.0017	0.0041
Cr	*Liza klunzingeri*	0.0008	0.0011	0.0026
Hg	*Liza klunzingeri*	0.055	0.073	0.19
Pb	*Liza klunzingeri*	0.0003	0.0005	0.0011

Abbreviations: THQ, target hazard quotient.

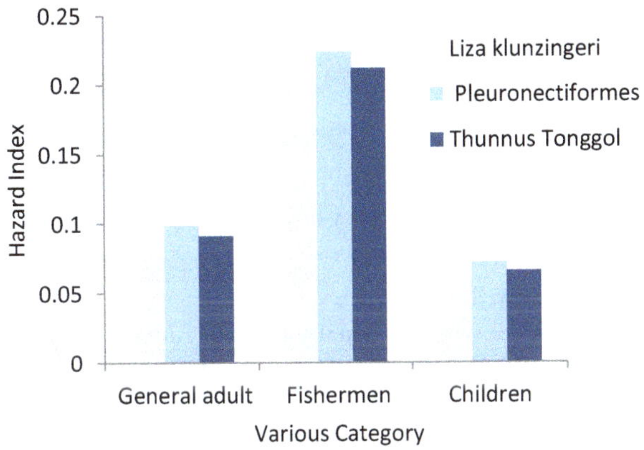

Figure 1. HI of heavy metals in three species (*Thunnus tonggol, Liza klunzingeri*, and Pleuronectiformes).

klunzingeri. In order of contamination levels of Pb, the fish species are ranked as *Liza klunzingeri* > Pleuronectiformes > *Thunnus tonggol*.

Thunnus tonggol was larger than the two other tested species of fish and average concentrations of Ni and Cr were higher in it than in the other two species (*Liza klunzingeri* and Pleuronectiformes). One reason for the higher contamination of Thunnus tonggol by Cr and Ni may be the result of their feeding patterns. Indeed, this species feeds on shrimp, crabs, and small fish and has a wider food chain than the other two species.

Pleuronectiformes was smaller than *Thunnus tonggol*, but its average concentration of Hg was higher. Maximum accumulation and storage of heavy metals occurred in benthic fish eaters, plankton eaters, and carnivores. The high concentration of Hg in Pleuronectiformes could be attributed to this reason.

The concentration of all studied heavy metals except Pb was lower in *Liza klunzingeri* than in the two other species. Heavy metals accumulation was lower in phytophagous species than in carnivorous species (34); the *Liza klunzingeri* species is small and phytophagous, and that could be a reason for its low heavy metals concentration. Statistical analysis showed a significant difference between the concentrations of Ni, Cr, and Pb in the tested fish species ($P<0.05$).

Comparison of results with guidelines and standards

The heavy metal concentrations of Ni, Cr, Hg, and Pb in three species of fish (*Thunnus tonggol, Liza klunzingeri*, and Pleuronectiformes) tested in this study were lower than the limits recommended by the guidelines of WHO, FAO, JECFA, the European Commission, and the standards formulated by the USFDA, the MAFF, and China's national regulations, which are shown in Table 1.

Comparison with other studies

The heavy metal concentrations for the species of *Thunnus tonggol, Liza klunzingeri*, and Pleuronectiformes in this study were compared with other studies (Table 7). Comparing the data in Tables 2 and 7 indicates that concentrations of Ni, Cr, and Pb were higher in fish of Lake Taihu, China and Tagus Estuary in Portugal than the fish in this study. The Cr concentration in fish of rivers in Italy placed in the Cr range in this study, but concentrations of Hg and Pb were higher than the range of this study.

Several factors are involved in the accumulation of met-

Table 6. Contribution of each species in food diet and permitted consumption levels for each of the seven groups and three new cases.

Consumer category	The ratio of species in food diet			Consumption permitted levels (kg per day)	
	Thunnus tonggol	Pleuronectiformes	*Liza klunzingeri*	Children	Adults
Just *Thunnus tonggol*	1	0	0	2.2	4
Just *Liza klunzingeri*	0	1	0	2.8	5.09
Just Pleuronectiformes	0	0	1	4.02	7.32
Thunnus tonggol and *Liza klunzingeri*	0.7	0.3	0	2.66	4.84
Thunnus tonggol and Pleuronectiformes	0.6	0	0.4	2.9	5.27
Liza klunzingeri and Pleuronectiformes	0	0.4	0.6	3.55	6.46
All tree species	0.5	0.2	0.3	3.05	5.55

Table 7. Comparison of measured heavy metal concentrations in tested fish with concentrations taken from the literature (µg/g)

Sampling site	Ni	Cr	Pb	Hg	Reference
Lake Taihu, China	0.64-2.59	0.7-0.98	1.73-12.87	-	(11)
Tagus estuary in Portugal	3.12-14.35	0-0.56	8.55-46.47	-	(35)
Italy rivers	-	0.04	0.04	0.34	(12)
Shadegan wetland, Iran	2.59-98.64	0.0-0.7	1.37-12.87	-	(36)
Moses estuary in Iran	2.35	48.14	1.32	-	(37)

als in different organs of fish, such as age, length, weight, gender, food habitat, ecological demands, metal concentrations in water and sediment, fish shelf life in contaminated environments, the fishing season, and the physical and chemical properties of water (17). Therefore, one cannot explain why the results of other studies are higher than those of this study. One major reason could be the noted differences, such as the feeding habits of the studied species and contamination in their living environments.

Estimated daily and weekly intake

Guidelines of the JECFA in 2010 designated 1.75 and 0.35 (μg/kg-week) as PTWI for Pb and Hg levels, respectively (7). In 2004, the PTWI for Ni was determined to be 35 (μg/kg-week) (6). Recommended PTWI amounts of Ni and Pb are higher than the calculated weekly absorption rate in this study. In the case of Hg, however, it is lower.

Risk assessment

The THQ of tested metals in the studied species and for different groups (children, normal adults, and fishermen) was less than 1. Thus, the danger caused by each of the studied metals alone does not threaten the health of consumers. The results of this study are consistent with those of Tao et al (29) in 2012 in Lake Taihu, China and the results of Taweel et al (38) in 2013 in Langat River and Engineering Lake, Malaysia. In contrast to the results of this study, Krishna et al reported in 2014 a THQ higher than 1 for Pb, Cr, and Hg in fish from the Machilipatnam coast in India (14). A HI of less than 1 was achieved in each of the studied species. Therefore, the current consumption rate of each species poses no danger to consumer health (considering the interference and resonance effects of the studied metals).

Conclusion

The order of mean heavy metal concentrations in fish is Ni > Hg > Cr > Pb. The highest mean concentrations of Ni and Cr were obtained in *Thunnus tonggol*. The highest mean concentration of Hg was obtained in Pleuronectiformes, and the highest mean concentration of Pb was obtained in *Liza klunzingeri*. There was a significant difference between the average concentrations of similar metals (except Hg) in different species. Only levels of Pb were significantly associated with the length of the *Liza klunzingeri* species. Obtained estimations of daily and weekly intakes (except for Hg) were much lower than the PTDI and PTWI. The THQ of all tested metals in the studied species and the HI of all three species were less than 1. Therefore, consumption of the studied fish has no health risk for consumers. The consumption of up to 4 kg/d of Thunnus tonggol and 7.32 kg/d of *Liza klunzingeri* causes no health problems for general adult consumers.

Acknowledgments

This research is derived from a Master's thesis and has been approved to be conducted at the Environmental Health Engineering Research Center. It was sponsored by the Vice-Chancellor for Research and Technology of Kerman University of Medical Sciences and supported by the Persian Gulf and Oman Sea (Bandar Abbas) Ecological Research Institute.

Competing interests

The authors declare that they have no competing interests.

Authors' contributions

All authors contributed equally and were involved in the design of the study, data collection, and article approval.

References

1. Foladifard R, Ebrahimi AA. Efficiency study of nickel (II) and cadmium (II) biosorption by powder of waste activated sludge from aqueous solutions. Iranian Journal of Health and Environment 2010; 3(4): 419-30. [In Persian].
2. Cruz GC, Din Z, Feri CD, Balaoing AM, Marie Gonzales E, Navidad HM. Analysis of toxic heavy metals (arsenic, lead, and mercury) in selected infant formula milk commercially available in the Philippines by AAS. E-International Scientific Research Journal 2009; 1(1): 40-51.
3. Askari SaryA, Khodadadi M, Mohammadi M. Concentration of heavy metal (Cd, Pb, Ni, Hg) in muscle, gill and liver tissues of Barbus xanthopterus in Karoon River. Iranian Scientific Fisheries Journal 2010; 19(4): 97-106.
4. Squadrone S, Prearo M, Brizio P, Gavinelli S, Pellegrino M, Scanzio T, et al. Heavy metals distribution in muscle, liver, kidney and gill of European catfish (Silurus glanis) from Italian Rivers. Chemosphere 2013; 90(2): 358-65.
5. Taghavi Jelodar H, Fazli H, Salman Mahiny A. Study on heavy metals (chromium, cadmium, cobalt and lead) concentration in three pelagic species of Kilka (Genus Clupeonella) in the southern Caspian Sea. Iranian Journal of Fisheries Sciences 2016; 15(1): 567-74.
6. Elnabris KJ, Muzyed SK, El-Ashgar NM. Heavy metal concentrations in some commercially important fishes and their contribution to heavy metals exposure in Palestinian people of Gaza Strip (Palestine). Journal of the Association of Arab Universities for Basic and Applied Sciences 2013; 13(1): 44-51.
7. Mortazavi MS, Sharifian S, Aghajari N. Risk estimation of heavy metals from consumption of silver pomfred and tiger tooth croaker in hormozgan province. Iranian Scientific Fisheries Journal 2013; 22(2): 127-36.
8. Sen I, Shandil A, Shrivastava VS. Study for determination of heavy metals in fish species of the River Yamuna (Delhi) by inductively coupled plasma-optical emission spectroscopy (ICP-OES). Advances in Applied Science Research 2011; 2(2): 161-6.
9. Ahmad AK, Sarah AA. Human health risk assessment of heavy metals in fish species collected from catchments of former tin mining. International Journal of Research Studies in Science, Engineering and Technology 2015; 2(4): 9-21.

10. Koki IB, Bayero AS, Umar A , Yusuf S. Health risk assessment of heavy metals in water, air, soil and fish. African Journal of Pure and Applied Chemistry 2015; 9(11): 204-10.
11. Zeng J, Yang L, Wang X, Wang WX, Wu QL. Metal accumulation in fish from different zones of a large, shallow freshwater lake. Ecotoxicol Environ Saf 2012; 86: 116-24.
12. Mortazavi A, Hatamikia M, Bahmani M, Hassanzadazar H. Heavy metals (mercury, lead and cadmium) determination in 17 species of fish marketed in Khorramabad city, West of Iran. Journal of Chemical Health Risks 2016; 6(1): 41-8.
13. Olusola JO, Festus AA. Assessment of heavy metals in some marine fish species relevant to their concentration in water and sediment from coastal waters of Ondo State, Nigeria. Journal of Marine Science: Research & Development 2015; 5(2): 1-6.
14. Krishna PV, Rao KM, Swaruparan IV, Rao DS. Heavy metals concentration in fish mugil cephalus from machilipatnam coast and possible health risks to fish consumers. British Biotechnology Journal 2014; 4(2): 126-35.
15. Akoto O, Bismark Eshun F, Darko G, Adei E. Concentrations and health risk assessments of heavy metals in fish from the Fosu Lagoon. Int J Environ Res 2014; 8(2): 403-10.
16. Bhupander K, DP M. Assessment of human health risk for arsenic, copper, nickel, mercury and zinc in fish collected from tropical wetlands in India. Advances in Life Science and Technology 2011; 2: 13-25.
17. Mashroofeh A, Bakhtiari AR, Pourkazemi M. Evaluation of cadmium, vanadium, nickel and zink concentrations in different tissues of beluga and stellate sturgeon and risk assessment regarding consuming their muscle tissue in south Caspian Sea. J Mazandaran Univ Med Sci 2012; 22(96): 90-7. [In Persian].
18. Al Sayegh Petkovšek S, Mazej Grudnik Z, Pokorny B. Heavy metals and arsenic concentrations in ten fish species from the Šalek lakes (Slovenia): assessment of potential human health risk due to fish consumption. Environ Monit Assess 2012; 184(5): 2647-62.
19. Sivaperumal P, Sankar TV, Viswanathan Nair PG. Heavy metal concentrations in fish, shellfish and fish products from internal markets of India vis-a-vis international standards. Food Chem 2007; 102(3): 612-20.
20. Rahman MS, Molla AH, Saha N, Rahman A. Study on heavy metals levels and its risk assessment in some edible fishes from Bangshi River, Savar, Dhaka, Bangladesh. Food Chem 2012; 134(4): 1847-54.
21. Yohannes YB, Ikenaka Y, Nakayama SM, Saengtienchai A, Watanabe K, Ishizuka M. Organochlorine pesticides and heavy metals in fish from Lake Awassa, Ethiopia: insights from stable isotope analysis. Chemosphere 2013; 91(6): 857-63.
22. Zhang Z, He L, Li J, Wu Z. Analysis of heavy metals of muscle and intestine tissue in fish – in Banan section of Chongqing from Three Gorges Reservoir, China. Polish J Environ Stud 2007; 16(6): 949-958.
23. Fathi HB, Othman MS, Mazlan AG, Arshad A, Amin SM, Simon KD. Trace metals in muscle, liver and gill tissues of marine fishes from Mersing, Eastern Coast of Peninsular Malaysia: concentration and assessment of human health risk. Asian Journal of Animal and Veterinary Advances 2013; 8(2): 227-236.
24. Arizhibowa F. A human health risk assessment of heavy metal bioaccumulation in Nile Tilapia Oreochromis niloticus from Lake Chivero, Zimbabwe [Thesis]. Zimbabwe: University of Zimbabwe; 2011.
25. Yi Y, Yang Z, Zhang S. Ecological risk assessment of heavy metals in sediment and human health risk assessment of heavy metals in fishes in the middle and lower reaches of the Yangtze River basin. Environ Pollut 2011; 159(10): 2575-85.
26. Hosseini SM, Mirghaffari N, Mahboobi Soofiani N, Hosseini SV. Risk assessment of mercury due to consumption of kutum of the Caspian Sea (Rutilus frisii kutum) in Mazandaran province. Iranian Journal of Natural Resources 2011; 64(3): 243-57. [In Persian].
27. Khosravi A, Najafi F, Rahbar MR, Motlagh MI, Kabir MJ. Landscape indicators of health in the Islamic Republic Iran.Tehran: Department of Health and Medical Education, Islamic Republic of Iran; 2009.
28. GDOFHP. Determination of per capita consumption of aquatic in level Hormozgan province. General Directorate of Fisheries Hormozgan province: coast sea researchers consulting engineers; 2010. http://www.hpf.ir/.
29. Tao Y, Yuan Z, Xiaona H, Wei M. Distribution and bioaccumulation of heavy metals in aquatic organisms of different trophic levels and potential health risk assessment from Taihu lake, China. Ecotoxicol Environ Saf 2012; 81: 55-64.
30. Hajizadeh S, Faghihzadeh S, Khaksari M, Boskabadi M, Pejhan A, Rostampour M, et al. Measurement of vital pulmonary parameters of sample iranian populations to determine reference equations for estimation of these parameters. Knowledge & Health 2009; 4(1): 1-7. [In Persian].
31. Zhao S, Feng C, Quan W, Chen X, Niu J, Shen Z. Role of living environments in the accumulation characteristics of heavy metals in fishes and crabs in the Yangtze River Estuary, China. Mar Pollut Bull 2012; 64(6): 1163-71.
32. Zhuang P, Li ZA, McBride MB, Zou B, Wang G. Health risk assessment for consumption of fish originating from ponds near Dabaoshan mine, South China. Environ Sci Pollut Res Int 2013; 20(8): 5844-54.
33. USEPA. Guidance for Assessing Chemical Contaminant Data for Use in Fish Advisories Volume 2 : Risk Assessment and Fish Consumption Limits. Washington, DC: USEPA; 2000.
34. Beheshti M, Askari Sari A, Velayatzadeh M. Assessment of heavy metals concentration of fish (Liza abu) in Karoon River, Khouzestan province. Water and wastewater. 2012; 3: 125-33 .[In Persian].
35. Caçador I, Costa JL, Duarte B, Silva G, Medeiros JP, Azeda C, et al. Macroinvertebrates and fishes as biomonitors of heavy metal concentration in the Seixal

Bay (Tagus estuary): Which species perform better? Ecol Indic 2012; 19: 184-90.

36. Hosseini Alhashemi A, Sekhavatjou MS, Hassanzadeh Kiabi B, Karbassi AR. Bioaccumulation of trace elements in water, sediment, and six fish species from a freshwater wetland, Iran. Microchem J 2012; 104: 1-6.
37. Parvaneh M, Khaivar N, Nikpour Y, Nabavi SM. Heavy metals (Hg, Cd, Pb, Ni, Cu) concentrations in Euryglossa orientalis and sediments from Khur-e-Musa Creek in Khuzestan Province. Iranian Scientific Fisheries Journal 2011; 20(2): 17-26. [In Persian].
38. Taweel A, Shuhaimi-Othman M, Ahmad AK. Assessment of heavy metals in tilapia fish (Oreochromis niloticus) from the Langat River and Engineering Lake in Bangi, Malaysia, and evaluation of the health risk from tilapia consumption. Ecotoxicol Environ Saf 2013; 93: 45-51.

10

Strategic management of solid waste in Tehran

Monireh Majlessi[1], Aida Vaezi[2*], Mohsen Mehdipour Rabori[3]

[1]Associate Professor, Department of Environmental Health Engineering, School of Public Health, Shahid Beheshti University of Medical Science, Tehran, Iran
[2]MA of Environmental Science, Islamic Azad University, Science and Research Branch, Tehran, Iran
[3]PhD Student of Environmental Health in SBMU, Member of Environmental Health Engineering Research Center, Kerman University of Medical Sciences, Kerman, Iran

Abstract
Background: Successful waste management programs are linked to internal factors, SW (strength and weakness points), and external factors, OT (opportunities and threats), and at the macro level are grounds for sustainable development. Studies may be related to David's study which presented a method to quantify strength, weakness, opportunity, threats (SWOT) analysis.
Methods: In this study, designed to investigate recycling and managing dry residues in District 1, threat factors were studied and internal factors evaluation (IFE) and external factors evaluation (EFE) matrixes were scored. Tehran Municipality District 1 has 10 subdistricts and 26 neighborhoods. The dry residues from Municipality District 1 that currently weigh approximately 2678 tons per day.
Results: Results showed that WO strategies for improving weaknesses and employing opportunities available in District 1 are ideal. After constructing the Quantitative Strategic Planning Matrix (QSPM), WO strategies were prioritized.
Conclusion: The evaluation matrix of internal and external factors in the city of Tehran indicated that waste management has weak internal factors. Meanwhile, evaluation points of external factors showed that, in the current situation, SWOT could achieve good results. Waste management systems are involved with different multi-disciplinary factors; therefore, trends in the development of waste treatment technologies have been led by various social, economic, and environmental drivers in Tehran.
Keywords: Solid waste management, SWOT, Strategic management

***Correspondence to:**
Aida Vaezi
Email: vaezei.ac@gmail.com

Introduction

Management, planning, and efforts to find optimal solutions for improving urban residue problems are inevitable, and city authorities seek optimal procedures to remedy urban problems and issues (1). Successful waste management programs are linked to internal factors (strength and weakness points) and external factors (opportunities and threats), and at the macro level are grounds for sustainable development (2). Studies on such subjects may be related to David's study in which the author presented a method to quantify SWOT. David (3) formed the matrix of internal & external factors and matrix-profile competition, but allocating a score to main factors having to be done mentally was a shortcoming of this method. Halla (4) studied the city of Dar al-Salam in Tanzania to analyze internal and external factors for urban management planning. Observations were made along with the participation and documented investigations in accordance with the available data.

Among the most important issues considered in this design are the collection, transfer, and disposal of environmentally friendly urban residues. Among other studies in this regard, the work of Abedin-Zadeh et al (2) can be noted. In this study, internal and external factors affecting waste management in the city of Tehran were identified, and analysis and strategy formulation using Quantitative Strategic Planning Matrix (QSPM) was performed. From other research papers, that of Zarei et al (5) can be noted, which attempted to study solid waste management in the city of Minab. Based on their results, WT strategies based on weaknesses and threats were chosen. Tehran Municipality District 1 has 10 regions and 26 neighborhoods. Currently, dry residues weighing approximately 2678 tons per day are collected from door fronts by a contractor company and 24 pick-up trucks. Seventeen recycling units have been established at the regional level so that residents can deliver their dry residues, which weigh about 103 tons per day, in person. The current study examined the strengths, weaknesses, opportunities, and threats of recycling and waste management of Tehran's dry residues in 2011 and prioritized relevant strategies.

Methods

Library information on recycling and urban waste management was collected by studying conducted internal and external research:

1. Data about regional waste from the waste management organization was collected, and
2. The internal and external factors evaluation (EFE) matrix and Quantitative Strategic Planning Matrix (QSPM) were created as described below.

After reviewing the internal and external factors, the most important items were listed. A number was then allocated to each coefficient: 0 (unimportant) to 1(very important). The number allocated to each coefficient is expressive of its significance. The sum of these coefficients had to equal one. Scores from 1 to 4 were allocated to each factor. A score of 1 showed fundamental weakness, 2 showed little weakness, 3 indicated strength, and 4 indicated high strength for the factor in question. The coefficient of each factor was then multiplied by the score, and so the overall scores for each factor and the final score was determined (4). To create the QSPM matrix, the first and second stages of the analysis framework (the evaluation matrix of internal and external factors) were used to objectively determine implementable strategies. In the QSPM matrix, important internal and external factors that influence design were considered. In this matrix, each factor was allocated an attractiveness score, which showed the attractions and the ability to deal effectively with the strategy of internal and external factors (5).

Results

Internal and External Factors Evaluation Matrix

After determining the strengths, weaknesses, opportunities, and threats of an organization in the area of Tehran, the internal and external factors matrix of the waste management system were set up in Tables 1, 2, 3, and 4 as follows:

Weights of dry residues from Tehran in the years 2007, 2009, and 2011 (kg) are shown in Table 5.

The final score for the internal factors evaluation (IFE) matrix is less than 2.5; therefore,the organization is weak at internal factors.

The final score for the EFE matrix final is 2.5; therefore, there are more organizational opportunities than threats, and the organization can take advantage of these opportunities to improve the trend of its activities.

Internal and External Matrix

To form this matrix, the scores created by the evaluation matrixes of internal and external factors already obtained are inserted in the horizontal and vertical dimensions of the matrix. Since the internal and external factor scores rate the situation of the organization as conservative (Figure 1), a conservative strategy (WO) is recommended. Since the organization is in a conservative position, strategies used are:

- WO_1: Development and promotion of recycling industries through financial support and marketing for

Table 1. Matrix of internal factors, recycling and management of dry residues from District 1

Descriptions	Weighted Score	Current Weight Score	Normalized weight	Internal Strategy Factors (Strong Points)
Waste separation scheme was conducted in District 1, but lacks proper quality.	0.12	3	0.04	1. Waste separation project in District 1
In some families, educated women have been effective in waste separation and recycling programs, but not enough.	0.15	3	0.05	2. Educated housewife awareness
Waste production volume is reduced through source separation and recycling.	0.12	3	0.04	3. Reduction of size and weight of residues
Using reflective clothing and uniforms has been effective in reducing accidents, especially at night.	0.08	4	0.04	4. Reflective clothing worn by municipal officers
Locating dry plastic residues into blue or black tanks at the regional level.	0.09	3	0.04	5. Locating dry residue tanks in District 1
The training of target groups by educators promotes citizen awareness and increases public participation.	0.21	3	0.07	6. Face to face training in source separation and recycling by educators in District 1
Existence of experienced experts in the field of urban services of municipal services and training is potentially possible.	0.24	4	0.06	7. Existence of specialists and experts in recycling and waste management
Employment opportunities are created through the creation of recycling industries and lateral affairs such as collection and recycling.	0.18	3	0.06	8. Employment opportunities through recycling industries
Studies and research to improve the system of collection, recycling, and disposal sites will be conducted.	0.28	4	0.07	9. Conducting research and studies in the field of waste management
The morning collection of dry residues by patrolling pick-up trucks plays an important role in enhancing cooperation and public participation.	0.12	3	0.04	10. Morning collection of dry residues by recycling office agents

Table 2. IFE matrix of recycling and waste management of District 1

Description	Weighted Score	Current Weight Score	Normalized Weight	Internal Strategy Factor (Weaknesses)
Municipal officers stand at the rear of the waste-carrying car and lack personal protective equipment such as gloves and mask which causes high risk, and conducted trainings are insufficient.	0.05	1	0.05	1. Failure of municipal officers to observe safety rules
Shortages in collection machinery exist, but essential measurements are being made in this respect.	0.1	2	0.05	2. Lack of adequate machinery for waste collection in District 1
Consumption patterns change and the willingness of citizens toward consumerism has increased waste volume; effective measures have not been made in this regard.	0.08	2	0.04	3. Changing consumption patterns and increasing volume of waste (consumerism)
The separation of waste from its origin is not performed principally; the necessary actions have not yet been taken.	0.06	1	0.06	4. Lack of segregation of waste at origin
Even if waste is to be separated at its point of origin, it will not be done during collection.	0.04	1	0.06	5. Porting mixed wastes by municipal officers
Citizens do not separate waste and the washing by agents is unprincipled, causing the vessel to be smelly.	0.04	1	0.04	6. Odors eliminated from waste tanks
Recycling industries require financing and face a lack of credit and investment risks.	0.12	2	0.06	7. Lack of financial resources; lack of independent funding defined in the recycling and waste management industries.
Creation of traffic and noise during waste loading and collection results in residents' unhappiness.	0.08	2	0.04	8. Road blocking and creation of traffic in District 1
Comprehensive information on all areas of the district has been targeted.	0.07	1	0.07	9. Lack of sufficient information to distinguish the origin of the waste
Non-compliance with waste collection schedule by officers has resulted in complaints from residents.	0.08	2	0.04	10. Noncompliance with the schedule of waste collection by officers
	2.42		1	Total internal factors

Abbreviation: IFE, internal factors evaluation.

Table 3. Recycling and waste management system of external evaluation matrix in District 1

Description	Weighted Score	Current Weight Score	Normalized Weight	External Strategy Factor (Opportunities)
Environmental laws and regulations are a suitable way to control the waste management process.	0.15	3	0.05	1. Environmental laws and regulations
Existence of traffic rules in order to reduce accidents during transportation of waste can be effective.	0.16	4	0.04	2. Traffic rules
Bread waste must be isolated and separated because of religious beliefs and optimal consumption.	0.12	3	0.04	3. Religious beliefs
The presence of NGOs can greatly help the management of waste materials.	0.24	4	0.06	4. NGOs presence in the district and their relationships with other NGOs
Given the importance of recycling policies and upstream financial support, there is a good opportunity to attract funding and Implement projects.	0.15	3	0.05	5. Acceptance and support of separation from the source by institutions and classes
Technology improvements in all aspects of waste management have not been carried out; in this respect, the experiences of other countries should be used.	0.15	3	0.05	6. Improving technology in waste collection
Potential and existing capacities have not been well utilized.	0.12	3	0.04	7. Presence of recycling industries
Greater possibilities for the participation of the private sector, public sector cuts, and attraction of private investors in this sector exist.	0.24	4	0.06	8. Presence of efficient and experienced contractors, and welcome from the private sector
Taking advantage of TV channels, especially provincial ones, and newspapers play an important role in informing citizens.	0.28	4	0.07	9. Use of advertising media to encourage waste separation and waste removal schedule implementation
The mentioned costs have been reduced to collecting and disposing of waste, and less need exists for waste disposal land.	0.15	3	0.05	10. Saving money in collection and deposal of wastes expenditures

Abbreviation: NGO; nongovernmental organization.

Table 4. Matrix of external evaluation in waste management for the dry residues recycling and management systems of District 1

Description	Weighted Score	Current Weight Score	Normalized weight	External strategy factor (Threats)
Paying salaries monthly leads to dissatisfaction, and the discontent of the people will influence both sides.	0.05	1	0.05	1. Obtain monthly salary from area residents 1
Many municipal activities take place at night, increasing the risk of accidents.	0.08	2	0.04	2. Traffic accidents due to non-compliance with traffic laws
Unauthorized or unlawful factors in processing industry and the absence of a formal market are issues.	0.05	1	0.05	3. Existence of black market and non-fixed prices
Population growth in the region increases waste volume and worn textures that prevent mechanized machines from moving easily.	0.08	2	0.04	4. Worn population structure and texture in some locales
Existing legislation on the environment (waste management) is not supported by implementation.	0.1	2	0.05	5. Lack of enforcement of environmental laws
A lack of coordination exists among various organizations and institutions such as the judiciary, environmental organizations, universities, and television.	0.12	2	0.06	6. Lack of external coordination of waste management
Advanced and up-to-date techniques for recovery of waste exist in the world but are not used or are used in inappropriate consumption ways.	0.14	2	0.07	7. Best techniques not used for waste recycling
Popular belief dictates that recycled containers are unhealthy; therefore, people do not use them.	0.1	2	0.05	8. Lack of a culture of using recycled goods
Mischievous animals cause bags to rupture the dispersion of waste. An animal removal plan is being conducted through health department cooperation.	0.08	2	0.04	9. Existence of stray animals and vermin in the region
The decay of automobiles leads to the distribution of environmental pollution and waste in the region.	0.08	2	0.04	10. State of decay of cars and vehicles, and falling of residual waste from their rear.
	2.64		1	Total internal and external factors

Table 5. Weight of dry residues in the years 2007, 2009, and 2011 (kg)

Year	District No.	Dry Bread	Plastic Types	Paper and Cardboard	Ferrous Metals (Light and Heavy)	Colored metals(Copper, Zinc, Aluminum, Brass)	Metals Types	PET	Textiles	Glass	Burial waste	Weights of Collected Dry Waste
2007	1	785730	948263	926113	-	-	562860	332735	3560	7728	40825.3	3660494
2009	1	3555465	5930495	6104452	4703955	1552880	6256835	2238488	1433488	1544970	253220	28047774
2011	1	5446659	7066368	7272274	4265258	1656875	5922133	2128522	1629319	1760914	401006	31627195

industrial products

- WO_2: Simultaneous training for workers (active forces) and citizens for development of mechanization and improved equipment.
- WO_3: Formation of regulations and guidelines for encouragement in the waste management sector, and employing private sector capital in this regard.

Quantitative Strategic Planning Matrix
Tables 6 and 7 show the WO strategies attractiveness score for each internal and external factor on the QSPM.

Discussion

Pollution resulting from the poor management of urban waste in the area of health and aesthetics imposes irreparable consequences upon society. Among them, the lack of waste management can lead to various diseases. The most important step in raising the level of understanding about the foundations of solid waste management is to have precise SWOT metrics.

In a study conducted in Tanzania, results suggested that the management plan in the city of Dar al-Salam acts stronger than a normal and ordinary plan; the comprehensive urban management plan for the city of Dar al-Salam is more general, and the city management plan is more specialized. Important factors in the stability and plan performance of this city were citizen participation in planning, coordinating groups working together, and

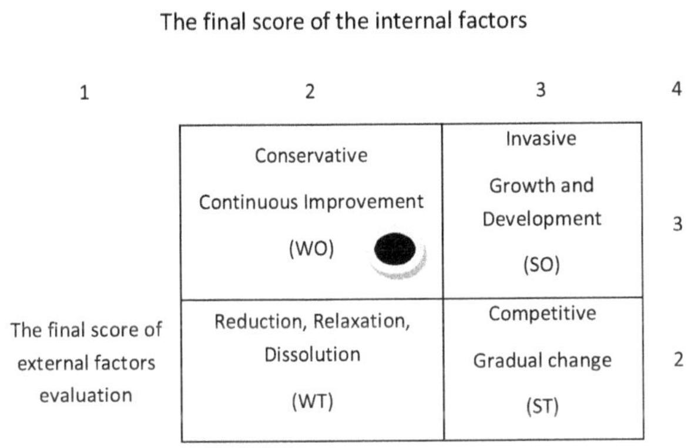

Figure 1. External and internal factors matrix.

providing equipment and resources for groups (4). Results of the evaluation matrix of internal and external factors in the city of Rasht indicate that waste management has weak internal factors. External factor evaluation points, however, show that in the current situation, strengthening opportunities and overcoming threats could achieve good results (2). Results obtained using quantitative strategic planning showed that implementing waste management law created the highest attractiveness, and education and proper method for pollutant reduction in burial sites created the least attractiveness (2).

Studies also indicate that Minab city's waste management strategy where WT strategies are based on weaknesses and threats is appropriate. Media and advertising strategies have the highest priority, and increasing the assignment of waste collection and recycling strategies to the private sector has the lowest priority (5). In this study, strategies were prioritized (Table 8) to determine the attractiveness of strategies using the QSPM.

In 2007, a study in Tehran ascertained that municipal solid waste comprised more than 97% of Tehran's solid waste, while 3 other types comprised less than 3%, namely hospital waste (1.0%), industrial waste (0.6%), and construction and demolition waste (0.5%). The contribution of household solid waste to total municipal solid waste was about 62.5% (6). The study also identified important challenges in solid waste management in Tehran including proper collection and management of hospital waste; public education aimed at reducing and separating household

Table 6. Quantitative strategic planning matrix (internal factors)

Internal Factors	Normalized Weight	WO_1		WO_2		WO_3	
		AS	TAS	AS	TAS	AS	TAS
Strengths							
S_1	0.05	3	0.15	4	0.16	3	0.12
S_2	0.05	1	0.05	3	0.15	1	0.05
S_3	0.04	2	0.08	3	0.12	3	0.12
S_4	0.04	1	0.04	2	0.08	1	0.04
S_5	0.04	2	0.08	3	0.12	2	0.08
S_6	0.07	3	0.21	4	0.28	3	0.21
S_7	0.06	4	0.24	3	0.18	4	0.24
S_8	0.06	4	0.24	3	0.18	3	0.18
S_9	0.07	4	0.28	4	0.28	4	0.28
S_{10}	0.04	2	0.08	2	0.08	2	0.08
Weaknesses							
W_1	0.05	1	0.05	3	0.15	3	0.15
W_2	0.05	3	0.15	3	0.15	3	0.15
W_3	0.04	2	0.08	2	0.08	2	0.08
W_4	0.06	3	0.18	3	0.18	3	0.18
W_5	0.04	2	0.08	3	0.12	2	0.08
W_6	0.04	1	0.04	3	0.12	2	0.08
W_7	0.06	4	0.21	3	0.18	4	0.24
W_8	0.04	1	0.04	1	0.04	1	0.04
W_9	0.07	3	0.21	4	0.28	3	0.21
W_{10}	0.04	1	0.04	2	0.08	2	0.08
Total Sum	1		2.53		3.01		2.69

Abbreviations: WO, weakness, opportunity; AS, available score; TAS, total available score.

Table 7. Quantitative Strategic planning matrix (external factors)

External Factors	Normalized Weight	WO_1		WO_2		WO_3	
		AS	TAS	AS	TAS	AS	TAS
Opportunities							
O_1	0.05	2	0.1	2	0.1	4	0.2
O_2	0.04	1	0.04	1	0.04	3	0.12
O_3	0.04	1	0.04	2	0.08	2	0.08
O_4	0.06	3	0.18	3	0.18	3	0.18
O_5	0.05	3	0.15	3	0.15	3	0.15
O_6	0.05	4	0.2	4	0.2	3	0.15
O_7	0.04	4	0.16	3	0.12	3	0.12
O_8	0.06	4	0.24	4	0.24	4	0.24
O_9	0.07	3	0.21	4	0.28	3	0.21
O_{10}	0.5	3	0.15	3	0.15	2	0.1
Threats							
T_1	0.05	1	0.05	2	0.1	2	0.1
T_2	0.04	1	0.04	1	0.04	3	0.12
T_3	0.05	4	0.2	1	0.05	2	0.1
T_4	0.04	1	0.04	2	0.08	2	0.08
T_5	0.05	3	0.15	2	0.1	4	0.2
T_6	0.06	3	0.18	3	0.18	3	0.18
T_7	0.07	4	0.28	3	0.21	3	0.21
T_8	0.05	4	0.2	1	0.05	1	0.05
T_9	0.04	1	0.04	2	0.08	1	0.04
T_{10}	0.04	2	0.08	4	0.18	3	0.12
Total Sum	1	2.73		2.64		2.75	

Abbreviations: WO, weakness, opportunity; AS, available score; TAS, total available score.

Table 8. Prioritized WO strategies

Scores Sum	External Factors	Internal Factors	Strategy
5.65	2.64	3.01	WO_2: Simultaneous training of workers and citizens for the development of mechanization and improvement of municipal services equipment
5.44	2.75	2.69	WO_3: Formulation of regulations and guidelines in order to encourage participation in waste management and attract private sector investment
5.26	2.53	2.73	WO_1: Upgrading recycling industries through financial support for the development and promotion of industry and marketing of recycled products

waste; educating municipal workers in order to optimize the waste collection system; and the participation of other related organizations and the private sector in solid waste management (7). The results of that study correspond with those of the current study.

Various studies have revealed that about 90% of municipal solid waste (MSW) is disposed of unscientifically in open dumps and landfills, creating problems to public health and the environment (8).

In another study conducted in Sweden, selected potential emerging waste treatment technologies were analyzed based on SWOT analysis, and the technologies were evaluated with a qualitative evaluation method based on waste handling capacity, development stage, and waste management problem solving capacity (9).

Generally, some causes of the failure in Tehran's municipality waste management and recycling can be stated as follows:

- Lack of coordination between the municipalities of different regions;
- Frequent changes in management and decision-making units;
- Lack of purpose and sustainable long-term plans to achieve goals;
- Lack of support for providing the necessary facilities, equipment, and resources to do the job right;
- Lack of incentives for private sector investment that will lead to 1-year contracts;
- Lack of motivation and confidence among citizens due to poor performance, lack of continuing education programs, and lack of collection of dry waste plans;
- Emergent and short-term projects not conducted fundamentally due to lack of time and money;
- Low quality of recycled products because of poor technology and lack of proper segregation of waste materials at the source, and
- Fragmentation and high number of recycling units of multiplicity, lack of oversight on their activities, and unauthorized agent activities.

Generally, recommendations that can be made to improve waste management include (*a*) defining "Recycling Day" as a symbolic movement in the country, (*b*) creating recycling exchange and offering recycling products, (*c*) gradual reduction in frequency of manually collecting waste in neighborhoods, (*d*) preventing unauthorized activities and badgers, (*e*) providing facts and figures about the problem of recycling waste and creating incentives to encourage more people to participate in this plan, and (*f*) avoiding parallel activities between parties related to the issue of waste management and recycling in order to reduce costs.

Conclusion

Waste management systems are dependent upon socio-economic issues such as population growth and gross domestic product (GDP) (10,11). Six waste management development drivers are categorized by Wilson in his study: (*a*) public health, (*b*) environmental protection, (*c*) resource value of waste closing the loop, (*d*) institutional development, (*e*) responsible issues, and (*f*) public awareness overtime (12). Waste management systems are involved with different multidisciplinary factors; therefore, trends in the development of waste treatment technologies have been led by various social, economic and environmental drivers in Tehran.

Encouraging nongovernmental organizations (NGOs), associations, urban and rural councils, and schools, universities and establishing programs for women, youth, and people interested in this issue with the collaboration of local authorities, municipalities, environmental agencies in order to gain public support for recycling waste through group focused activities should be considered. An important consideration is that, in the field of waste management, only by using the synchronized and simultaneous "education and culture making," "applying law" and "employing superior technologies and promoting them" can we be hopeful that waste management laws will take the right step toward optimal management of waste materials (13). The development of waste technologies also involves other externalities, like shifting personal and social viewpoints on waste such as 'waste' to 'resource'. A number of studies have been conducted by different researchers on the 'zero waste' concept (14).

Competing interests

The authors declare that they have no competing interests.

Authors' contributions

Monireh Majlessi, Aida Vaezi, and Mohsen Mehdipour conceived and designed the study. Mohsen Mehdipour performed the literature search and wrote the manuscript. All authors participated in data acquisition, analysis and interpretation. All authors critically reviewed, refined, and approved the manuscript.

Acknowledgments

It is necessary to thank all those who participated in this research, especially the authorities of District 1 of Tehran Municipality.

References

1. Abbasvand M. Examining urban waste management process of Golestan province. In: 4th National Conference on Waste Management; 2008; Mashad, Iran.
2. Abedinzadeh N, Abedinzadeh F, Abedi T. Examining strategic factors in waste management of Rasht city using SWOT analysis and QSPM matrix. Journal of Ecology 2011; 37(57): 93-104.
3. David FR. Strategic Management. Parsaeiyan A, Arabi M, trans. 21st ed. Tehran: Culture and Management Publications; 2011. p. 687.
4. Halla F. A SWOT analysis of strategic urban development planning: the case of Dar al Salaam city in Tanzania. Habitat International 2007; 31(1): 130-42.
5. Zarei M, Raeisi A. Documenting a strategic management plan for Minab city's waste. In: 6th National Conference on Waste Management; 2012; Mashad, Iran.
6. Damghani AM, Savarypour G, Zand E, Deihimfard R. Municipal solid waste management in Tehran: Current practices, opportunities and challenges. Waste Manag 2008; 28(5): 929-34.
7. Karbasi A, Monavari M, Mogouyi R. Strategic Management in the Environment. Kavosh Ghalam Publications; 2007.
8. Sharholy M, Ahmad K, Mahmood G, Trivedi RC. Municipal solid waste management in Indian cities – A review. Waste Manag 2008; 28(2): 459-67.

9. Zaman AU. Identification of waste management development drivers and potential emerging waste treatment technologies. Int J Environ Sci Technol (Tehran) 2013; 10(3): 455-64.
10. Mazzanti M, Zoboli R. Waste generation, waste disposal and policy effectiveness: Evidence on decoupling from the European Union. Resources, Conservation and Recycling 2008; 52(10): 1221-34.
11. EEA indicator fact sheet. http://www.eea.europa.eu/data-and-maps/indicators. Accessed April 22, 2009. Published 2008.
12. Wilson DC. Development drivers for waste management. Waste Manag Res 2007; 25(3): 198-07.
13. Fallahi Gilani R, Darvish A. Awareness and importance of waste separation and environmental protection: case study from Tehran's 22 municipality regions. In: 5th National Conference of Environmental Engineering; 2011; Terhan, Iran.
14. Zaman AU, Lehmann S. Challenges and opportunities in transforming a city into a 'Zero Waste City'. Challenges 2011; 2(4): 73-93.

11

Performance evaluation of wastewater stabilization ponds in Yazd-Iran

Mahdi Farzadkia[1], Mohammad Hassan Ehrampoush[2], Shahram Sadeghi[3], Majid Kermani[4], Mohammad Taghi Ghaneian [5], Vahid Ghelmani[6], Ehsan Abouee Mehrizi[7*]

[1]Associate Professor, Department of Environmental Health Engineering, Iran University of Medical Sciences, Tehran, Iran
[2]Professor, Department of Environmental Health Engineering, Shahid Sadoughi University of Medical Sciences, Yazd, Iran
[3]MSc Student, Environmental Health Engineering, Student Research Committee, Kurdistan University of Medical Sciences, Sanandaj, Iran
[4]Assistant Professor, Department of Environmental Health Engineering, Iran University of Medical Sciences, Tehran, Iran
[5]Associate Professor, Department of Environmental Health Engineering, Shahid Sadoughi University of Medical Sciences, Yazd, Iran
[6]Bachelor of Civil Engineering, Manager of Wastewater Treatment Plant in Yazd, Iran
[7]PhD Student, Environment Research Center, Iran University of Medical Sciences, Department of Environmental Health Engineering, Faculty of Public Health, North Khorasan University of Medical Sciences- Bojnurd, Bojnurd, Iran

Abstract

Background: Yazd waste stabilization pond facilities consist of three stabilization pond systems, module 1, module 2 and module 3 that AWSP module 1 has started its operation. The existing facilities have had several problems in their operation. The objectives of this research were to evaluate the performance of stabilization ponds in wastewater treatment of the city of Yazd, due to several problems in their operation, and to prepare a scheme of its upgrading, if necessary.

Methods: During the period from December to June 2010, data analysis were carried out for both raw and treated wastewater.

Results: Results of these investigations showed that the average effluent concentrations of Biochemical Oxygen Demand 5 (BOD_5), Chemical Oxygen Demand (COD) and Suspended Solid (SS) taken from anaerobic pond and secondary facultative ponds of module 1 were 306.9, 135.18, 139.75 and 136.75, 69.025, 136.5 mg/L, respectively.

Conclusion: These results indicated that the effluent of the anaerobic pond of module 1 was complied with the Iranian treated wastewater standards for agricultural reuse in terms of BOD_5 and COD concentrations; hence the secondary facultative ponds could be changed to other primary facultative ponds in order to increase the capacity of wastewater treatment plant.

Keywords: Biological wastewater treatment, Stabilization pond, Yazd

***Correspondence to:**
Ehsan Abouee Mehrizi
Email: Ehsan.abouee@gmail.com

Introduction

Wastewater Stabilization Pond (WSP) is considered as the most appropriate system to treat the increasing flows of urban wastewater in tropical and subtropical regions of the world. WSPs are commonly used as efficient means of wastewater treatment relying on little technology and minimal, albeit regular maintenance. Their low capital and hydraulic loads have been valued for years in rural regions and in many countries wherever suitable land is available at reasonable cost (1-5). They generally consist of a series of ponds where the wastewater has around twenty days retention time and usually a depth from one to three meters depending on the type of pond (6). These ponds have been used for treatment of municipal, agricultural, and industrial wastewater. The primary function of anaerobic lagoons is stabilization and breakdown of the high concentrations of organic pollutants contained in wastewater and not necessarily production of a high-quality of effluents. Most often anaerobic lagoons are operated in series with facultative and aerobic lagoons (7,8). Although, stabilization ponds are effective methods of wastewater reclamation and reduction of the Biochemical Oxygen Demand 5 (BOD_5) and coliform as efficient, high concentrations Suspended Solids (SSs) exceeding 100 mg/L in their effluents is one of the major disadvantages of these systems (9,10). Several techniques are used to treat domestic wastewater. These can be classified into two groups: conventional and non-conventional treatment plants. The former has high-energy requirements. The later is solely dependent on natural purification processes. The conventional systems of wastewater treatment include trickling filters, activated sludge systems, biodisc rotators, and aerated lagoons. The non-conventional systems, which are also called eco-technologies include constructed wetlands and WSPs. Among these technologies, the widely recommended ones for developing countries are the WSPs (11). Oxidation ponds are also called stabilization ponds or lagoons and serve mostly small rural areas, where land is readily available at relatively low cost (12). WSPs are biological treatment systems, the processes and operations of which are highly dependent on the environmental conditions, such as temperature, wind speed and light intensity which are highly variable, and any given combination of these environmental parameters is usually unique to a given location (13). There are many advantages of using this kind of biological treatment like ease of operation, low energy requirement, less equipment maintenance, and better sludge thickening. However, the effluent quality from fixed-film system is relatively poorer than suspended growth systems in terms of BOD_5 and SS (14). If pond systems are correctly designed and managed in order to cultivate anaerobic and aerobic bacteria and green micro-algae, then such systems will decompose waterborne organic wastes effectively and efficiently, and will help in reducing some of the problems associated with the treatment and disposal of wastewater. In addition, about 90% of the ponds in the United States are used in small communities with less than 10,000 residents and are very effective in wastewater treatment (13). The city of Yazd is located in the central part of Iran, with a population of around 900,000 people and many small and large industries. Municipal and industrial wastewater of this city is conducted to a wastewater treatment plant through sewer. The basic wastewater treatment process in Yazd is stabilization pond. However, due to inappropriate design and consideration of both biological process and physical aspects of the ponds, the existing facilities suffer serious malfunctioning problems. Hence, a program was developed during the period from December to June 2010 with a case study on the existing facilities. The main objectives of the program were to train the personnel to monitor, and evaluate the pond performance and effluent quality of the stabilization ponds, and to propose a scheme for upgrading and expanding WSPs, if necessary, depending on the results obtained. Similar programs have been developed in many parts of the world (3,15-17).

Methods

Site specifications

The wastewater treatment plant of Yazd is located in the north of the city, close to the main road of Yazd airport (Figure 1). The latitudinal location of the Yazd WSPs is about 34.08 N, the longitude is around 49.70 E, and the pond's altitude is 1710 m above sea level. Yazd treatment plant consists of three WSP systems as AWSP module 1 (M1), AWSP module 2 (M2) and AWSP module 3 (M3). The M1, M2 and M3 facilities are parallel with each other and AWSP module 1 have started their operation in 1993 and 2006, for the equivalent population of 25,000 and 80,000, respectively. As can be seen in Figure 2, the studied WSP systems are the same as classical pond configurations with anaerobic and facultative ponds. The studied wastewater treatment plants in Yazd have a pretreatment unit that includes screens followed by the WSP systems. Table 1 presents the physical and operational characteristics of the AWSP systems. The M1 AWSP comprises one Anaerobic Pond (AP) in parallel followed by a distribution tank that distributes the APs effluent into one parallel Primary Facultative Pond (PFP), followed by two Secondary Facultative Ponds (SFPs) in parallel (Figure 2). The treated wastewater of M1 facilities is used for agricultural reuse. As pointed out by Mara *et al* (18), the current reuse of wastewater for agricultural purposes is attractive to many local authorities, especially to those in water-scarce regions. It is known that agriculture is responsible for more than 80% of total world water consumption (19).

Sampling

This was a descriptive cross-sectional study carried out during 6 months including the cold-season months (from mid-January to mid-March) and warm-season months

Figure 1. Overview of treatment plant

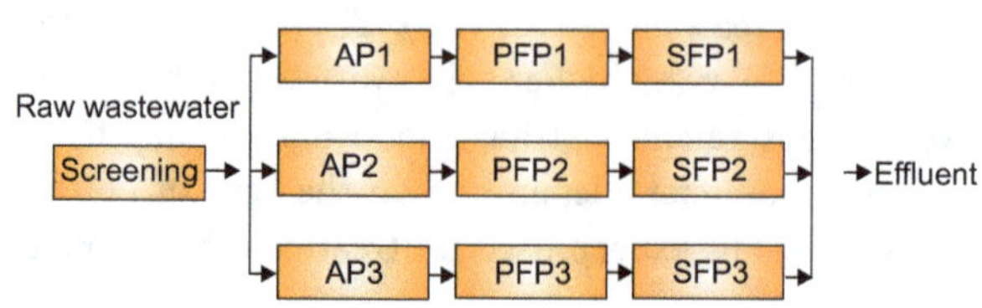

AP= Anaerobic Pond
PFP= Primary Facultative Pond
SFP= Secondary Facultative Pond

Figure 2. Schematic flow diagram for M1 AWSP

Table 1. Physical and operational characteristics of the M1 AWSP system

Parameter	AP	PFP	SFP
Hydraulic retention time (day)	5	4	4
The depth of the pool basin	5	5	5
Free height (m)	0.75	0.75	0.75
Area (m^2)	24617	43981	43981
Size Bottom of Ponds (m^2)	19015	39676	39676
Slope walls	1 to 5 (Vertical to horizontal)	1 to 5 (Vertical to horizontal)	1 to 5 (Vertical to horizontal)
Percentage BOD_5 reduction in anaerobe ponds	40	70	70

AP= Anaerobic Pond; PFP= Primary Facultative Pond; SFP= Secondary Facultative Pond

(from mid-May to early August in 2010-11). The warm and cold months of the year were determined through the weather data of the previous years. Wastewater samples were taken monthly at the inlet and outlet of each pond. The collected samples were composite samples taken over a period of 48 hours. The samples were taken directly by means of 2 L beaker glass. Each sample of 2 L taken at a wastewater depth of 1 m was directly transferred to a 30 L sample container and fixed for physicochemical analysis (17). Sampling was conducted from December to June 2010.

Climate

Yazd has a relatively cold and dry climate. The maximum temperature may rise up to +38 °C in summer and fall down to -10 °C in winter. The average temperature in the coldest month is -7.48 °C. The average precipitation is around 300 mm, and the annual relative humidity is 50%.

Analyzed parameters

Total BOD_5, Chemical Oxygen Demand (COD) and SS were determined for both influent and effluent of the module. The measurement of flow was carried out by means of a Partial flume located at the inlet channel. Analytical approaches were based on the standard methods (20).

Results

Total performance evaluation system

The results obtained for each stage and for the total systems of M1 AWSP, are presented in Tables 2, 3 and 4 respectively.

The averages of raw wastewater flow rates entering the systems were 4,300 and 13,500 m^3/d for the AWSP system of M1, respectively, which were equivalent to the expected design. The measured average of BOD_5 and COD concentrations of raw wastewater, as around 272.08 and 577.13 mg/L, were also near the expected design concentrations of 250 and 550 mg/L, respectively, for BOD_5 and COD. However, the average SS concentration for raw wastewater, around 258.66 mg/L was upper than expected design concentration of 220 mg/L. Thus, the raw wastewater in Yazd could be classified as medium to strong, in terms of BOD_5, COD, and SS (14).

Analysis of pond performance parameters AWSP system M1

As Table 3 indicates, the removal efficiencies of BOD_5, COD, and SS for the APs with HRT=5 days and the PFPs

Table 2. The average removal efficiencies of parameters in the stabilization pond treatment

Parameter	Input (mg/l)	Output (mg/l)	Removal efficiency (%)
BOD_5	272.08	69.025	74.6
COD	577.13	136.75	76.31
TSS	258.6	136.5	47.2

BOD_5: Biochemical Oxygen Demand 5; COD: Chemical Oxygen Demand; TSS: Total Suspended Solids

Table 3. The average removal efficiency parameters of the anaerobic ponds

Parameter	Input to the anaerobic ponds (mg/l)	Anaerobic ponds output (mg/l)	Removal efficiency (%)
BOD_5	272.08	135.18	50.31
COD	577.13	307	46.24
TSS	258.66	139.75	45.97

BOD_5: Biochemical Oxygen Demand 5; COD: Chemical Oxygen Demand; TSS: Total Suspended Solids

Table 4. The mean change in TSS (mg/l) during summer and cold during the whole period of stabilization pond system

Sampling time		Raw wastewater input to anaerobic (mg/l)	The first optional input to wastewater pond (mg/l)	Effluent treatment plants (mg/l)
Three months of warm	June	352	174	215
	August	197	100.5	98
	July	240	104	133
	Average period of warm	263	126.16	148.6
Three months of cold	January	202	131	87
	February	301	189	155
	March	260	140	131
	Average period of cold	254.33	153.33	124.33
Average total study period		258.66	139.745	136.5

with HRT=11.3 days, were 50.31%, 46.24%, and 45.97%, respectively. The SFPs with the HRT=8 days had the removal efficiencies of 74.6%, 76.31%, and 47.2% for BOD_5, COD, and SS, respectively.

Based on meteorological studies, the average temperature of Yazd in the coldest and the hottest months are -0.3 °C and +33.6 °C, respectively. The average precipitation is around 300 mm and the annual relative humidity is 50 %. Table 5 summarizes the climatic conditions in Yazd, Iran (2010).

Discussion

M1 AWSP system

Wastewater stabilization pond system has been operated for municipal and industrial wastewater treatment worldwide. These systems are relatively affordable, especially in the tropics that construction and operation is simple and low cost. With respect to the effluent quality of the PFPs and SFPs and in comparison with the Iranian treated wastewater standards for agricultural irrigation that has indicated BOD_5, COD, and SS concentrations should be less than 100, 200, and 100 mg/L, respectively, the results indicated that the average effluent concentrations of BOD_5, COD, and SS were 135.18, 307, and 139.75 mg/L, respectively, for PFPs, and 69.025, 136.75, and 136.5 mg/L, respectively, for SFP. The effluent of the studied PFPs complied with the considered standards in terms of BOD_5 and COD concentrations. As shown in Figures 3, 4 and 5, although the average effluent concentrations of BOD_5 and COD of the PFPs were lower, the average concentration of effluent SS was higher than the concentration of the effluent SS of the SFPs. The main constraint in the WSPs is the high SS in the effluents, which is primarily due to high concentrations of algal cells in the effluent (21). Thus in practice, the SFPs would not be required and could be replaced with other PFPs, in parallel with the existing PFPs, to enhance the quantity of treated wastewater in forthcoming years and to optimize the treated wastewater quality.

Untreated wastewater can create many environmental problems. Low-tech wastewater treatment systems con-

Table 5. Climatic condition in Yazd, Iran

Parameters	Annual Mean
Temperature (°C)	19.9
Sun Light Hours (h/month)	274.9
Evaporation (mm/month)	255.2
Wind Speed (m/s)	9.5

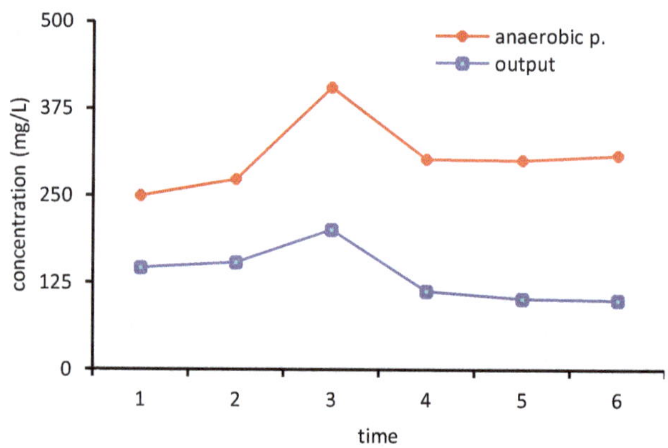

Figure 3. COD variations for M1 AWSP

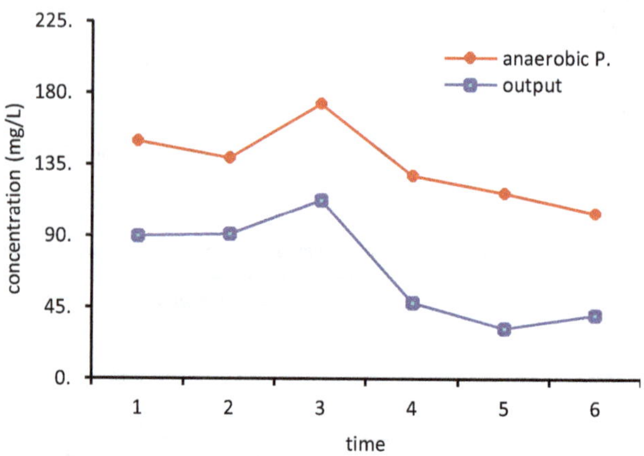

Figure 4. BOD_5 variations for M1 AWSP

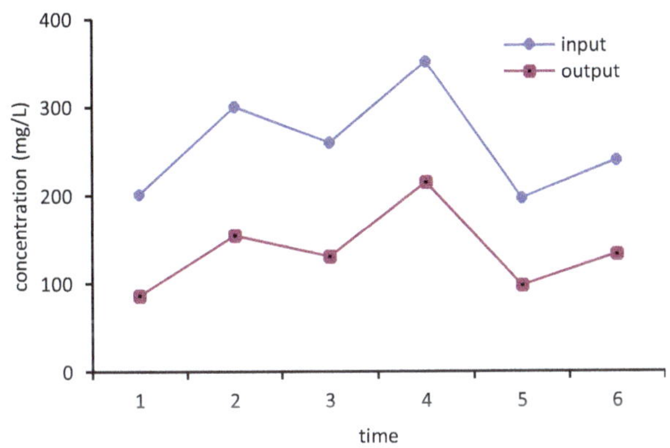

Figure 5. TSS variations for M1 AWSP

suming no energy or low-consuming systems improve our environment in addition to reduce economic costs. The range of BOD_5 concentrations of SPFs for M1 was less than the results obtained in a study conducted for stabilization ponds in Egypt (4). The removal efficiency of Yazd facility for BOD_5 was higher than the removal efficiency of another study that was conducted in Spain, as 54% (22). However, the removal efficiency of COD of that study was about the same as that in Yazd (about 70%). In a study that was carried out in Tanzania, the rate of COD removal was 66% for PFP, 68% for SFP1, 71% for Maturation Pond (MP), and the overall COD removal rate was about 94%, (23), much higher than that in Yazd which were 76.31% for M1. For conclusion, the TFP of M2 can be used as a serial SFP in order to increase Yazd wastewater plant capacity and effluent quality enhancing by population growth. In another way for enhancing effluent quality of Yazd facility, it could be practical to put some baffles in SPFs of both M1 to optimize HRT and plug flow condition of wastewater, and consequently, enhance removal efficiencies of BOD_5, COD and SS. The results show that the total amount of pollutants removed in this system is in good and acceptable situation compared to expensive mechanical systems with high energy consumption and operational problems. The share of the anaerobic pond removal was determined by stabilization pond system. Altogether, the removal of the BOD_5, COD parameters is significant in anaerobic ponds. High organic matter removal in the stabilization pond can be by a suitable retention time and high temperature because anaerobic pond performance significantly increases with increasing temperature, in a study was conducted by Pena on the anaerobic pond, removal of BOD_5, and COD was reported 59%, 68% respectively. In another study, the rate of BOD_5 removal at temperatures above 20 °C and retention time of more than 2.5 days, were reported 60%. They reported that the main mechanism in the anaerobic pond is the removal of SSs through the settling. Anaerobic pond is the first pond and its share is up in the removal. Therefore, the above removal of the anaerobic ponds substantially reduces the concentration into facultative ponds. Naddafi *et al* (24) reported that anaerobic pond of the first module of Arak stabilization pond has efficiency of 40% and 47% for BOD_5 and COD, respectively, higher than that of AP in YWSPs. Employment of skilled operators is expensive, and energy consumption due to its related problems from environmental and economical point of view is important worldwide in recent years, and Iran does not have a suitable state in energy consumption, therefore, YWSPs upgrading programs should be considered based on fundamentals with no need for energy resources and no expert operators. As mentioned above, the problems of YWSPs are related to operational and design parameters, and the quality of effluent from this wastewater treatment plant may be improved by controlling entered wastewater and supply of distributed flows in ponds. Determination of hydraulic regime and its relationship with pond geometry is another important issue which should be considered in YWSPs. Also, training personnel to monitor and evaluate the pond performance and response to upgrading effluent quality is important and should be recognized.

Conclusion

Based on this research, it can be concluded that the stabilization pond system is one of the appropriate techniques used for the treatment of various types of wastewater worldwide. These systems can be constructed with local materials and operated without the need for skilled workers. The results showed that the variation of organic load, pH, Electrical Conductivity (EC) and season variation had no effects on organic matter removal, and the removal of BOD_5 was approximately constant.

Acknowledgements

The authors highly appreciate the sponsorship of Yazd Water and Wastewater Company and Iran University of Medical Sciences.

Competing interests

The authors declare that they have no competing interests.

Authors' contributions

MF, MK, MHE and EAM conceived and designed the study. EAM and MTG performed the literature search and wrote the manuscript. All authors participated in the data acquisition, analysis and interpretation. All authors critically reviewed, refined and approved the manuscript.

References

1. Nameche TH, Vasel JL. Hydrodynamic studies and modelization for aerated lagoons and waste stabilization ponds. Water Res 1998; 32(10): 3039-45.

2. Agunwamba JC. Effect of tapering on the performance of waste stabilization ponds. Water Res 2001; 35(5): 1191-200.
3. Nelson KL, Cisneros BJ, Tchobanoglous G, Darby JL. Sludge accumulation, characteristics, and pathogen inactivation in four primary waste stabilization ponds in central Mexico. Water Res 2004; 38(1): 111-27.
4. Hamdy A, Rabia N, Hamdy S. Study of waste stabilization pond geometry for wastewater treatment efficiency. Ecol Eng 2006; 28 (1): 25-34.
5. Kaya D, Dilek FB, Gokcay CF. Reuse of lagoon effluents in agriculture by post-treatment in a step feed dual treatment process. Desalination 2007; 215 (1-3): 29-36.
6. Toumi A, Nejmeddine A, El Hamouri B. Heavy metal removal in waste stabilization ponds and high rate ponds. Water Sci Technol 2000; 42(10-11): 17-21.
7. Crites RW, Middlebrooks EJ, Reed CS. Natural Wastewater Treatment Systems. Boca Raton, Florida: CRC Press, Taylor & Francis Group; 2006.
8. International Water Association. Pond Treatment Technology. IWA Publishing; 2006. p. 218-20.
9. Middlebrooks EJ. Upgrading pond effluents: an overview. Water Sci Technol 1995; 31(12): 353-68.
10. Kimwaga RJ, Mashauri DA, Mbwette TSA, Katima JHY, Jorgensen SE. Use of coupled dynamic roughing filters and subsurface horizontal flow constructed wetland system as appropriate technology for upgrading waste stabilisation ponds effluents in Tanzania. Phys Chem Earth 2004; 29(15-18): 1243-51.
11. Awuah E. Pathogen removal mechanisms in macrophyte and algal waste stabilization ponds. CRC Press; 2007.
12. Bitton G. Wastewater microbiology. Department of Environmental Engineering Sciences, University of Florida, Gainesville, Florida; 2005.
13. Gray NF. Biology of wastewater treatment. 2th ed. World Scientific Publishing Company; 2004
14. Metcalf L, Eddy HP, Tchobanoglous G, Burton FL, Stensel HD. Wastewater Engineering: Treatment, Disposal and Reuse. 4th ed. New York: McGraw-Hill; 2004.
15. Escalante V, Moeller G, Rivas A. Pond evaluation and upgrading in Mexico. Water Sci Tech 2000; 42(10): 43-50.
16. Oakley SM, Pocasangre A, Flores C, Monge J, Estrade M. Waste stabilization pond use in Central America: the experiences of El Salvador, Guatemala, Honduras and Nicaragua. Water Sci Tech 2000; 42(10): 51-8.
17. Yaghoubi M, Echihabi L, Foutlane A, Bourchich L, Jellal J, Wittland C, *et al.* The performance of the waste stabilization pond system at Boujaad, Morocco. Water Sci Tech 2000; 42 (10): 9-11.
18. Mara DD, Alabaster GP, Pearson HW, Mills SW. Waste stabilization ponds: A design manual for Estern Africa. Leeds: Lagoon Technology International; 1992.
19. Valencia E. Potencialidad del Reuse del Efluente de una laguna facultative en Irrigacion: Coparacion de la produccion utilizando dos Hortalizas Regadas con el Efluente y Agua subterranean [MSc Dissertation] Instituto Cinara, Universidad del Valle, Cali; 1998.
20. American Public Health Association (APHA), American Water Works Association (AWWA), Water Environment Federation (WEF). Standard Methods for the Examination of water and wastewater. 22th edition. Washington, DC: American Public Health Association; 2012.
21. Esen II, Puskas K, Banat IM, Al-Daher R. Algae removal by sand filteration and reuse of filter material. Waste Manag 1991; 11(1-2): 59-65.
22. Travieso L, Sanchez E, Borja R, Benitez F, Raposo F, Rincon B, *et al.* Evaluation of a laboratory- scale stabilization pond for tertiary treatment of distillery waste previously treated by a combined anaerobic filter– aerobic trickling system. Ecol Eng 2006; 27(2): 100-8.
23. Kayombo S, Mbwette TS, Mayo AW, Katima JH, Jorgrnsen SE. Diurnal cycles of variation of physical–chemical parameters in waste stabilization ponds. Ecol Eng 2002; 18(3): 287-91.
24. Naddafi K, Hassanvand MS, Dehghanifard E, Faezi Razi D, Mostofi E, Kasaee N, *et al.* Performance evaluation of wastewater stabilization ponds in Arak-Iran. Iran J Environ Health Sci Eng 2009; 6(1): 41-6.

12

Nitrate removal from water using alum and ferric chloride: A comparative study of alum and ferric chloride efficiency

Ali Ahmad Aghapour[1], Sepideh Nemati[2], Amir Mohammadi[2*], Heshmatollah Nourmoradi[3], Sima Karimzadeh[2]

[1]Assistant Professor of Environmental Health Engineering, Department of Environmental Health Engineering, School of Health, Urmia University of Medical Sciences, Urmia, Iran
[2]MSc of Environmental Health Engineering, Student Research Committee, School of Health, Urmia University of Medical Sciences, Urmia, Iran
[3]Assistant Professor of Environmental Health Engineering, Department of Environmental Health Engineering, School of Health, Ilam University of Medical Sciences, Ilam, Iran

Abstract
Background: Nitrate is an acute and well-known hazardous contaminant, and its contamination of water sources has been a growing concern worldwide in recent years. This study evaluated the feasibility of nitrate removal from water using the traditional coagulants alum and ferric chloride with lower concentrations than those used in the conventional coagulation process.
Methods: In this research, two coagulants, alum and ferric chloride, were compared for their efficiency in removing nitrate in a conventional water treatment system. The removal process was done in a batch system (jar test) to examine the effects of coagulant dosages and determine the conditions required to achieve optimum results.
Results: The results revealed that ferric chloride at an initial dose rate of 4 mg/L reduced nitrate concentration from 70 mg/L to less than the World Health Organization (WHO) guideline value (50 mg/L N-NO_3). However, the removal efficiency of alum was not salient to significant nitrate reduction.
Conclusion: In conclusion, ferric chloride was more effective than alumin removing NO^-_3, even in common dosage range, and can be considered a cost-effective and worthy treatment option to remediate nitrate-polluted water. Furthermore, the removal of nitrate by coagulation can be simple and more economical than other treatment alternatives.
Keywords: Nitrate, Water treatment, Coagulation, Alum, Ferric chloride

***Correspondence to:**
Amir Mohammadi
Email: amahammadi@gmail.com

Introduction

Nitrate (NO_3^-) is found naturally in the environment and is an important plant nutrient (1). However, the contamination of water sources by nitrate has been a growing concern worldwide in recent years (2). Nitrate in water sources is related to nitrogen fertilizer and domestic wastewater (3). This problem is found in both developed and developing countries. In European Union (EU) countries, nitrate concentrations in 20% of groundwater resources were more than 50 mg NO^-_3/L between the years 1996 and 1998 (4,5). In several Middle Eastern countries, including some places in Iran (6), Iraq (7), Saudi Arabia (8), and the United Arabic Emirates (9), high nitrate concentrations have been detected in ground water. Nitrate causes disease and negatively affects health with such issues as methemoglobinemia, goiter, thyroid disorder, stomach cancer, cytogenetic defects, and birth defects (10,11). The maximum concentration level (MCL) of nitrate is 50 mg NO_3/l as set by the World Health Organization (WHO) standards for drinking water, and the Environmental Protection Agency (EPA) and EU have determined the MCL of nitrate to be 10 and 11.3 mg N/L, respectively (12-14). Thus, the removal of nitrate from contaminated water for drinking purposes is important.

There are numerous physical and chemical methods to remove nitrate from water, including ion exchange, reverse osmosis, and electrodialysis. These methods are useful but produce concentrated brine as a waste by-product which should be treated or disposed of, and this requires expensive processes (6,12).

Nitrogen and phosphorus compounds are typically removed from wastewater with biological processes. Biological denitrification is an extensively used procedure for the reduction of nitrate from water and wastewater (15).

The most significant problem with the denitrification process is that it requires disinfection in order to remove bacteria. Furthermore, this process is very sensitive to some parameters, such as temperature, pH, conductivity, and toxic compounds (10,12,16). Other biological treatment options include treatment using fluidized beds or a fixed bed and membrane biofilm reactor. Chemical reduction methods using media such as zero valent iron (ZVI) and sulfur modified iron (SMI) have also been used recently (17,18).

Coagulation and flocculation may be broadly described as chemical and physical processes that mix coagulating chemicals and flocculation aids with water. Coagulation destabilizes particles and enables them to become attached to other particles so that they may be removed in subsequent processes (19). Three main mechanisms, electrostatic coagulation (reduction of electrostatic forces and separation of particles), sweep coagulation (entrapment of particles by coagulant metal-hydroxides), and adsorptive coagulation (destabilization by adsorption of polymers or long hydroxide chains to the particle surface), are considered for the coagulation/flocculation process (20).

In summary, these processes are low–cost, simple methods for the destabilization of suspended particles and the removal of turbidity, natural organic matter (NOM), disinfection by product precursors or color in raw water.

Al(III) and Fe(III) coagulants are two main inorganic compounds used in water treatment plants. Ferric sulphate, aluminum sulphate, and ferric chloride are the most common coagulants (21,22).

The main source of drinking water in most of Iran's cities is ground water, which generally contains higher concentrations of nitrate. The current study aimed to evaluate the feasibility of nitrate removal from drinking waters with conventional coagulation treatments by applying alum and ferric chloride in a batch system. Furthermore, the efficiency rates of nitrate removal by alum and ferric chloride were compared, and the effects of factors such as initial nitrate concentration and coagulant dosage were studied.

Methods

Containers and synthetic water preparation

All containers were glass. They were washed with diluted acid (H_2SO_4, 10%) and then rinsed with tap water and deionized water. Synthetic water samples in concentrations of 10, 30, 50, 70, and 100 mg/L NO^-_3-N were prepared from nitrate stock solution (dissolving KNO_3 [Merck Co., 99.0%] in tap water). All samples were stored at temperatures below 4°C, and experiments were carried out within 24 hours of sample preparation. The main source of tap water was groundwater with no special treatment except chlorination. The physico-chemical characteristics of the tap water are shown in Table 1.

Table 1. Physico-chemical characteristics of tap water

Parameter	Values
pH	7 ± 0.2
EC (µS/cm)	690
Temperature (°C)	20 ± 2
Turbidity (NTU)	0.5
Alkalinity (mg/L)	317
TDS (mg/L)	448
SO_4^- (mg/L)	58
Cl^- (mg/L)	21
Total Hardness (mg/L)	285
Mg (mg/L)	25
Ca (mg/L)	72
Na (mg/L)	33
K (mg/L)	4

Experimental procedure

In this study, the most common coagulants, i.e. alum (aluminum sulfate, $Al_2(SO_4)_3$) and ferric chloride ($FeC1_3$), were applied to determine the nitrate removal efficiency of a conventional water treatment system.

The coagulation-flocculation process was performed in a batch system (jar test) with a six-paddle stirrer (Hach), and the chemicals, dosages, and conditions required to achieve optimum results were evaluated (ASTM, 1995). The alum and ferric chloride solutions at desired concentrations were prepared by diluting the stock solution [10 g/L] immediately before use. Coagulant dosages for the removal of nitrate were 2.5, 3, 3.5, 4, and 4.5 mg/L.

Conventional coagulation by alum depends on pH and temperature (23); in this study, the pH values of solutions were around 7±0.2, and tests were conducted at room temperature which is recommended for $FeCl_3$ and alum usage. Acidic pH may lead to the formation of weak HNO_3 ions or the complexation of NO_3^- ions with dissolved aluminum (24). Therefore, in this study, acidic pH was not applied. Furthermore, as reported in the literature, a pH above 9 is not efficient in the coagulation process for nitrate removal (25).

Rapid mixing was simulated for 1 minute at 120 rpm and flocculation was done by slow mixing at 20 rpm for 15 minutes. The flocs were then allowed to settle for 20 minutes without mixing.

Sample analyses

The samples were analyzed for residual nitrate concentration using a UV spectrophotometer apparatus (HACH DR/5000) at a wavelength of 220 nm according to the 1998 standard methods for the examination of water and wastewater (4500-NO_3^- B) (5).

The pH values of samples were measured with a pH meter WTW 3110 at the end of the settling period.

Statistical analysis

The data acquired from the comparative surveys was analyzed by SPSS 16.0 for Windows (Chicago, IL, USA). Due to the lack of normal data distribution, the non-parametric statistics analysis Mann-Whitney U Test was used to further analyze the data. All data was analyzed with Microsoft Excel 2007.

Results

The effect of initial nitrate concentration on the efficiency of the conventional coagulation-flocculation process was investigated at different nitrate concentrations (10, 30, 50, 70, and 100 mg/L N-NO_3). The removal percentages of different concentrations of nitrate using ferric chloride and alum as a function of coagulant dosages are separately presented in Figure 1. Results of the Mann-Whitney U tests showed significant differences in the abilities of ferric chloride and alum to remove nitrate ($P<0.01$).

Clearly, ferric chloride exhibited a greater removal efficiency than aluminum sulfate. As can be seen, the nitrate uptake by both ferric chloride and alum decreased when the initial nitrate concentrations were increased from 10 to 100 mg/L N-NO_3.

Figure 2 shows the typical operating ranges for the coagulants. Taking into account the primary pH values and optimum dosages of coagulants, the predominant removal mechanisms for nitrate could be attributed to adsorption and sweep precipitation.

Discussion

Effect of initial nitrate concentration and coagulant dose

As seen in Figure 1, ferric chloride removed nitrate with a higher efficiency than that of alum. Results of the Mann-Whitney U test also showed significant differences in ferric chloride and alum abilities for nitrate removal ($P<0.01$).

The highest removal efficiency rate using each one of the applied coagulants was obtained for an initial nitrate concentration of 10 mg/LN-NO_3, and minimum removal was found for a 100 mg/LN-NO_3 concentration (see Figure 1). When initial nitrate concentration was 10 and 30 mg/LN-NO_3, the removal fraction using $Fecl_3$ was more than 90%; however, for 70 mg/LN-NO_3, it was reduced to less than 35% even when applying the highest coagulant dose (4 mg/L). Thus, it can be concluded that when the concentration of the contaminant is increased, removal efficiency is decreased. This is true for alum as well.

Koparal and Öğütveren achieved similar results in the removal of nitrate from water by an electrochemical process and reported that removal efficiency was inversely related to initial nitrate concentration (26). It seems that increasing nitrate concentration is an important factor effecting the removal of nitrate from water.

The experimental results also showed removal restrictions in concentrations above 50 mg/L of pollutant, and ferric chloride was more effective in the removal of low ranges of N-NO^-_3 (<30 mg/L). Similarly, another study which investigated nitrate removal from water by a coagulation process reported removal efficiency as 85% when 0.3 g of hydrotalcite-like compounds were used in the NO_3^- solution with an initial concentration of 10 mg/L (27). Nonetheless, the optimum dosage of ferric chloride in the current study was 4 mg/L, which is less than the typically used coagulation dosage in the water treatment industry. In the case of alum, applied dosages ranged from 2.5 to 4 mg/L, which is also less than the typically used dosages for the coagulation process. The maximum removal efficiency was attained in higher ranges of alum (3.5-4 mg/L), but in nitrate concentrations of more than 30 mg/L, it had a negligible efficiency. Ayyasamy et al studied the removal of remained nitrate from groundwater samples using chemical coagulation after a one-stage biological treatment. They reported that the optimum dosage of alum for the removal of nitrate at a concentration of 50 mg/L was 150 mg/L and removal efficiency of the process was 74.3% (10). In this dosage of alum, the residual aluminum concentration in water treated by the coagulation process may be above the guidelines of the WHO and the admissible limit set by the Institute of Standard and Industrial Research of Iran, and

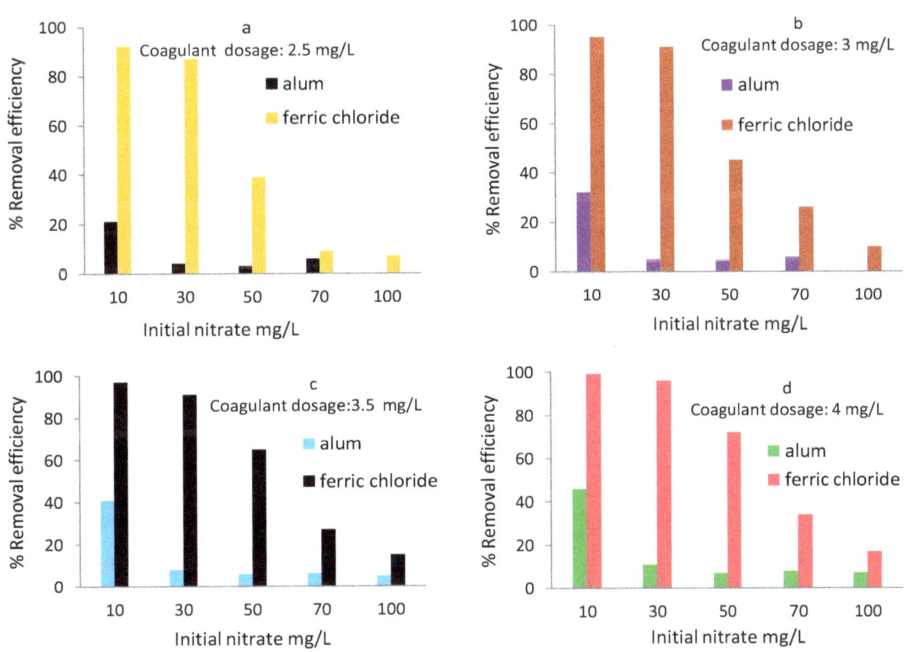

Figure 1. Efficiency of alum and ferric chloride in nitrate removal.

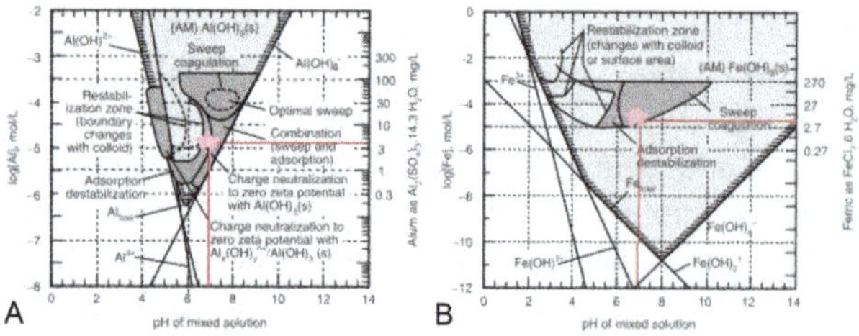

Figure 2. Typical operating ranges for coagulants: (A) alum and (B) ferric chloride (24).

it is suspected in Alzheimer disease (21).

Nitrate removal mechanism

To determine the nitrate removal mechanism using alum and ferric chloride coagulants, direct chemical reactions and adsorption mechanisms were appraised. As reported in the literature, for a chemical reaction between aluminum and nitrate or iron and nitrate, pH values should be raised above 9 (24,25,28). However, due to the high stability and solubility of the chemical products of this reaction in water, nitrate removal was not expected in pH values above 9 (24,25).

In pH values less than 9, anions of Cl^- and SO_4^- could compete with nitrate for reduction by coagulants ($Cl^- > SO_4^{-2} > NO_3^-$) (27). According to equations 1 and 2, adding alum and ferric chloride in the range of 3 to 4 mg/L could lead to the production of HCl and H_2SO_4; therefore, the range of the operating region diminished for iron and especially for aluminum hydroxide.

Eq. (1): $Al_2(SO_4)_3.14H_2O \rightarrow 2Al(OH)_3 + 3H_2SO_4 + 8H_2O$

Eq. (2): $FeCl_3.6H_2O \rightarrow Fe(OH)_3 + 3HCl + 3H_2O$

Since the initial pH was 7, nitrate removal could not be achieved by a chemical reaction. Hence, bulk reactions for nitrate removal by ferric chloride and alum were adsorption and sweep precipitation (see Figure 2) (24).

Comparison of alum and ferric chloride abilities in removing nitrate

Based on the results, an initial ferric chloride dose of 4 mg/L induced a reduction in nitrate concentration from 70 mg/L to less than the WHO guideline value (50 mg/L as NO_3) (13). Alum efficiency, however, was not sufficient. As previously stated, ferric ions are better than aluminum ions for adsorption and sweep coagulation because a lower dosage is needed.

Comparing the solubility of alum and ferric chloride showed that ferric chloride is more insoluble than aluminum in a wide range of pH values. Moreover, adding alum in ranges of 3.5 to 4 mg/L could lead to the production of H_2SO_4 and decreased pH. Therefore, the optimum range of pH for alum is less than that for ferric chloride, and decreases in pH values are greater with alum than with ferric chloride. Therefore, the application of ferric chloride could be more effective than alum for the removal of nitrate in a water treatment plant.

The high stability and solubility of nitrate in water makes the removal of this anion from water sources with high nitrate concentrations difficult (11,29). As such, various chemical and biological methods have been examined for their ability to remove this soluble anion. Some types of treatment methods should be considered highly experimental since there are currently no full-scale applications, such as biological treatment, subsurface biological treatment, phytoremediation, and nanotechnology (e.g., using nanoscale ZVI, SiO_2–FeOOH, polyvinyl alcohol/polymer composites, etc.) (30-33).

Although conventional treatment methods such as ion exchange, reverse osmosis, electrodialysis, and biological denitrification are feasible, effective options for nitrate removal from potable water, their applications have some limitations.

Conclusion

The main objective of the current study was to evaluate the ability of conventional water treatment processes to reduce nitrate contamination. Although the studied method is not a specialized or ideal technology for the treatment of nitrate, it may be used in some circumstances to reduce the health risks associated with nitrate contamination of drinking water. As the results of this study showed, the conventional coagulation process using alum and ferric chloride could be effective in removing nitrate at concentrations less than 50 mg-NO_3/L; ferric chloride exhibited a greater removal efficiency than alum. Therefore, the use of ferric chloride for water treatment, even in common dose ranges, can play a basic role in the removal of nitrate from water.

Acknowledgements

The authors appreciate the support provided by Urmia University of Medical Sciences as project numbers 92-01-34-1126 and 92-01-34-1187.

Competing interests

The authors declare that they have no competing interests.

Authors' contributions

All authors contributed equally and were involved in the study design, data collection, and article approval.

Reference

1. Loganathan P, Vigneswaran S, Kandasamy J. Enhanced removal of nitrate from water using surface modification of adsorbents – a review. J Environ Management 2013; 131: 363-74.
2. Ota K, Amano Y, Aikawa M, Machida M. Removal of nitrate ions from water by activated carbons (ACs)—influence of surface chemistry of ACs and coexisting chloride and sulfate ions. Appli Surf Sci 2013; 276: 838-42.
3. Bhatnagar A, Kumar E, Sillanpää M. Nitrate removal from water by nano-alumina: Characterization and sorption studies. Chem Eng J 2010; 163(3): 317-23.
4. Della Rocca C, Belgiorno V, Meriç S. Overview of in-situ applicable nitrate removal processes. Desalination 2007; 204(1-3): 46-62.

4. Della Rocca C, Belgiorno V, Meriç S. Overview of in-situ applicable nitrate removal processes. Desalination 2007; 204(1-3): 46-62.
5. Khorsandi H, Mohammadi A, Karimzadeh S, Khorsandi J. Evaluation of corrosion and scaling potential in rural water distribution network of Urmia, Iran. Desalination and Water Treatment 2015; 57(23): 10585-92.
6. Nanbakhsh H, Mohammadi A, Ebrahimi A. Investigating of nitrate and nitrite concentration of drinking water wells in villages around of the industrial park. Health System Research 2010; 6: 881-88. [In Persian].
7. Barbooti MM, Bolzoni G, Mirza IA, Pelosi M, Barilli I, Kadhum R, et al. Evaluation of quality of drinking water from Baghdad, Iraq. Sci World J 2010; 5(2): 35-46.
8. Alabdula'aly AI, Al-Rehaili AM, Al-Zarah AI, Khan MA. Assessment of nitrate concentration in groundwater in Saudi Arabia. Environ Monit Assess 2010; 161(1-4): 1-9.
9. Rizk ZS. Inorganic chemicals in domestic water of the United Arab Emirates. Environ Geochem Health 2009; 31(1): 27-45.
10. Ayyasamy PM, Shanthi K, Lakshmanaperumalsamy P, Lee SJ, Choi NC, Kim DJ. Two-stage removal of nitrate from groundwater using biological and chemical treatments. J Biosci Bioeng 2007; 104(2): 129-34.
11. Song H, Zhou Y, Li A, Mueller S. Selective removal of nitrate from water by a macroporous strong basic anion exchange resin. Desalination 2012; 296: 53-60.
12. McAdam E, Judd S. A review of membrane bioreactor potential for nitrate removal from drinking water. Desalination 2006; 196(1-3): 135-48.
13. World Health Organization. Guidelines for Drinking-Water Quality. Geneva: WHO; 2008.
14. Drinking Water Contaminants – Standards and Regulations. http://www.epa.gov/dwstandardsregulations. 2016.
15. Yousefi Z, Motallebi R, Takdastan A. Water desalination by membrane technology (RO) in southern Iran (Jask city). Environmental Health Engineering and Management 2014; 1(1): 13-8.
16. Sharma SK, Sobti RC. Nitrate removal from ground water: a review. J Chem 2012; 9(4): 1667-75.
17. Jensen VB, Darby JL, Seidel C, Gorman C. Drinking water treatment for nitrate Davis. Center for Watershed Sciences, University of California; 2012.
18. Mosaferi M, Nemati S, Khataee A, Nasseri S, Hashemi AA. Removal of Arsenic (III, V) from aqueous solution by nanoscale zero-valent iron stabilized with starch and carboxymethyl cellulose. J Environ Health Sci Eng 2014; 12: 74.
19. Baruth EE. Water Treatment Plant Design (American Water Works Association, and American Society of Civil Engineers). 4th ed. New York: McGraw-Hill; 2005.
20. Zhou T. Removal of organic micropollutants by coagulation in wastewater treatment. Delft University of Technol; 2011.
21. Aghapour AA, Mohammadi A. Investigating the coagulation efficiency of PACI in removal of water turibidty in Shahrchay river in Orumia and its health effects. Health System Research. 2011; 6(4): 762-9. [In Persian].
22. Jiang JQ, Lloyd B. Progress in the development and use of ferrate (VI) salt as an oxidant and coagulant for water and wastewater treatment. Water Res 2002; 36(6): 1397-408.
23. Pernitsky DJ, Edzwald JK. Selection of alum and polyaluminum coagulants: princip applic. Journal of Water Supply: Research and Technology - AQUA 2006; 55(2): 121-141.
24. Crittenden JC, Trussell RR, Hand DW, Howe KJ, Tchobanoglous G. MWH's Water Treatment: Principles and Design. Wiley; 2012.
25. Murphy AP. Chemical removal of nitrate from water. Nature 1991; 350: 223-5
26. Koparal AS, Öğütveren ÜB. Removal of nitrate from water by electroreduction and electrocoagulation. J Hazard Mater 2002; 89(1): 83-94.
27. Bhatnagar A, Sillanpää M. A review of emerging adsorbents for nitrate removal from water. Chem Eng J 2011; 168(2): 493-504.
28. Lacasa E, Cañizares P, Sáez C, Fernández FJ, Rodrigo MA. Removal of nitrates from groundwater by electrocoagulation. Chem Eng J 2011; 171(3): 1012-7.
29. Murkani M, Nasrollahi M, Ravanbakhsh M, Bahrami P, Jaafarzadeh Haghighi Fard N. Evaluation of natural zeolite clinoptilolite efficiency for the removal of ammonium and nitrate from aquatic solutions. Environmental Health Engineering and Management 2015; 2(1): 17-22.
30. Selecky M, Adair J, Clifford D. Nitrate Treatment Alternatives for Small Water Systems. http://www.doh.wa.gov/ehp/dw. Published 2005.
31. Khalil AM, Eljamal O, Jribi S, Matsunaga N. Promoting nitrate reduction kinetics by nanoscale zero valent iron in water via copper salt addition. Chem Eng J 2016; 287: 367-80.
32. Ensie B, Samad S. Removal of nitrate from drinking water using nano SiO 2–FeOOH–Fe core–shell. Desalination 2014; 347: 1-9.
33. Rajeswari A, Amalraj A, Pius A. Adsorption studies for the removal of nitrate using chitosan/PEG and chitosan/PVA polymer composites. Journal of Water Process Engineering 2016; 9: 123-34.

Feasibility of electrochemical oxidation process for treatment of saline wastewater

Kavoos Dindarloo[1], Hamza Ali Jamali[2*], Parvin Lakbala[1], Hamid Mahmoodi[3], Fatemeh Kazemi[4]

[1]Assistant Professor, Environmental Health Engineering Research Center, Instructor Department of Health, Hormozgan University of Medical Sciences, Bandar Abbas, Iran
[2]Assistant Professor, Department of Environmental Health Engineering, Qazvin University of Medical Sciences, Qazvin, Iran
[3]MSc, Environmental Health Engineering Research Center, Instructor Department of Health, Hormozgan University of Medical Sciences, Bandar Abbas, Iran
[4]MSc, Environmental Health Engineering Research Center, Hormozgan University of Medical Sciences, Bandar Abbas, Iran

Abstract
Background: High concentration of salt makes biological treatment impossible due to bacterial plasmolysis. The present research studies the process of electrochemical oxidation efficiency and optimal levels as important factors affecting pH, salt concentration, reaction time and applied voltage.
Methods: The sample included graphite electrodes with specifications of 2.5 cm diameter and 15 cm height using a reactor with an optimum capacity of 1 L. Sixty samples were obtained with the aid of the experiments carried out in triplicates for each factor at 5 different levels. The entire experiments were performed based on standard methods for water and waste water treatments.
Results: Analysis of variance carried out on effect of pH, salt concentration, reaction time and flow intensity in elimination of chemical oxygen demand (COD) showed that they are significant factors affecting this process and reduce COD with a coefficient interval of 95% and test power of 80%. Scheffe test showed that at optimal level, a reaction time of 1 hour, 10 g/L concentration, pH = 9 and 15 V electrical potential difference were obtained.
Conclusion: Waste waters containing salt may contribute to the electro-oxidation process due to its cations and anions. Therefore, the process of electrochemical oxidation with graphite electrodes could be a proper strategy for the treatment of saline wastewater where biological treatment is not possible.
Keywords: Saline wastewater, Graphite electrode, Salinity removal, Electrochemical oxidation.

***Correspondence to:**
Hamzah Ali Jamali
Email: jamalisadraei@yahoo.com

Introduction

Saline wastewater contains high amounts of organic compounds and soluble inorganic salts which are produced from industries such as fish processing, sea food packing, tanning and petrochemicals (1). In biological treatment of wastewaters with high salt concentrations especially in the of aerobic activated sludge phase process, the efficiency of decontamination would be very low due to bacterial plasmolysis in high concentration of salt (over 1%) (2,3). Desalination is required before treatment process. Therefore, to treat the waste waters containing total dissolved solids (TDS), other advanced treatment methods such as ultra-filtration (4), nano-filtration (5), reverse osmosis (6), electro-Fenton processes (7), and photochemical electrodialysis (8) have been studied.

In the methods of membrane filtration of saline wastewaters where some solution is used for demineralization, the influent organic matter to the membrane should be zero level (9). Waste waters containing high salt appeared to have high electrical conductivity due to high concentration of cations and anions. Hence, this seems to be an appropriate method in combination with electrochemical methods in the treatment of such waste waters (10).

Many studies have been conducted on electro oxidation of organic compounds where various materials are used as anode. This method has been successfully employed in the treatment of wastewaters from industries such as textile, tanning, distillation and leachate industries as well as urban wastewater. Different anodes have been employed for the treatment of industrial wastewaters. Recently, electrode graphite has been widely used for the decomposition of organic matter because it is economical. Graphite has a large surface area; this feature enables it to increase the rate of organic matter removal through adsorption and electro-oxidation. These electrodes are more efficient compared with other electrodes. Hydroxyl radicals are dominant because of the physical adsorption of oxygen on to the graphite electrode where they cause complete decomposition of organic materials. However, one of the considerable disadvantages of graphite electrode is its relative short life due to corrosion of the its surface especially when electric oxidation occurs in high potential difference (11).

Detoxification of tannery wastewater through electrolysis by the use of Ti/Pt anode and SS304 cathode was studied. The results showed a 52% decrease in chemical oxygen demand (COD) using electro oxidation process (11). In a study by Rao et al (12), using Ti/Pt electrode in the successful oxidation of tannery waste water, the study showed that the material of the electrode has an effect on COD removal. Kargi Dincer reported that biological treatment of saline wastewater by activated sludge resulted in low COD removal efficiencies as a result of plasmolysis of cells caused by high salt content (13). Szpyrkoicz et al suggested the electro oxidation process as a supplementary treatment after biological treatment as a more economical process of ammonium removal (14). In electrochemical oxidation, the organic pollutants are eliminated by chlorine and hypochlorite. During the oxidation process, the reaction of electrographite in the presence of sodium chloride produces chlorine (14).

Anode reactions:

$$2Cl^- \rightarrow Cl_2 + 2e^- \quad (1)$$

$$4OH^- \rightarrow O_2 + 2H_2O + 4e^- \quad (2)$$

Cathode reactions

$$2H_2O + 2e^- \rightarrow H_2 + 2OH^- \quad (3)$$

Reactions at the anode (Equations 1 and 2) and at the cathode (Equations 3) for man integrated cell at the anode. Hydroxide is produced at the cathode and reacts with chloride and hypochlorite. The compounds of hypochlorite and free chlorine are chemically reactive which decomposes organic pollutants into carbon dioxide and water, hence HOCl is produced:

HOCl decomposition reaction:

$$Cl_2 + H_2O \rightarrow H^+ + Cl^- + HOCl \quad (4)$$

$$HOCl \leftrightarrow H^+ + OCl^- \quad (5)$$

Hypochlorite is a leading organic matter. The final reaction in the electro oxidation process is expressed according to Equation 6 below:

$$\text{Organic matter} + OCl^- \rightarrow \text{interstitial material} \rightarrow CO_2 + Cl^- + H_2O \quad (6)$$

The present study was carried out to find out the factors such as salt concentration, reaction time, the intensity of voltage and pH affecting the treatment of saline waste water through electrooxidation process and determination of optimal conditions for each of the factors.

Methods

This research was a descriptive-analytical study conducted in a cubic-rectangular batch reactor made of Plexiglas with an optimum capacity of 1/L with dimensions of 14 cm × 7 cm ×15 cm. Graphite columns with diameter of 2.5 cm and length of 15 cm were interconnected by pieces of wire and embedded in the reactor as 2 series of parallel electrode. Each series of graphite electrode connected to direct current was supplied by AC/DC convertor (PAYANIK, RN-3003D). The experimental setup for the electrooxidation studies is shown in Figure 1.

2.5 g bovine albumin serum and hydrochloric acid were used to prepare synthetic wastewater as well as required salt in the process and concentration of COD at the required level of saline wastewater. To balance the pH level, caustic 1 N and hydrochloric acid solutions made by Merck Company were used. All experiments were performed according to the standard methods for the analysis of water and wastewater (15). Experimental design was carried out based on the book for experimental design (16). The optimal level of each factor – pH, primary concentration of salt, current intensity and time – was determined separately. Since the present study aimed at comparing the mean percent of COD removal at different levels of the factors, graphs were used for choosing sample size. The frequency of each test was selected at a confidence interval of 95% and a power of test of 80%. A mean difference of 20% or higher for COD removal between the 2 levels of each factor in the experiment showed that it was statistically significant (17).

To determine the frequency of each test and the level of each factor, the standard deviation estimated to be 10% was evaluated based on the formula 7:

$$Q = \frac{1}{\sigma}\sqrt{\frac{\sum T_j^2}{r}} = \frac{1}{10} \times \sqrt{\frac{5 \times 15^2}{5}} = 1.5 \quad (7)$$

Where:

r = Number of groups

σ = Equal to Q

T^2 = Mean difference between the two concentrations.

Using the following formula, Ω coefficient was calculated at 0.5 error levels and 80% power of test, and the ratio of the numbers of pH factor levels in the sample graphs. According to above mentioned book, 3 experiments were repeated for each level. As the results of the calculations, the total numbers of the experiments was estimated 60 samples (16).

After the experiments were conducted and laboratory data for the related parameters were collected, graphs were constructed using Excel and SPSS version 16. To identify optimal level of each variable, analysis of variance (ANOVA) and Scheffe test were used.

Results

Effects of pH

Tables 1 & 2 and Figure 2 show the results of wastewater COD changes (790 mg/L primary COD) at 10 g/L salt concentration, 1 hour reaction time, 15 V potential difference and different pH.

Also, ANOVA results for the effect of pH on COD removal at 10 g/L salt concentration,1 hour reaction time, and 15 V potential difference showed that pH had a significant ($F = 44.24$, $P < 0.001$) effect on the process.

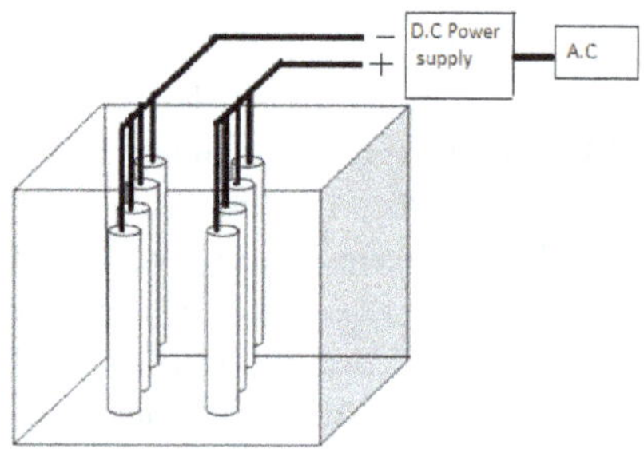

Figure 1. Schematic diagram of the reactor used in this study

Table 1. Data of reduction of COD at different pH levels with 10 g/L salt concentration, 1 hour reaction time, and 15 V difference

pH	No. of experiments	Mean	SD	SE	95% CI
3	3	42	3	1.7	34.54-49.45
5	3	52.3	6.6	3.8	35.79-68.87
7	3	63	6.5	3.7	46.71-79.28
9	3	80.33	4.5	2.6	69.13-91.53
11	3	87.33		2.3	77.29-97.37

Abbreviations: SD, standard deviation; SE, standard error.

Table 2. Comparison test for different levels of pH factor in reducing COD at 10 g/L salt concentration, 1 hour reaction time, and 15 V potential difference at 95% CI

pH level compared	Other pH levels	Mean difference	SD	*P*
	3	38.33 (SE)	4.2	<0.001
9	5	28 (SE)	4.2	<0.001
	7	17.33 (SE)	4.2	<0.029
	11	-7	4.2	<0.615

Abbreviation: SD, standard deviation.

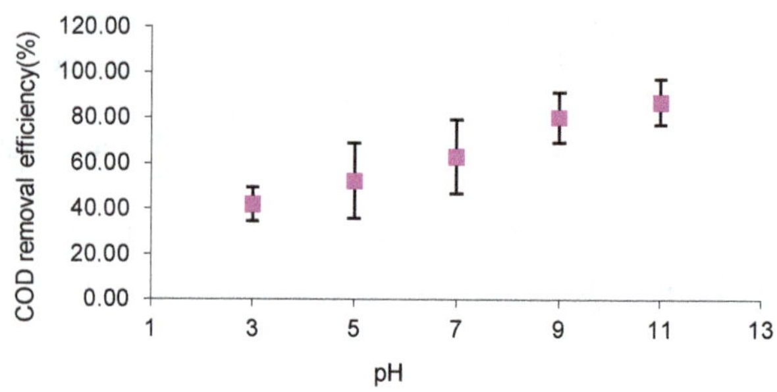

Figure 2. The graph of COD removal changes at different levels of pH at 10 g/L salt concentration, 1 hour reaction time, and 15 V potential difference.

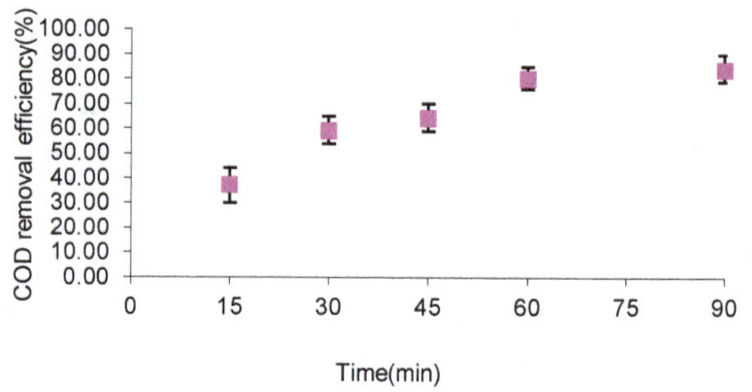

Figure 3. The graph of changes in COD removal at different levels of reaction time, 10 g/L salt concentration, pH = 9 and 15 V potential difference.

Table 3. Data description in reduction percent of COD at different reaction time with 10 g/L salt concentration, pH = 9 and 15 V difference

Time	No. of experiments	Mean	SD	SE	95% CI
15	3	37.23	2	1.2	30-44
30	3	59.33	5.5	3.17	54-65
45	3	64.33	5.5	3.17	59-70
60	3	80.33	4.5	2.6	76-85
90	3	84	1	0.57	79-90

Abbreviations: SD, standard deviation; SE, standard error.

Table 4. Comparisons of different levels of reaction time in reducing COD for the determination of optimal level at 95% CI

pH level compared	Other pH levels	Mean difference	SD	*P*
	15	43(SE)	3.39	<0.001
16	30	21 (SE)	3.39	<0.002
	45	16 (SE)	3.39	<0.013
	90	-3.66	3.39	<0.877

Abbreviation: SD, standard deviation.

Effect of reaction time

Tables 3 & 4 and Figure 3 show the results of wastewater COD changes in this study with 10 g/L salt concentration, pH = 9 and 15 V potential difference at different reaction time.

ANOVA results for the effects of reaction time on COD removal at 10 g/L salt concentration, pH = 9, and 15 V potential difference showed that reaction time had a significant ($F = 60.55$, $P < 0.001$) effect on this process.

Effect of salt concentration

Tables 5 & 6 and Figure 4 show the results of wastewater COD changes at 1 hour reaction time, pH = 9 and 15 V potential difference time at different salt concentrations. ANOVA results for the effects of salt concentration on COD removal at 1 hour reaction time, pH = 9, and 15 V potential difference showed that salt concentration had a significant effect on this process ($F = 0.4.72$, $P < 0.05$).

Influence of applied voltage

Tables 7 & 8 and Figure 5 show the results of wastewater COD changes at 1 hour reaction time, pH = 9 and 10 g/L salt concentration, at different potential differences.

ANOVA results for the effects of potential differences on COD removal with a 1 hour reaction time, pH = 9, and 10 gr/L salt concentration showed that potential difference had a significant effect on the process ($F = 8.75$, $P < 0.05$).

Table 5. Data description of percentage reduction of COD at different salt concentrations, 1 hour reaction time, pH = 9 and 15 V potential difference

Saline concentration (g/L)	No. of experiments	Mean	SD	SE	CI %95
10	3	80.33	4.5	2.6	69.13-91.53
15	3	83.66	3.2	1.8	75.68-91.65
20	3	84.66	3.7	2.18	75.26-94.07
25	3	88	3	1.73	80.54-95.45
30	3	91	1.5	0.88	87.53-95.12

Abbreviations: SD, standard deviation; SE, standard error.

Table 6. Comparison test for different levels of voltage percentage on COD reduction at different salt concentrations with 1 hour reaction time, pH = 9, and 15 V difference

Saline concentration level compared	Other time levels	Mean difference	SD	*P*
	15	-3.33	2.7	0.82
10	20	-4.33	2.7	0.65
	25	-7.66	2.7	0.17
	30	-11 (SE)	2.7	0.03

Abbreviation: SD, standard deviation.

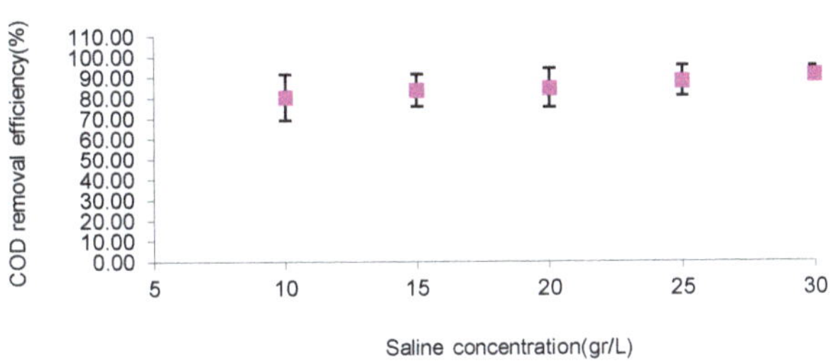

Figure 4. COD removal changes at different levels of salt concentration, with a 1 hour reaction time, pH = 9 and 15 V potential difference.

Table 7. Data description in COD reduction percentage at different potential differences, 1 hour reaction time, pH = 9 and 10 g/L salt concentration

Voltage	No. of experiments	Mean	SD	SE	CI 95%
2	3	64.66	4.5	2.6	53.46-75.8
5	3	70.66	3	1.7	63.7-78.25
8	3	73.66	3.21	1.8	65.68-81.65
10	3	77	1.7	1	72.69-81.3
15	3	80.33	4.5	2.6	69.13-91.53

Abbreviations: SD, standard deviation; SE, standard error.

Table 8. Comparison test for different levels of potential differences on COD reduction at 1 hour reaction time, pH = 9 and 10 g/L salt concentration

Voltage level compared	Other voltage levels	Mean difference	SD	*P*
	2	15.66 (SE)	2.9	0.005
15	5	6.66	2.9	0087
	8	3.33	2.9	0.329
	10	-11 (SE)	2.9	0.852

Abbreviation: SD, standard deviation.

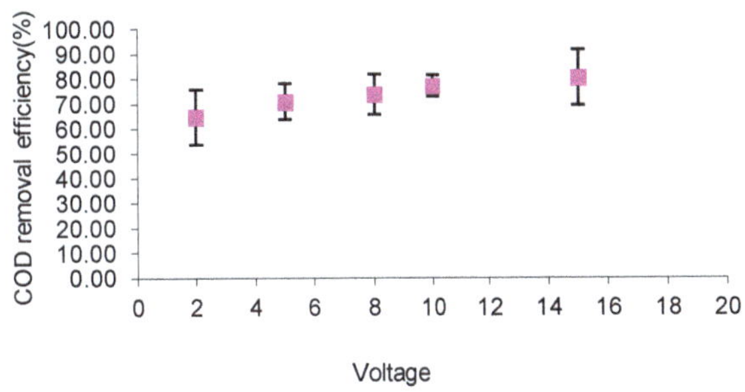

Figure 5. COD removal changes at different levels of potential difference, at 1 hour reaction time, pH = 9 and 15 potential difference.

Discussion

ANOVA results for the effects of pH on COD removal at 10 g/L salt concentration, 1 hour reaction time, and 15 V potential difference showed that pH had a significant effect on this process ($F = 44.24$, $P < 0.001$). The results showed that an increase in pH from 3 to 11, resulted in a COD decrease from 87% to 42%. Scheffe test for the comparison of different levels of pH factor was significant and showed that as pH increases, removal of COD increases as well which could be due to hydroxyl radical production in the electro-oxidation process. The findings of this study are in accordance with those of Sundarapandiyan et al (2) which showed that an increase in pH from 5 to 11, causes a COD decreases from 89% to 57%. According to Equations 1-5, the reason for this is that in acidosis state, hypochlorite ions, CLO^-, remain unstable and in combination with hydrogen ion tends to form HCLO. The hypochlorite which is formed in an alkaline state accounts for the most important factor inorganic matter oxidation. However, Yan et al (18) reported a higher removal of COD in acidic conditions using the Fenton process a sit is more efficient in acidic conditions.

ANOVA results for the effects of reaction time on COD removal at 10 g/L salt concentration, pH = 9, and 15 V difference showed that reaction time had a significant effect on this process ($F = 60.55$, $P < 0.001$). The mean COD removal of was 37.33% at a 95% CI and 80% power of test at 15 minutes reaction time (Table 3). Scheffe test for the comparison of different levels of reaction time factor was significant and showed that with increase in reaction time, COD removal also increases. In comparing different levels of time, it is observed that 60 minutes interval is significantly different from 15, 30, and 45 minutes intervals. There was, however, no significant difference at 90 minutes time interval where COD reduction rate increased by 3% which does not seem reasonable with the amount of energy consumed. Therefore, a reaction time of 60 minutes is suggested to be an optimal level. The finding of this study is in accordance with those of Zhang et al (19)and Yazdanbakhsh et al (20). However, Yan et al and Javid et al showed a higher time for COD decrease which could be due to the difference in potential difference used in the process (18,21).

ANOVA results for the effects of salt concentration on COD removal at 1 hour reaction time, pH=9, and 15 V potential difference showed that salt concentration had a significant effect on the process ($F=4.72$, $P<0.05$). The mean COD removal was 80.33% at 95% CI and 80% power of test at 10 g/L salt concentration (Table 5). The reason is likely due to an increase in the production of hypochlorite ions which occurs through the addition of NaCl. This is consistent with Sundarapandiyan et al study (2). Scheffe test for the comparison of different levels of salt concentration factor was only significant at 10 and 30 g/L, however it was insignificant at other levels. This shows that although salt acts as a facilitator, its increase in the environment has no effect on the efficiency of the process. ANOVA results for the effects of potential differences on COD removal at 1 hour reaction time, pH=9, and 10 g/L salt concentration showed that it had a significant effect on the electrochemical oxidation process ($F=8.75$, $P<0.05$).The mean COD removal was 64.66% at 95% CI and 80% power of test at 15 minutes reaction time (Table 7). Seems that rendering increase of voltage results in hypochlorite ion increase; this is consistent with Sundarapandiyan study (2). Also, Al-Qaim et al (22) reported that with the increase of voltage the removal efficiency of caffeine by electrochemical oxidation increased to 95% at 10 V from 75% at 6 V and 10% at 3 V. Also, the study by Duan et al (23) showed that phenol removal efficiency is almost 100% at 3 V after 300 minutes of polarization, with only 40% and 83% of phenol removed at 1.5 and 2 V, respectively (23). Scheffe test was used for the comparison of different levels of reaction time which was significant; although potential difference is effective in the electrochemical oxidation process considering the useful capacity of the reactor, an 8-15 Voltage difference seems to be sufficient. An increase in potential differences with this capacity causes foam formation and discoloration due to destruction of graphite electrode. The study of Yan et al showed an increase in COD removal with an increase in voltage. It indicates that some organic compounds are decomposed following voltage increase (18,20).

Conclusion

ANOVA analysis showed that pH, salt concentration, reaction time, and flow intensity are significant in COD removal at a CI 95% and test power of 80%. Scheffe test showed the optimal levels for the factors thus; reaction time 1 hour,10 g/L concentration, pH=9 and 15 V electrical potential difference. The electrochemical oxidation process with graphite electrodes could be a proper strategy for the treatment of saline wastewater where biological treatment is not possible.

Acknowledgements

The authors appreciate the cooperation of the Vice chancellor of Research and Technology, Hormozgan University of Medical Sciences.

Competing interests

The authors declare that they have no competing interests.

Authors' contributions

All the authors contributed equally to this study and therefore reviewed and approved the final draft of the manuscript.

References

1. Lefebvre O, Moletta R. Treatment of organic pollutant in industrial saline wastewater: a literature review. Water Res 2006; 40(20): 3671-82.
2. Sundarapandiyan S, Chandrasekar R, Ramanaiah B, Krishnan S, Saravanan P. Electrochemical oxidation and reuse of tannery saline wastewater. J Hazard Mater 2010; 180(1-3): 197-203.
3. Kargi F, Dincer AR. Biological treatment of saline wastewater by bed batch operation. J Chem Technol Biotechnol 1997; 69(2): 167-72.
4. Brites Alves AM, Norberta de Pinho M. Ultrafiltration for colour removal of tannery dyeing wastewaters. Desalination 2000; 130(2): 147-54.
5. Aouni A, Fersi C, Ben Sik Ali M, Dhahbi M. Treatment of textile wastewater by a hybrid electrocoagulation/nanofiltration process. J Hazard Mater 2009; 168(2-3): 868-74.
6. Cassano A, Molinari R, Romano M, Drioli E. Treatment of aqueous effluents of the leather industry by membrane processes. J Memb Sci 2001; 181(1): 111-26.
7. Kurt U, Apaydin O, Gonullu MT. Reduction of COD in wastewater from an organized tannery industrial region by electro-Fenton process. J Hazard Mater 2007; 143(1-2): 33-40.
8. Rodrigues MA, Amado FD, Xavier JL, Streit KF, Bernardes AM, Ferreira JZ. Application of photoelectrochemical-electrodialysis treatment for the recovery and reuse of water from tannery effluents. J Clean Prod 2008; 16(5): 605-11.
9. MänttäriaM, ViitikkobK, NyströmaM. Nanofiltration of biologically treated effluents from the pulp and paper industry. J Memb Sci 2006; 272(1-2): 152-60.
10. Grimm J, Bessarabov D, Sanderson R. Review of electro-assisted methods for water purification. Desalination 1998; 115(3): 285-94.
11. Vlyssides AG, Israilides CJ. Detoxification of tannery waste liquors with an electrochemical system. Environ

Pollut 1997; 97(1-2): 147-52.
12. Rao NN, Somasekar KM, Kaul SN, Szpyrkowicz L. Electrochemical oxidation of tannery wastewater. J Chem Technol Biotechnol 2001;76(11): 1124-31.
13. Kargi F, Dincer AR. Use of halophilic bacteria in biological treatment of saline wastewater by fed-batch operation. Water Environ Res 2000; 72(2): 170-4.
14. Szpyrkowicz L, Kaul SN, Neti RN, Satyanarayan S. Influence of anode material on electrochemical oxidation for the treatment of tannery wastewater. Water Res 2005; 39(8): 1601-13.
15. APHA, AWWA, WPCF. Standard Methods for the Examination of Water and Wastewater. 21st ed. Washington, DC: American Public Health Association; 2005.
16. Douglas C. Montgomery, Design and Analysis of Experiments. 8th ed. Wily Publication; 2012.
17. Neter J, Wasserman W, Kutner M H. Applied linear statistical models: Regression, analysis of variance, and experimental designs. Homewood, IL: Irwin; 1974.
18. Yan L, Ma H, Wang B, Wang Y, Chen Y. Electrochemical treatment of petroleum refinery wastewater with three –dimensional multi-phase electrode. Desalination 2011; 276(1-3): 397-402.
19. Zhang J, Zhang Y, Quan X. Electricity assisted anaerobic treatment of salinity wastewater and its effects on microbial Communities. Water Res 2012; 46(11): 3535-43.
20. Yazdanbakhsh AR, Kermani M, Komasi S, Aghayani E, Sheikhmohammadi A. Humic acid removal from aqueous solutions by peroxi-electrocoagulation process. Environ Health Eng Manag J 2015; 2(2): 53-8.
21. Javid AH, Malakootian M, Mehdipour M. Investigation of ozone and coagulant material's (aluminum sulfate, ferric chloride, poly aluminum chloride and lime) efficiency in "Kerman Kork" industry wastewater treatment. Environ Health Eng Manag J 2015; 2(1): 1-6.
22. Al-Qaim FF, Mussa ZH, Othman MR, Abdullah MP. Removal of Caffeine from aqueous solution by indirect electrochemical oxidation using a graphite-PVC composite electrode: A role of hypochlorite ion as an oxidizing agent. J Hazard Mater 2015; 300: 387-97.
23. Duan F, Li Y, Cao H, Wang Y, Crittenden JC, Zhang Y. Activated carbon electrodes: electrochemical oxidation coupled with desalination for wastewater treatment. Chemosphere 2015; 125: 205-11.

14

Adsorption of copper, lead and cadmium from aqueous solutions by activated carbon prepared from saffron leaves

Shidvash Dowlatshahi[1], Ahmad Reza Haratinezhad Torbati[2], Mahshid Loloei[1*]

[1]Instructor, Environmental Health Engineering Research Center and Department of Environmental Health, School of Public Health, Kerman University of Medical Sciences, Kerman, Iran
[2]M.Sc Student of Environmental Health Engineering, Department of Environmental Health, School of Public Health, Kerman University of Medical Sciences, Kerman, Iran

Abstract
Background: Industrial development has caused the release of various pollutants including heavy metals into the environment. These toxic compounds are extremely dangerous to living beings and the environment due to their non-biodegradability, severe toxicity, carcinogenicity, the ability to be accumulated in nature and the ability to contaminate groundwater and surface water. The aim of the present research was to provide an appropriate and cost-effective adsorbent to remove heavy metals from aqueous solutions.
Methods: The activated carbon was produced from the dried. Batch experiments were performed on real and synthetic samples at room temperature. The effect of pH, adsorbent dose, initial concentration, and contact time were studied, and the adsorption isotherms of heavy metals were determined. The removal efficiency was evaluated on real wastewater.
Results: The maximum removal efficiency of heavy metals (copper, cadmium and lead) by activated carbon adsorbent prepared from saffron leaves was obtained in pH 7. The optimum amount of adsorbent was 0.6 g, and the optimum contact times were 45 min for copper and cadmium ions and 90 min for lead ion, respectively. In these optimum conditions the removal efficiencies were 76.36%, 91.25% and 97.5%, respectively. The removal efficiencies of heavy metals from actual samples (copper industry and the battery industry) in the optimum conditions were 82.25%, 69.95% and 91.23%, respectively. The results obtained showed the highest correlation with Langmuir isotherm model.
Conclusion: Based on the results obtained, the activated carbon produced from saffron leaves has a good capability in removal of the metal ions from the aqueous solutions. Considering the availability of saffron leaves in Khorasan, its cost-effectiveness, and high uptake capacity, it can be applied as a proper absorbent to remove the heavy metals from industrial wastewater.
Keywords: Adsorption, Heavy metal, Activated carbon, Saffron leaves

***Correspondence to:**
Mahshid Loloei
Email: mloloei@gmail.com

Introduction

Increasing pollution of municipal and industrial sewages by toxic compounds due to industry development and growth is a concerning issue. Heavy metals because of the nature of their non-degradability, sever toxicity, ability to be accumulated, and carcinogenicity not only endanger aquatic life, but also cause the acceptor waters to be inappropriate for various uses such as drinking (1). Discharging sewage of many industrial processes like plating, photography, nuclear power plants, and petrochemical industries into water resources causes water pollution. Heavy metals like lead and cadmium, even at low concentrations are serious threats to living organisms through their being stimulating, property aggregation ability, carcinogenicity, and mutagenesis (2). These metals are non-degradable and very resistant against biodegradation. The adverse effects of their presence in body can be diarrhea, abdominal pains, severe vomiting, fracture, sterilization, and they can damage the central nervous system, body immune systems, and also cause psychological disorders, possible damage to DNA, and cancer (3-8).

There are a variety of methods such as sedimentation, filtration, oxidation and reduction, ion exchange, and membrane separation for heavy metals removal. These methods are often ineffective or expensive in very low concentrations (3,4).

Adsorption is a process during which a soluble material accumulates on the surface of another substance (9). In fact, adsorption is the function of transferring from liquid phase to solid phase. Adsorption is an important phenomenon in many physical, chemical, and biological processes in the nature. Adsorption is also widely used in water and wastewater treatment (7). One of the most widely used adsorbents for this purpose is activated carbon. Its high cost of activation stages and its low regenerative capacity has limited the use of this material on a large scale. In recent years, using inexpensive adsorbents has attracted the attention of many researchers. These adsorbents are frequently found and are available, and their preparation costs are low (6). The use of bioadsorbents to remove metal ions from aqueous solutions is a cost-effective and environment-friendly method. Therefore in recent decades, it has attracted the attention of many researchers. Successful use of natural materials such as activated carbon produced from olive kernel in the removal of heavy metals like copper, lead and cadmium was experienced by Bohli *et al* (8) in 2013, in Tunisia.

Activated carbon produced from tassel to remove lead was one of the applied adsorbents used by Moyo *et al* (5). Also Hadoun *et al* (10) investigated the removal of cadmium from the aqueous solutions using activated carbon produced from palm stem in Algeria. *Yahaya* et al (11) investigated the removal of copper using activated carbon produced from rice husk. Amuda *et al* studied the heavy metals removal by charcoal made from coconut shell that resulted in a very effective absorbent from cheap row material (11). The study by Song *et al* (12) on the removal of lead by activated carbon made of coconut skin activated by potash had good results. Nale *et al* (13) performed studies of kinetics and copper absorption equilibrium from aqueous solutions on activated carbon produced from corn cob. Studies related to kinetics and cadmium removal equilibrium from aqueous solutions by activated carbon produced of apple skin were carried out by Sartape *et al* (14). Another study was conducted by Arivoli *et al* (15) titled "The dynamic of copper adsorption by low-cost activated carbon". Saffron with the scientific name of Crocus sativus, is a plant from the family of gladiola. There was 57,000 hectares of cultivated area in Razavi Khorasan which was equivalent to 77% total production of country in 2012.

The aim of this study was to investigate the removal efficiency of heavy metals (copper, cadmium and lead) with activated carbon prepared from leaves of saffron in the synthetic and real solutions as an available and inexpensive absorbent in the region. The effective parameters on absorption process such as contact time, pH, and initial concentration of metal ions were studied. Also the adsorption equilibriums were studied.

Methods

In this practical study, the efficacy of activated carbon prepared from saffron leaves for the removal of metal ions (copper, cadmium and lead) from the aqueous solutions was studied.

The chemical materials made by Merch company with high purity were used. The stock solutions of metal ions with concentration of 1000 mg/l were prepared. In all stages of the experiment, the 0.1 N normal solutions of NaOH and H_2SO_4 were used to adjust pH. pH meter (HANNA instruments HI 110 Model) was used to measure the pH of the solution

Adsorbent preparation

Sorbent preparation was performed as follows. Saffron leaves were harvested from the cultivated fields around Torbat Heydarie city located in Razavi Khorasan and collected in the suitable containers and moved to laboratory. The leaves were washed by distilled water, then dried at 60 °C, and were powered after crushing. In order to improve the adsorption capacity, 100 g of obtained powder ere mixed with a special volume of phosphoric acid with a mass ratio of 1:5, and this mixing alternatively continued for one hour to obtain a better contact of acid and saffron leaves. Then the modified powder was washed by distilled water for the removal of outputs. After removing the additional water, the obtained materials in a metal reactor in diameter of 50 mm and length of 100 mm were placed into an electric furnace for one hour in a temperature of 750 °C (16). This carbon was washed by distilled water until pH reached about 6.5 and then was dried in the oven for one hour at 100 °C, and then the two sieves with mesh sizes of 100 and 200 were used to separate the absorbent particles between 0.85 and 0.15 mm. Memert oven, which was made in Germany, and 1200 °C Electric Furnace were used to provide required temperatures.

Determining the characteristics of the adsorbent

In order to investigate the properties of the initial carbon and activated carbon prepared from saffron leaves, several experiments were conducted to determine the percentages of cellulose, lignin, volatile material, humidity, density, and iodine number (17). Morphological properties of activated carbon prepared from saffron leaves were studied by Scanning Electronic Microscope (SEM).

Adsorption tests

The effect of various parameters on the absorption process including the initial concentration of pollutants, the amount of adsorbent, contact time, and the optimum pH was studied. To determine the optimum pH, the value of 0.1 g activated carbon prepared from saffron leaves was poured in 100 ml of the synthetic solution with concentration of 30 ppm. Then experiment was conducted at contact times of 30, 45, 60, and 90 min at various pHs (3, 5, 7, and 9) separately, and then adsorbent was contacted with the metal ions of pollutant in the ambient temperature. Then the samples were filtered using filter paper in size of 0.45 μ and the remaining metal ions were measured by atomic absorption apparatus (YOUNGLTN ASS 8020 Model). The process was separately repeated for all the metal ions.

In the next step considering optimum pH determined,

the amounts of adsorbent of activated carbon (0.3, 0.6, 1, and 1.3 g), initial concentration of pollutants (20, 30, 50, and 40 mg/L), and the contact time (30, 45, 60, and 90 min) were separately tested for individual pollutants. The removal efficiency was calculated using the following equation:

$$\text{Removal}(\%) = (C_0 - C_t)/C_0 \times 100 \quad (1)$$

Where C_0 is the initial concentration (mg/l), and C_t is metal ion concentration after each adsorption step.

Langmuir adsorption isotherm

The assumptions of Langmuir equation include the followings: a) maximum absorption occurs when the adsorbent surface is covered by a single molecular layer of soluble material. b) The absorption energy is fixed and identical at all the points. c) The molecules of adsorbed material cannot move in the adsorbent surface.

$$Q_0 = \frac{Q_{max}\ BC}{1+BC} \quad (2)$$

Where C is the concentration of soluble material in the steady state mg/L.

Q_{max} is the maximum absorption capacity (mg/g) and B is Langmuir equation constant (L/mg).

Freundlich adsorption isotherm

The Freundlich equation expresses adsorption on a heterogeneous surface in terms of adsorption, and it can be expressed as follows:

$$Q = KC_f^{1/n} \quad (3)$$

Where n is Freundlich intensity parameter, and K is Freundlich capacity factor.

The Freundlich equation is an empirical one, and it is applicable for interpreting the experimental data (7).

With respect to determining optimal conditions in synthetic samples, the efficiency of metals removal from the sewage of Copper Complex Industries was studied to provide copper ion and the battery-making for electronics power (for cadmium and lead ions) was sampled. After determining the characteristics of sewage [pH, concentrations of copper, lead, cadmium ions, Chemical Oxygen Demand (COD), Biochemical Oxygen Demand (BOD), and Total Dissolved Solids (TDS)], adsorption experiments were conducted by produced active carbon, and the removal efficiency was calculated (18).

Results

The results of studying the characteristics of activated carbon produced from saffron leaves are shown in the Table 1. The results showed that the density of activated carbon produced is lower than the other initial materials, and the amount of lignin is almost similar to the almond skin and walnut shell. The amount of cellulose is lower than pistachio bark and almond skin and is equal to walnut shell, but the value of the iodine number is less than these materials.

Figure 1 shows the results of the photos provided by the electron microscopy of coal and activated carbon prepared from saffron leaves.

In studying different pHs in the adsorption of metal ions, the optimum pH was 7. In the other pHs, the amount of adsorption was decreased.

According to efficiency results the absorption of copper and cadmium during contact time of 45 min was the highest with 83.34 % and 86.64 %, respectively. The lead adsorption was balanced at 90 min. At this time, the highest observed removal amount was 90.22%. Table 2 shows the results.

To investigate the amount of optimal absorbent, the adsorption experiments at optimum pH 7, the initial concentration of pollutants with 50 ppm, and the optimum contact times from the previous stage in the different values of adsorbents were tested and the results are shown in Table 3.

The results of adsorption experiments of heavy metals by activated carbon prepared from saffron leaves showed that in the optimum amount of adsorbent of 0.6 g and the equilibrium contact time of 45 min, the removal efficiency of cadmium ion is more than that of copper ion (82.52% against 75.56%). Removal efficiency of lead ion in the amount of adsorbent of 0.6 and 90 min was 95.55%. Increasing amount of adsorbent caused no significant change in absorption rate.

In the following research, to investigate the effect of adsorbent prepared on the real sample, samples of the effluents of two copper and battery-making industries were studied. The results related to study the raw wastewater quality of copper and battery-making industries are shown in the Table 4.

Testing optimized results on the actual samples related to the copper and cadmium ions showed that in optimum conditions (absorbent 0.6 g and the equilibrium contact time of 45 min) the removal efficiencies were equal to 68.12% and 78.95%, respectively. The experimental results related to the real lead sample showed that in the optimum contact time of 90 min and adsorbent amount

Table 1. The characteristics of activated carbon prepared from saffron leaves

Ingredients	Iodine number, mg/g	Percent of lignin	Percent cellulose	Density kg/m³	Percentage of total ash	Percentage of volatile material	Moisture (%)	References
Saffron leaves	501	30	34	160	7.84	1.05	5.4	This study
Almond Skin	830	27	39	470	1.5	-	-	Kaghzchi (17)
Walnut shell	806	33	34	457	6	-	-	Kaghzchi (17)
Pistachio bark	953	13	42	448	1.2	-	-	Kaghzchi (17)

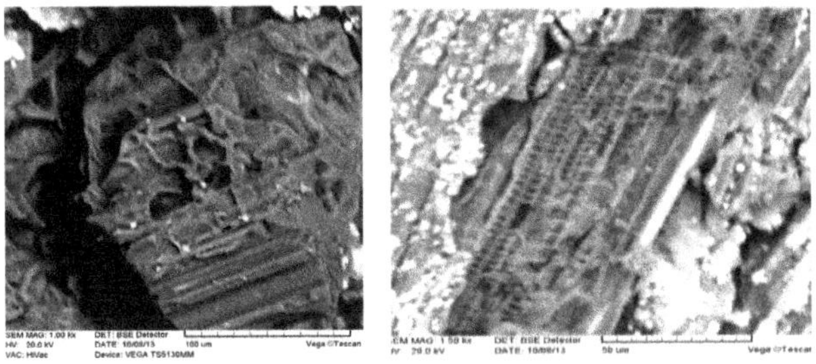

Figure 1. SEM image of charcoal (Left) prepared from saffron leaves without activating and activated carbon (Right).

of 0.6 g, the removal efficiency of this ion was 88.96%. To study the behavior of the adsorbent, Langmuir and Freundlich adsorption isotherms were plotted, and the results are shown in Figures 2 and 3. On the basis of the obtained correlation coefficients, it can be concluded that the adsorption of ions studied on prepared adsorbent given to the correlation coefficient higher than 0.99 follows Langmuir isotherm. Isothermal parameters have been shown in Table 5.

Table 2. The removal efficiency of the heavy metals (%) at different pHs and contact times (initial concentration 30 mg/L of pollutant and adsorbent dose 0.1 g).

Contact time(min)	Metal ion											
	Cd^{2+}				Cu^{2+}				Pb^{2+}			
	30	45	60	90	30	45	60	90	30	45	60	90
pH= 3	33.30	36.65	43.36	46.67	40.25	40.36	53.36	50.58	60.31	63.32	66.96	66.68
pH= 5	43.32	46.60	50.25	60.36	56.63	63.69	56.68	53.32	66.63	70.10	73.30	76.65
pH= 7	82.25	83.34	81.95	70.65	73.35	86.64	73.30	70.11	83.32	86.65	86.66	90.22
pH= 9	59.25	63.56	63.26	66.63	50.22	63.37	56.69	60.11	76.65	80.35	80.35	83.36

Table 3. The removal efficiency of heavy metals at different amounts of adsorbent (optimum contact time, optimum pH 7, initial concentration of 50 ppm)

Heavy metals	Optimum contact time (min)	The amount of adsorbent (g)	The final concentrations of heavy metals (ppm)	Absorption efficiency in synthetic solution (%)
Pb^{+2}	90	0.3	4.9	90.20
	90	0.6	2.22	95.55
	90	1	2.33	95.34
	90	1.3	2.27	95.45
Cu^{+2}	45	0.3	14.21	71.58
	45	0.6	12.22	75.56
	45	1	12.72	74.56
	45	1.3	12.37	75.25
Cd^{+2}	45	0.3	15.61	68.78
	45	0.6	8.75	82.50
	45	1	9.27	81.45
	45	1.3	9.21	81.58

Table 4. Characteristics of the sample of raw wastewater in the copper and battery-making industries

Type of industry	pH	Lead Concentration (ppm)	Cadmium concentration (ppm)	Copper concentration (ppm)	COD mg/L	BOD mg/L	TDS mg/L
Copper	4.21	1.4	0.4	5.88	185	60	260
Battery-making	5.9	53.56	20.32	2.8	350	180	1100

COD: Chemical Oxygen Demand; BOD: Biochemical Oxygen Demand; TDS: Total Dissolved Solids

Table 5. Adsorption isotherm parameters for Cu, Cd and Pb by activated carbon prepared from saffron leaves

Metal ion	Freundlich isotherms			Langmuir isotherm		
	R^2	1/n	K (mg/g)	R^2	B (L/g)	Q_{max} (mg/g)
Copper	0.96	0.476	1.74	0.99	1.90	62.96
Lead	0.96	0.476	1.50	0.99	1.44	79.6
Cadmium	0.98	0.44	1.77	0.99	16.66	68.75

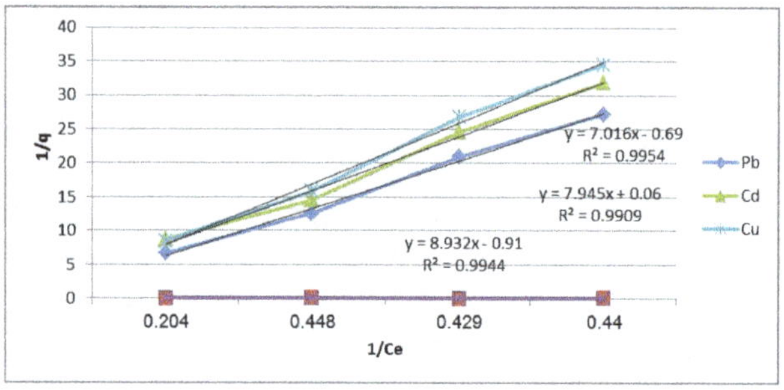

Figure 2. Langmuir adsorption isotherm of copper, cadmium, and lead

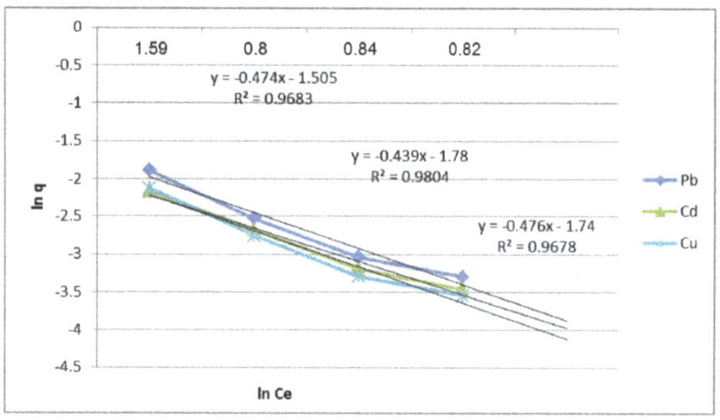

Figure 3. Freundlich adsorption isotherm of copper, cadmium, and lead

Discussion

The experimental results related to the prepared activated carbon shown that the way of activation stages have a significant impact on the quality of activated carbon prepared.

The effect of pH on adsorption efficiency of lead, cadmium, and copper by the activated carbon prepared from saffron leaves

pH is an important parameter in the adsorption process. In this study the effect of pH was tested by changing pH of solution in the range of 3-9 in concentration of 30 ppm from metal ions. Metal ions adsorption indicated more dependence from pH of solution (Table 2). Increasing solution pH increases the rate of adsorption because of electrostatic forces. This process continued until pH reached to 7 after that, increasing pH did not show positive effect on increasing absorption. This is because of reduced mobility of ions resulting from changing the properties and its load .The highest absorption was obtained in pH 7. In most of the studies, best pH for removal heavy metals was between 5-7. pH in terminal experiment was decreased by 1 value.

The results of this study were consistent with those of Amouei *et al*'s (19) research titled "Cadmium removal from canola plant", Madhava Rao *et al* (20) research in India about copper and cadmium removal from the aqueous solutions by activated carbon produced from Ceiba pentandra, and also and also with Venkatesan *et al* (21) about cadmium removal by activated carbon produced from Derris Indica wood. Nwabanne and Igbokwe (22) obtained some similar results related to the copper removal from aqueous solutions by activated carbon produced from various agricultural products in 2012.

The properties of the metal ion and also selectivity of the adsorbent are important factors in the adsorption process. In this study, the orders of ions absorption were observed Pb>Cd>Cu, respectively. This trend has also been the same in other studies (23). Pb ion occupies the adsorption places sooner than Cu and Cd due to its larger ionic radius. Some various trends have been also in the other studies (23).

The effect of adsorbent dose on the adsorption efficiency of copper, cadmium, and lead by activated carbon prepared from leaves of saffron

The dose of adsorbent is another important parameter in the adsorption process. The number of sites available for adsorption of metal ions depends on the amount of adsorbent used. In Table 3, it is observed that the adsorption of metal ions increases by increasing the amount of adsorbent to 0.6 g through holding the other parameters constant in the initial concentration of 50 mg/L. This increase is due to an increase in the surface area of adsorbent. At this value, the adsorbent surface is saturated with metal ions and after that the amount of adsorption reduces (24).

The results of this study regarding the effect of the adsorbent on the adsorption efficiency were consistent with Venkatesan *et al* research (21) related to the removal of cadmium by activated carbon produced from Derris Indica wood ,and also with the results of Madhava Rao *et al* research (20) on copper and cadmium removal from aqueous solutions by activated carbon prepared from Ceiba pentandra and were consistent with Amouei *et al* (19) studies about cadmium removal from water through activated carbon prepared from canola plant. In the study of Awwad and Salem about adsorption of copper and lead from aqueous solutions by modified adsorbent, the optimum amount of adsorbent was obtained 0.5 g (25).

The effect of contact time on adsorption efficiency of lead, cadmium, and copper by activated carbon produced from saffron leaves

The results related to ions adsorption of Cu (II) and Cd (II) on prepared adsorbent indicated that 45 min contact time was enough to reach equilibrium (Table 2). Adsorption does not change with an increase in contact time due to the saturation of adsorbent surface. Thus, this obtained contact time was used in other adsorption studies. The results indicated that this time is less than the contact time of these metal ions on some other adsorbents (25,26).

The results of this research on the effect of contact time were similar to those of Patnukao *et al* research (27) titled "Removal of metal ions (copper and lead) from synthetic wastewater by activated carbon produced from Eucalyptus Camaldulensis Dehnh shell related to copper metal". Also, Madhava Rao *et al* research (20) related to removal of copper and cadmium from aqueous solutions by activated carbon prepared from Povak tree skin was consistent with studies of Nale *et al* (13) on kinetics and equilibrium of lead adsorption from aqueous solutions on activated carbon produced from maize cobs (2012) with equilibrium contact time of 90 min. Additionally, these were associated with the other study by Arivoli *et al* (15) titled "Dynamic of copper adsorption by inexpensive activated carbon".

The isotherm study in the adsorption process of copper, lead, and cadmium from aqueous solutions by activated carbon prepared from saffron leaves

In the present study, the Freundlich and Langmuir isotherm models were studied to describe the relationship between the level of adsorption and the equilibrium concentration in liquid phase. This relationship and adsorption parameters for each model are shown in Figures 1, 2 and Table 5.

The amount of the correlation coefficient (R^2) shows that Langmuir model states a better relationship in this adsorption process. The maximum adsorption capacity for copper, lead, and cadmium ions is 62.92, 79.6, and 68.75 mg/g, respectively. These values represent high potential of the produced absorbent in the adsorption of metal ions studied from aqueous solutions because of its high adsorption capacity.

The results of this study on correspondence of adsorption isotherms are consistent with those of Saeedi *et al* (26) research related to cadmium removal by activated carbon produced from walnut and almond shell and Moreno-Pirajàn *et al* research (28) about the study of kinetic and removal of Mn, Ni, Cu, Fe ions from wastewater by activated carbon produced from coconut shell. Also, Nikazar and Noorbakhsh's (29) study on the removal of heavy metals Cd, Cr, and pb from aqueous solutions by activated carbon produced from agricultural wastewater (rice bran, wheat bran and straw) was consistent with these studies. Amouei *et al* (19) studied the Freundlich isotherm model related to cadmium adsorption using canola plant.

Comparing the absorption efficiency of heavy metals (copper, cadmium, and lead) from synthetic and real wastewater by activated carbon prepared from saffron leaves

Also in a research done by Elamin *et al* (30) about activated carbon produced from almond shell for the absorption of heavy metals of contaminated water, removal efficiencies of copper, and lead and cadmium in the synthetic model were 98%, 94%, and 49%, respectively and in the real model were 73%, 75%, and 20%, respectively.

The effect of initial concentration had an important effect on the absorption studies because removal efficiency decreased with rising initial concentration. Increasing the initial concentration of metal from 20 to 50 mg/L reduced the absorption efficiency of copper from 76.36% to 75.56% mg/L and those of lead absorption from 97.5% to 95.5% and cadmium absorption from 91.25% to 82.5%.

Conclusion

The results of experiments related to heavy metals absorption including copper, lead, and cadmium on activated carbon prepared from saffron leaves indicated a significant efficiency. The absorption level of Pb>Cd>Cu in the amount of adsorbent for all metals were 0.6 g, and the optimum contact times were 45, 45, and 90 min, respectively.

In these conditions the highest absorption efficiencies were observed 76.36%, 91.25%, and 97.5%, respectively. Also comparing the results of the study on synthetic and real models showed that the level of absorption in real wastewater is less than synthetic wastewater which is justifiable given analyzing the real wastewater of target industries and existence of other competing compounds, such as heavy metals and the other effective materials in absorption.

Due to the abundances of leaves and undesired use, saffron can be used as appropriate absorbent in the heavy metals removal from aqueous solutions.

Acknowledgments

This paper is the result of the students' thesis and it is conducted under the supervision of the Engineering Research Center for Environmental Health, Kerman University of Medical Sciences and sponsored by deputy of research and technology. Thereby, I thank them for their assistance to conduct this research.

Competing interests
The authors declare that they have no competing interests.

Authors' contributions
All authors participated equally in data acquisition, analysis and interpretation. All authors reviewed and approved the manuscript.

References

1. Ulmanu M, Marañón E, Fernández Y, Castrillón L, Anger I, Dumitriu D. Removal of copper and cadmium ions from diluted aqueous solutions by low cost and waste material adsorbents. Water Air Soil Pollut 2003; 142(1-4): 357-73.
2. Holan ZR, Volesky B. Biosorption of lead and nickel by biomass of marine algae. Biotechnol Bioeng 1994; 43(11): 1001-9.
3. Lee HS, Suh JH. Continuous biosorption of heavy metal ions by Ca-loaded laminaria japonica in fixed bed column. Korean J Chem Eng 2000; 17(4): 477-9.
4. Monteiro A, Boaventura R, Dinis MA. Treatment Aguas De Residuais: O Papel Das Microalgas. Revista from Faculdade de Ciencia and Technology 2004; 1: 41-54.
5. Moyo M, Chikazaza L, Nyamunda BC, Guyo U. Adsorption batch studies on the removal of Pb (II) using maize tassel based activated carbon. J Chem 2013; (2013): 1-8.
6. Sharma YC. Thermodynamics of removal of cadmium by adsorption on an indigenous clay. Chem Eng J 2008; 145(1): 64-8.
7. Weber WJ. Physicochemical processes for water quality control. New York: Wiley Interscience; 1972.
8. Bohli T, Villaescusa I, Ouederni A. Comparative study of bivalent cationic metals adsorption Pb (II), Cd (II), Ni (II) and Cu (II) on olive stones chemically activated carbon. J Chem Eng Process Technol 2013; 4(4): 1-7.
9. Carey RO, Migliaccio KW. Contribution of wastewater treatment plant effluents to nutrient dynamics in aquatic systems: a review. Environ Manage 2009; 44(2): 205-17.
10. Hadoun H, Belmedani M, Sadaoui Z, Toumert I. Removal of cadmium from aqueous solution by adsorption onto activated carbon prepared from date stems. J Chem Chem Eng 2013; 7: 979-84.
11. Amuda OS, Giwa AA, Bello IA. Removal of heavy metal from industrial wastewater using modified activated coconut shell carbon. Biochemical Engineering Journal 2007; 36: 174-81.
12. Song C, Wu S, Cheng M, Tao P, Shao M, Gao G. Adsorption studies of coconut shell carbons prepared by KOH activation for removal of Lead (II) from aqueous solutions. Sustainability 2014; 6(1): 86-98.
13. Nale B, Kagbu J, Uzairu A, Nwankwere E, Saidu S, Musa H. Kinetic and equilibrium studies of the adsorption of Lead (II) and Nickel (II) ions from aqueous solutions on activated carbon prepared from maize cob. Der Chemica Sinica 2012; 3(2): 302-12.
14. Sartape AS, Mandhare AM, Salvi PP, Pawar DK, Kolekar SS. Kinetic and equilibrium studies of the adsorption of Cd (II) from aqueous solutions by wood apple shell activated carbon. Desalination and Water Treatment 2013; 51(22-24): 4638-50.
15. Arivoli S, Nandhakumar V, Saravanan S, Nagarajan S. Adsorption dynamics of copper ion by low cost activated carbon. Arabian Journal of Science and Engineering 2009; 34: 1-12.
16. Owlad M, Aroua MK, Daud WA, Baroutian S. Removal of hexavalent chromium-contaminated water and wastewater: a review. Water Air Soil Pollut 2009; 200(1-4): 59-77.
17. Yenisoy-Karakas S, Aygun A, Gunes M, Tahtasakal E. Physical and chemical characteristics of polymer-based spherical activated carbon and its ability to adsorb organics. Carbon 2004; 42(3): 477-84.
18. APHA, AWWA, WEF. Standard methods for the examination of water and wastewater. 20th edition. Washington DC: American Public Health Association, American Water Work Association, Water Environment federation; 1998.
19. Amouei A, Ehrampoush MH, Ghaneian MT, Asgharzadeh F, Mousapour A, Parsian H. Removing cadmium from aqueous solutions by the Canola residuals. J Mazand Univ Med Sci 2014; 23(110): 153-64.
20. Madhava Rao M, Ramesh A, Purna Chandra Rao G, Seshaiah K. Removal of copper and cadmium from the aqueous solutions by activated carbon derived from Ceiba pentandra hulls. Journal of Hazardous Materials 2006; 129(1-3): 123-9.
21. Venkatesan G, Senthilnathan U. Adsorption batch studies on the removal of cadmium using wood of derris Indica based activated carbon. Research Journal of Chemistry and Environment 2013; 17(5): 19-24.
22. Nwabanne J, Igbokwe P. Mechanism of copper (II) removal from aqueous solution using activated carbon prepared from different agricultural materials. International Journal of Multidisciplinary Sciences and Engineering 2012; 3(7): 46-52.
23. Kazemipour M, Ansari M, Tajrobehkar S, Majdzadeh M, Kermani HR. Removal of lead, cadmium, zinc, and copper from industrial wastewater by carbon developed from walnut, hazelnut, almond, pistachio shell, and apricot stone. Journal of Hazardous Materials 2008; 150(2): 322-7.
24. El-Sherif IY, Ashmawy A, Badr S. Biosorption of cadmium and nickel by nile water algae. Journal of

Applied Sciences Research 2008; 4(4): 391-6.
25. Awwad AM, Salem NM. Biosorption of copper (II) and lead (II) ions from aqueous solutions by modified loquat (Eriobotrya japonica) leaves (MLL). J Chem Eng Mater Sci 2012; 3(1): 7-17.
26. Saeedi M, Jamshidi A, Abessi A. Removal of dissolved cadmium by adsorption onto walnut and almond shell charcoal: comparison with Granular Activated Carbon (GAC). Water and Wastewater 2009; 20(70): 16-22.
27. Patnukao P, Kongsuwan A, Pavasant P. Batch studies of adsorption of copper and lead on activated carbon from Eucalyptus camaldulensis dehn. bark. J Environ Sci (China) 2008; 20(9): 1028-34.
28. Moreno-Piraján JC, Garcia-Cuello VS, Giraldo L. The removal and kinetic study of Mn, Fe, Ni and Cu ions from wastewater onto activated carbon from coconut shells. Adsorption 2011; 17(3): 505-14.
29. Nikazar M, Noorbakhsh N. Removal of heavy metals ((Cr (VI), Pb (II), Cd (II) from aqueous solution by active carbon prepared from agricultural waste (rice bran, wheat bran, straw). Science and Technology Environment 2006; 8(1): 34-43.
30. Elamin A, Reddy M, Rehrah D. Activated carbon from almond shells to adsorb the heavy metals from contaminated water. Int J Chem Environ & Tech 2013; 1(3): 1-8.

15

Growth and lipid accumulation in response to different cultivation temperatures in *Nannochloropsis oculata* for biodiesel production

Mohammad Malakootian[1*], Behnam Hatami[2], Shidwash Dowlatshahi[3], Ahmad Rajabizadeh[3]

[1]Professor, Environmental Health Engineering Research Center and Department of Environmental Health, Kerman University of Medical Sciences, Kerman, Iran
[2]MSc, Department of Environmental Health, Kerman University of Medical Sciences, Kerman, Iran
[3]Instructure, Environmental Health Engineering Research Center, Department of Environmental Health, Kerman University of Medical Sciences, Kerman, Iran

Abstract
Background: Microalgal lipid is a promising feedstock for biodiesel production. The aim of the present study was to investigate the effects of cultivation temperature on the growth and lipid accumulation properties of *Nannochloropsis oculata* microalgae.
Methods: *Nannochloropsis oculata*can grow in a wide range of temperatures (5 ~ 35°C). Late in the stationary growth phase of microalgae, biomass production and lipid accumulation were measured. The methanol-chloroform extraction method was used to extract total lipids from dried cells. The direct esterification method was used to measure fatty acids. Constituents were identified by gas chromatography.
Results: The results show that the maximum specific growth rate at 20°C was 0.1569 day^{-1}, and the maximum biomass production of microalgae at 25°C was 2.2667 g/L. The highest percentage of biomass conversion into lipid (35.71%) occurred at 30°C. Maximum lipid productivity was seen at temperatures of 15°C, 20°C, and 25°C, but the analysis of fatty acids in the three temperatures shownare maximum accumulations of triglycerides in the microalgae cells at 20°C and 25°C.
Conclusion: In the cultivation of *Nannochloropsis oculata*, the optimal temperature range for maximum efficiency in biodiesel production from lipids is 20°C to 25°C.
Keywords: Freshwater microalga, *Nannochloropsis oculata*, Cultivation temperature, Lipid accumulation

***Correspondence to:**
Mohammad Malakootian
Email: m.malakootian@yahoo.com

Introduction

Finding sufficient supplies of clean energy for the future is one of society's most daunting challenges and is intimately linked with global stability, economic prosperity, and quality of life (1). Fuels represent around 70% of total global energy requirements, particularly in transportation, manufacturing, and domestic heating. At present, electricity accounts for only 30% of global energy consumption (2). In the European Union (EU), the transport sector is responsible for almost one quarter of greenhouse gas emissions (3). Concerns about decreasing fossil fuel sources, especially oil and natural gas, along with climate changes and global warming necessitate giving more attention to renewable energy (2, 4). Recently, biodiesel (green fuel) as a renewable and non-toxic energy which produced low amounts of pollution has been proposed as a replacement for fossil fuels. Its use leads to a decrease in the harmful emissions of carbon monoxide, hydrocarbons, and particulate matter and to the elimination of SO_x emissions, with a consequent decrease in the greenhouse effect, in line with the Kyoto Protocol agreement (5-8). Microalgae are photosynthetic organisms that have an extraordinary potential for cultivation as energy crops. They are also able to produce a wide range of commercially interesting by-products such as fats, oils, sugars, pigments, and many other functional bioactive compounds (9-12). The microalgal lipid productivity/biomass is about 15–300 times that of conventional crops (6). Production and accumulation of lipids by microalgae is one of the important sources of biodiesel (13). Microalgal oil is mostly accumulated as triglycerides, and its reaction with simple alcohols (a chemical reaction known as "transesterification") leads to the formation of a chemical compound known as an alkyl ester, more generically known as biodiesel (10,14,15). Several types of microalgae are able to accumulate a large amount of lipids in their cells. *Nanno-*

*chloropsis oculata*is a marine unicellular alga of the class Eustigmatophyceae and is most often considered (16-19). The growth rate, cell density, and lipid content of the microalgae are essential parameters contributing to lipid productivity and reducing the total cost of biodiesel production (6,20). Therefore, determining the optimum microalgae culture conditions helps increase lipid productivity, promoting biodiesel investigation levels and its usage on an industrial scale (20,21). The quantity and proportion of different types of lipids within *Nannochloropsis* cells under a variety of growth conditions (salinity, light intensity) and medium specifications (nitrogen density, phosphate, and iron) have been determined (14,22,23). One of the important factors of microalgae cultivation is temperature; it affects the structural components of the cells (particularly lipids and proteins), controls the basic rates of all chemical reactions, and causes significant changes in metabolic regulatory mechanisms, specificity of enzyme reactions, cell permeability, and cell composition (18). This typicality will ultimately be effective on biomass and lipid production rates. There is no consistent pattern in relation to the effect of temperature on lipid production. Increasing culture temperature in some microalgae causes an increase in lipid rates; however, low temperatures cause greater lipid production in *Nibshiapleacea* microalgae (19).

Studies done on *Nannochloropsis oculata* have only investigated the effects of various factors on biomass production or cellular lipid rate; less attention has been given to lipid productivity. The current study investigated the effect of temperature on lipid productivity as one of the effective factors in economical biodiesel productivity. Moreover, this study analyzed the lipid extracted from microalgae for the quality and quantity of its compositions, and its applicability for biodiesel production was investigated.

Methods

The current study was conducted during the second half of 2012 at the Environmental Health Engineering Research Center of Kerman University of Medical Science using laboratory-scale pilot plants.

Nannochloropsis oculata

Stock culture of *Nannochloropsis oculata* was obtained from the Research Institute for Aquaculture in the south of Iran. This microalgae is a eukaryotic photosynthetic microorganism and, given its simple structure, has a fast growth rate (22). An image of *Nannochloropsis oculata* is shown in Figure 1. In their study conducted in China, Chen et al used high density microalgae (25×10^6 cell/ml) to determine cell density on microalgae growth and lipid composition (9).

Culture conditions

Stock culture was cultivated in a liquid Walne medium ($MnCl_2$, $CoCl_2$, $CuSo_4$, $FeCl_3$, $(NH_4)_6Mo_7O_{24}$, $ZnCl_2$, $NaNO_3$, H_3BO_3, Na_2EDTA, NaH_2PO_4), and simulated sea water was attained by dissolving sea salt in sterile water.

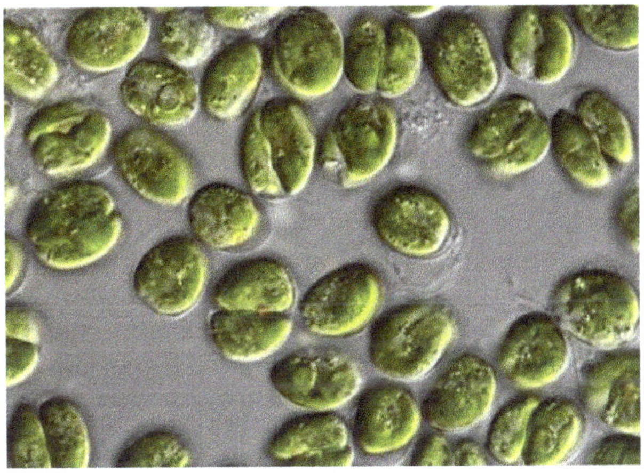

Figure 1. Image of *Nannochloropsis oculata* used in this study.

All tests were repeated three times in a ficolab equipped with temperature and light intensity controls under sterile conditions and in a 2-liter Meyer flask. In this study, the effect of temperature on microalgae was investigated at seven temperatures (5°C, 10°C, 15°C, 20°C, 25°C, 30°C, and 35°C) that were controlled with a thermometer. According to the studies of Sen et al and Banerjee et al, continuous illumination at 70 $\mu E\ m^{-2}\ S^{-1}$ with a 12-hour light cycle and medium salinity of 25 ppt were considered in this study (24,25). Fluorescent lamps (FL20D, OSRAM, Korea) were used as the light source for microalga growth. Sterile-air containing 2% (v/v) CO2 and filtered using glass fiber was aerated into the column through an air sparger at the bottom of the column at a flow rate of 0.5 v/v.

Microalgae cell counting and dry weight

Cell density (cells mL^{-1}) was measured using a UV/Visible spectrophotometer (Shimadzu Corporation). Absorbance of cell density in wavelengths 100 to 800 nm was measured. Maximum absorbance was observed at wavelength 680 nm. Each sample was diluted to give an absorbance in the range of 0.1–1.0, if optical density was greater than 1.0. Microalgae dry weight per liter (g L^{-1}) was measured according to the method previously reported. Microalgae cells were collected by centrifugation of wet biomass for 30 minutes at 15°C with 3000 RPM. The dry weight of marine microalgae samples was affected by the salt absorbed on the cell surface, and its presence in the intercellular water ensured error in estimating the amount of biomass. This explains the differentiations in the amount of dry cell weight in various papers. Hence, before gravimetric analysis was performed, the salts were removed by again solving the centrifuged cells in 200 ml ammonium formate (0.5M, pH 8.0, adjusted with 1M NaOH) and centrifuging under the mentioned conditions. Microalga pellets were dried at 100°C for 4 hours for dry weight measurements. The specific growth rate of microalgae in the logarithmic phase was calculated as follows:

$$\mu = \frac{lnN_f - lnN_i}{t_2 - t_1} \quad \text{(Eq. 1)}$$

where μ (day^{-1}) is the specific growth rate, lnN_i and lnN_f

are cell densities (cell/ml) at the beginning and end of the logarithmic growth phase, respectively, and t is the time (day) (26).

Extraction and measurement of lipid content and TAG

The methanol-chloroform (1/1, V/V) extraction method was used to extract total lipids from the dried cells. To remove residual microalgae, the extraction lipids were filtered on membranes with 0.45 µm mean pore size. After washing twice with methanol and its complete evaporation, gravimetric analysis was done, and part of the lipid fraction was expressed as the percentage of dry cell weight. Lipid productivity was calculated by:

$$P_{Lipid}\left(gl^{-1}day^{-1}\right) = \frac{(C_f \times DCW_f)-(C_i \times DCW_i)}{T} \quad \text{(Eq. 2)}$$

Where P_{Lipid} is lipid productivity; C_f and DCW_f are lipid content (g/g) and biomass (g/l) of the microalgae in the final stationary growth phase, respectively; Ci and DCW_i are lipid content (g/g) and biomass (g/l) of the microalgae in the initial stationary growth phase, respectively; and T is cultivation time (day). After total lipid amount was measured, the dried lipid was solved in 0.4 ml of isopropyl alcohol, and TAG was estimated by an enzymatic colorimetric method using a commercial kit from BHKT Clinical Reagent Co., Ltd., No. 2400076 (27).

Determination of fatty acid profiles

The direct esterification method was used to measure fatty acid properties. A mixture of 100 mg lyophilized microalgae and 8 ml KOH was sonicated for 3 minutes. For saponification, the mixture was heated to 100°C for 15 minutes and then cooled to room temperature. For esterification, 8 ml of 0.7 NHCl in methanol and BF_3/CH_3OH was added to the mixture (14% V/V), and it was again heated to 100°C for 15 minutes. After cooling to room temperature, emulsification was avoided by adding 2 ml of a saturated solution of sodium chloride to the mixture. FAMEs were extracted by adding aliquots of n-hexane. The FAMEs in the hexane layer were analyzed using standard gas chromatography (Agilent technologies 7890A-5975c) with a capillary column and a flame ionization detector. Nitrogen was used as the carrier gas and delivered at a rate of 1.5 mL min^{-1}. The temperature was programmed to increase from 130°C to 180°C at a rate of 10°C min^{-1} and thereafter ramped to 210°C at a rate of 2°C min^{-1}. The injector and detector were kept at 220°C and 250°C, respectively (27).

Statistical analysis

Data was analyzed using one-way analysis of variance (ANOVA).To determine the statistical difference between temperatures, the Tukey test was used. A value of $P<0.05$ was considered statistically significant (16,28).

Results

Effects of temperature on cell growth rate of *Nannochloropsis oculata*

Figure 2 shows the cell growth rate of *N. oculata* in different cultivation temperatures. No increase was observed in the number of microalgae cells at temperatures of 5°C and 35°C. At 20°C and 25°C, the microalgae cells entered into the logarithmic growth phase after one day, and the maximum cell growth rate was seen at the end of the stationary growth phase. At temperatures of 10°C and 15°C, the microalgae cells entered into the logarithmic growth phase after three days and at the beginning of the fourth day. The results also showed that microalgae cells had very low growth at the temperature of 30°C. As seen in Figure 2, maximum cell growth rates were seen at the temperatures of 20°C and 25°C. The results of one-way ANOVA showed that the differences between cell growth rates at different temperatures were significant ($P<0.05$). Tukey test results confirmed that the cells were not put in homogeneous groups and growth rates at temperatures of 20°C and 25°C had significant differences with temperatures of 10°C ($P=0.001$, $P=0.000$) and 15°C ($P=0.043$, $P=0.01$). There was no significant statistical difference in growth rates between temperatures of 20°C and 25°C ($P=0.599$), nor between 10°C and 15°C ($P=0.725$).

Effects of temperature on biomass and lipid production

As seen in Figure 3, the results of tests of temperature effects on biomass production in *Nannochloropsis oculata* at the end of the stationary growth phase showed that the maximum amount of biomass was produced at the temperature of 25°C. Because of unsufficient microalgae growth, there were no differences between the initial and final biomass amounts at 5°C, 30°C, and 35°C. The results of one-way ANOVA showed that the differences between

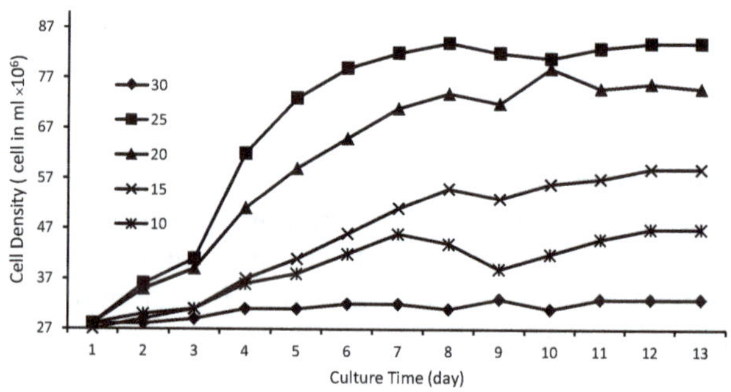

Figure 2. Cell growth rate of *Nannochloropsis oculata* in different cultivation temperatures.

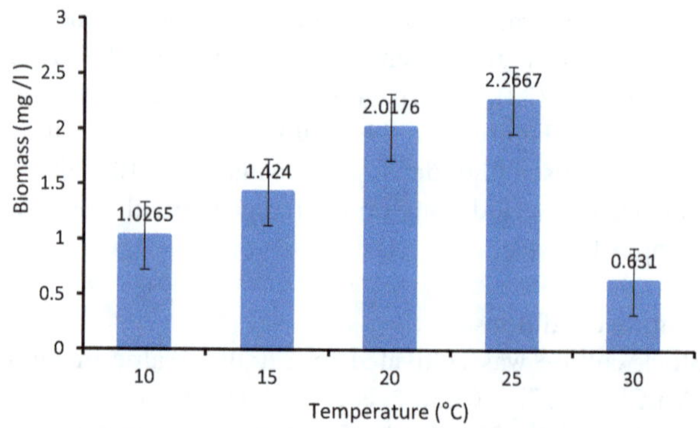

Figure 3. Biomass production by Nannochloropsis oculata in different cultivation temperatures.

biomass production rates at different temperatures were significant ($P=0.000$), and Tukey test results confirmed that the cells were not put in homogeneous groups.

Figure 4 shows that, at the 20°C and the end of the stationary growth phase, despite the large amount of biomass production, the percentage of converting biomass to lipid was minimum. Moreover, at 30°C, despite the small amount of biomass production, the percentage of converting biomass to lipid was maximum. In this study, the results of one-way ANOVA showed that lipid production in the stationary growth phase and at 20°C was significantly different from other temperatures ($P=0.000$); the minimum lipid content was seen at this temperature (16.93%), and the maximum amount of biomass to lipid conversion was seen at 30°C.

Effect of temperature on fatty acids composition

The results of the fatty acids composition of *N. oculata* grown in different cultivation temperatures are shown in Table 1.

Discussion

Effects of temperature on cell growth rate

Temperature is a very sensitive factor for microalgal growth and metabolic activities in microalgal cells. It is also an easy-control factor in the practical operation of microalgae cultivation (6,29). In this study, a different temperature regime was used and temperature was considered as an independent variable. Most microalgae respond to increases in temperature with increased exponential growth rates until reaching their optimum temperature, after which point loss of structural integrity tends to lead to sharp declines in growth rates (30). In this study, cell density increased along with temperature increases from 10°C to 25 °C and reached its maximum growth at 25°C

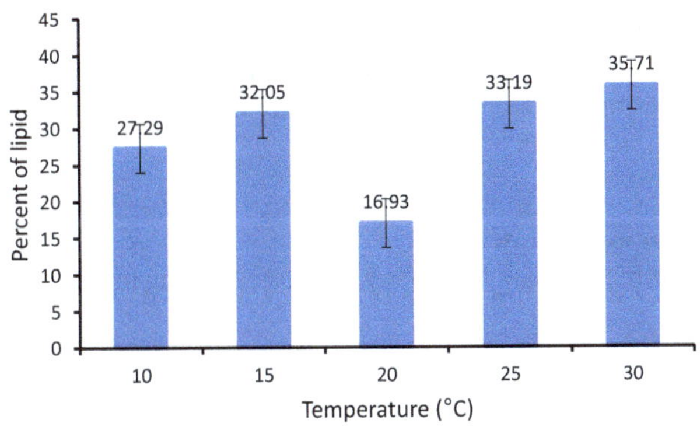

Figure 4. Lipid production by *Nannochloropsis oculata* in different cultivation temperatures

(Figure 2). Abu-reza et al referred to 24°C to 26°C as the optimal temperature for *Nannochloropsis* growth (31). Sukenik et al in their study on *Nannochloropsis* sp. referred to the temperature of 25°C as the optimal temperature for growth (32). Wagenen et al also studied *Nannochloropsis salina* and referred to the temperature of 23°C as the optimal temperature for growth; this result matches the results of the current study (30).

Investigations of the special growth rate in the studied temperature regime showed that the maximum special growth rate in the logarithmic growth phase was at 25°C and was 0.1569 day^{-1}; for 20°C it equaled 0.1388 day^{-1}, for 15°C it equaled 0.1016 day^{-1}, and for 10°C, it equaled 0.0645 day^{-1}. In their studyon *Nannochloropsis oculata*, Converti et al found that by decreasing the temperature from 20°C to 15°C, the special growth rate decreased from 0.13 to 0.06 day^{-1}. By increasing temperature to 25°C, the special growth rate decreased to 0.07 day^{-1}; this result matched the findings of the current study (22). In a study by Chiu et al, after 6 to 8 cultivation days of *Nannochloropsis oculata* in F/2 medium, light intensity of 300 $\mu E\ m^{-2}\ S^{-1}$, and a temperature of 26°C, the special growth rate reached 0.571 day^{-1} (15). Banerjee et al cultivated *Nannochloropsis oculata* in Walne medium with a light intensity of 150 $\mu E\ m^{-2}\ S^{-1}$ and a temperature of 20°C. They attained a special growth rate of 0.004 day^{-1} (25). The differences in special growth rates between studies could have been caused by differences in cultivation type, light intensity, previous microalgae medium, photo-bio reactor type or medium salinity (14,17).

Effects of temperature on biomass production

Microalgae are unicellular photosynthetic organisms that use light energy and carbon dioxide with a higher photosynthetic efficiency than plants for the production of biomass (20). In this study, as shown in Figure 3, the maximum amount of biomass was attained in the stationary growth phase at a temperature of 25°C (2.2667 g/L). Chiu et al, after 6 to 8 cultivation days of *Nannochloropsis oculata* in F/2 medium and using 2% carbon dioxide at 26°C, could produce 1.277 g of biomass per liter (15). In the study of Su et al on a single-stage batch culture of *Nannochloropsis oculata* in Walne medium and at a temperature of 25°C through the logarithmic growth phase, the biomass density was about 0.58 g/L which increased in the stationary growth phase to 0.654 g/Ll (33).

Effects of temperature on lipid productivity

Environmental factors such as temperature, light, pH, salinity, and nutrient status of the culture medium not only

Table 1. Fatty acid composition in dry weight, percentage of *Nannochloropsis oculata* grown in 15, 20, 25°C and end of stationary growth phases

Fatty acid	Scientific name	15°C	20°C	25°C
C14:0	Myristic acid	2.16± 0.04	6.02 ± 0.03	6.17 ± 0.19
C16:0	Palmitic acid	28.66 ± 0.84	33.57 ± 0.85	39.93 ± 0.09
C16:1 n-7	Palmitoleic acid	35.17 ± 0.29	34.92 ± 0.1	29.56 ± 0.41
C18:0	Stearic acid	1.87 ± 0.01	1.87 ± 0.00	1.86 ± 0.00
C18:1 n-9	Oleic acid	5.53 ± 0.13	4.07 ± 0.05	4.13 ± 0.01

affect the photosynthesis and productivity of algal cells, but also influence the pattern, pathway, and activity of cellular metabolism and, thus, dynamic cell composition (20). Temperature is one of the effective factors on lipid accumulation in cell and fatty acid percentages in lipids. The results of this study showed that the highest biomass to lipid conversion was achieved at 30°C. These results match those of Sukenik et al (32). In a study by Converti et al on *Nannochloropsis oculata*, lipid content was increased from 7.9% to 14.9% by increasing cultivation temperature from 20°C to 25°C (22). These results also match those of the current study in which increasing the temperature from 20°C to 25°C increased lipid content from 16.93% to 33.19%. In the study by Enebo and Iwamoto on *Rhodotorula gracilis*, however, increasing cultivation temperature from 27°C to 35°C decreased the lipid content (34). That result showed that the effects of temperature on different types were not similar. Increasing lipid accumulation with increasing cultivation age (reaching the stationary growth phase) was caused by cell division and the diversion of photosynthetic energy into TAG production (14).

The results showed that more cell biomass production by microalgae did not cause higher lipid production. Thus, lipid content or biomass amount was not singly a proper scale for microalgae lipid productivity. The roles of both factors should have been considered beside each other using Eq. 2 (19). The results of the current study showed that lipid productivity at 30°C, 25°C, 20°C, 15°C, and 10°C was, sequentially, 0.01647, 0.0395, 0.0394, 0.03841, and 0.0206 gl^{-1} day^{-1}. These results show that the maximum lipid productivity was at 25°C, 20°C, and 15°C and was significantly more than lipid productivity at 10°C and 30°C. Furthermore, there was no difference in lipid productivity at 15°C, 20°C, and 25°C. These results match those of Converti et al (22), Griffiths and Harrison (7), and Huerlimann (14).

Effects of temperature on fatty acid composition

An important characteristic of microbial fat production is the possibility of changing the fatty acid composition by altering culture conditions. Onefactor affecting fatty acid composition is temperature. Under unfavorable conditions, microalgal growth is arrested, photosynthetic activity decreases, and excess energy might be stored in the form of lipids and/or carbohydrates (28). The microalgae oil in the form of triglycerides is convertible to biodiesel, and vegetable oils currently used for biodiesel are mainly C16 and C18 (7). In a study by Olofsson et al (5), the fatty acids appropriate for producing biodiesel were identified as myristic acid, palmitic acid, palmitoleic acid, stearic acid, and oleic acid, which comprise more than 45%-78% of total fatty acids. As seen in Table 1, in this study, these compositions comprised more than 70%-80% of fatty acid compositions at the temperatures of 15°C, 20°C, and 25°C. Furthermore, at all studied temperatures, the palmitic acid and palmitoleic acid comprised more than 60%-70% of fatty acids; these results match those of Kleinschmidt et al (29).

The TAG of *Nannochloropsis* consist mainly of saturated and monounsaturated fatty acids and are stored in vacuoles within the cell (5, 28). Most microalgae accumulate very little TAGs during exponential growth, but substantial amounts of TAGs can accumulate during the stationary phase (5). As the result, in this study, the fatty acid composition analysis was done in the microalgae stationary growth phase. According to Table 1 of this study, TAG content at the end of the stationary growth phase equaled 73.39% at 15°C and was significantly less than the TAG amount at 20°C (80.45%) and 25°C (81.65% fatty acid). In a study on *Nannochloropsis* sp., Wang and Wang achieved a TAG amount of 41.1% of fatty acids (35). The EN14214 standard for biodiesel quality attached the limitation of less than 12% for the composition of linolenic acid. This composition in the current study at all temperatures was less than one percent of fatty acids.

Conclusion

The amount of biomass and lipid produced in a cell was not alone a proper criterion to choose microalgae for economical biodiesel production. However, the maximum *Nannochloropsis oculata* lipid productivity was achieved at temperatures of 15°C, 20°C, and 25°C. The results of fatty acid analyses at the aforementioned temperatures showed that the triglyceride amount was significantly more at the temperatures of 20°C and 25°C than at 15°C. Therefore, temperatures of 20°C to 25°C were the best temperatures for lipid productivity and biodiesel production from *Nannochloropsis oculata* microalgae.

Acknowledgements

The current work was financially supported by the Research and Technology Assistance Division of Kerman University of Medical Sciences, Iran. This work was also partially supported by the Environmental Health Engineering Research Center of Kerman University of Medical Sciences, Iran.

Competing interests

Authors declare that they have no competing interests.

Authors' contributions

SD and BH conceived and designed the study. MM and AR performed the literature search and wrote the manuscript. All authors participated in data acquisition, analysis, and interpretation. All authors critically reviewed, refined, and approved the manuscript.

References

1. Taylor RL, Rand JD, Caldwell GS. Treatment with algae extracts promotes flocculation, and enhances growth and neutral lipid content in Nannochloropsis oculata--a candidate for biofuel production. Mar Biotechnol (NY)

2012; 14(6): 774-81.
2. Xiaodong D, Yajun L, Xiaowen F. Effects of selective medium on lipid accumulation of chlorellas and screening of high lipid mutants through ultraviolet mutagenesis. African Journal of Agricultural Research 2011; 6(16): 3768-74.
3. Gouveia L, Oliveira AC. Microalgae as a raw material for biofuels production. J Ind Microbiol Biotechnol 2009; 36(2): 269-74.
4. Rajasri Y, Ramgopal RS, Rao CS. Lipid accumulation studies in chlorella pyrenoidosa using customized photobioreactor-effect of nitrogen source, light intensity and mode of operation. International Journal of Engineering Research and Applications 2012; 2(3): 2446-53.
5. Olofsson M, Lamela T, Nilsson E, Bergé JP, del Pino V, Uronen P, et al. Seasonal variation of lipids and fatty acids of the microalgae nannochloropsis oculata grown in outdoor large-scale photobioreactors. Energies 2012; 5(5): 1577-92.
6. Li X, Hu HY, Zhang YP. Growth and lipid accumulation properties of a freshwater microalga Scenedesmus sp. under different cultivation temperature. Bioresour Technol 2011; 102(3): 3098-102.
7. Griffiths MJ, Harrison ST. Lipid productivity as a key characteristic for choosing algal species for biodiesel production. J Appl Phycol 2009; 21(5): 493-507.
8. Li Y, Horsman M, Wang B, Wu N, Lan C. Effects of nitrogen sources on cell growth and lipid accumulation of green alga Neochloris oleoabundans. Appl Microbiol Biotechnol 2008; 81(4): 629-36.
9. Chen Y, Wang J, Liu T, Gao L. Effects of initial population density (IPD) on growth and lipid composition of Nannochloropsis sp. J Appl Phycol 2012; 24(6): 1623-7.
10. Karampudi S, Chowdhury K. Effect of Media on Algae Growth for Bio-Fuel Production. Notulae Scientia Biologicae 2011; 3(3): 33-41.
11. Banskota AH, Stefanova R, Gallant P, McGinn PJ. Mono- and digalactosyldiacylglycerols: potent nitric oxide inhibitors from the marine microalga Nannochloropsis granulata. J Appl Phycol 2013; 25(2): 349-57.
12. Nigam S, Rai MP, Sharma R. Effect of nitrogen on growth and lipid content of chlorella pyrenoidosa. Am J Biochem Biotechnol 2011; 7(3): 124-9.
13. Lee SJ, Yoon BD, Oh HM. Rapid method for the determination of lipid from the green alga Botryococcus braunii. Biotechnology Techniques 1998; 12(7): 553-6.
14. Huerlimann R, de Nys R, Heimann K. Growth, lipid content, productivity, and fatty acid composition of tropical microalgae for scale-up production. Biotechnol Bioeng 2010; 107(2): 245-57.
15. Chiu SY, Kao CY, Tsai MT, Ong SC, Chen CH, Lin CS. Lipid accumulation and CO2 utilization of Nannochloropsis oculata in response to CO2 aeration. Bioresour Technol 2009; 100(2): 833-8.
16. Durmaz Y. Vitamin E (α-tocopherol) production by the marine microalgae Nannochloropsis oculata (Eustigmatophyceae) in nitrogen limitation. Aquaculture 2007; 272(1-4): 717-22.
17. Spolaore P, Joannis-Cassan C, Duran E, Isambert A. Optimization of Nannochloropsis oculata growth using the response surface method. J Chem Technol Biotechnol 2006; 81(6): 1049-56.
18. Sandnes JM, Källqvist T, Wenner D, Gislerød HR. Combined influence of light and temperature on growth rates of Nannochloropsis oceanica: linking cellular responses to large-scale biomass production. J Appl Phycol 2005; 17(6): 515-25.
19. Hu H, Gao K. Response of growth and fatty acid compositions of Nannochloropsis sp. to environmental factors under elevated CO2 concentration. Biotechnol Lett 2006; 28(13): 987-92.
20. Huang X, Huang Z, Wen W, Yan J. Effects of nitrogen supplementation of the culture medium on the growth, total lipid content and fatty acid profiles of three microalgae (Tetraselmis subcordiformis, Nannochloropsis oculata and Pavlova viridis). J Appl Phycol 2013; 25(1): 129-37.
21. Simionato D, Sforza E, Corteggiani Carpinelli E, Bertucco A, Giacometti GM, Morosinotto T. Acclimation of Nannochloropsis gaditana to different illumination regimes: Effects on lipids accumulation. Bioresour Technol 2011; 102(10): 6026-32.
22. Converti A, Casazza AA, Ortiz EY, Perego P, Del Borghi M. Effect of temperature and nitrogen concentration on the growth and lipid content of Nannochloropsis oculata and Chlorella vulgaris for biodiesel production. Chemical Engineering and Processing: Process Intensification 2009; 48(6): 1146-51.
23. Srinivas R, Ochs C. Effect of UV-A Irradiance on Lipid Accumulation in Nannochloropsis oculata. Photochem Photobiol 2012; 88(3): 684-9.
24. Sen B, Koser M, Alp M, Erbas H. Studies on growth of marine microalgae in batch cultures: III. Nannochloropsis Oculata (Eustigmatophyceae). Asian Journal of Plant Science 2005; 4(6): 642-4.
25. Banerjee S, Hew WE, Khatoon H, Shariff M, Fatimah M. Growth and proximate composition of tropical marine Chaetoceros calcitrans and Nannochloropsis oculata cultured outdoors and under laboratory conditions. Afr J Biotechnol 2011; 10(8): 1375-83.
26. Malakootian M, Hatami B, Dowlatshahi S, Rajabizadeh A. Evaluation of factors affecting on lipid extraction for recovery of fatty acids from Nannochloropsis oculata micro-algae to biodiesel production. Environmental Health Engineering and Management Journal 2014; 1(1): 19-24.
27. Malakootian M, Hatami B, Dowlatshahi S, Rajabizadeh A. Optimization of culture media for lipid production by Nannochloropsis oculata for Biodiesel production. Environmental Health Engineering and Management Journal 2015; 2(3): 141-7.
28. Pal D, Khozin-Goldberg I, Cohen Z, Boussiba S. The effect of light, salinity, and nitrogen availability on lipid production by Nannochloropsis sp. Appl Microbiol Biotechnol 2011; 90(4): 1429-41.
29. Kleinschmidt M, McMahon VA. effect of growth temperature on the lipid composition of Cyanidium caldarium I. Class separation of lipids. Plant Physiol 1970; 46(2): 286-9.
30. Wagenen J, Miller T, Hobbs S, Hook P, Crowe P, Huesemann M. Effects of Light and Temperature on Fatty Acid Production in Nannochloropsis Salina. Energies 2012; 5(3): 731-40.
31. Abu-Rezq T, Al-Musallam L, Al-Shimmari J, Dias P. Optimum production conditions for different high-quality marine algae. Hydrobiologia 1999; 403: 97-107.
32. Sukenik A, Zmora O, Carmeli Y. Biochemical quality of marine unicellular algae with special emphasis on lipid composition. II. Nannochloropsis sp. Aquaculture 1993; 117(3): 313-26.
33. Su CH, Chien LJ, Gomes J, Lin YS, Yu YK, Liou JS, et al. Factors affecting lipid accumulation by Nannochloropsis oculata in a two-stage cultivation process. J Appl Phycol 2011; 23(5): 903-8.
34. Enebo L, Iwamoto H. Effects of Cultivation Temperature on Fatty Acid Composition in Rhodotorula gracilis. Acta Chem Scand 1966; 20(2): 439-43.
35. Wang G, Wang T. Characterization of Lipid Components in Two Microalgae for Biofuel Application. J Am Oil Chem Soc 2012; 89(1): 135-43.

Identification of clinically antibiotic resistant genes *Aac(3)-IIa* and *Aac(6')-Ib* in wastewater samples by multiplex PCR

Naser Samadi[1], Rahim Aali[2*], Esrafil Asgari[3], Hamed Mirhosaeini[4], Ali Shahriari[5], Farhad Mahmoodi[6], Fardin Nouri[7], Asad Hamdi[7], Farzad Fanaie[7], Shahla Yosefi[7], Saied Dadashi[7]

[1]MSc of Cellular and Mulecular Research Center, Faculty of medicine, Urmia University of Medical Sciences, Urmia, Iran
[2]Assistant Professor of Department of Environmental Health Engineering, Faculty of Nursing and Health, Urmia University of Medical Sciences, Urmia, Iran
[3]Instructor of Department of Environmental Health Engineering, Faculty of Nursing and Health, Urmia University of Medical Sciences, Urmia, Iran
[4]Instructor of Medical Microobiology, Department of Environmental Health Engineering, Faculty of Health, Arak University of Medical Sciences, Arak, Iran
[5]Assistant Professor of Department of Environmental Health Engineering, Environmental Health Research Center, Golestan University of Medical Sciences, Gorgan, Iran.
[6]PhD student of Department of Environmental Engineering, Ahvaz Branch, Islamic Azad University, Ahvaz, Iran
[7]BSc of Environmental Health, Student Research Committee, Urmia University of Medical Sciences, Urmia, Iran

Abstract

Background: Aminoglycoside antibiotics are widely used in medical centers, particularly to treat infections. The resistance developed against these agents is a huge concern in health care. A number of researchers have reported that hospital and municipal wastewaters are among the most important dissemination sources of these agent into the environment. Some, however, do not agree with this opinion. In the present study, the prevalence of aminoglycoside resistance genes was investigated in raw and effluent wastewater from hospital and municipal wastewater treatment plants.

Methods: To conduct this descriptive-analytical study, 30 samples were taken according to sampling principles and cold cycle and transferred to the molecular laboratory. DNA was extracted by the freeze-thaw method using a kit (Promega). The genes *aac(3)-IIa* and *aac(6')-Ib* which code aminoglycoside resistance were examined in this study.

Results: The results indicated that the studied genes are present in 35% of urban and hospital wastewaters, and their frequency percentage is higher in hospital wastewater (52%) than urban wastewater (48%). The studied genes were identified in 61% of raw hospital wastewater samples; however, they were not detected in the output wastewater from the studied treatment plants.

Conclusion: Although, the studied genes were not detected in the final effluent, there is a high potential for their release into the environment. The current study demonstrated that the coding genes of aminoglycoside antibiotic resistance are present in raw urban and hospital wastewaters. In the case of improper exploitation of wastewater treatment plants, the output water can contaminate other environmental sections, such as soil and water resources, and result in the emission of these contaminants.

Keywords: Antibiotic resistance, Aminoglycosides, Urban and hospital wastewaters

***Correspondence to:**
Rahim Aali
Email: Aali@hlth.mui.ac.ir

Introduction

Excessive use of antibiotics in the treatment of humans and animals leads to antibiotic resistance (1-3). Antibiotic resistant bacteria (ARB) and antibiotic resistant genes (ARGs) cause serious problems in the treatment of infectious diseases. They enter the environment in different ways. Urban and hospital wastewaters have a high potential for the release of resistant bacteria and genes into the environment (4,5). Such release increases concerns about public health (4). Antibiotic resistance in the environment can be transferred to nonpathogenic bacteria from pathogenic bacteria and increase pressure on the aquatic ecology through physiological changes and population mobility (4-6). Research has shown that a large number of ARB and

ARGs, in particular hospital strains, have been found in regions where hospital wastewaters have been discharged (7). Importantly, some of these contaminated water resources are used for public water supplies. Some investigations have revealed that discharge of hospital wastewater is associated with the widespread dissemination of resistant bacteria in the environment (1). Other studies have demonstrated that urban wastewaters are important means through which resistant bacteria and genes are discharged into the environment (8). Some researchers have shown that resistant bacteria and genes are rarely removed from wastewater in treatment plant facilities (3,4). Some studies have even indicated that the number of resistant bacteria increases in such facilities (9).

Aminoglycosides are a group of medications that have similar chemical, antimicrobial, pharmacological, and toxic properties. Currently, this group includes streptomycin, gentamicin, tobramycin, kanamycin, amikacin, netilmicin, neomycin, sizomycin, and others (10). Aminoglycoside antibiotics are widely used in medical centers to treat life-threatening infections caused by gram-negative bacteria (10,11). Resistance to aminoglycosides is often caused by drug enzymatic changes by phosphorylation, adenylation, and acetylation of amine and hydroxyl groups, the most common mechanism of antibiotic resistance in bacteria (12). Among enzymes that change aminoglycosides, the *aac(6')-Ib* is predominant and causes resistance to many antibiotics. This gene was first identified in 1986 in *Klebsiella pneumoniae* isolates (13). Although the production of aminoglycoside-changing enzymes is the most common mechanism for producing resistance to aminoglycosides, the gram-negative bacteria that produce 16S rRNA methylase have become increasingly prevalent in recent years (14), and such widespread resistance to aminoglycosides has become a huge challenge to the medical profession (15). Filipova et al (16) and Chandrakanth et al (17) reported aminoglycoside resistant bacteria isolated from clinical environments. The *aac* gene is distributed in different microorganisms isolated from polluted waters (18-20). Szczepanowski et al (21) reported that aminoglycoside resistance genes prevailed in clinical compartments and can be spread to the environment through hospital wastewater. Considering the risk of these pollutant genes entering the environment, this study examined the prevalence of the *aac(3)-IIa* and *aac(6')-Ib* resistant genes in raw and wastewater effluent and determined the role of urban wastewater treatment plants in removing/reducing *aac* genes.

Methods

Preparation of samples

A certain volume of the main samples (50 ml) was taken and centrifuged for 10 minutes at 6000 rpm. The sediment was extracted, 300 μl of distilled water was added, and the solution was subjected to freeze and thaw using liquid nitrogen and boiling water, each for 1 minute periodically at 4-5 stages (22).

DNA extraction

DNA was extracted using a Promega Wizard Genomic DNA purification kit, (Madison, WI) in 3 stages. In the lysing stage of the cell, pipetting was performed well after the addition of cell lyse solution (600 μl). The solution was then incubated at 80°C for 5 minutes. Finally, RNase was added, and the solution was incubated again for 15-60 minutes at 37°C. At the protein precipitation stage, 200 μl of precipitation solution was added and incubated for 5 minutes on ice, followed by centrifugation at 13000-16000 rpm for 3 minutes. At the DNA precipitation stage, the top liquid was transferred to a tube containing 600 μl of isopropanol at room temperature and mixed. Afterwards, 600 μl of 70% ethanol was added at room temperature and the mixture was centrifuged for 2 minuts at 13 000-16 000 rpm. The ethanol was then removed and the tubes dried under ambient conditions. DNA was suspended in 100 μl of distilled water and kept for 1 night at 4°C; it was ultimately maintained at -20°C until molecular experiments were performed. The concentration of the extracted DNA was read by spectrophotometry (Nanodrop ND 1000).

PCR method

The pair of primers used to amplify the *aac(3)-IIa* and *aac(6')-Ib* genes are listed in Table 1. PCR amplification was conducted at a volume of 25 λ containing 2.5 μl buffer 10X PCR, 1 μl of each primer, 0.75 μl of $MgCl_2$, 0.5 μl of dNTP, 0.35 μl of tag polymerase, 2 μl of the DNA template, and 14.9 μl of deionized water. All PCR experiments had positive and negative controls. The process was performed using the initial denaturation stage for 5 minutes at 94°C, 35 cycles at 94°C for 45 seconds (denaturation), at 53°C for 30 seconds (annealing), and at 72°C for 1 minute (extension). The final extension was performed at 72°C for 10 minute. PCR products were investigated using electrophoresis (1.5%) and DNA safe stain. They were then de-

Table 1. Characteristics of primers

Primers	Target gene	Sequence (5'-3')	Amplified size (bp)	Annealing temperature (°C)	References
F R	*Aac*(3)-IIa	F: 5-CGGAAGGCAATAACGGAG- 3 R: 5 -TCGAACAGGTAGCACTGAG -3	749 bp	53°C	(23)
F R	*Aac*(6')-Ib	F: 5'-ATG ACT GAG CAT GAC CTT G-3' R: 5'-AAGGGTTAGGCAACACTG-3'	524 bp	53°C	(23)

tected on Transluminar UV.

Data analysis
To compare the presence and absence of genes among various sites, the chi-square test was employed. McNemar test was utilized before and after each treatment plant.

Results
Amplified multiplex genes are illustrated in Figure 1. The results indicated that *aac(3)-IIa* and *aac(6')-Ib*genes were found in 35% of the samples (Figure 2A). The *aac(3)-IIa* (37%) was found in a larger number of samples than the *aac(6')-Ib* gene (33%) (Figure 2B). Distinctive analysis of urban and hospital wastewaters indicated that the detection percentage of the studied genes was higher in hospital wastewaters (52%) than in urban wastewaters (48%) (Figure 2C).
The results revealed that the *aac(3)-IIa* gene was present in 50% of the hospital wastewater samples, while *aac(6')-Ib* was observed in 42% of the samples. The studied genes were found in 28% of the urban wastewater samples. Investigations of the input and output of urban wastewaters indicated that, generally, 56% of the studied genes were found in the input. *aac(3)-IIa* and *aac(6')-Ib* were presented in the raw samples 56% and 44%, respectively. The studied genes were found in none of the effluent of urban wastewater treatment plants, but in 61% of the wastewater samples from hospitals. One of the studied hospitals had its own wastewater treatment plant; the results demonstrated that 83% of samples from this plant's raw wastewater contained the studied genes, but none of the genes were found in the effluent.

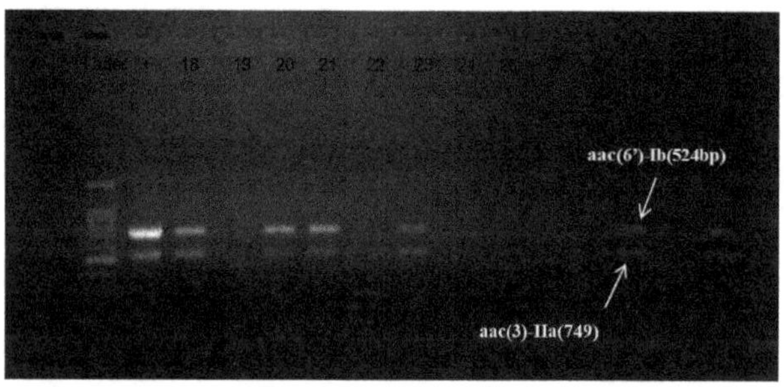

Figure 1. Aminoglycoside resistance genes, *aac(3)-IIa* (749) bp and *aac(6')-Ib*(524 bp).

Discussion
Aminoglycoside antibiotics are widely used in medical clinics, especially to treat life-threating infections caused by gram-negative bacteria (24-26). Because of the widespread incidence of aminoglycoside resistance and the resulting complications for the medical field (24), the present study investigated two aminoglycoside-resistance genes. Clinical genes resistant to aminoglycosides were found to be present in municipal wastewaters. In their studies, Guardabassi et al (27), Heuer et al (18), and Aali et al (8) also indicated that the genes resistant to aminoglycosides exist in environmental resources. The results of this study revealed the lowest percentage of clinically genes. The results of other studies also confirm these findings. Jakobsen et al (28) found *aac* genes in wastewater isolates and showed a possible dissemination of *aac* from hospital to wastewater. Aali et al (8) reported an association between the lowest number of bacteria and aminoglycoside-resistance genes. The reason can be attributed to less use of aminoglycoside antibiotics. Recently, the consumption of some members of the aminoglycoside family (such as

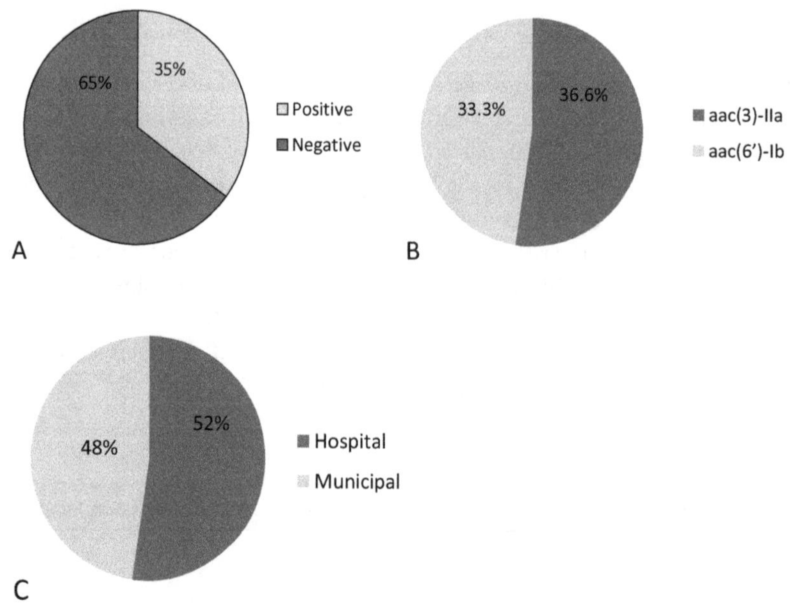

Figure 2. (A) Overall detection of ARGs, (B) total detection of *aac(3)-Iia* and *aac(6')-Ib*, and (C) differences between municipal and hospital wastewaters.

gentamycin) has been limited because of side effects; however, studies have indicated that resistance can develop even in low antibiotic concentrations (29). Heuer et al (18) attributed the diversity of the genes that code resistance to gentamycin (including *Aac3-I*) in different environments to the increased frequency of genetic transfer elements. The primary mechanism involved in resistance to gentamycin has been reported as enzyme deactivation, and over 50 types of enzyme have been identified. These enzymes, based on their biochemical action, include acetyltransferases, phosphotransferases, and nucleotidyl transferases (adenylyl tranferases) (16,17,30). These enzymes are coded by various genes, including *Aac*, *Aph*, and *Ant* (*Aad*). The mentioned enzymes and the coder genes have been identified in a wide range of isolated bacteria (*Aeromonas, Escherichia, Vibrio, Salmonella,* and *Listeria*) from hospital environments and contaminated waters (18-20). The *Aac3* gene has been identified as the coder of aminoglycoside-N-acetyl transferases in the microbial populations isolated from wastewater treatment facilities, river water, and agricultural regions (31,32). This can also be applicable to this study, yet complementary research is necessary to investigate the role of key factors in genetic transfer in the studied wastewaters.

The findings of the current study revealed that the percentages of *aac(3)-IIa* and *aac(6')-Ib* genes are higher in hospital wastewaters than in urban wastewaters. These results are consistent with those of other researchers. The high rate of aminoglycoside-resistance genes in hospitals can be caused by common unnecessary prescriptions for patients. This also demonstrates the aminoglycoside resistant mobile genetic agents' movements between clinical and environmental bacteria (18). Plasmid and integrons have main roles in the dissemination of these agents (30). In their study of the *Aac3-1* gene, Aali et al (8) reported that the genes and bacteria resistant to aminoglycosides are more prevalent in hospital than urban wastewaters. Their low percentage in urban wastewaters can be related to process design and operation. The *aac(3)-IIa*gene was identified in hospital wastewaters more than the *aac(6')-Ib* gene. This can be due to bacteria carrying or better expression of the *aac(3)-IIa* gene.

In the current study, contrary to other studies, neither *aac(3)-IIa* nor *aac(6')-Ib* genes were detected in effluent wastewater samples from different wastewater treatment plants. Szczepanowski et al (21) identified many aminoglycoside resistance genes (eg, *aacA*, *aacA1*, *aacA4*, *aacA7*, *aacA29b*, *aacC1*, *aacC2*, *aacC3*, *aacC4*, *aac(3)-Id*, and *aac[69]-Im*) in the final effluent of wastewater treatment plants. The absence of the studied resistance genes in the effluent wastewater can be due to the effect of the plant's disinfection unit, the shock caused by a huge organic load, or the specific conditions of the wastewater treatment plants.

Conclusion

Human contact with resistant bacteria and genes has been a health concern associated with the management of diseases. This contact can occur through various means such as environmental resources. Hospital and urban wastewater plays a great role in the spread of resistant genes and bacteria. Wastewater treatment plants are very important as a controlling barrier against these contaminants entering environmental resources. Therefore, the design, operation, and continual monitoring of the effluent from treatment plants together with the monitoring of design parameters can be essential in the success of wastewater treatment plants in controlling these pollutants. Various parameters of treatment together with gene expression and the way genes are affected in the wastewater treatment process require further broad and extensive research. The presence of *aac(3)-IIa* and *aac(6')-Ib* genes in the wastewater samples of treatment plants indicates that conventional treatment approaches cannot guarantee the absence of these genes; thus it is suggested that more advanced methods be applied to control antibiotic-resistant genes.

Acknowledgments

The authors would like to thank the staff of the Cellular and Molecular Laboratory (Urmia Medicine Faculty) and Dr. Nikaeen, Head of the Advanced Environmental Molecular Laboratory (Department of Environmental Health Engineering), Isfahan University of Medical Sciences.

Competing interests

The authors declare that they have no competing interests.

Authors' contributions

RA designed the study. NS (PCR setup), RA (DNA extraction and PCR setup), FF (testing), FN (testing), AH (testing), SD (testing) and SHY (testing) did the lab work, and ASH, HM, and FM performed the literature search and wrote the manuscript. All authors critically reviewed, refined, and approved the manuscript.

References

1. Kümmerer K. The presence of pharmaceuticals in the environment due to human use--present knowledge and future challenges. J Environ Manageme 2009; 90(8): 2354-66.
2. Tao R, Ying GG, Su HC, Zhou HW, Sidhu JP. Detection of antibiotic resistance and tetracycline resistance genes in Enterobacteriaceae isolated from the Pearl rivers in South China. Environ Pollut 2010;

158(6): 2101-9.
3. Kim S, Aga DS. Potential ecological and human health impacts of antibiotics and antibiotic-resistant bacteria from wastewater treatment plants. J Toxicol Environ Health B Crit Rev 2007; 10(8): 559-73.
4. Munir M, Wong K, Xagoraraki I. Release of antibiotic resistant bacteria and genes in the effluent and biosolids of five wastewater utilities in Michigan. Water Res 2011; 45(2): 681-93.
5. Gao P, Munir M, Xagoraraki I. Correlation of tetracycline and sulfonamide antibiotics with corresponding resistance genes and resistant bacteria in a conventional municipal wastewater treatment plant. Sci Total Environ 2012; 421-422: 173-83.
6. Pruden A, Larsson DG, Amézquita A, Collignon P, Brandt KK, Graham DW, et al. Management Options for Reducing the Release of Antibiotics and Antibiotic Resistance Genes to the Environment. Environ Health Perspect 2013; 121(8): 875.
7. Rahube TO, Yost CK. Antibiotic resistance plasmids in wastewater treatment plants and their possible dissemination into the environment. Afr J Biotechnol 2010; 9(54): 9183-90.
8. Aali R, Nikaeen M, Khanahmad H, Hassanzadeh A. Monitoring and comparison of antibiotic resistant bacteria and their resistance genes in municipal and hospital wastewaters. Int J Prev Med 2014; 5(7): 887-94.
9. Aali R, Nikaeen M, Khanahmad H, Hejazi Z, Kazemi M, Hassanzadeh A. Occurrence of tetracycline resistant bacteria and resistance gene(tetW) in hospital and municipal wastewaters. Fresenius Environmental Bulletin 2014; 23(10a): 2560-6.
10. Klare I, Konstabel C, Badstübner D, Werner G, Witte W. Occurrence and spread of antibiotic resistances in Enterococcus faecium. Int J Food Microbiol 2003; 88(2-3): 269-90.
11. Bradfor PA. Extended-spectrum beta–lactamases in the 21st century: characterization, epidemiology, and detection of this important resistance threat. Clin Microbiol Rev 2001; 14(4): 933-51.
12. Wright GD. Antibiotic resistance in the environment: a link to the clinic? Curr Opin Microbiol 2010; 13(5): 589-94.
13. Savic M, Lovric J, Ilic Tomic T, Vasiljevic B, Conn LG. Determination of the target nucleosides for members of two families of 16S rRNA methyltransferases that confer resistance to partially overlapping groups of aminoglycoside antibiotics. Nucleic Acids Res 2009; 37(16): 5420-31.
14. Yamane K, Wachino JI, Suzuki S, Shibata N, Kato H, Shibayama K, et al. 16S rRNA Methylase producing, Gram-negative Pathogens. Emerg Infect Dis 2007; 13(4): 642-6.
15. Jawetz E, Melnick JL, Adelberg EA. Review of Medical Microbiology. 22nd ed. Los Altos: Appleton & Lange; 2007.
16. Filipova M, Bujdakova H, Drahovska H, Lišková A, Hanzen J. Occurrence of aminoglycoside-modifying-enzyme genesaac (6′)-aph (2″), aph (3′), ant (4′) andant (6) in clinical isolates ofEnterococcus faecalis resistant to high-level of gentamicin and amikacin. Folia Microbiol (Praha) 2006; 51(1): 57-61.
17. Chandrakanth RK, Raju S, Patil S. Aminoglycoside-resistance mechanisms in multidrug-resistant Staphylococcus aureus clinical isolates. Curr Microbiol 2008; 56(6): 558-62.
18. Heuer H, Krögerrecklenfort E, Wellington EM, Egan S, van Elsas JD, van Overbeek L, et al. Gentamicin resistance genes in environmental bacteria: prevalence and transfer. FEMS Microbiol Ecol 2002; 42(2): 289-302.
19. Tennstedt T, Szczepanowski R, Braun S, Pühler A, Schlüter A. Occurrence of integron-associated resistance gene cassettes located on antibiotic resistance plasmids isolated from a wastewater treatment plant. FEMS Microbiol Ecol 2003; 45(3): 239-52.
20. Tennstedt T, Szczepanowski R, Krahn I, Pühler A, Schlüter A. Sequence of the 68,869 bp IncP-1[alpha] plasmid pTB11 from a waste-water treatment plant reveals a highly conserved backbone, a Tn402-like integron and other transposable elements. Plasmid 2005; 53(3): 218-38.
21. Szczepanowski R, Linke B, Krahn I, Gartemann KH, Gützkow T, Eichler W, et al. Detection of 140 clinically relevant antibioticresistance genes in the plasmid metagenome of wastewater treatment plant bacteria showing reduced susceptibility to selected antibiotics. Microbiology 2009; 155: 2306-19.
22. Asghari FB, Nikaeen M, Mirhendi H. Rapid monitoring of Pseudomonas aeruginosa in hospital water systems: a key priority in prevention of nosocomial infection. FEMS Microbiol Lett 2013; 343(1): 77-81.
23. Samadi N, Hosainzadegan H, Akbari Qomi M. Assessment of aac(3')-IIa and aac(6')-Ib genes incidence involved in aminoglycoside resistance in Klebsiella.pneumoniae isolated from Clinical samples in urmia hospitals. Jundishapur Journal of Microbiology 2014: in press.
24. Yu FY, Yao D, Pan JY, Chen C, Qin ZQ, Parsons C, et al. High prevalence of plasmid-mediated 16S rRNA methylase gene rmtB among Escherichia coli clinical isolates from a Chinese teaching hospital. BMC Infect Dis 2010; 10: 184.
25. Bogaerts P, Galimand M, Bauraing C, Deplano A, Vanhoof R, De Mendonca R, et al. Emergence

of ArmA and RmtB aminoglycoside resistance 16S rRNA methylases in Belgium. J Antimicrob Chemother 2007; 59(3): 459-64.
26. Zhang Y, Zhou H, Shen XQ, Shen P, Yu YS, Li LJ. Plasmid-borne armA methylase gene, together with blaCTX-M-15 and blaTEM-1 in a Klebsiella oxytoca isolate from China. J Med Microbiol 2008; 57(Pt 10): 1273-6.
27. Guardabassi L, Lo Fo Wong DM, Dalsgaard A. The effects of tertiary wastewater treatment on the prevalence of antimicrobial resistant bacteria. Water Res 2002; 36(8): 1955-64.
28. Jakobsen L, Sandvang D, Hansen LH, Bagger-Skjøt L, Westh H, Jørgensen C, et al. Characterisation, dissemination and persistence of gentamicin resistant Escherichia coli from a Danish university hospital to the waste water environment. Environ Int 2008; 34(1): 108-15.
29. Bouki C, Venieri D, Diamadopoulos E. Detection and fate of antibiotic resistant bacteria in wastewater treatment plants: a review. Ecotoxicol Environ Saf 2013; 91: 1-9.
30. Zhang XX, Zhang T, Fang H. Antibiotic resistance genes in water environment. Appl Microbiol Biotechnol 2009; 82(3): 397-414.
31. Park J, Lee J, Oh J, Jeong Y, Cho J, Joo H, et al. Antibiotic selective pressure for the maintenance of antibiotic resistant genes in coliform bacteria isolated from the aquatic environment. Water Sci Technol 2003; 47(3): 249-53.
32. Szczepanowski R, Krahn I, Linke B, Goesmann A, Pühler A, Schlüter A. Antibiotic multiresistance plasmid pRSB101 isolated from a wastewater treatment plant is related to plasmids residing in phytopathogenic bacteria and carries eight different resistance determinants including a multidrug transport system. Microbiology 2004; 150(11): 3613-30.

17

Prediction of the waste stabilization pond performance using linear multiple regression and multi-layer perceptron neural network

Maryam Khodadadi[1], Alireza Mesdaghinia[2], Simin Nasseri[2], Mohammad Taghi Ghaneian[3], Mohammad Hassan Ehrampoush[4], Mahdi Hadi[5*]

[1]Ph.D. Student of Environmental Health, International Campus, Shahid Sadoughi University of Medical Sciences, Yazd, Iran
[2]Professor, Center for Water Quality Research (CWQR), Institute for Environmental Research (IER), Tehran University of Medical Sciences, Tehran, Iran
[3]Associate professor, Department of Environmental Health, School of Public Health, Shahid Sadoughi University of Medical Sciences, Yazd, Iran
[4]Professor, Department of Environmental Health, School of Public Health, Shahid Sadoughi University of Medical Sciences, Yazd, Iran
[5]Ph.D. student, Center for Water Quality Research (CWQR), Institute for Environmental Research (IER), Tehran University of Medical Sciences, Tehran, Iran

Abstract

Background: Data mining (DM) is an approach used in extracting valuable information from environmental processes. This research depicts a DM approach used in extracting some information from influent and effluent wastewater characteristic data of a waste stabilization pond (WSP) in Birjand, a city in Eastern Iran.

Methods: Multiple regression (MR) and neural network (NN) models were examined using influent characteristics (pH, Biochemical oxygen demand [BOD_5], temperature, chemical oxygen demand [COD], total suspended solids [TSS], total dissolved solid [TDS], electrical conductivity [EC] and turbidity) as the regression input vectors. Models were adjusted to input attributes, effluent BOD_5 (BODout) and COD (CODout). The models performances were estimated by 10-fold external cross-validation. An internal 5-fold cross-validation was also used for the training data set in NN model. The models were compared using regression error characteristic (REC) plot and other statistical measures such as relative absolute error (RAE). Sensitivity analysis was also applied to extract useful knowledge from NN model.

Results: NN models (with RAE = 78.71 ± 1.16 for BODout and 83.67 ± 1.35 for CODout) and MR models (with RAE = 84.40% ± 1.07 for BODout and 88.07 ± 0.80 for CODout) indicate different performances and the former was better ($P < 0.05$) for the prediction of both effluent BOD_5 and COD parameters. For the prediction of CODout the NN model with hidden layer size (H) = 4 and decay factor = 0.75 ± 0.03 presented the best predictive results. For BODout the H and decay factor were found to be 4 and 0.73 ± 0.03, respectively. TDS was found as the most descriptive influent wastewater characteristics for the prediction of the WSP performance. The REC plots confirmed the NN model performance superiority for both BOD and COD effluent prediction.

Conclusion: Modeling the performance of WSP systems using NN models along with sensitivity analysis can offer better understanding on exploring the most significant parameters for the prediction of system performance. The findings of this study could build the foundation for prospective work on the characterization of WSP operations and optimization of their performances with a view to conducting statistical approaches.

Keywords: Data mining, Multiple regression, Neural network, Waste stabilization pond

***Correspondence to:**
Mahdi Hadi
Email: hadi_rfm@yahoo.com

Introduction

In recent decades, the application of computer modeling techniques has been introduced in many environmental issues (1). Given the significance of wastewater treatment and its role in reducing environmental pollution, the control and proper operation of a wastewater treatment system is the most important environmental issue. The waste stabilization ponds (WSPs) are commonly recommended wastewater treatment systems used in arid and semi-arid developing countries. WSPs are valuable treatment systems due to suitable climatic conditions and availability of land (2). A stabilization pond is the most simple, reliable and cost-effective process with low maintenance requirements that can be used as an appropriate alternative for wastewater treatment by reducing biological oxygen demand (BOD_5) (3). Inappropriate operation of a waste water treatment systems may result in severe environmental and public health problems, as its effluent to a receiving water body can cause or spread various diseases to human beings (4). The performance of a WSP is often affected by various factors such as physical and biological factors (5). Moreover the performance efficiency of a WSP is greatly controlled by raw wastewater characteristics because of the variations in raw wastewater characteristics, its strengths and flow rates which is attributed to the changing and complex nature of the treatment process (6). In a WSP treatment system there are certain basic explanatory variables which can be used in explaining the plant performance. Among these variables, chemical oxygen demand (COD) and BOD_5 are the two important. Thus, predicting the COD and BOD_5 in the effluent, as the WSP's performance indices, depending upon the influent raw wastewater quality will aid the operator to discover the foremost effective factors on treatment efficiency and take necessary safety measures before the occurrence of any challenge. Furthermore, given the required time needed to measure BOD_5 (5 days) and COD (2 hours), and also their procedures which involve the use of several dangerous chemicals, the prediction of these parameters values instead of their measurement may be an environmentally and economically safe approach (7).

Attention has been diverted from manually treatment plant effluent to mathematically techniques due to these objectives (8). As a result, modeling of the effluent characteristics of treatment system is imperative for the prediction of plant performance and operation. Artificial neural network (ANN) technique is a non-parametric mathematical modeling technique which can be used for modeling such processes. It can be employed for better prediction of the process performance owing to their high accuracy, adequacy and quite promising applications in engineering (9-11). The ANN models have been used with the prediction objectives in several fields including air quality (12-14), water treatment (15), wastewater treatment (16), atmospheric sciences (17) among others.

The emphasis of this study is the analysis of a WSP treatment system data using data mining (DM) approaches. The data were collected from a WSP treatment system in Birjand, a city in the east of Iran. The purpose of this study is the prediction of the WSP performance in terms of COD and BOD_5 as the main performance related parameters. Several analyses were performed by considering and comparing two DM techniques (i.e. multiple regression [MR] and neural network [NN]).

Methods

WSP description

The studied stabilization pond is located in Birjand, the capital city of Southern Khorasan province in east of Iran. It is located at a latitude of 32°86' N and longitude of 59°21' E and about 1490 m above sea level. Birjand city has a cold and dry climate. The average annual temperature is 16.35°C with the warmest time in June (average 27.5°C) and the coldest in February (average 3.2°C). The Birjand's WSP has been constructed with a capacity of 10 500 cubic meter per day for a population of 64 000 people (18). The WSP configuration is divided into anaerobic, facultative and maturation ponds (Figure 1). According to Figure 1, this treatment system has a pretreatment unit which includes grit chamber and screens that is followed by the WSP systems. The wastewater samples were taken from the influent raw wastewater and maturation ponds effluent within a period of one year. The analysis of the influent raw wastewater composite samples characteristics (including pH, temperature, BOD_5, COD, total suspended solids [TSS], total dissolved solid [TDS], electrical conductivity [EC] and turbidity) were carried out according to the standard procedures (19). Only two main distinct parameters namely COD and BOD were analyzed for the effluent treated wastewater.

Data preparation

All performance parameters of Birjand's WSP including pH, COD, BOD_5, TSS, TDS, EC and Turbidity were measured in Birjand wastewater treatment plant (water and wastewater laboratory). In this study, pH, EC and TDS were measured using portable devices. COD, BOD and TSS were determined according to standard methods (19). Data analyses were performed using statistical packages in R (20).

The descriptive statistics of the raw data of the influent and effluent wastewater characteristics are summarized in Table 1.

A dataset of the operation parameters of Birjand's WSP was used where the objective was to estimate the effluent COD and BOD using influent wastewater characteristics (eight continuous attributes). The data comprised 96 raw instances, some of which are outlier values. Thus it is necessary to preprocess the data before fitting the DM models. This process includes operations such as choosing the data (e.g. attributes or examples) or dealing with missing values and outliers. Outlier data is defined as data with considerably distance from the normal distribution. However, in some instances, these outlier values may be correct because they are as a result of the natural product of the distribution of variables (21). All examples with missing values were deleted. The box-and-whisker plot was applied in order to detect outliers. Samples beyond the whiskers of the plot were considered as outliers (Figure 2).

Figure 2 summarizes each raw variable by four components as follows: a central line in each box is the sample median to specify the central tendency; a box (with edges of 25th and 75th percentiles) to indicate variability around the central tendency; whiskers around the box to show the range of the variable; and the observations beyond the

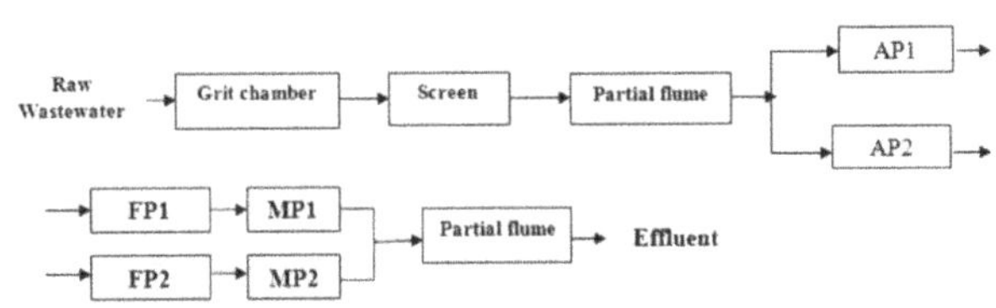

Figure 1. Schematic flow diagram for Birjand's WSP. AP, anaerobic pond; PF, facultative pond; MP, maturation pond; WSP, waste stabilization pond (Adapted from reference 18).

Table 1. Raw data descriptive statistics

Parameter	Unit	Mean	SD	SE	Max	Min	UB	LB
pHin	-	7.91	0.13	0.03	8.20	7.50	7.94	7.89
Tin	°C	23.40	3.11	0.63	28.70	17.10	24.03	22.77
BODin	mg/L	510.64	101.63	20.59	726.00	210.00	531.23	490.04
CODin	mg/L	815.55	154.66	31.34	1167.00	516.00	846.89	784.22
TDSin	mg/L	1299.16	158.45	32.11	1680.00	962.00	1331.26	1267.05
TSSin	mg/L	238.73	210.52	42.66	2217.00	115.00	281.39	196.07
ECin	μ mohs/cm	3100.94	312.53	63.33	3810.00	2400.00	3164.26	3037.61
NTUin	NTU	211.22	113.00	22.90	1172.00	69.00	234.11	188.32
BODout	mg/L	116.85	20.80	4.22	182.00	67.00	121.07	112.64
CODout	mg/L	259.04	51.32	10.40	392.00	168.00	269.44	248.65

whisker length are marked as outliers displayed with a circle sign that fall within the distance defined by quartile ±1.5× interquartile range (IQR). Where the IQR, or midspread or middle fifty, is a measure of statistical dispersion, being equal to the difference between the upper and lower quartiles, IQR = $Q_3 - Q_1$. In current study, using a normal distribution, the date falls out of three standard deviations (SDs) of the mean was considered as outliers. Thus, the data that were more than μ±3SD, were regarded as outliers. Figure 3 summarizes the Box-and-Whisker plot for each processed variable. The descriptive statistics of the pre-processed influent and effluent wastewater characteristics are summarized in Table 2.

The last step in the data preparation procedure was the data scaling. The objective here was to ensure that the statistical distribution of the values for each model input and output variable is approximately uniform. Therefore all attributes were standardized to a zero mean and one standard deviation (22).

Statistical modeling

The modeling and evaluation of the model which is an iterative process are the key procedures in DM approach (23). This research addresses these steps with an emphasis on the use of NN and MR functional models to solve the regression goals for the prediction of the performance of a WSP treatment system. All experiments reported in this study were conducted using R statistical environment (20). R tool is an open access with a set of software packages that allows the manipulation of data, performing of calculations, drawing of graphics and conducting statistical analysis. Taking the advantage of open access, Cortez (24) developed the *Rminer* package that aids the use of DM techniques in classification and regression tasks. In the present work, the *Rminer* package was utilized in performing the statistical modeling. The regression dataset is made up of *k*∈*(1,...,N)* examples, each mapping an input vector ($x_1^k,...,x_n^k$) to a given target y_k. The error is given by $e_k = y_k - \bar{y}_{pk}$, where $\bar{y}_{pk}$ represents the average of the pre-

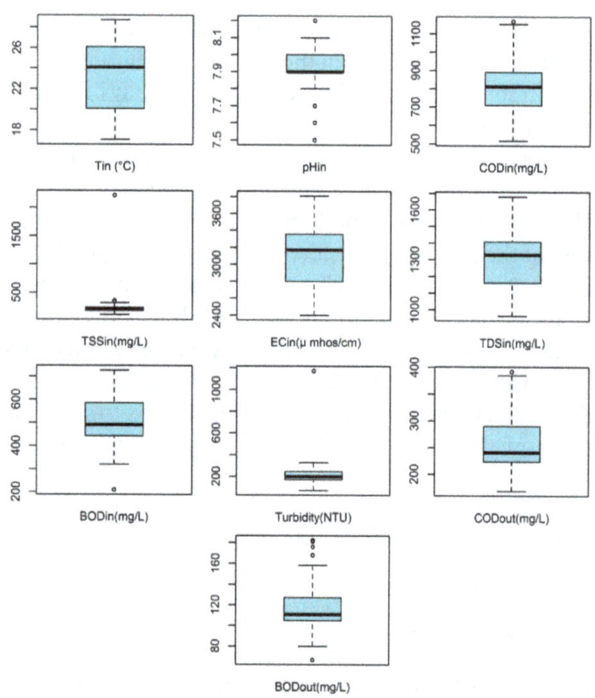

Figure 2. Box-and-Whisker Plots for the raw data of WSP characteristics

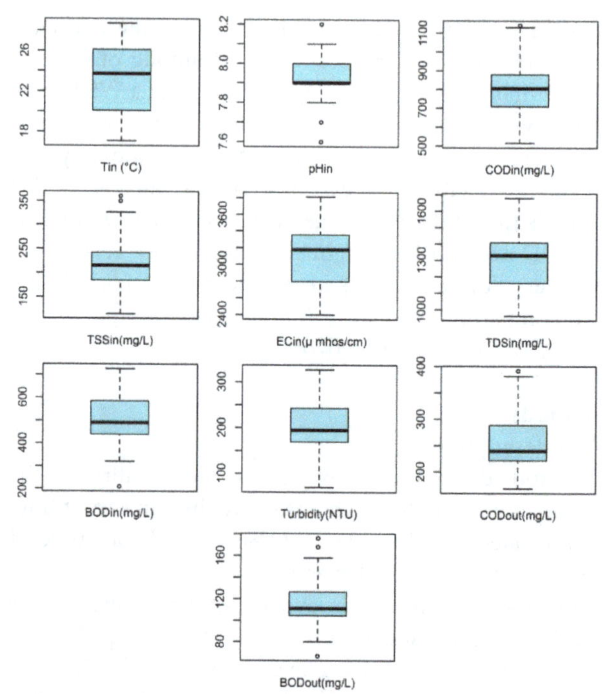

Figure 3. Box-and-Whisker Plots for the pre-processed data of WSP characteristics.

Table 2. Processed data descriptive statistics

Parameter	Unit	Mean	SD	SE	Max	Min	UB	LB
pHin	-	7.92	0.12	0.03	8.20	7.60	7.95	7.90
Tin	°C	23.36	3.11	0.65	28.70	17.10	24.00	22.71
BODin	mg/L	506.12	99.73	20.65	726.00	210.00	526.77	485.47
CODin	mg/L	807.25	147.92	30.63	1141.00	516.00	837.88	776.62
TDSin	mg/L	1297.57	158.83	32.89	1680.00	962.00	1330.46	1264.67
TSSin	mg/L	215.45	50.45	10.45	360.00	115.00	225.89	205.00
ECin	µ mohs/cm	3100.98	315.17	65.27	3810.00	2400.00	3166.25	3035.71
NTUin	NTU	200.04	54.84	11.36	327.00	69.00	211.40	188.69
BODout	mg/L	115.51	18.88	3.91	176.00	67.00	119.42	111.60
CODout	mg/L	257.27	50.09	10.37	392.00	168.00	267.64	246.90

Abbreviations: SD, standard deviation; SE, standard error; UB, upper bound of 95% confidence interval for the mean; LB, lower bound of 95% confidence interval for the mean.

dicted value for the k input pattern.

Here, we selected eight wastewater treatment plant's influent characteristics (pH, BODin, Tin, CODin, TSSin, TDSin, ECin and NTUin) as the regression dataset input vectors. Both models were assessed with a supervised learning, where each model was adjusted to a dataset with examples that map input attributes into a given target (BODout and CODout).

The generalization performance of the models is often estimated by the holdout validation (i.e. train and test split). However, in some conditions the k-fold cross-validation is also used as the more robust approach (22). The latter approach is a more powerful method which requires k times more computation since k models are fitted in this approach. In this study, 10-fold cross-validation approach was used for the model's generalization performance estimation. The multilayer perceptron network was utilized for NN model. This network includes one hidden layer of H neurons with logistic functions. The overall model is given in the form:

$$y_i = f_i(w_{i,0} + \sum_{j=I+1}^{I+H} f_j(\sum_{n=1}^{I} x_n w_{m,n} + w_{m,0}) w_{i,n}) \quad (1)$$

where y_i is the output of the network for node i, $w_{i,j}$ is the weight of the connection from node j to i and f_j is the activation function for node j. In regression method, the output neuron is usually a linear function. The *Rminer* use the *nnet* package for modeling the NN. In *nnet* the optimization is done through the BFGS (Broyden, Fletcher, Goldfarb and Shanno)method of *optim* package. The BFGS is a quasi-Newton method, specifically published simultaneously in 1970 by Broyden, Fletcher, Goldfarb and Shanno (25). This uses function values and gradients to build up a structure of the surface to be optimized. For regression tasks, the algorithm minimizes the squared error (22). To solve this issue, the solution adopted is to train *Nr* different networks and then select the NN with the lowest error (22). In *Rminer*, this option is set using *model = 'mlp.'* The performance of NN model depends significantly on the number of nodes in hidden layer. NN in simplest form has *H = 0*, while some more complex NN may use a high values for *H* (24). Optimizing the network structure is a crucial step in the design of NNs. The NN structure must be optimized to minimize computer processing and obtain a good performance to avoid overfitting (10). There is no way to determine the best number of hidden units without training several networks and estimating the generalization error of each (26). If the hidden units are few, then high training error and high generalization error due to under-fitting may result. Conversely, if many hidden units are used, low training error can be achieved at the expense of network generalization which degrades overfitting (27). Since the NN network was not too complex in this study, the number of hidden nodes (*H*) was estimated by using the formula, *H=I/2*, where *I* is the number of nodes in input layer. However, the weight decay factor hyper-parameter was optimized with a 10-range grid-search between zero and one. To avoid overfitting for the training data set, an internal 5-fold cross-validation was used. In this study all input and output attributes were standardized with zero mean and standard deviation of unity using setting *scale* argument in *fit* function on *all* value.

After selecting the best parameters, the model was retrained with all training data. To quantify the importance of input variables in the model, the sensitivity analysis was applied after the training phase in order to analyze the model responses when a given input is changed. The weight of each input variable was measured by varying its value through its full range while the other input variables remained with their mean values (28). If the analyzed input variable is very important, its variance in the model output will be high. Therefore, the most important variable in the model output is the input variable that has a higher variance (24). The overall performance of each model was computed by the global metrics, namely the mean absolute deviation (MAD), root mean squared error (RMSE) and relative absolute error (RAE), which can be computed as (29):

$$MAD = \frac{\sum_{i=1}^{N} |y_i - \hat{y}_i|}{N} \quad (1)$$

$$RMSE = \sqrt{\frac{\sum_{i=1}^{N} (y_i - \hat{y}_i)^2}{N}} \quad (2)$$

$$RAE = \frac{\sum_{i=1}^{N} |y_i - \hat{y}_i|}{\sum_{i=1}^{N} |y_i - \bar{y}_i|} \times 100\% \quad (3)$$

where N denoted the number of considered cases, y_i the observed value, $\hat{y}_i$ the predicted value and $\bar{y}_i$ the average of observed value.

In all three metrics, lower values result in better predictive models. Nevertheless, the RMSE showing the overall accuracy of the model is more sensitive to high errors. Another approach used in comparing regression models is the regression error characteristic (REC) curve (30), which plots the error tolerance given in terms of the absolute deviation, versus the percentage of predicted points. An independent variable is expected to presents low predictive error and high REC area. In this study, the comparison of NN and MR models was examined with a statistical test. The Welch's two-sample t test that handles inequality in variance by adjusting degrees of freedom was used to compare RAE of the models.

Results

Figure 4 shows the typical three-layered feed-forward ANN. Eight input nodes corresponding to eight independent attributes, four hidden layer nodes and one output node are estimating BODout (Figure 4a) and CODout (Figure 4b) concentrations. Connections between nodes are presented by solid lines, which are associated with synaptic weights adjusted during the training procedure. The bias nodes were also shown, with 1 as their output value. The NN model (H = 4, $decay$=0.75±0.03) achieved the best predictive results for the prediction of effluent COD (CODout). Similar results was found for BODout in which an NN model (H = 4, $decay$ = 0.73±0.03) predicts this parameter with lower error in comparison with MR regression. Table 3 shows the MAE, RAE and RMSE for the NN and MR models. The RMSE is the root of mean squared difference between outputs and targets.

The Welch's two sample t test was also carried out in order to compare the performance of NN and MR models for the prediction of BODout and CODout in terms of influent wastewater characteristics. The results are summarized in Table 4. The REC curves of NN models for CODout and BODout were shown in Figures 5a and 5b, respectively.

The results of the sensitivity analysis (Figure 6a and 6b) procedure are useful for knowledge discovery for NN models. In this way, it is possible to quantify the contribution of a given attribute for the model.

The diagnostic plots of fitted NN models for BODout and CODout are shown in Figures 7 and 8, respectively. For indicating the models performance, the actual vs. predicted values was visualized in Figure 7a and Figure 8a. It was observed that the targets were well tracked by the output. Figure 7b and Figure 8b show the residual histograms for the BODout and CODout NN models, respectively. The fitted normal distribution curve on the data set indicated that the residual of the models followed a normal

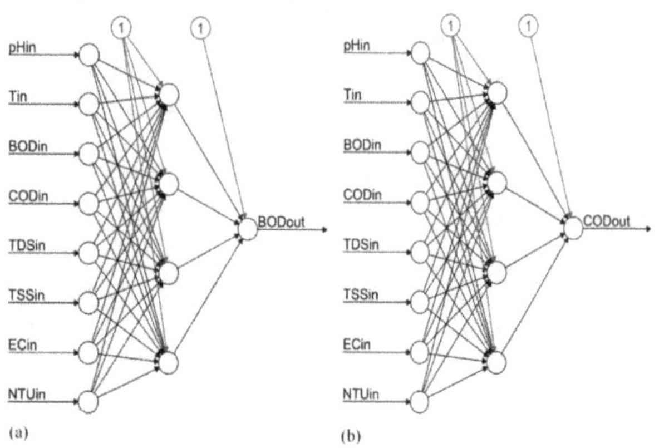

Figure 4. Structure of the constructed three-layer feed-forward ANN to predict BODout (a) and CODout (b) .

Table 3. Global metrics for the NN and MR models

Mertic	BODout		CODout	
	NN	MR	NN	MR
MAE	11.41±0.16	12.24±0.15	33.72±0.54	35.50±0.32
RAE	78.71±1.16	84.40±1.07	83.67±1.35	88.07±0.80
RMSE	16.37±0.21	16.81±0.22	44.75±0.74	45.55±0.36

Table 4. Welch's two sample t test results for comparing NN and MR models

Parameter	Model	RAE	t statistic	df	P value*
BODout	NN	78.71±1.16	-8.1337	17.89	0.0000
	MR	84.40±1.07			
CODout	NN	83.67±1.35	-6.3215	14.64	0.0000
	MR	88.07±0.80			

*Ho = no difference between means; H1 = true difference in means is not equal to 0; 95% CI.

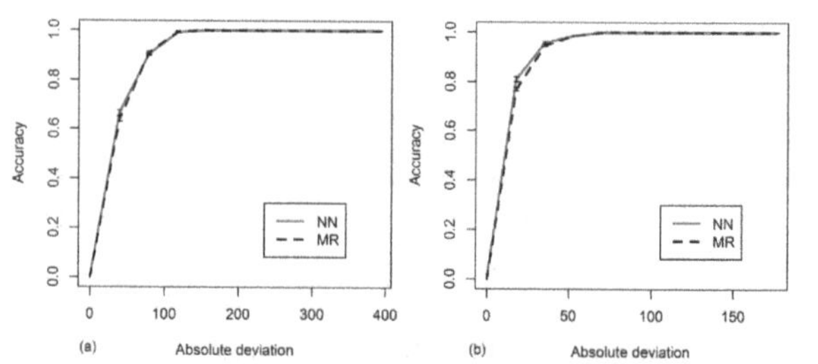

Figure 5. The REC curves confirm the NN performance superiority for the prediction of CODout (a) BODout (b).

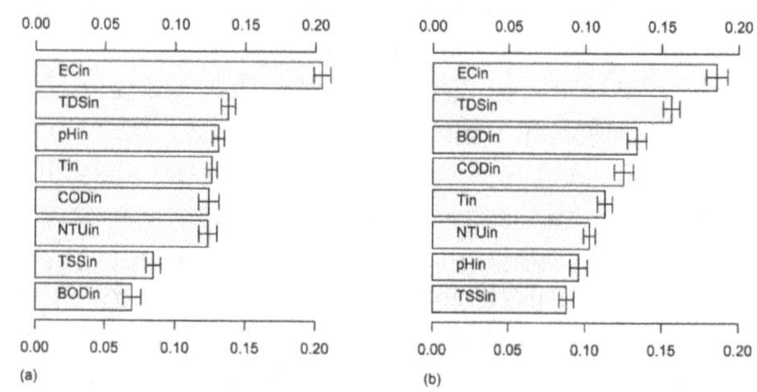

Figure 6. The relative input importance of the NN model for CODout (a) and BODout (b) in the order of importance shows the ECin and TDSin as the most relevant inputs.

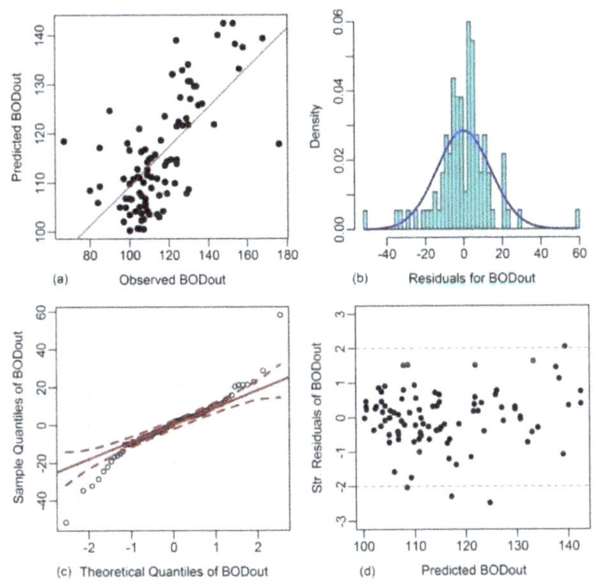

Figure 7. Result of NN model to predict BODout with a 10-fold external and a 5-fold internal cross-validation approach. Scatter plot of estimated values vs. observed values (R^2 = 0.56) (a). Histogram of residuals with normal adjustment curve (mean = 0.003 ± 2.9) in the data set (b). Q-Q plot of the agreement between the residual quantiles and normal quantiles (c). Relationship between standardized residuals and values predicted by the model (d).

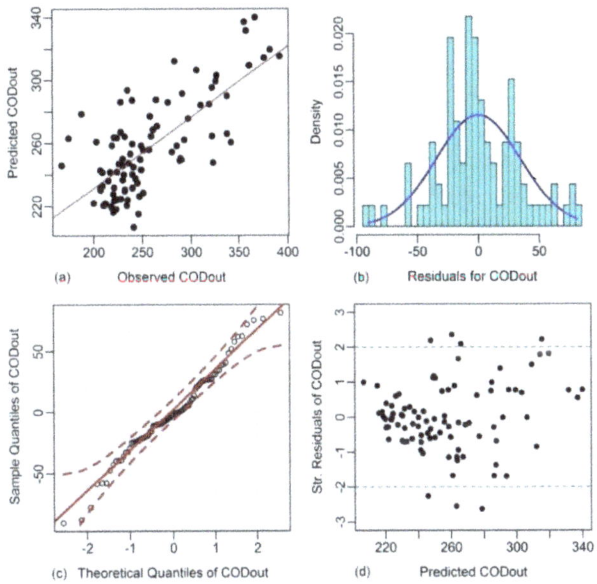

Figure 8. Result of NN model to predict CODout with a 10-fold external and a 5-fold internal cross-validation approach. Scatter plot of estimated values vs. observed values (R^2 = 0.54) (a). Histogram of residuals with normal adjustment curve (mean = 0.071 ± 7.15) in the data set (b). Q-Q plot of the agreement between the residual quantiles and normal quantiles (c). Relationship between standardized residuals and values predicted by the model (d).

distribution.

Figure 7c shows the residuals normal Q-Q plot of the BODout model. This plot helps to confirm the rationality behind the conclusion that the residuals are fairly normally distributed. Figure 8c shows the residuals normal Q-Q plot of the CODout model.

The relationship between standardized residuals and values predicted by the models were shown in Figure 7d and Figure 8d.

Discussion

According to Figure 6a and 6b which present the relative input importance of the NN model for CODout and BODout, respectively, the EC and TDS which are the measures of dissolved materials in wastewater are seen as the most important inputs. Conversely, the influent TSS has a minimal impact on the predicted variations of BODout and CODout. In other words, less than 10% of the BODout or CODout originated from the suspended solids entering the WSP and most of the influent suspended materials have a biodegradable nature and probably were converted or deposited through the treatment process. Thus the most fractions of the suspended biodegradable materials entering the WSP may be removed by treatment process, while some part of biodegradable materials, which are mainly dissolved materials, were leaved the system without being remarkably treated. This may be as a result of low considerable effect of the treatment process on the elimination of soluble materials entering the WSP. The lower metric values for NN models revealed the higher performance of NN models in comparison with MR models.

Welch's two sample *t* test was found to be statistically significant, indicating that the NN and MR have different performances and the former is better ($P<0.05$) for the prediction of both BODout and CODout parameters. The obtained average RAE = 78.71 ± 1.16 for BODout and 83.67 ± 1.35 for CODout in the case of NN model are better when statistically compared with MR model results (RAE = 84.40% ± 1.07 for BODout and 88.07 ± 0.80 for CODout). The REC curve plots in Figures 5a and 5b also confirm the NN model performance superiority. The whiskers in all graphs represent the 95% *t* student CI. According to the findings of this study, both the REC curves and global metrics, present the best performance to be the NN model followed by MR model. In fact, the NN model for both BODout and CODout responses have lower errors and higher area under the REC curve.

The degradability of WSP's influent wastewater as the ratio of BOD_5 to COD is called Biodegradability Index (BI). Generally, the BI Index ranges between 0.4 to 0.8 for domestic wastewater. If BOD_5/COD is >0.6 then the waste is biodegradable fairly and can be effectively treatable using biological treatment methods If BOD_5/COD is >0.6 then the waste is fairly biodegradable and can be excellently treated using biological treatment. If BOD_5/COD ratio is between 0.3 and 0.6, then there is need to seed the wastewater in order to treat it biologically. In cases where the BOD_5/COD is <0.3, then the wastewater is not biologically treatable (13,31). From data in Table 2, the BI index was obtained to be 0.62 and 0.44 for influent and effluent of WSP, respectively. These values indicate that the influent wastewater can be classified as degradable and the effluent has no further biodegradability potential. This is consistent with Figure 6a where the BODin has a minimal impact on the predicted variations of CODout. In other words, most fractions of materials in the effluent are dissolved and non-biodegradable materials.

The coefficient of determination (R^2) between observed and predicted BODout and observed and predicted CODout were 0.56 and 0.54, respectively. Despite the slightly weak correlation values, the points are almost well aligned on the acceptable prediction diagonal of coordinates 1:1. These values which indicate that 56% and 54% of the variance of the BODout and CODout variables, respectively, can be explained by using the network input attributes. The remaining 44% and 46% of their variances can be attributed to unknown, lurking variables, or inherent variability (11). The Histogram of the residuals can be used to determine if the variance is normally distributed. A symmetric bell-shaped histogram which is evenly distributed around zero shows that the normality assumption of the residuals is likely to be valid (32,33).

According Figure 7b and Figure 8b the residuals are close to a normal distribution around a mean value of 0.003 ± 2.9 (SD=14.00) and 0.071 ± 7.15 (SD=34.54) for BODout and CODout NN models, respectively. A Q-Q (Quantile-Quantile) plot is another graphic method for testing if the residuals of the models follow the normal distribution. The residuals are said to follow a normal distribution if all the scatter points are close to the reference line (34). The point pattern in the middle of Q-Q plot is fairly linear. Chambers (35) and Fowlkes (36) discuss the interpretations of commonly encountered departures from linearity. When the left end of pattern is below the line and right end of pattern is above the line, then there are long tails at both ends of the data distribution. No curvature was seen in the middle in Figure 7c thus showing that there is no skewness in the residuals distribution. It is noted that at the left and right ends of the plot, the circles are somewhat farther away from the line than elsewhere in the plot. But it is common for points at either end of the plot to be farther from the line than elsewhere, even when the data are normal (37). When the distribution of the residuals is skewed and their variance is found not to be constant, a transformation on the response variable may be quite advantageous (38). For CODout model (Figure 8c) the circles or points all lie quite close to the line and within the 95% CI zone; as such is safe to say that these residuals emanated from a normal distribution. Consequently, the normality assumption of residuals for both BODout and CODout models may be appreciated. In the standardized residuals rescale residual values by the regression standard error, if the residuals are found to be distributed normally, about 95% should fall within 2σ around the fitted curve. Consequently, 95% of the standardized residuals will also fall between -2 and +2 in the residual plot (33). Figure 7d and Figure 8d show a random scatter around zero with only a few points outside the ± 2 limits. Therefore, the points are properly distributed on both sides of the horizontal line of zero ordinate representing the standardized average of the residuals.

In recent years, modeling of wastewater treatment plant (WWTP) or constructed wetlands performances through NN for the prediction of wastewater characteristic parameters has gained enormous interest. The ANN application of these articles was summarized in Table 5.

As can be seen in Table 5 the ANN models performance based on coefficient of determination for the prediction of effluent BOD_5 varies between 31% and 84 %. For effluent COD the models performances varies between 39 and 98%. The values of 0.56 for BOD_5 and 0.54 for COD obtained in this study are within the range of performance predictions reported in literature.

According to Table 5, it can be established that DM approaches were most commonly used for the performance prediction of the conventional WWTP systems. Indeed, DM approach were rarely utilized for the performance prediction of natural treatment system such as WSPs and constructed wetlands (CWs) which have been proven to be effective substitutes for treating wastewater (45). However, results of this study could be the basis for better understanding of DM approach using ANN models for the prediction of WSPs performance.

Conclusion

The real-world DM application case for the prediction of a WSP performance presented in this study indicates the possibility of analyzing data by using the R environment and *rminer* package. It was shown that the influent WSP indices could be applied to the prediction of effluent quality. Two DM techniques were explored: MR and NN. Overall, Welch's two-sample *t* test on global metrics and

Table 5. Summary ANN application of researcher's studies in modeling of WWTPs and WTPs

Predicted parameters	Location	R^2	References
BOD_5, COD, TSS	WWTP, El-Agamy	Reaching up to 0.9	(39)
COD, TSS	Paper mill WWTP, China	0.98 for COD ; 0.96 for SS	(40)
BOD_5, COD, TSS	WWTP, Doha	0.39-0.84 for COD; 0.31-0.84 for BOD_5; 0.54-0.96 for TSS	(10)
COD_5, TSS, pH	WWTP, Taiwan industrial park	0.92 for SS; 0.86 for COD; 0.90 for pH	(41)
NH_3, BOD_5, TSS, COD, TN	WWTP, Denmark	More than 0.95	(42)
pH, TDS, turbidity, TN, TP	Industrial WWTP, Iran	0.85-0.92 for pH; 0.23-0.53 for Turbidity; 0.07-0.45 for TP; 0.22-0.51 for TN; 0.67-0.82 for TDS	(43)
BOD_5	Industrial WWTP, Govindpura	0.64-0.87 for BOD_5	(8)
BOD_5	Constructed wetlands (CWs), Greece	0.52-0.68 for BOD_5	(44)
BOD_5, COD	WSPs, Birjand, Iran	0.56 for BOD_5; 0.54 for COD	This study

REC curve analysis confirmed NN model as the best predictive model for the prediction of both BOD and COD parameters. The sensitivity analysis of the WSP influent characteristics for the selected NN model can offer guidance which will be very important for the prediction of WSP performance based on the effluent BOD and COD concentrations. EC and TDS were found to be the most descriptive influent wastewater characteristics for the prediction of the WSP performance.

Acknowledgments
The authors are thankful to the Center for Water Quality Research (CWQR) at the Institute for Environmental Research (IER) of Tehran University of Medical Sciences for providing facilities and supports for this research.

Competing interests
The authors declare that they have no competing interests.

Authors' contributions
MK, MTG and MHE participated in raw data provision. AM and SN contributed to drafting and editing the manuscript. MH contributed in the design of study, data analysis and drafting the manuscript.

References
1. Maier HR, Dandy GC. Neural network based modelling of environmental variables: a systematic approach. Math Comput Model 2001; 33(6): 669-82.
2. Mozaheb S, Ghaneian M, Ghanizadeh G, Fallahzadeh M. Evaluation of the stabilization ponds performance for municipal wastewater treatment in Yazd, Iran. Middle-East Journal of Scientific Research 2010; 6(1): 76-82.
3. Wiley PE, Brenneman KJ, Jacobson AE. Improved algal harvesting using suspended air flotation. Water Environ Res 2009; 81(7): 702-8.
4. Peterson JD, Murphy RR, Jin Y, Wang L, Nessl MB, Ikehata K. Health effects associated with wastewater treatment, reuse, and disposal. Water Environ Res 2011; 83(10): 1853-75.
5. Shammas NK, Wang LK, Wu Z. Waste stabilization ponds and lagoons. In: Wang LK, Pereira NC, Yung-Tse Hung YT, eds. Biological Treatment Processes. Springer; 2009. p. 315-70.
6. Saqqar M, Pescod M. Modelling the performance of anaerobic wastewater stabilization ponds. Water Sci Technol 1995; 31(12): 171-83.
7. Al-Asheh S, Mjalli FS, Alfadala HE. Forecasting influent-effluent wastewater treatment plant using time series analysis and artificial neural network techniques. Chemical Product and Process Modeling 2007; 2: 3.
8. Vyas M, Modhera B, Vyas V, Sharma A. Performance forecasting of common effluent treatment plant parameters by artificial neural network. J Eng Appl Sci 2011; 6(1): 38-42.
9. Hanbay D, Turkoglu I, Demir Y. Prediction of wastewater treatment plant performance based on wavelet packet decomposition and neural networks. Expert Syst Appl 2008; 34(2): 1038-43.
10. Mjalli FS, Al-Asheh S, Alfadala H. Use of artificial neural network black-box modeling for the prediction of wastewater treatment plants performance. J Environ Manage 2007; 83(3): 329-38.
11. Solaimany Aminabad M, Maleki A, Hadi M. Application of artificial neural network (ANN) for the prediction of water treatment plant influent characteristics. J Adv Environ Health Res 2014; 1(2): 89-100.
12. Rostami Fasih Z, Mesdaghinia A, Nadafi K, Nabizadeh Nodehi R, Mahvi AH, Hadi M. Forecasting the air quality index based on meteorological variables and autocorrelation terms using artificial neural network. Razi Journal of Medical Sciences 2015; 22(137): 31-43. [In Persian].
13. Motesaddi S, Nowrouz P, Alizadeh B, Khalili F, Nemati R. Sulfur dioxide AQI modeling by artificial neural network in Tehran between 2007 and 2013. Environmental Health Engineering and Management Journal 2015; 2(4): 173-8.
14. Shakerkhatibi M, Mohammadi N, Zoroufchi Benis K, Behrooz Sarand A, Fatehifar E, Asl Hashemi A. Using ANN and EPR models to predict carbon monoxide concentrations in urban area of Tabriz. Environmental Health Engineering and Management Journal 2015; 2(3): 117-22.
15. Khataee AR, Kasiri MB. Artificial neural networks modeling of contaminated water treatment processes by homogeneous and heterogeneous nanocatalysis. Journal of Molecular Catalysis A: Chemical 2010; 331(1): 86-100.
16. Hamed MM, Khalafallah MG, Hassanien EA. Prediction of wastewater treatment plant performance using artificial neural networks. Environ Model Softw 2004; 19(10): 919-28.
17. Gardner MW, Dorling S. Artificial neural networks (the multilayer perceptron)—a review of applications in the atmospheric sciences. Atmos Environ (1994) 1998; 32(14): 2627-36.
18. Hayati H, Doosti M, Sayadi M. Performance evaluation of waste stabilization pond in Birjand, Iran for the treatment of municipal sewage. Proceedings of the International Academy of Ecology and Environmental Sciences 2013; 3(1): 52-8.
19. Rice EW. Standard methods for the examination of water and wastewater. Washington, DC: American Public Health Association; 2012.
20. R Core Team. R: A language and environment for statistical computing. Vienna, Austria: R Foundation for Statistical Computing; 2013. Available from: http://www.R-project.org.
21. Masters T. Practical neural network recipes in C++. California: Morgan Kaufmann; 1993.
22. Hastie T, Tibshirani R, Friedman J, Franklin J. The elements of statistical learning: data mining, inference and prediction. The Mathematical Intelligencer 2005; 27(2): 83-5.
23. Chapman P, Clinton J, Kerber R, Khabaza T, Reinartz T, Shearer C, et al. CRISP-DM 1.0 Step-by-step data mining guide. 2000. Available from: https://www.the-modeling-agency.com/crisp-dm.pdf.
24. Cortez P. Data mining with neural networks and support vector machines using the R/rminer tool. In: Perner p, ed. Advances in Data Mining Applications and Theoretical Aspects. Springer; 2010. p. 572-83.

25. Fletcher R. A new approach to variable metric algorithms. Comput J 1970; 13(3): 317-22.
26. Nguyen N, Cripps A. Predicting housing value: A comparison of multiple regression analysis and artificial neural networks. Journal of Real Estate Research 2001; 22(3): 313-36.
27. Geman S, Bienenstock E, Doursat R. Neural networks and the bias/variance dilemma. Neural Comput 1992; 4(1): 1-58.
28. Kewley RH, Embrechts MJ, Breneman C. Data strip mining for the virtual design of pharmaceuticals with neural networks. IEEE Trans Neural Netw 2000; 11(3): 668-79.
29. Witten IH, Frank E. Data Mining: Practical Machine Learning Tools and Techniques. Amsterdam: Morgan Kaufmann; 2005.
30. BI J, Bennett P. Regression error characteristic curves. Twentieth International Conference on Machine Learning (ICML-2003); Washington, DC; 2003.
31. Rim-Rukeh A, Agbozu L. Impact of partially treated sewage effluent on the water quality of recipient Epie Creek Niger Delta, Nigeria using Malaysian Water Quality Index (WQI). J Appl Sci Environ Manag. 2013; 17(1): 5-12.
32. Montgomery DC, Runger GC, Hubele NF. Engineering Statistics. New York: John Wiley & Sons; 2009.
33. Graphic Residual Analysis. Available from: http://www.originlab.com/doc/Origin-Help/Residual-Plot-Analysis.
34. Thode HC. Testing for normality. Boca Raton: CRC Press; 2002.
35. Chambers JM. Graphical Methods for Data Analysis. New York: Champman & Hall; 1983.
36. Fowlkes EB. A folio of distributions: A collection of theoretical quantile-quantile plots. New York: Marcel Dekker; 1987.
37. Brown DE. Bro. Brown's statistics reference pages: examples of interpreting Q-Q plots 2014. Available from: http://emp.byui.edu/BrownD/Stats-intro/dscrptv/graphs/qq-plot_egs.htm.
38. Neter J, Kutner MH, Nachtsheim CJ, Wasserman W. Applied Linear Statistical Models. Chicago: Irwin; 1996.
39. Nasr MS, Moustafa MA, Seif HA, El Kobrosy G. Application of artificial neural network (ANN) for the prediction of EL-AGAMY wastewater treatment plant performance-EGYPT. Alexandria Engineering Journal 2012; 51(1): 37-43.
40. Wan J, Huang M, Ma Y, Guo W, Wang Y, Zhang H, et al. Prediction of effluent quality of a paper mill wastewater treatment using an adaptive network-based fuzzy inference system. Appl Soft Comput 2011; 11(3): 3238-46.
41. Pai T, Yang P, Wang S, Lo M, Chiang C, Kuo J, et al. Predicting effluent from the wastewater treatment plant of industrial park based on fuzzy network and influent quality. Appl Math Model 2011; 35(8): 3674-84.
42. Raduly B, Gernaey KV, Capodaglio A, Mikkelsen PS, Henze M. Artificial neural networks for rapid WWTP performance evaluation: methodology and case study. Environ Model Softw 2007; 22(8): 1208-16.
43. Naser M, Hamed H, Mohammad TJ, Hamidreza H, Hamid A. Simulation of low TDS and biological units of Fajr industrial wastewater Treatment plant using artificial neural network and principal component analysis hybrid method. Journal of Water Resource and Protection 2012; 6(4): 370-6.
44. Akratos CS, Papaspyros JN, Tsihrintzis VA. An artificial neural network model and design equations for BOD and COD removal prediction in horizontal subsurface flow constructed wetlands. Chem Eng J 2008; 143(1): 96-110.
45. Kayombo S, Mbwette T, Katima J, Ladegaard N, Jørgensen S. Waste stabilization ponds and constructed wetlands -design manual. UNEP-IETC with the Danish International Development Agency (Danida); 2005.

18

Comparison of hybrid regression and multivariate regression in the regional flood frequency analysis

Seyed Hashem Hosseini[1], Mohammad Reza Khaleghi[1*], Hasan Jami[2], Sajjad Baygi[2]

[1]Department of Range and Watershed Management, Torbat-e-Jam Branch, Islamic Azad University, Torbat-e-Jam, Iran
[2]Department of Civil Engineering, Torbat-e-Jam Branch, Islamic Azad University, Torbat-e-Jam, Iran

Abstract

Background: Magnitude, rate and frequency of the stochastic and unexpected events are of great significance and importance in hydrology. Nowadays, for economic planning of the projects, the use of analytical methods of unexpected events in hydrology is unavoidable. The aim of this study was to compare hybrid regression and multivariate regression to estimate flood peak discharge in the province of Khorasan Razavi and in the selected water measured stations.

Methods: For this purpose, 19 hydrometric stations were selected and analyzed. In the first step, the rate of peak discharge was estimated with different return periods and by selecting the best regional distribution (lognormal distribution type III). In the next step, independent and important variables including area, mean annual rainfall, the average height of the watershed and its slope were determined using functional analysis and using SPSS software version 22. Then, two hydrologically homogeneous regions were determined by homogeneity test using cluster analysis, and accordingly, two models were presented for the whole area and also for homogeneous areas. To compare and evaluate the accuracy and efficiency of the estimated models, the rates of discharges were estimated and compared with observational rates using three control watersheds. To compare models, it was used from the average absolute values of the relative error index.

Results: It was revealed that the hybrid method was more accurate than the multivariate regression method in the return period of 50 years and provides better results of flood discharges for the area. Homogenous areas had a higher coefficient of determination (R^2) and lower relative standard error (RSE) compared to the whole area. It was also revealed that with increase of return period, the rates of R^2 decreased but the rates of relative standard error increased.

Conclusion: The accuracy of multivariate regression and hybrid methods was the same in the 25-year return period. In the present study, the importance and necessity of homogenous areas compared with the model of the whole area are completely evident.

Keywords: Floods, Regression analysis, Khorasan Razavi province, Statistical distributions, Hybrid regression

***Correspondence to:**
Mohammad Reza Khaleghi
Email: drmrkhaleghi@gmail.com

Introduction

Flood is one of the most important natural disasters in the world that makes main problems for the development and reclamation of many countries. This natural phenomenon has been converted to a continuous phenomenon that can destroy capital and financial facilities of countries. Flood management in a basin will not be successful unless the hydrological behaviors of catchments are predicted (1-4). Hence, the determination of flood peak discharge with different return periods is essential for conducting flood control plans, designing hydraulic structures, preventing erosion and designing storages in an area (5-8). For this reason, different methods such as regional flood analysis have been presented. All these methods are made based on the recognition of factors affecting flood. To estimate flood frequency at ungauged drainage basins, flood frequency analysis using regression analysis is performed (9). In regional methods for flood prediction, the estimation of the index flood is a crucial point in order to attain reliable predictions in ungauged basins

(10). In recent years, the regression methods have been successfully employed and demonstrated satisfactory results (11-13). Kjeldsen and Rosbjerg compared different methods for estimating T-year events. Their simulation study revealed that in regional index-flood estimation, the method of probability-weighted moments is preferable to the method of moment estimation with regard to bias and RMSE (14). Kuldkepp in a study on watersheds located in arid areas of Nevada State, presented a new method called hybrid method to estimate peak flood discharge in areas with deficient hydrologic statistics (15). Kjeldsen et al, in a flood frequency analysis in South Africa, suggested that the effective factors in the peak flood discharge like the physiographic and climatic characteristics including longitude, latitude, watershed area, mean annual rainfall, rainfall concentration and soil compaction coefficient (16). Eng et al estimated the peak flow of 50 years using hybrid regression in the Southeast of the United States and concluded that this method has less error compared with geographic methods or methods based on the predictive variables (17). Song and Singh performed flood frequency analysis in Illinois by comparing curves of flood frequency obtained from flood simulation series with those obtained from equations of regional regression. They found that the curves obtained from flood simulation series have a higher accuracy compared to those obtained from the equations of regional regression (18). Eng et al used a hybrid region-of-influence (HRoI) regression method (as a hybrid geographic region-of-influence approach) to estimate the 50-year peak flow (Q50) in the Southeastern United States (17). They concluded that by increasing the accuracy of the geographic proximity of the stations, the accuracy of the model also increases. Marofi showed that Gumbel method is the best-fitted distribution in 10 regions of Hamadan province (19). Shukla et al dealt with the probabilistic estimates of extreme maximum rainfall using the regional generalized extreme value (GEV) distribution and concluded that the GEV model satisfied the selection criteria (20). Haberlandt and Radtke evaluated a hybrid rainfall model for the flood frequency analysis and reported a good performance in capturing average and extreme rainfall characteristics (21). Liu et al in downscaling some data-driven methods to forecast the daily precipitation and daily maximum and minimum temperature series concluded that time-lagged feed forward neural network (TLFFNN) and evolutionary polynomial regression (EPR) have more efficiency than statistical downscaling model (SDSM) (22). Smith and Karr investigated the flood frequency analysis using the Cox regression model and considered the assessment of the relative importance of physical processes such as snowmelt or soil moisture storage on flood frequency at an area and derivation of time-varying flood frequency estimates (23). Speyrer and Ragas examined the impact of flood risk and mandatory flood insurance on property values (24). Several studies performed the popular SPARROW model using a hybrid approach combining conventional regression methods with spatial data based on landscape characteristics and stream properties to predict continuous water quality from point observations (25-27). Firat compared different models for daily river flow forecasting and concluded that adaptive neuro-fuzzy inference system (ANFIS) has higher accuracy and reliability than other models (28). Soong and Soong compared flood-frequency curves derived from two different methods and concluded that flood-frequency curves derived from regional regression equations was slightly sloppy than those derived from the simulated flood-frequency curves (29). The aim of this study was to determine the physical and climatic factors affecting regional flood analysis, to find the best statistical distribution for determining the maximum flood discharges and also to compare two models of multivariate regression and hybrid method in the regional flood frequency analysis.

Material and Methods

Study area

The study area was located in Northeast of the country and from geological view, the sedimentary basin of Hezar Masjed- Kopedagh and Binalood as two separate tectonic, was the main part of the study area. Based on Domartan, the climate of these areas is arid and semi-arid. Figure 1 shows the location of the study stations in Khorasan Razavi.

Selection of the most suitable regional frequency distribution

After testing and data reconstruction, the flood frequency analysis included Pearson type Π distribution, Log Pearson type III, two-parameter gamma, three-parameter Lognormal, two-parameter normal and Lognormal, were fitted with peak discharge of each station by HYFA software. To select the best statistical distribution, Chi-square test was used, so that its minimum value was considered $\chi^2=1$ and its maximum value was considered $\chi^2=6$.

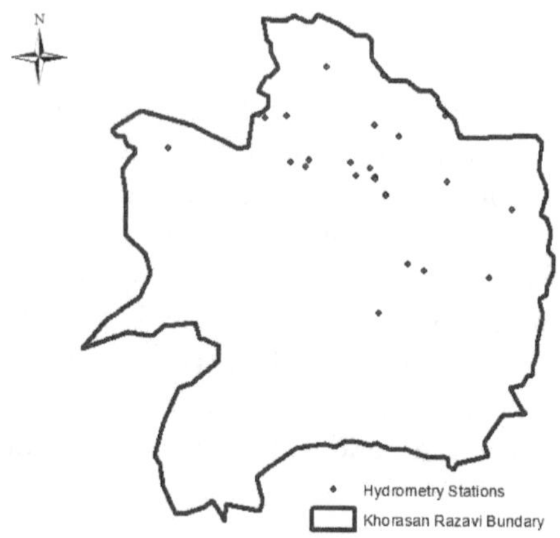

Figure 1. Location of the study stations in Khorasan Razavi.

Selection of the independent variables and limiting variables

In this study, a method of factor analysis was performed for 17 variables in the selected watersheds using SPSS software. Variables include different characteristics of watershed such as area, perimeter, average slope, the slope of the main river, length of the watershed, length of the main river, average height, average rainfall, maximum 24-hour precipitation, drainage density, maximum and minimum height and finally Gravelius, Horton, Miller and Schumm coefficients. As a unit of measurement, each variable differed from another one, therefore, for proper comparison of the variables with each other, all of them were standardized to be the same unit. As the results obtained from factor analysis was complex at first and did not produce an optimal solution, so to maximize the variance of each factors and to simplify the expression of variables of functional structure, functional axis has been rotated using a varimax rotation, one of the conventional methods, to be as an independent factor. Meanwhile, the naming of factors is done on the basis of the rotated functional values. Then, the matrices of functional scores of the stations were extracted using regression estimation. To confine the number of factors, Kaiser-Meyer-Olkin (KMO) measure was used to determine the rate of proportionality of a number of selective factors. Nonessential variables were extrapolated using the anti-image correlation matrix by measurement system analysis (MSA). These variables were diameter elements of the correlation matrix. In the elimination of data, the rate of KMO and percentage of the variance should be considered, because by eliminating one variable, KMO value and percentage of variance increase or decrease. The coefficient values of about 0.9 were considered very suitable, 0.8, suitable, 0.7, moderate, 0.6, medium, and less than 0.5 were not suitable. After determination of essential variables (12 factors), functional analysis was performed based on these variables.

Identification of homogeneous regional classes

The first step in the analysis of flood-prone area is to determine the homogeneous areas that were done using multivariate regression and index flood methods by cluster analysis. For this purpose, in the first step independent variables were determined by applying the process of factor analysis among input variables. In the next step, the homogenous areas were identified by cluster analysis using cumulative classes in SPSS software version 22. In the hybrid method, homogeneous areas were partitioned based on the most important parameters of the watershed (obtained from factor analysis), that affect the flood characteristics.

Hybrid method

Hybrid method is on the basis of the station-year method. In this method, all statistics available to the stations were used to overcome the lack of statistics. In the hybrid method, standardization was done by dividing each statistical data by a power of hydrological parameter of the watershed and then statistics of each class was combined with each other. The hybrid method included two sections: In the first section, the area was partitioned into at least three homogenous areas on the basis of the most important parameter of the watershed, and the flood peak statistics in each class were combined with each other. In the second section, statistics of annual flood peak discharge were standardized using an approximate factor, and then at the end of the hybrid process, an approximate factor of the standard was corrected using a technique combined from regression and flood frequency analysis. The process of standardization was done on using one of the parameters of the watershed in each replication. The model used in the hybrid method (equation 1), is like those that are used in different methods of regional flood analysis.

$$Q_T = aAbBcCd... \qquad (1)$$

Where Q_T is peak discharge in return period of T years, A, B, C are independent parameters of the watershed, B, c, d, symbols in the regression equation and A is the fixed rate of the equation.

Partitioning of homogenous areas was done on the basis of the most important parameter of the watershed that affects flood characteristics. So, the area of the watershed, as the most important parameter of the watershed, was used in the first process. Partitioning must be done in a way that in each class, total data would be at least 100 classes and sum of a number of data in different could be near to each other and otherwise, by weighing to class data could be near as far as possible. A maximum number of classes with respect to the hypothesis of at least 100 data in each class were determined using equation 2:

$$J \leq \frac{Nf}{100} \qquad (2)$$

Where J is the maximum number of classes while its minimum value is three classes and Nf is the number of total statistical data in each class. Then, the average weight of the catchment area (first parameter) was determined according to equation 3:

$$\overline{A}_j = \text{antilig} \frac{\sum_{j=1}^{g}\sum_{k=1}^{h} \log A_{ijk}}{gh} \qquad (3)$$

Where $\overline{A}_J$ is average weight of the area of the watershed in class I, A_{ijk}, area of the watershed in class i and at the station j and i=1, 2, 3…f, number of classes, J=1, 2, 3… g, number of stations in ith class, and K, number of years in the jth station. It should be noted that the number of classes should be determined on the basis of the second and nth physiographic parameters in order and the average weight of the area in classes should be also calculated.

Standardization of peak discharge was replicated from equation 4 and it was performed through dividing peak discharge by the average weight of the area in each class; in the first replication, b was considered equal to 1.

$$S_{ijk} = \frac{Q_{ijk}}{A_{ijk}^{b}} \quad (4)$$

Then, S_{ijk} of each class was fitted by the curve of intermittent flood or by an empirical equation. For each class, a suitable probability distribution was determined, then, S_{ijk} with different return periods for each class, was obtained using a suitable distribution (Sti). To obtain peak discharge with a return period of T years, the rates of S_{ti} was non-standardized as equation 5:

Qti=Sti (Āi) b (5)

The average weight of the area was fixed along with the replication but the values of b vary to a constant value. The new exponent in each return period was obtained as equation 6:

$$b_t = \frac{\sum_{i=1}^{F} \bar{A}_i Q_{ti} - \sum_{i=1}^{F} A_i \sum_{i=1}^{F} Q_{ti}}{\sum_{i=1}^{F} \bar{A}_i^2 - \frac{\left[\sum_{i=1}^{F} A_i\right]^2}{F}} \quad (6)$$

In fact, if a regression equation is plotted between Q_{ti} and A_i, b_t is the slope of the equation, in other words, it is the coefficient variation of A_i. By the new exponent of calculation, the next stage was replicated so that the rates of Q_{ijk} was standardized through dividing A_{ijk} by exponent of b_t, and stages of determination of a suitable distribution and calculation of S_{ti} were replicated to get a newer b_t. This replication was continued until the variation of b_t reached less than 1% . After application of a parameter, if b was not fixed, it might be indicated that there was no linear relationship between parameter and peak discharge, so it was not included in the model. When the most important parameter of the watershed (area) was included in the model and fixed exponent and final b_t were calculated, then, it was time to determine the second parameter of the watershed, for example, the annual rainfall of the watershed. After determination of the second parameter, all stages performed for the first parameter, were also replicated for the selected parameter with a difference that for standardization of discharges of each rainfall class, Q_{ijk} of each station was divided by (A^b P^c) to obtain b. Exponent of c was also equal to 1 in the first replicate. After standardizing the discharges, the steps for determining the S_{ti} and C_t are repeated. The repetition process continues as long as the power of each new parameter is fixed. The rate of coefficient A was determined based on the correlation.

Method of multivariate regression

In this method, some ratios are presented between discharges in different return periods and physiographic characteristics of the watershed (12). The general relation of multiple regression is as equation 7:

$$Q_T = F\left(A^a,\ B^b,\ C^c,\ \ldots,\ Z^z\right) \quad (7)$$

Estimation of the regional distribution parameters

In this method, regression equations were developed between characteristics of watershed and parameters of the probability distribution. Firstly, after frequency analysis, suitable parameters were obtained for each station, then; regression equations were developed if necessary, to estimate the parameters of the probability distribution. In fact, using characteristics of the watershed, distribution parameters were obtained for different areas with deficient statistics, then, by application of obtained distribution parameters, different Q_{Ts} were calculated (1,4).

Results

Functional analysis

In this stage, essential variables were determined based on the value of KMO = 0.721. After determination of the essential variables (14 factors), the functional analysis was performed based on these variables. Table 1 shows the rates of invisible matrix root and variance percentage of the factors.

By application of functional analysis, from input variables, four factors including area, mean annual rainfall, the average height of watershed and slope of the main river, were identified as the most important factors. The total variance of these factors was considered 91.18%.

Determination of the homogenous areas

In the study area, the cluster analysis method was carried out through cumulative classes. Factors included area, mean annual rainfall, the average height of watershed and pure slope of the river was done using hierarchical clustering method proposed by Wardez. By considering the maximum coefficient of similarity, two groups were obtained as follows:

The homogenous group 1: Watersheds of Golmakan, Dolat Abad, Mooshang, Sarasiab, Chakaneh, kooshk bala, Golestan, Hesar, anderkh, Zoshk, Zarandeh, Taghoon, Dorood, Nashib.

The homogenous group 2: Band Sarooj, Sang Divar, Ghara Tikan, Kalateh Rahman, Kartian, Agh Darband, Bagh Abbasi, Jaafar Mashhadi, Chahchaheh, Kardeh, Kashafrood, and Hatiteh.

The abovementioned processes are shown by a tree diagram (Figure 2).

Hybrid method

In the study area, 19 watersheds that had at least 10 statistical years of peak flood discharge, were selected. These 19 watersheds had totally 328 years of discharge statistics. After functional analysis, the factor of area was identified as the most important factor and also the factor of rainfall was identified as the second effective factor on the peak discharge. So, in the first step, the

Table 1. Values of invisible matrix root and variance percentage of the factors

Component (Factor)	Special primary values			Total	Variance percentage (%)	Cumulative variance
	Total	Variance percentage (%)	Cumulative variance			
1	6.581	54.839	54.839	4.401	36.357	36.67
2	1.966	16.385	71.223	2.766	23.05	59.72
3	1.313	10.941	82.164	2.536	21.137	80.86
4	1.083	9.023	91.187	1.239	10.32	91.187
5	0.371	3.09	94.276			
6	0.268	2.235	96.512			
7	0.213	1.77	98.29			
8	0.088	0.234	99.025			
9	0.044	0.354	99.379			
10	0.036	0.302	99.681			
11	0.029	0.242	99.992			
12	0.009	0.077	100			

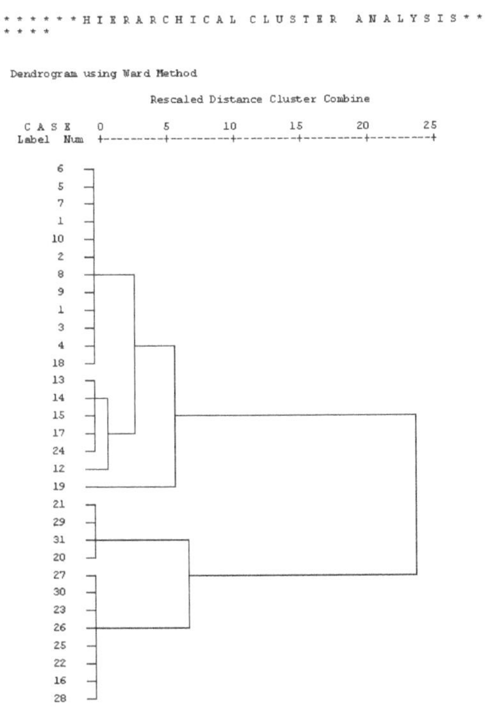

Figure 2. Tree diagram for area, mean annual rainfall, and mean height of watershed and pure slope of the river.

area was homogenized based on the basin area, then, the average weight of area for each class was calculated using equation 3. In the process of the hybrid method, In the process of hybrid, the instantaneous flood observations of the studied classes were standardized according to equation 4. Also the amount of b in the first replication was equal to 1. In other words, peak flood discharge of each station was divided by the exponent of its area to obtain the standardized discharges. As in the subsequent replication, the area exponent (b) would change, thus, for regional flood frequency analysis in each class; a long-term standardized compound statistic was obtained.

Then, based on the distribution of the three-parameter Lognormal (the best regional distribution), standard floods with return periods of 2, 5, 10, 25, 50 and 100 years were extracted for each class. After this stage, peak discharge of t years in each class (Q_{ti}) was determined as equation 5.

The second replication of the hybrid method was begun from equation 4. In this stage, instead of b that in the first stage was equal to 1, the rates of Table 2 that have been determined for each return period was used and replication process continued until exponent of b was fixed and if the power (b) is not fixed, the corresponding variable is abandoned and it indicating that there was no linear relationship between the related factor and discharge. In this research, the parameter b in the third replication means that the fourth step was fixed. These rates have been shown in Table 3.

After the factor of watershed area, mean annual rainfall was used. In the first step, the number of classes was determined and the stages done for the area, were replicated. As shown in equation 4, the method used for obtaining standard discharges for the first factor (area) and the second one (mean annual rainfall) is different. For subsequent agents, this process is repeated.

After determination of regression coefficients of the area in different return periods, the stages for subsequent important factors were replicated to complete the abovementioned relations. For discharges with different return periods, the abovementioned stages for the factor

Table 2. The rates of regression of watershed area factor (b) for the first replication in the study area

Return Period (y)	2	5	10	25	50	100
B-value	0.1612	0.2747	0.3389	0.4089	0.4534	0.4889

Table 3. The rates of regression coefficient of watershed area factor (b) for various replications in the study area.

Return Period (y)	2	5	10	25	50	100
First step	1	1	1	1	1	1
Second step	0.1612	0.2747	0.3389	0.4089	0.4534	0.4889
Third step	0.1373	0.2498	0.3391	0.4587	0.5339	0.6158
Fourth step	0.1401	0.2518	0.3391	0.4461	0.5184	0.5788

of mean annual rainfall were replicated and when the regression coefficient of these factors was fixed, the repetition was stopped and coefficients were replaced in the equation. In this study, the rates of c in different return periods were fixed after four steps (Table 4).

After the factor rainfall factors, average height and slope of watershed were determined and as the regression coefficients of these factors did not reach a fixed and logical value, they were not included in the model and put aside. After determination of c and b and other factors (average slope of the watershed, the average height of watershed) that could not include in the model, the fixed rate of the factor was determined. To determine regression fixed value (a) in each station, first, the rates of Q_t were obtained using the dominant distribution of the area, and simultaneously, with respect to physiographic characteristics included in the model, the rates of $(A^b P^c L^d)_t$ in different return periods for each station were calculated separately. Then, the ratio of Q_t to $(A^b P^c L^d)$ was determined for each station, finally, the median of these rates in different return periods, as regression fixed value (a), was included in the model. These rates are presented in the Table 5.

Multivariate regression

All models used for the estimation of peak flood discharge in a return period of 2, 5, 10, 25, 50 and 100 years for the whole area and homogenous areas of watersheds, are presented in Tables 6 and 7, respectively.

Discussion

The first step in the analysis of regional flood is to determine homogenous areas that were done using multivariate regression method by cluster analysis. For homogenizing the areas, firstly, functional analysis was performed on the input variables, and four factors including area, mean annual rainfall, average height of the watershed and slope of the main river, were identified as the most important factors. The total variance of these factors was considered 91.18%.

In the next stage, homogeneity of the watershed was determined by cluster analysis and two homogenous areas were determined. Also, with respect to four main variables, two series of models were estimated: one in the whole area and another one in the homogenous areas. Homogenous areas had a higher coefficient of determination (R^2) and less relative standard errors (RSE) compared to the models for the whole area. It was also revealed that with increase of return period, the rates of R^2 decreased but the rates of RSE increased. Figure 3 shows the mean relative error in the whole area and in the homogenous areas in the multivariate regression method.

Kjeldsen and Rosbjerg performed a flood frequency analysis in South Africa and suggested that the physiographic and climatic characteristics of the area including the area of the watershed, mean annual rainfall, rainfall concentration, the average slope of the watershed, altitude, edaphic characters and finally soil compaction coefficient, are effective in maximum flood discharges (14). Ouarda et al introduced some effective characteristics of the watersheds in peak discharges like area, length of the canal, the slope of the main canal and the mean annual rainfall (30). Honarbakhsh, in the analysis of regional flood in watersheds of Salt Lake,

Table 4. The rates of regression coefficient of mean annual rainfall factor (b) for various replications in the study area

Return Period (y)	2	5	10	25	50	100
First step	1	1	1	1	1	1
Second step	0.0716	0.0436	0.0716	-0.0089	-0.0146	-0.0124
Third step	0.0908	0.0806	0.057	0.01	0.0036	0.0012
Fourth step	0.09	0.0709	0.0578	0.0095	0.0032	0.001

Table 5. Regression fixed rates of hybrid model in various return periods

Return Period (years)	2	5	10	25	50	100
Regression fixed	54.17	57.96	44.49	35.9	111.7	100.9

Table 6. Multivariate regression models of peak discharge for the whole area

Return Period	Model	SE	R^2
2	Q = 480.9 + 0.154 A – 0.151 H	0.17	0.88
5	Q = 975.6 + 0.284 A – 0.321 H	0.18	0.88
10	Q = 1303 + 0.380 A – 0.439 H	0.2	0.87
25	Q = 1725.9 + 0.51 A – 0.594 H	0.19	0.86
50	Q = 2048 + 0.154 A – 0.716 H	0.2	0.8
100	Q = 2237 + 0.154 A – 0.841 H	0.21	0.73

Table 7. Multivariate regression models of peak discharge for the homogenous areas

Return Period	Model	SE	R^2
2	Q = 1085 + 0.126 A – 0.372 H	0.17	0.87
5	Q = 1856 + 0.216 A – 0.63 H	0.17	0.89
10	Q = 2335 + 0.277 A – 0.79 H	0.18	0.88
25	Q = 2915 + 0.355 A – 0.981 H	0.19	0.86
50	Q = 3335 + 0.413 A – 1.12 H	0.19	0.82
100	Q = 3752 + 0.47 A – 1.25 H	0.24	0.75

introduced physiographic characteristics like the area of the watershed, average height and the average slope of the watershed, annual rainfall and length of the watershed as the most important parameters affecting the flood (31). In this study, according to the functional analysis, area, mean annual rainfall, the average height of watershed and slope of the main river, were identified as the most important factors. Figure 3 shows that SE of the model of the whole area was higher than that of the homogenous areas and R^2 was also lower than it. Therefore, it is concluded that models of homogenous areas have higher efficiency compared to those of the whole area.

Conclusion

The aim of this study was to determine the physiographic and climatic characteristics affecting analysis of regional flood, to find the best statistical distribution for determining the maximum discharges and also to compare two models of multivariate and hybrid in the analysis of regional flood. Many studies have shown the importance of hydrologically homogenous areas in higher accuracy and efficiency of the regional analysis models. In his study, the importance and necessity of homogenous areas compared with the whole area were completely evident. Generally, homogenizing in most cases makes effective factors in the flow, to show higher accuracy in the model. The comparison showed that in the hybrid method, by increasing the return period, the rate of error decreased at first and then increased. Hybrid method in the return period of 50 years had a higher accuracy

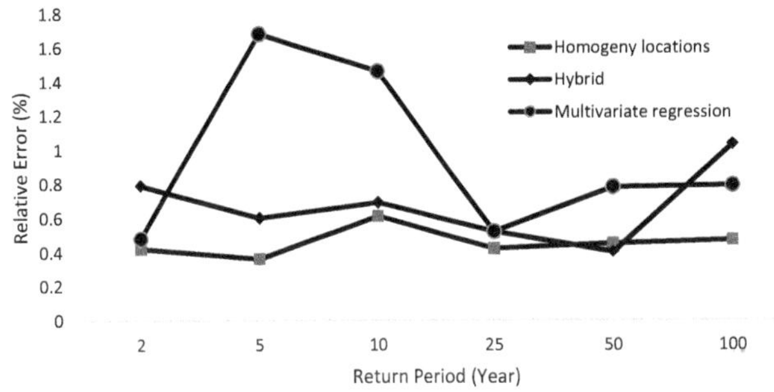

Figure 3. Comparison of the mean relative error in the multivariate regression (whole area) and the multivariate regression (homogenous areas) and hybrid regression.

compared to the multivariate regression method and presented better results of flood discharge for the area. Methods of multivariate regression and hybrid in a 25-year return period have similar accuracy.

Acknowledgements

The authors would like to gratitude the Khorasan Razavi Regional Water Company for providing and preprocessing the data on rainfall and flood hydrograph.

Competing interests

The author declares that he has no competing interests.

Authors' contributions

The author was involved in the suggestion of the problem, design of experiments, data collection, and article approval.

References:

1. Bhadra A, Panigrahy N, Singh R, Raghuwanshi NS, Mal BC, Tripathi MP. Development of a geomorphological instantaneous unit hydrograph model for scantily gauged watersheds. Environ Model Softw 2008; 23(8): 1013-25. doi: 10.1016/j.envsoft.2007.08.008.
2. Khaleghi MR, Gholami V, Ghodusi J, Hosseini H. Efficiency of the geomorphologic instantaneous unit hydrograph method in flood hydrograph simulation. Catena 2011; 87(2): 163-71. doi: 10.1016/j.catena.2011.04.005.
3. Khaleghi MR, Ghodusi J, Ahmadi H. Regional analysis using the geomorphologic instantaneous unit hydrograph (GIUH) method. Soil Water Res 2014; 9(1): 25-30.
4. Khaleghi MR. The influence of deforestation and anthropogenic activities on runoff generation. J For Sci 2017; 63(6): 245-53. doi: 10.17221/130/2016-JFS
5. Jain V, Sinha R. Derivation of unit hydrograph from GIUH analysis for a Himalayan River. Water Resources Management 2003; 17(5): 355-76. doi: 10.1023/a:1025884903120.
6. Gholami V, Khaleghi MR. The impact of vegetation on the bank erosion (case study: the Haraz River). Soil Water Res

2013; 8(4): 158-64.
7. Khaleghi MR, Varvani J. Simulation of relationship between river discharge and sediment yield in the semi-arid river watersheds. Acta Geophys 2018; 66(1): 109-19. doi: 10.1007/s11600-018-0110-9.
8. Khaleghi MR, Varvani J. Sediment rating curve parameters relationship with watershed characteristics in the semiarid river watersheds. Arab J Sci Eng 2018; 43(7): 3725-37. doi: 10.1007/s13369-018-3092-7.
9. Senaratne S, Cunnane C. Assessment of model error in regression estimates of floods at an ungauged site. Integrated Water Resources Management; 2000 Apr; LAHS, California; 2000.
10. Gioia A, Fiorentino M, Iacobellis V, Margiotta MR. Theoretical derivation of the index flood. Adv Geosci 2005; 2: 249-53. doi: 10.5194/adgeo-2-249-2005.
11. Yu PS, Chen CJ, Chen SJ, Lin SC. Application of Grey Model toward Runoff forecasting. J Am Water Resour Assoc 2001; 37(1): 151-66. doi: 10.1111/j.1752-1688.2001.tb05482.x.
12. Trivedi HV, Singh JK. Application of grey system theory in the development of a Runoff prediction model. Biosyst Eng 2005; 92(4): 521-6. doi: 10.1016/j.biosystemseng.2005.09.005.
13. Chou CM. Applying multi-resolution analysis to differential hydrological grey models with dual series. J Hydrol 2007; 332(1-2): 174-86. doi: 10.1016/j.jhydrol.2006.06.031.
14. Kjeldsen TR, Rosbjerg D. Comparison of regional index flood estimation procedures based on the extreme value type I distribution. Stoch Environ Res Risk Assess 2002; 16(5): 358-73. doi: 10.1007/s00477-002-0104-6.
15. Kuldkepp M. Rudolf Kjellén: geopolitiken och konservatismen edited by Ragnar Björk, Bert Edström and Thomas Lundén Stockholm, Hjalmarson & Högberg, 2014.337 pp., 198 kr, ISBN 978-91-7224-196-1. J Balt Stud 2017; 48(4): 549-51. doi: 10.1080/01629778.2017.1394699.
16. Kjeldsen TR, Smithers JC, Schulze RE. Regional flood frequency analysis in the KwaZulu-Natal province, South Africa, using the index-flood method. J Hydrol 2002; 255(1-4): 194-211. doi: 10.1016/S0022-1694(01)00520-0.
17. Eng K, Milly PC, Tasker GD. Flood regionalization: a hybrid geographic and predictor-variable region-of-influence regression method. J Hydrol Eng 2007; 12(6): 585-91. doi: 10.1061/(ASCE)1084-0699(2007)12:6(585).
18. Song S, Singh VP. Meta-elliptical copulas for drought frequency analysis of periodic hydrologic data. Stoch Environ Res Risk Assess 2010; 24(3): 425-44. doi: 10.1007/s00477-009-0331-1.
19. Marofi M. Investigation of the probability distributions fitting to extreme values of the flood. Geophysical Research Abstracts, EGS - AGU - EUG Joint Assembly; 6-11 April 2003; Nice, France.
20. Shukla RK, Trivedi M, Kumar M. On the proficient use of GEV distribution: a case study of subtropical monsoon region in India. Anale. Seria Informatică 2010; 83(1):81-92.
21. Haberlandt U, Radtke I. Hydrological model calibration for derived flood frequency analysis using stochastic rainfall and probability distributions of peak flows. Hydrol Earth Syst Sci 2014; 18(1): 353-65. doi: 10.5194/hess-18-353-2014.
22. Liu X, Coulibaly P, Evora N. Comparison of data-driven methods for downscaling ensemble weather forecasts. Hydrol Earth Syst Sci 2008; 12(2): 615-24. doi: 10.5194/hess-12-615-2008.
23. Smith JA, Karr AF. Flood frequency analysis using the cox regression model. Water Resour Res 1986; 22(6): 890-6. doi: 10.1029/WR022i006p00890.
24. Speyrer JF, Ragas WR. Housing prices and flood risk: An examination using spline regression. The Journal of Real Estate Finance and Economics 1991; 4(4): 395-407. doi: 10.1007/bf00219506.
25. Smith RA, Schwarz GE, Alexander RB. Regional interpretation of water-quality monitoring data. Water Resour Res 1997; 33(12): 2781-98. doi: 10.1029/97WR02171.
26. Lyon SW, Seibert J, Lembo AJ, Steenhuis TS, Walter MT. Incorporating landscape characteristics in a distance metric for interpolating between observations of stream water chemistry. Hydrol Earth Syst Sci 2008; 12(5): 1229-39. doi: 10.5194/hess-12-1229-2008.
27. Samadi MT, Sadeghi S, Rahmani A, Saghi MH. Survey of water quality in Moradbeik river basis on WQI index by GIS. Environ Health Eng Manag 2015; 2(1): 7-11.
28. Firat M. Comparison of artificial intelligence techniques for river flow forecasting. Hydrol Earth Syst Sci 2008; 12: 123-39.
29. Soong TW, Straub TD. Flood-Frequency Curves Estimated by Regional Regression Equations and Simulated Flood Series in the Ungaged Areas of Blackberry Creek Watershed, Illinois. World Water and Environmental Resources Congress; 2003 Jun 23-26; ASCE, Philadelphia; 2003. doi: 10.1061/40685(2003)357.
30. Ouarda T, Girard C, Cavadias GS, Bobee B. Regional flood frequency estimation with canonical correlation analysis. J Hydrol 2001; 254(1-4): 157-73. doi: 10.1016/S0022-1694(01)00488-7.
31. Honarbakhsh A. Regional analysis of flood in Namak lack watershed [dissertation]. Tehran: University of Tehran; 1996. [In Persian].

19

Removal of Congo red dye from aqueous solutions by a low-cost adsorbent: activated carbon prepared from Aloe vera leaves shell

Yusef Omidi Khaniabadi[1], Mohammad Javad Mohammadi[2,3], Mojtaba Shegerd[4], Shahram Sadeghi[5], Sedigheh Saeedi[6], Hassan Basiri[7,8*]

[1]MSc of Environmental Health, Health Care System of Karoon, Ahvaz Jundishapur University of Medical Sciences, Ahvaz, Iran
[2]PhD, Abadan School of Medical Sciences, Abadan, Iran
[3]PhD, Student Research Committee, Department of Environmental Health Engineering, School of Public Health and Environmental Technologies Research Center, Ahvaz Jundishapur University of Medical Sciences, Ahvaz, Iran
[4]BSc of Occupational Health, Health Center of East, Ahvaz Jundishapur University of Medical Sciences, Ahvaz, Iran
[5]MSc of Environmental Health, Environmental Health Research Center, Kurdistan University of Medical Sciences, Sanandaj, Iran
[6]MSc of Environmental Health, Sama Technical and Vocatinal College, Islamic Azad University, Izeh Branch, Izah, Iran
[7]MSc of Environmental Health, Rural Water and Wastewater Company, Semnan, Iran
[8]MSc of Environmental Health, Department of Environmental Health Engineering, School of Health, Lorestan University of Medical Sciences, Lorestan, Iran

Abstract
Background: Synthetic dyes have several harmful effects on human health as well as aquatic life. In this study, activated carbon (AV-AC), based on Aloe vera leaf shells, was used as a novel agricultural adsorbent, one that is low-cost and available for the removal of Congo red (CR) as a carcinogenic dye from aqueous solutions.
Methods: In the batch system, the influence of different parameters like contact time, pH, adsorbent dosage, and initial CR concentration were examined on the dye removal from liquid medium. The experimental data were fitted by pseudo-first-order and pseudo-second-order kinetics, and also Langmuir and Freundlich isotherms models.
Results: The optimum contact time and pH for the uptake of CR were obtained at 20 minutes and acidic pH of 2. The maximum uptake capacity of CR dye by AV-AC was 1850 mg/g. The results showed that the experimental data were well-fitted by the pseudo-second-order kinetic model ($R^2 > 0.99$) and Freundlich isotherm model ($R^2 > 0.99$).
Conclusion: According to the results of our study, the AV-AC is a low-cost, non-toxic, and effective adsorbent for the uptake of CR dye from aqueous media.
Keywords: Aloe Vera, Congo red, Kinetic, Charcoal

***Correspondence to:**
Hassan Basiri
Email: h.basiri29@gmail.com

Introduction

Water pollution caused by discharge of wastewater as a result of industrial activities has been considered one of the grave environmental issues in the world, especially in developing countries (1-3). The discharge of many organic and inorganic pollutants into water media by different industries, as a global environmental problem, has stimulated worldwide attention because of their harmful effects on the environment and also human health (4,5). One of the most important of these pollutants is dye substances. Various industries generate large volumes of dye containing effluents, such as the textiles, paper, rubber, leather, cosmetics, and food industries. The disposal of these coloured wastewaters without acceptable treatment into aquatic bodies will adversely influence the natural aquatic environment as well as have carcinogenic and mutagenic effects on human beings (6,7). Thus, removal of these dyes is essential before these are discharged into the water bodies (8). Among different kinds of anionic dyes, Congo red (CR) (4-aminonaphthalene-1-sulphonic acid) is a benzidine-based anionic diazo dye, which is generated by some industries, including textiles, printing and dyeing, paper, rubber and plastic industries. It is toxic for many organisms and is a doubted carcinogen and mutagen dye (6). CR is very stable to biological degradation; it has a complex aromatic structure (5,9,10). Several physical, chemical and microbial methods, including membrane separation, ion exchange, biological degradation, coagulation-flocculation, electrochemical techniques, chemical oxidation, and adsorption, have been used to remove CR from coloured water (11,12). Among these processes, the adsorption technique is current due to its simplicity, high sorption capacity, environmental friendliness, non-toxicity, and availability of a wide range of sorbents. Adsorption onto activated carbon because of its high surface area and high uptake capacity has been commonly used as an adsorbent for the remove of organic pollutants, but it is high-priced, and is hard to regenerate (2,13). In addition to activated carbon, some adsorbents—such as montmorillonite (14), rice husk (15), bentonite (16), rice hull ash (17), *Azadirachta indica* leaf (18), Jujuba seeds (19), red mud (20), polypyrrole–polyaniline nanofibres (21), anion exchange membrane (22), and ball-milled sugarcane bagasse (23)—have been employed for the removal of CR dye. Today, more attention is being given to the use of low-cost adsorbents. Aloe vera grows in warm tropical areas such as the United States, Africa, Australia, South America, India and Iran (2). The original material of Aloe vera is used in the production of latex substance and drugs, but the Aloe vera leaves shell are by-products of the agricultural industries (2,3,24). Omidi-Khaniabadi et al used the Aloe vera leaf wastes as a novel adsorbent for the adsorption of pollutants such as 4-chlorophenol (2), methylene blue (24), aniline and methyl orange (6) from aqueous solutions. In all the abovementioned contaminants, the results obtained were of interest for their removal. In this study, Aloe vera leaves shells were used as a cheap adsorbent to produce the activated carbon for CR dye uptake from synthetic wastewater.

Materials and Methods

Materials

Aloe vera leaves were collected from suburban farms of Ahvaz city in Khuzestan province, Iran. The chemical substances—which included CR dye, H_2SO_4 and NaOH—were acquired from Merck Co, Germany. The molecular structure of CR dye is presented in Figure 1. Dye stock solution (1000 mg/L) was prepared by dissolving 1 g of CR dye in 1 L deionized water and different concentrations were made by dilution of the stock solution. A rotary shaker (Behdad-Rotomix model) was used to mix the suspension of dye and AV-AC at 150 rpm.

Preparation of adsorbent

All the waste was carefully washed by deionized water to eliminate the pollutants and impurities. It was then dried in an oven at 105°C for 24 hours. The dried shells were crushed through a laboratory mill and were then sieved. The particle sizes with a range of 300–500 μm were carbonized in a furnace at 550°C for 20 minutes in a without-air condition and prepared activated carbon was again crushed and sieved (mesh no = 40) for the uptake experiments.

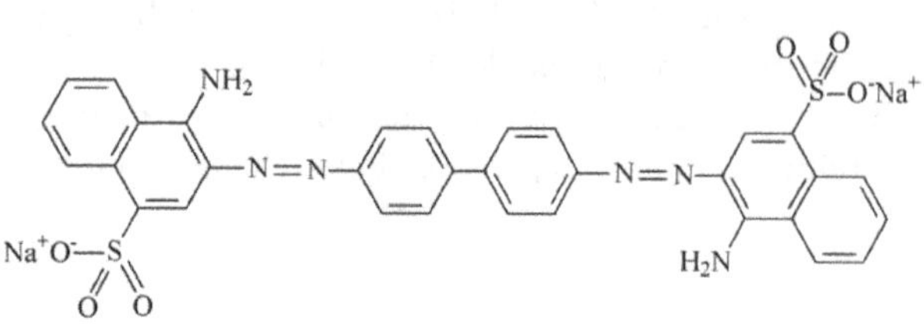

Figure 1. Molecular structure of Congo red (CR).

Characterization and analysis

For elemental analysis of AV-AC samples, a Heraeus Elemental Analyser (Jobin-Yvon Ultima ICP-AES) was used. The surface morphology of the adsorbent was characterized (before and after CR uptake) by a scanning electron microscope (SEM, Jeol Model Jsm-T330). The remaining concentration of CR in the solution was measured through an UV-Vis spectrophotometer (PG Instrument Limited Model) at a maximum absorbance wavelength of 495 nm (λ max).

Zero point of charge (pHzpc) of the adsorbent was determined by preparing 50 mL 0.01 M NaCl solution in a 100 mL Erlenmeyer flask. The initial pH in each flask was adjusted in the range of 2 to 12 by 0.1 N H_2SO_4 or NaOH solutions. In each flask containing NaCl solution, 0.1 g of sorbent was poured, and then the suspensions were mixed for 24 hours by a rotary shaker. After mixing stage, the final pH of solutions was determined to identify pHzpc.

Adsorption experiments

The batch adsorption experiments were conducted to identify the effects of various factors including contact time (0–100 minutes), pH (2–12), adsorbent dosage (0.5–5 g/L), and adsorbate content (100–500 mg/L) on the sorption of CR by using AV-AC. In all, 44 samples were taken, and the results analysed and the figures drawn in Excel software. All the experiments were performed at laboratory temperature (25°C) and were then shacked at 150 rpm in a 250 mL Erlenmeyer flask. After agitation time, the suspension was filtered through a fibreglass paper (0.45 μm) and the absorbance of the clear sample was analysed to determine the remaining CR. The amounts of CR adsorbed on AC-AV and the sorption percentage (%) were computed by the following equations:

$$q_e = \frac{(C_0 - C_e)V}{m} \tag{1}$$

$$Sorption(\%) = \frac{(C_0 - C_e)}{C_0} \times 100 \tag{2}$$

where q_e (mg/g) is the uptake capacity the AV-AC. C_0 and C_e are respectively the initial and equilibrium dye concentrations in the solution (mg/L). V and m are respectively the volume of solution (L) and adsorbent mass (mg).

Results

Characterization

The elemental analysis of chemical composition of the AV-AC illustrated that Ca, O, K, and Mg are the majors that formed 92.7% of the weight (wt. %) of the adsorbent.

Na (5.92%) and Cl (1.35%) are the minors comprising the adsorbent (2,3). Figure 2A and 2B respectively explain the morphology of the surface of activated carbon before and after the CR uptake.

Effect of contact time

The effects of different contact times on the sorption of CR dye by Aloe vera leaves shell-based activated carbon (AV-AC) at the time intervals of 0–100 minutes in 100 mL dye solution and 1 g/L of AV-AC at the initial pH of solution were investigated. The results are shown in Figure 3A.

Kinetics study

The experimental data of CR dye by AV-AC was fitted by three adsorption kinetics, including pseudo-first-order, pseudo-second-order, and intraparticle diffusion models to obtain better understanding of the sorption process. The pseudo-second-order kinetic model and parameters of the uptake of CR sorption by AV-AC are illustrated in Figure 3B and Table 1 respectively. As can be seen in Table 1, the highest correlation coefficient (R^2) of CR adsorption was related to the pseudo-second-order kinetic. Therefore, the adsorption process of CR dye by AV-AC was followed by the pseudo-second-order kinetic model.

Effect of solution pH and adsorbent dosage

The effect of solution pH (2–12) on the uptake of CR (initial concentration = 100 mg/L) by AV-AC has been shown in Figure 4A. As is obvious, by increment the solution pH, from 2 to 12, the uptake capacity of the adsorbent was decreased. The maximum uptake of CR (100 mg/g) by AV-AC was obtained in pH 2. Thus, pH 2 has been selected as the optimum pH for the next stages. As seen in Figure 4A, the isoelectric point of the activated carbon from natural adsorbent was observed at pH of 11.3. There were positive charges on the AV-AC surface at the lower pH values of the isoelectric point, which will promote a reaction with CR as an azo dye.

The effect of different adsorbent dosages of AV-AC in an initial CR concentration (100 mg/L) at 25°C was examined. The results are explained in Figure 4B.

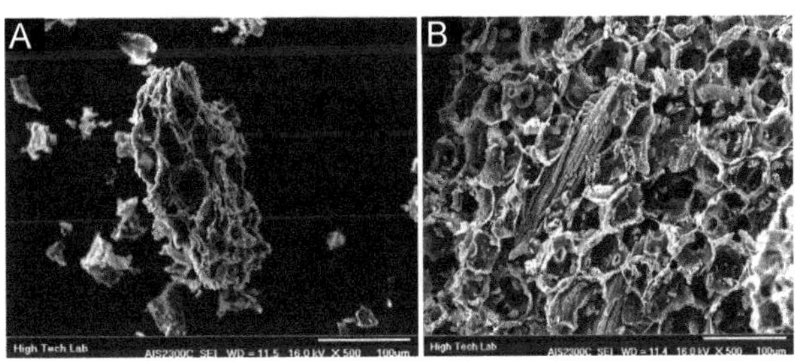

Figure 2. SEM image the AV-AC (A) before and (B) after the uptake process of Congo red (CR).

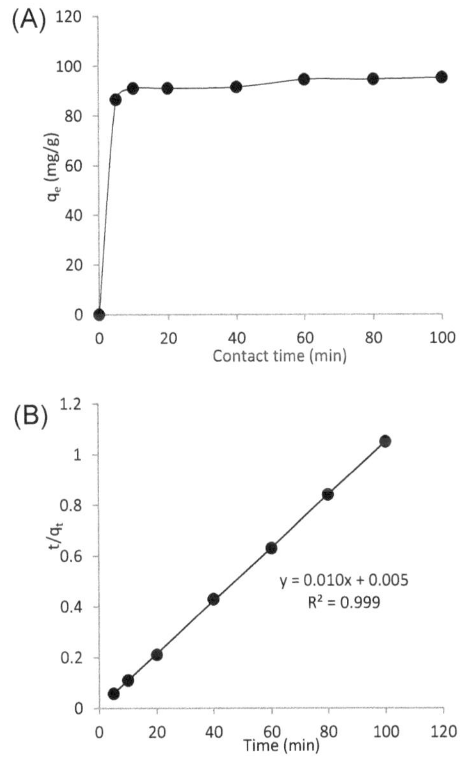

Figure 3. (A) The effect of contact time on the sorption of Congo red (CR) dye by AV-AC (CR con. = 100 mg/L, pH = 7 and adsorbent dosage = 1 g/L) and (B) pseudo-second-order kinetic model.

Effect of initial dye concentration

The influence of initial content (100 to 500 mg/L) on the uptake capacity of CR by AV-AC was evaluated. As shown in Figure 5A, the adsorption value of AV-AC at the beginning was quickly increased by the rise in CR concentration in the dye solution.

Isotherm study

Two uptake isotherm models, including Langmuir and Freundlich, were applied to fit the experimental data of CR dye removal by AV-AC.

As can be seen in Figure 5B, the amount of K_f and n were obtained from the cut-off and gradient of liner plotting of ln q_e vs. ln C_e respectively. The uptake isotherm parameters of the Langmuir and Freundlich models are presented in Table 2. As is obvious, the value of the correlation coefficient (R^2) of the Langmuir ($R^2 > 0.96$) model was lower than Freundlich ($R^2 > 0.99$) isotherm model.

Discussion

As seen in Figure 2A, before the uptake of CR dye, the surface morphology of adsorbent had irregular cavities

Table 1. Calculated kinetic parameters of Congo red (CR) dye sorption onto AV-AC

Adsorbate	Pseudo-first-order			Pseudo-second-order			Intraparticle diffusion		
	q_e(mg/g)	K_1 (1/min)	R^2	q_e (mg/g)	K_2 (g/mg.min)	R^2	C (mg/g)	K (mg/g min$^{1/2}$)	R^2
Congo red	1.03	2×10^{-4}	0.84	100	2.4×10^{-3}	0.99	2.52	8×10^{-2}	0.97

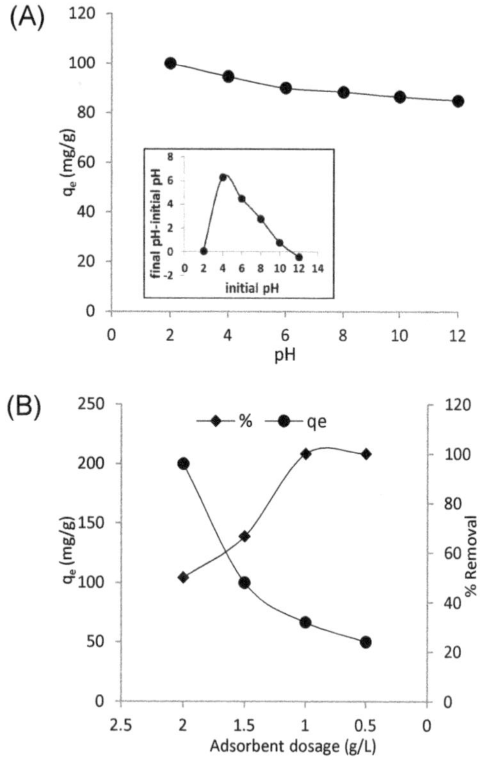

Figure 4. (A) The effect of pH on the uptake of CR dye by AV-AC (contact time=20 min, CR solution con.=100 mg/L and activated carbon dosage=1 g/L) and (B) the effect of adsorbent dosage on the sorption (contact time=20 min, CR solution con.=100 mg/L and pH=2).

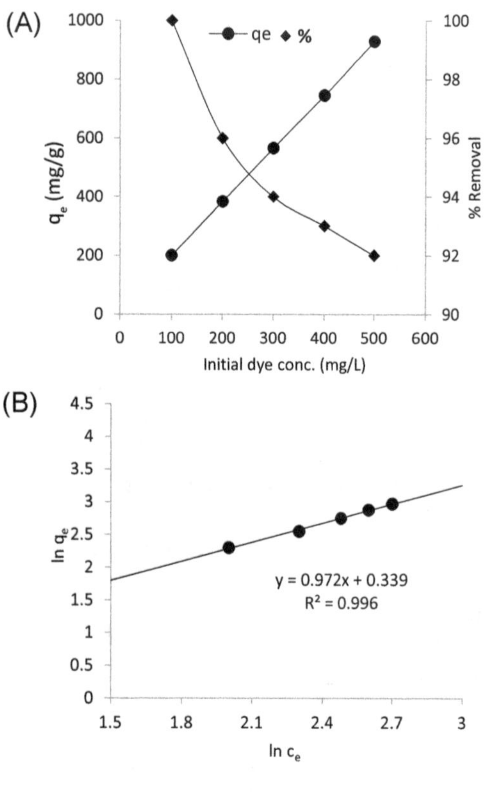

Figure 5. (A) The effect of initial concentration on the sorption of CR dye by AV-AC (contact time=20 min, pH=2 and activated carbon dosage=0.5 g/L) and (B) Freundlich isotherm model.

via fine open holes. After adsorption, as can be seen in Figure 2B, a smoother configuration, regular surface, and expanded pores were formed, after the uptake of adsorbate onto activated carbon (3). The expansion of the pores can be because of the uptake of CR onto the sorbent surface that filled these pores.

The contact time of the uptake process is an important parameter in the context of the economic aspects of the treatment of polluted water (3). The plot of contact time illustrates that the sorption of CR dye was rapidly increased at the first of the uptake process due to the availability of more vacant sites at the adsorbent surface and, with increases in contact time, these vacant sites were filled by CR. Thus, adsorption capacity was gradually increased. The equilibrium was acquired at contact time of 20 minutes (q_e= 91 mg/g) and then slowly reached a fixed state during the residual time up to 100 minutes. Thus, the contact time of 20 minutes was chosen as the optimum for the subsequent stages.

Table 3 illustrates the uptake capacity and equilibrium time of CR uptake by various adsorbents. As can be seen, the comparison of low-cost adsorbent (AV-AC) with other sorbents of CR illustrates that this adsorbent has a high uptake capacity (91.0 mg/g) and its sorption is approximately 3 to 5 times higher than that of chitosan (19.15 mg/g), valoria bryopsis (10.52 mg/g), and pinus brutia (35.3 mg/g), obtained from other studies. This may be due to the unique characteristics of AV-AC (2).

The pseudo-first-order kinetic is expressed by Eq. (3):

$$ln\,(q_e - q_t) = ln\, q_e - K_1 t \tag{3}$$

where q_e (mg/g) and q_t (mg/g) are respectively the amounts of CR dye adsorbed onto the AV-AC surface at the equilibrium and at t (min). The parameters of K_1 (uptake rate constant, 1/min) and q_e were determined from linear plotting of $ln\,(q_e - q_t)$ against t (min), which were acquired from the gradient and cut-off of the plot respectively (2). The experimental data of the CR adsorption were also analysed by the pseudo-second-order kinetic. This sorption model can be expressed by the Eq. (4):

$$\frac{t}{q_t} = \frac{1}{k_2 q_e^2} + \frac{t}{q_e} \tag{4}$$

where q_e and q_t have similar definitions to the parameters of pseudo-first-order kinetic mentioned above. Kinetic rate constant (K_2, g/mg. min) and adsorption capacity at equilibrium state (q_e, mg/g) were obtained from the intercept and slope of t/q_t against t (min) respectively. The intraparticle diffusion model is explained by Eq. (5):

$$q_t = k_{id}\sqrt{t} + C \tag{5}$$

where K_{id} (g/mg.min$^{1/2}$) is the kinetic model constant rate and C (mg/g) is associated with the thickness of the boundary layer. The amounts of C and K_{id} are achieved from the intercept and slope of the liner plot of q_t versus $\sqrt{t}$ respectively.

Cheng et al illustrated the pseudo-second-order kinetic well followed the uptake of CR dye via activated carbon/

Table 2. Calculated isotherm parameters of Congo red (CR) dye sorption onto AV-AC

Adsorbate	Langmuir			Freundlich		
	Q_m (mg/g)	b (L/mg)	R^2	K_f	n	R^2
Congo red	1850	8.5×10^{-4}	0.96	4.6×10^{-2}	1.02	0.99

Table 3. Comparison of various adsorbents in respect of sorption of Congo red (CR) dye from aqueous media

Adsorbent	C_0 (mg/L)	Equilibrium time (min)	q_e (mg/g)	Ref.
EBTAC	50	22h	21.73	(22)
CB-CTAB	20	4h	19.15	(25)
Valoria bryopsis	30	180	10.52	(26)
Pinus brutia	100	120	35.3	(27)
Maghemite nanoparticles	20	30	20.0	(28)
Eichhornia crassipes	104.45	90	4.97	(29)
PPy–PANI NFs	50	120	50.50	(30)
AV-AC	100	20	91.0	This study

surfactant than other kinetics (5). In addition, according to the amount of C (mg/g) for the uptake of CR, the liner plot of this kinetic model did not pass through the origin. Thus, the intraparticle diffusion model is not the step of rate limiting in the adsorption of CR dye by AV-AC.

The solution pH changes the surface properties of the adsorbent and the level of adsorbate ionization (2,3). Furthermore, the uptake of adsorbate is affected by the solution pH due to the sorption of H^+ and OH^- ions in the liquid (10,25). This outcome can be due to the change in the surface charge of the activated carbon. The amounts of positive charges onto the adsorbent surface were enhanced through decrease in initial solution pH, subsequently causing increase in the sorption capacity in more acidic solution pH. The electrostatic repulsions between the negatively charged AV-AC and anionic dye of CR were increased in the basic pH.

It was identified that with the increase in the AV-AC dosage from 0.5 to 2 g/L, the adsorption capacity was reduced from 200 to 50 mg/g. Decrease in the sorption amounts of upper dosages may be due to the inaccessibility of the dye molecules that cannot cover all the surface sites of the adsorbent. In other words, a huge number of the active sites on the AV-AC surface cannot attain saturation state at higher dosages of the activated carbon. So, dose of 0.5 g/L of AV-AC was chosen as the optimum dose for the subsequent stages of the sorption process.

The quantity of CR adsorbed was incremented from 200 to 929.88 mg/g by the increase in the adsorbate content from 100 to 500 mg/L. This finding can be due to the fact that the adsorbed polluted created van der Waal's driving force to overcome the mass transmission resistance of CR dye to the active surface sites of the AV-AC. Thus, the above-mentioned power increase through an increase in the adsorbate content, therefore, caused an enhancement of CR sorption capacity.

Adsorption isotherms provide helpful requirements to identify the uptake mechanism and characteristics of the sorbent surface for the design of sorption systems (26,27). The Langmuir model assumes that the monolayer uptake occurred onto the homogeneous surface of the AV-AC. This model is widely used to the uptake of pollutants from aqueous media. The Langmuir isotherm model is exhibited by Eq. (6):

$$\frac{C_e}{q_e} = \frac{C_e}{Q_m} + \frac{1}{bQ_m} \tag{6}$$

where q_e (mg/g) and C_e (mg/L) are the uptake capacity of the adsorbent and initial CR concentration at equilibrium time respectively. The values of Q_m (mg/g, the maximum uptake capacity of monolayer sorbent) and b (L/mg, constant rate of Langmuir model) were obtained from the gradient and cut-off of liner plotting of C_e/q_e versus C_e respectively. The Freundlich isotherm model is generally used for multilayer adsorption on heterogeneous surface of the adsorbent. This uptake model can be illustrated by the following equation:

$$\ln q_e = \ln k_f + \frac{1}{n} \ln C_e \tag{7}$$

where K_f (L/g) and n are the constant rates of isotherm model respectively associated with the uptake capacity and adsorption intensity.

It can be found that the experimental data of CR dye onto AV-AC were well-fitted to the Freundlich isotherm model. In general, the adsorption bonds between activated carbon and adsorbed CR dye were suitably powerful, since the n value achieved from the Freundlich model was further than that one. Rouf et al showed that the experimental data of CR dye by chitosan were well followed through the Freundlich isotherm model (28).

Comparison with other studies

The amount of maximum monolayer uptake capacity of CR dye by AV-AC was obtained at 1850 mg/g. This amount has been compared with Q_m achieved from the other studies for the removal of CR. Table 4 shows the maximum uptake capacity from the Langmuir model by various sorbents. It can be noted that the maximum of CR

Table 4. Comparison of the maximum uptake capacity of AV-AC with that of other adsorbents

Adsorbents	Q_m (mg/g)	Ref.
Eichhornia crassipes	1.58	(29)
Activated carbon (laboratory grade)	1.88	(34)
Cashew nut shell	5.18	(35)
Kaolin (clay materials)	5.44	(36)
Activated carbon prepared from coir pith	6.70	(37)
Acid activated red mud	7.08	(20)
Bagasse fly ash	11.89	(34)
Montmorillonite	12.70	(14)
Cattail root	38.79	(9)
Ball-milled sugarcane bagasse	39.8	(23)
Neem leaf powder	41.20	(38)
Magnetic core–manganese oxide shell	42.0	(39)
Chitosan/montmorillonite nanocomposite	54.52	(14)
Chitosan beads modified with CTAB	94.39	(25)
Ca-bentonite	107.41	(40)
Mesoporous activated carbon	189.0	(41)
Maghemite nanoparticles	208.33	(28)
Cone of Pinus brutia biomass	329.6	(27)
CTAB modified chitosan beads	352.2	(42)
AC/DDAC	769.23	(5)
AV-AC	1850	This study

uptake capacity of the AV-AC is noticeably further than the maximum capacity of other sorbents.

Conclusion

In this study, CR was removed from aqueous media by Aloe vera leaf shells-based activated carbon (AV-AC). The influence of different parameters like contact time, pH, adsorbent dosage, and CR content were evaluated on the uptake capacity of the adsorbent. The optimum uptake capacity of the AV-AC (91.0 mg/g) was achieved within 20 minutes at pH of 2. The maximum monolayer adsorption capacity was obtained equal to 1850 mg/g. The results showed that the experimental data were well-followed from the second-order-kinetic and Freundlich isotherm models. This natural adsorbent has several advantages, such as low cost, environmental friendliness, high uptake capacity, and non-toxicity. Thus, it can be considered an effective adsorbent on the removal of CR dye from aqueous solutions.

Acknowledgments

The authors wish to thank the Lorestan University of Medical Sciences.

Competing interests

The authors declare that they have no competing interests.

Authors' contributions

Study concept, design, and critical revision of the manuscript for important intellectual content: YOK; drafting of the manuscript and advisor, and performing experiments: YOK, HB, MJM, MS, ShS, and SeS.

References

1. Nourmoradi H, Nikaeen M, Khiadani (Hajian) M. Removal of benzene, toluene, ethylbenzene and xylene (BTEX) from aqueous solutions by montmorillonite modified with nonionic surfactant: Equilibrium, kinetic and thermodynamic study. Chem Eng J 2012; 191: 341-8. doi: 10.1016/j.cej.2012.03.029
2. Omidi-Khaniabadi Y, Jafari A, Nourmoradi H, Taheri F, Saeedi S. Adsorption of 4-chlorophenol from aqueous solution using activated carbon synthesized from aloe vera green wastes. J Adv Environ Health Res 2015; 3(2): 120-9.
3. Basiri H, Nourmoradi H, Moghadam FM, Moghadam KF, Mohammadian J, Khaniabadi YO. Removal of aniline as a health-toxic substance from polluted water by aloe vera waste-based activated carbon. Der Pharma Chemica 2015; 7(11): 149-55.
4. Jourvand M, Shams Khorramabadi G, Omidi-Khaniabadi Y, Godini H, Nourmoradi H. Removal of methylene blue from aqueous solutions using modified clay. J Bas Res Med Sci 2015; 2(1): 32-41.
5. Cheng Z, Zhang L, Guo X, Jiang X, Li T. Adsorption behavior of direct red 80 and Congo red onto activated carbon/surfactant: process optimization, kinetics and equilibrium. Spectrochim Acta A Mol Biomol Spectrosc 2015; 137: 1126-43. doi: 10.1016/j.saa.2014.08.138.
6. Omidi Khaniabadi Y, Heydari R, Nourmoradi H, Basiri H, Basiri H. Low-cost sorbent for the removal of aniline and methyl orange from liquid-phase: Aloe vera leaves wastes. Journal of the Taiwan Institute of Chemical Engineers 2016; 68: 90-98. doi: 10.1016/j.jtice.2016.09.025.
7. Omidi Y, Basiri H, Jafari A, Saeedi S, Goudarzi G, Taheri F, et al. Removal of textile dye methylene blue from liquid-phase by activated carbon from aloe vera wastes. Jundishapur J Health Sci 2016; In Press. doi: 10.17795/jjhs-38242.
8. Wasti A, Awan MA. Adsorption of textile dye onto modified immobilized activated alumina. Journal of the Association of Arab Universities for Basic and Applied Sciences 2016; 20: 26-31. doi: 10.1016/j.jaubas.2014.10.001.
9. Hu Z, Chen H, Ji F, Yuan S. Removal of Congo red from aqueous solution by cattail root. J Hazard Mater 2010; 173(1-3): 292-7. doi: 10.1016/j.jhazmat.2009.08.082.
10. Chatterjee S, Lee MW, Woo SH. Adsorption of Congo red by chitosan hydrogel beads impregnated with carbon nanotubes. Bioresour Technol 2010; 101(6): 1800-6. doi: 10.1016/j.biortech.2009.10.051.
11. Pavan FA, Dias SLP, Lima EC, Benvenutti EV. Removal of Congo red from aqueous solution by anilinepropylsilica xerogel. Dyes Pigm 2008; 76(1): 64-9. doi: 10.1016/j.dyepig.2006.08.027.
12. Khodadadi M, Saghi MH, Azadi NA, Sadeghi S. Adsorption of chromium VI from aqueous solutions onto nanoparticle sorbent: chitozan-Fe-Zr. J Mazandaran Univ Med Sci 2016; 26(141): 70-82.
13. Chen H, Zhao J. Adsorption study for removal of Congo red anionic dye using organo-attapulgite. Adsorption 2009; 15(4): 381-9. doi: 10.1007/s10450-009-9155-z.
14. Wang L, wang A. Adsorption characteristics of Congo red onto the chitosan/montmorillonite nanocomposite. J Hazard Mater 2007; 147(3): 979-85. doi: 10.1016/j.

jhazmat.2007.01.145.

15. Han R, Ding D, Xu Y, Zou W, Wang Y, Li Y, et al. Use of rice husk for the adsorption of Congo red from aqueous solution in column mode. Bioresource Technology 2008; 99 (8): 2938-46. doi: 10.1016/j.biortech.2007.06.027.
16. Bulut E, Özacar MO, Şengil İA. Equilibrium and kinetic data and process design for adsorption of Congo Red onto bentonite. J Hazard Mater 2008; 154(1-3): 613-22.
17. Chou KS, Tsai JC, Lo CT. The adsorption of Congo red and vacuum pump oil by rice hull ash. Bioresour Technol 2001; 78(2): 217-9.
18. Shankar S, Rai A, Ahmad A, Sastry M. Rapid synthesis of Au, Ag, and bimetallic Au core–Ag shell nanoparticles using Neem (*Azadirachta indica*) leaf broth. J Colloid Interface Sci 2004; 275(2): 496–502. doi: 10.1016/j.jcis.2004.03.003.
19. Reddy MC, Sivaramakrishna L, Reddy AV. The use of an agricultural waste material, Jujuba seeds for the removal of anionic dye (Congo red) from aqueous medium. J Hazard Mater 2012; 203-204: 118-27. doi: 10.1016/j.jhazmat.2011.11.083.
20. Tor A, Cengeloglu Y. Removal of Congo red from aqueous solution by adsorption onto acid activated red mud. J Hazard Mater 2006; 138(2): 409-15. doi: 10.1016/j.jhazmat.2006.04.063.
21. Bhaumik M, Maity A, Srinivasu V, Onyango M. Removal of hexavalent chromium from aqueous solution using polypyrrole-polyaniline nanofibers. Chem Eng J 2012; 181-182: 323–33. doi: 10.1016/j.cej.2011.11.088.
22. Khan MA, Akhtar S, Zafar S, Shaheen A, Khan MA, Luque R, et al. Removal of Congo red from aqueous solution by anion exchange membrane (EBTAC): adsorption kinetics and themodynamics. Materials 2015; 8(7): 4147-61. doi: 10.3390/ma8074147.
23. Zhang Z, Moghaddam L, O'Hara IM, Doherty WOS. Congo Red Adsorption by Ball-Milled Sugarcane Bagasse. Chem Eng J 2011; 178: 122-8. doi: 10.1016/j.cej.2011.10.024
24. Omidi-Khaniabadi Y, Jafari A, Basiri H, Saeedi S, Taheri F. A natural sorbent for the removal of textile dye methylene blue from liquid-phase: Aloe vera wastes. Int J Chem Environ Eng 2016; 7(1): 53-9.
25. Rouf S, Nagapadma M, Ramakoteswara Rao R. Removal of harmful textile dye Congo red from aqueous solution using chitosan and chitosan beads modified with CTAB. Int J Eng Res Appl 2015; 5(3): 75-82.
26. Jayaraj R, Jeyasingh Thanaraj P, Thillai Natarajan S, Martin Deva Prasath P. Removal of Congo red dye from aqueous solution using acid activated eco-friendly low cost carbon prepared from marine algae valoria bryopsis. J Chem Pharm Res 2011; 3(3): 389-96.
27. Bouguettoucha A, Chebli D, Mekhalef T, Noui A, Amrane A. The use of a forest waste biomass, cone of *Pinus brutia* for the removal of an anionic azo dye Congo red from aqueous medium. Desalination and Water Treatment. 2015; 55(7): 1956-65. doi: 10.1080/19443994.2014.928235.
28. Afkhami A, Moosavi R. Adsorptive removal of Congo red, a carcinogenic textile dye, from aqueous solutions by maghemite nanoparticles. J Hazard Mater 2010; 174(1-3): 398-403. doi: 10.1016/j.jhazmat.2009.09.066.
29. Wanyonyi WC, Onyari JM, Shiundu PM. Adsorption of Congo red dye from aqueous solutions using roots of *Eichhornia crassipes:* kinetic and equilibrium studies. Energy Procedia 2014; 50: 862-9. doi: 10.1016/j.egypro.2014.06.105.
30. Bhaumik M, McCrindle R, Maity A. Efficient removal of Congo red from aqueous solutions by adsorption onto interconnected polypyrrole–polyaniline nanofibres. Chem Eng J 2013; 228: 506-15. doi: 10.1016/j.cej.2013.05.026.
31. Du Q, Sun J, Li Y, Yang X, Wang X, Wang Z, et al. Highly enhanced adsorption of congo red onto graphene oxide/chitosan fibers by wet-chemical etching off silica nanoparticles. Chem Eng J 2014; 245: 99-106. doi: 10.1016/j.cej.2014.02.006.
32. Mohd Din AT, Hameed BH, Ahmad AL. Batch adsorption of phenol onto physiochemical-activated coconut shell. J Hazard Mater 2009; 161(2-3): 1522-9. doi: 10.1016/j.jhazmat.2008.05.009.
33. Omidi-Khaniabadi Y, Kamarehei B, Nourmoradi H, Goudarzi G, Jourvand M, Basiri H, et al. Hexadecyl trimethyl ammonium bromide-modified montmorillonite as a low-cost sorbent for the removal of methyl red from liquid-medium. IJE Transactions A: Basics 2016; 29(1):60-7. doi: 10.5829/idosi.ije.2016.29.01a.09.
34. Mall ID, Srivastava VC, Agarwal NK, Mishra IM. Removal of Congo red from aqueous solution by bagasse fly ash and activated carbon: kinetic study and equilibrium isotherm analyses. Chemosphere 2005; 61(4): 492-501. doi: 10.1016/j.chemosphere.2005.03.065.
35. Senthil Kumar P, Ramalingam S, Senthamarai C, Niranjanaa M, Vijayalakshmi P, Sivanesan S. Adsorption of dye from aqueous solution by cashew nut shell: studies on equilibrium isotherm, kinetics and thermodynamics of interactions. Desalination 2010; 261(1-2): 52-60. doi: 10.1016/j.desal.2010.05.032.
36. Vimonses V, Lei S, Jin B, Chow C, Saint C. Kinetic study and equilibrium isotherm analysis of Congo red adsorption by clay materials. Chem Eng J 2009; 148(2-3): 354-64. doi: 10.1016/j.cej.2008.09.009.
37. Namasivayam C, Kavitha D. Removal of Congo red from water by adsorption onto activated carbon prepared from coir pith, an agricultural solid waste. Dyes Pigm 2002; 54(1):47-58. doi: 10.1016/S0143-7208(02)00025-6.
38. Bhattacharrya KG, Sharma A. Azadirachta indica leaf powder as an effective biosorbent for dyes: a case study with aqueous Congo red solutions. J Environ Manage 2004; 71(3): 217-29. doi: 10.1016/j.jenvman.2004.03.002.
39. Zhai Y, Zhai J, Zhou M, Dong S. Ordered magnetic core-manganese oxide shell nanostructures and their application in water treatment. J Mater Chem 2009; 19: 7030-5. doi: 10.1039/B912767D.
40. Lian L, Guo L, Guo C. Adsorption of Congo red from aqueous solutions onto Ca-bentonite. J Hazard Mater 2009; 161(1): 126-31. doi: 10.1016/j.jhazmat.2008.03.063.
41. Lorenc-Grabowska E, Gryglewicz G. Adsorption characteristics of Congo red on coal based mesoporous activated carbon. Dyes Pigm 2007; 74(1):34-40. doi: 10.1016/j.dyepig.2006.01.027.
42. Jain R, Sikarwar S. Photocatalytic and adsorption studies on the removal of dye Congo red from wastewater. Int J Environ Pollut 2006; 27(1-3): 158-78. doi: 10.1504/IJEP.2006.010460.

20

Baffle and fixed media effects on coliform removal and bacterial die-off rate coefficient in waste stabilization ponds

Hossein Sasani[1*], Naser Mehrdadi[2], Behnoush Aminzadeh[3], Afshin Takdastan[4]

[1]PhD Student of Environmental Engineering, Department of Environmental Engineering, School of Environment, University of Tehran, Tehran, Iran
[2]Assistant Professor of Environmental Engineering, Department of Environmental Engineering, Graduate Faculty of Environment, University of Tehran, Iran
[3]Assistant Professor of Environmental Engineering, Department of Environmental Engineering, School of Environment, University of Tehran, Tehran, Iran
[4]Associate Professor of Environmental Health Engineering, Department of Environmental Health Engineering, School of Public Health, Ahvaz Jundishapur University of Medical Sciences, Ahvaz, Iran

Abstract
Background: The use of waste stabilization ponds is one of the cheapest wastewater treatment processes. This study evaluated the effects of baffles and fixed media on coliform removal in facultative lagoons.
Methods: In this study, the settled wastewater from four pond systems of the city of Ahvaz was used as input. Each system was composed of two ponds that were connected to each other serially. Three of them were equipped with two, three, and four baffles. Packages of fixed media were installed in the first baffled pond equal to the number of baffles. During a 12-month sampling period from March 2016 to February 2017, the capability of each system to remove coliform with different detention times was studied.
Results: The control system with no baffle and no media reduced the coliform index by an average of 67% in a detention time of 6 to 12 days. Increasing the baffles and fixed media in the ponds improved the coliform removal efficiency; systems with two, three, and four baffles achieved coliform removal in the amounts of 77%, 81%, and 83%, respectively. The coliform die-off coefficients (K_b) were higher in the attached growth systems than in the control system. The coefficients were determined to be 0.21, 0.26, 0.29, and 0.31 d^{-1} for the second ponds of the control, two-, three-, and four-baffle systems, respectively.
Conclusion: This method can be used to upgrade the existing waste stabilization pond and to design new ponds with at least two baffles in the facultative lagoons.
Keywords: Waste stabilization ponds, Baffle, Fixed media, Die-off coefficient, Coliform removal

***Correspondence to:**
Hossein Sasani
Email: sasanihossein@ut.ac.ir

Introduction

Waste stabilization ponds (WSPs) system is one of the important natural processes in developing countries, especially in tropical areas, because this process does not require electrical or mechanical equipment. WSPs treat a variety of wastewaters from domestic to complex industrial wastewater, and they function under a wide range of weather conditions, from tropical to arctic (1). Pollutants such as biochemical oxygen demand (BOD), chemical oxygen demand (COD), nitrogen, phosphor, and pathogens can be removed by WSPs. One benefit of WSPs is pathogen reduction. The most important factors that lead to pathogen removal in the ponds are temperature, solar radiation, pH, food shortage, predator organisms, and toxins (2). The main advantage of this process is its lower costs compared with other wastewater treatment processes in terms of construction and operational costs (3). These ponds are easily operated and need no electrical or mechanical equipment as oxygen is supplied by micro-algae in the facultative lagoons. However, a larger area of land is necessary compared with energy-based processes such as activated sludge (4). In addition, occasionally, a high concentration of suspended solids is seen in the effluent that refers to the algae cells. Odor production and water loss due to evaporation are its other disadvantages (5). The concentration of pollutants in pond effluent is a function of the ponds number in series, geometry, hydraulic retention times (HRT), and the number of internal and external factors of ponds (6). To improve the quality of the effluent, it is recommended that processes such as coagulation and flocculation, microstrainer, supplementary ponds, and rock filters be used in the end of the ponds (7). Wetland can also be added to supplement the treatment process (8). Another way to improve the treatment process is to put baffles in the ponds to improve system operations, increase retention time, and reduce the dispersion number (9,10). In addition to baffles, putting the fixed media in the ponds provides more surface area for the growth of microbial films and makes it receive more hydraulic loads (11,12). Attached growth process was introduced in England in 1893 and provided a trickling filter. Conventional attached growth processes include rotating biological discs, bioreactors with submerged beds, floating biofilm reactors, and integrated attached and suspended processes (13). Nielson et al studied the performance of three baffled ponds compared to an unbaffled model in four organic and hydraulic loading rates on the laboratory scale and reported that a considerable effect of the baffling was observed in lower hydraulic detention times (14). Pearson et al studied the effects of configuration, depth of the facultative lagoon and supplementary pond, and baffles on pond operations in pollutant removal and concluded and demonstrated that installing baffles in a supplementary pond improved its efficiency in removing organic materials and pathogens (15). Mattamara and Puetpaiboon evaluated the performance of baffles in WSPs on the laboratory scale and showed that baffled ponds gave higher COD, nitrogen, and coliform removal compared with normal ponds (16). Abbas et al studied the effects of length to width and number of baffles on the performance of ponds and showed that a ratio of 4:1 with two and three baffles is the most appropriate geometric shape for stabilization ponds (17). Babu highlighted four pilots in which one was without a baffle and three others had 15 baffles with similar surfaces and different configurations and reported that the pond equipped with baffles with vertical flows was the most effective pond in terms of reducing short-circuit current (18). In a system in which rock media was placed in the facultative lagoon with three baffles, Al-Sa`ed et al reported the more suitable performance of a baffled attached growth system relative to a control system in the elimination of COD, total suspended solids (TSS), NH4, and coliforms (19). Ouali et al studied a wastewater treatment plant the process of which was conventional activated sludge followed by three maturation ponds in a series acting as a tertiary treatment. The results showed that the installation of two baffles in the first maturation pond improved coliform removal efficiency (20). This research studied the effects of number of baffles and fixed media packages on the performance of WSPs in the removal of total coliforms of settled wastewater from Choneibeh wastewater treatment plant in the city of Ahvaz.

Methods

Wastewater quality

Choneibeh wastewater treatment plant is the only wastewater treatment plant of the city of Ahvaz. Its process is conventional activated sludge with an area of 13 hectares, covers 140 000 people, and treats 390 liters of wastewater per second (21). This study used the effluent of primary sedimentation as the pilot raw wastewater. Settled wastewater quality parameters in the sampling period (from March 2016 to February 2017) are presented in Table 1.

Pilot specifications

As shown in Figure 1, the pilot of the research was composed of a control system and three other systems. The control system included two facultative lagoons connected serially. The dimensions of the first and second ponds were (4 m × 1 m × 1 m) and (4 m × 0.8 m × 1 m), respectively. The other systems had the same sizes and dimensions as the control system. The first and second ponds of each system had two, three, and four baffles. Wooden vertical baffles were installed in the pond so that the baffle width was 70% of the pond width. For sealing ponds, there are many sealers that classified into three major categories: synthetic, cement, and natural.[1] In this research, first, the inner surface of the pilot was covered with cement sealer. Then a synthetic sealer was used as the final lining. It was made from an asphalt component that is usually applied to house roofs as waterproofing (LA4 model made in Iran Glass Wool Company). Mineral shells with a specific surface area of approximately 100 m^2/m^3 and average diameter of 5 cm was put in plastic boxes and used as fixed attached growth media. In the first pond of each baffled system, three boxes were installed over each other and beside each baffle according to Figure 1.

Pilot start up and operation

To evaluate the capability of ponds to remove pathogens, settled wastewater entered the pilot at a flow rate of 0.6 m^3/d. At this stage, since the ponds had high detention times to prevent the formation of anaerobic conditions in the pilot, instead of settled wastewater, the ponds were filled with a mixture of piping network water and effluent from the wastewater treatment plant. Then, settled wastewater was discharged into the ponds by a submerged open-propeller pump made in China by the Leo factory. Over a month, the ponds were monitored for color and for dissolved oxygen concentration which showed the amount of 1.5 to 2 mg/L. In this condition, grab samples were taken from the effluents of eight ponds at 10:00 o'clock at the end of each month (three times). In the second and third phases, the same four flows entered the pilot as shown in Figure 2.

Table 1. Quality of settled wastewater in sampling period

Parameter	Unit	March	April	May	June	July	Aug.	Sep.	Oct.	Nov.	Dec.	Jan.	February	Average	SD
BOD_5	mg/L	167	132	124	171	142	118	107	117	103	160	134	152	135.6	19.0
COD	mg/L	274	236	172	252	218	168	155	174	159	290	235	254	215.6	41.7
TSS	mg/L	86	62	63	68	55	81	75	69	78	84	63	71	71.3	8.0
TP	mg/L	2.8	2.4	3.9	2.3	2.6	1.8	2.2	2.0	2.5	2.9	2.2	2.7	2.5	0.4
TKN	mg/L	35	33.4	29.7	31	32	28	30	36	37	28	35	32	32.3	2.5
TC	MPN/100 m	8.5×10^{12}	8.7×10^{12}	7.9×10^{12}	7.8×10^{12}	8.2×10^{12}	8.8×10^{12}	7.6×10^{12}	7.1×10^{12}	7.8×10^{12}	8.9×10^{12}	7.3×10^{12}	8.6×10^{12}	7.8×10^{12}	5.1×10^{11}
pH	-	7.7	7.3	7.2	7.8	7.5	7.8	7.3	7.6	7.2	7.5	7.8	7.6	7.5	0.2

Abbreviations: BOD_5: biochemical oxygen demand in 5 days; COD: chemical oxygen demand; TSS: total suspended solids, TP: total phosphor, TKN: total Kjeldahl nitrogen, TC; total coliform.

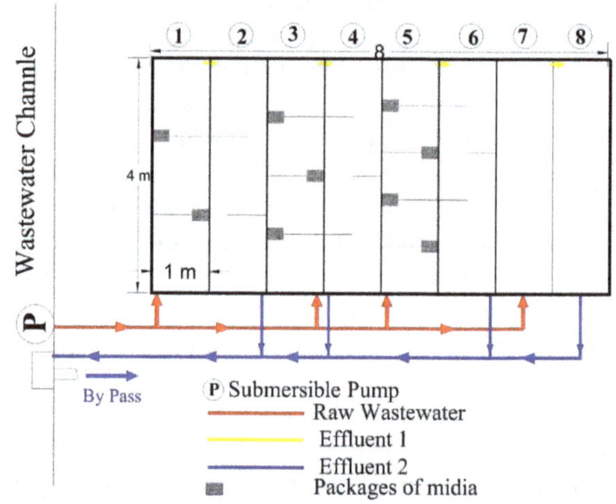

Figure 1. Plan and flow diagram of pilot.
Pilot was composed of four systems: 3 baffled systems and a control system that had no baffle and no fixed media. Ponds were numbered 1 to 8. Settled wastewater entered the first ponds of all systems and are shown with the numbers 1, 3, 5, and 7. The second ponds of all systems are shown with the numbers 2, 4, 6, and 8.

Experiments procedure

All experiments were carried out based on the standard method (22). Grab samples were taken from effluents of the ponds and analyzed using the following methods: 5210 B for BOD, 5220 B for COD, 2540 D for TSS, 4500-NorgC for Total Kjeldahl nitrogen (TKN), 4500-P D for total phosphate (TP), and 9221 B for total coliforms (multiple tube fermentation technique). DO was measured using a DO meter made by the Hack Company (HQ40d model), and pH was measured using a pH meter made by the WTW Company (720 model).

Statistical analysis

Data was analyzed using XLSTAT software version 2016.5. Student's *t* test was applied to determine statistical differences. The level of significance was considered to be 95% and an error rate of less than 5% was acceptable (*P* value < 0.05).

Calculation of coliform die-off coefficient

Coliform die-off coefficient (K_b) in ponds is usually modelled assuming first-order kinetics (23). According to such an equation, K_b is directly proportional to the concentration of indicator organisms. These equations are different from the point of view of the pond hydraulic regime. There are three types of hydraulic regime: plug flow, complete mix, and dispersed flow. In fact, the hydrodynamic behavior of the ponds is not consistent with ideal plug flow or complete mix, but it does correspond to the dispersed flow. Thus, to determine the coliform die-off coefficient, the Wehner-Wilhelm equation was used (5).

$$N = N_0 \frac{4ae^{1/2d}}{(1+a)^2 e^{a/2d} - (1-a)^2 e^{-a/2d}} \tag{1}$$

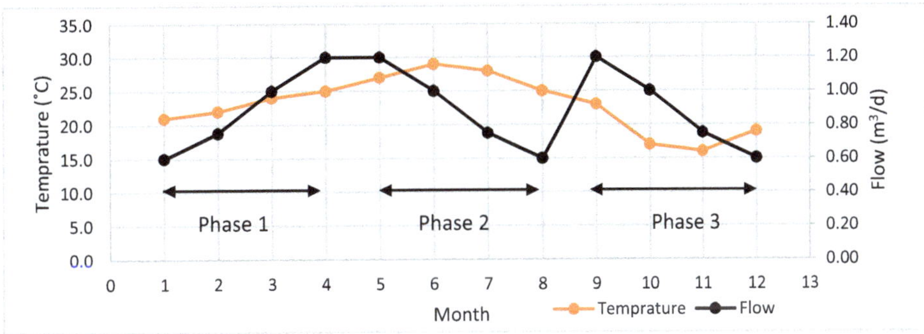

Figure 2. Flow and wastewater temperature in sampling duration.
The pilot operation was conducted in three phases. Each phase was composed of four months. Phase 1: 1st to 4th months, phase 2: 5th to 8th months, and phase 3: 9th to 12th months. In each month, settled wastewater entered each system with a fixed flow. In the duration of each phase, the wastewater flow was changed.

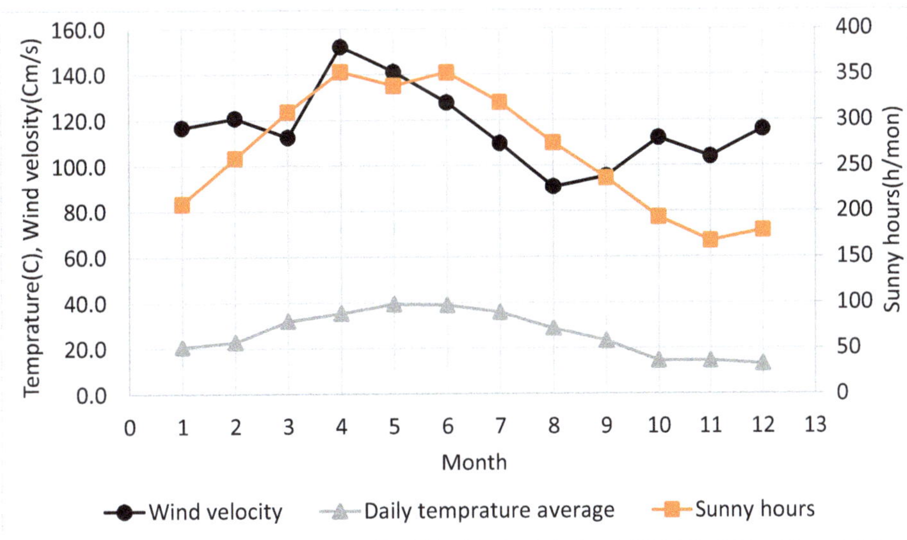

Figure 3. Climatological factors in sampling duration (24).

$$a = \sqrt{1 + 4K_b.t.d} \quad (2)$$

In equation 1, d is the dispersion number. Values of d close to zero indicate plug flow regime and in large amounts indicate complete mix in the reactors. According to previous research, the dispersion number can be calculated using the following formula (6):

$$d = \frac{1}{\frac{L}{B}} \quad (3)$$

For ponds in which the number of baffles installed parallel to the pool width is n, the ratio of (L/B) can be calculated by the following equation (5):

$$\frac{L}{B} = \frac{B}{L}(n+1)^2 \quad (4)$$

L: pond length
B: pond width
n: The number of baffles

With the Wehner-Wilhelm model and using data from the ponds, it can be used to design the facultative lagoons and supplementary ponds (5). For this purpose, the first index (L/B) was calculated based on pond dimensions and number of baffles. Then the dispersion number (d) was determined using equation 3. After that, the K_b value was assumed and the theoretical HRT determined. Finally, the ratio of output coliform to input coliform (N/N_0) was calculated using the Wehner–Wilhelm equation. According to the values measured in the sampling, measured (N/N_0) can also be determined. If the measured ratio is equal to the calculated ratio, the assumed values will be correct; if not, the previous steps must be repeated to achieve the correct result by trial and error method. Since K_b is a function of temperature, in this study, it was assumed to be 20°C and modified to the real wastewater temperature using equation 5.

$$K_T = K_{b20} \times 1.07^{(T-20)} \quad (5)$$

Results

Coliform and BOD_5 concentrations

Average values of the main climatological parameters recorded for the investigated region are summarized in Figure 3. Figure 4A shows that effluent coliform concentrations of all systems in each month were higher than the permissible limits recommended by Iran's Department of Environment (DOE) for discharge to surface water and agricultural usage (1000 MPN/100 mL). Moreover, during the first phase, the coliform concentration had an ascending procedure that was related to increased HRT. The HRT of each system in the first month was 12 days, but it decreased to 6 days at the end of the process. As shown in Figure 2, wastewater temperature during this phase increased considerably from 21°C to 25°C. The results of sampling showed that during the first 2 months, the effects of HRT decrement dominated the effect of temperature increment for coliform removal. The coliform concentration increment happened in the effluent of each system (Figure 4A); however, it decreased in the third month. This may be related to the increment of wastewater temperature as 4°C and sunny hours (Figure 3). In the fourth month, despite the increase in climatological factors, the coliform concentration decreased. It may refer to the HRT decrement as shown in Figure 2 that dominated the other factors. In the second phase, as shown in Figure 3, the warmest months of the year occurred; thus, hydraulic loading was applied in a descending procedure unlike the first phase. As shown in Figure 4A, coliform concentration decreased in this phase. In the third phase, the arrangement of hydraulic loads was the same as that of the second phase; thus, the coliform concentration decreased like in the previous phase, except in the 10th month. It seems that the considerable decrease in temperature and the sunny hours in this month dominated the effect of increased HRT (Figure 3). Another point is the improvement of effluent microbial quality of baffled systems compared to the control system. Figure 4A shows that the average coliform concentration in the two-, three-, and four-baffle systems decreased 29.2%, 41.7%, and 50% more than the control system, respectively. This refers to the role of the number of baffles and the amount of fixed media that increased real HRT and attached growth area, respectively, and resulted in better effluent quality (16,18). Figure 4B shows the BOD_5 concentrations of effluents from all systems. According to this figure, the average BOD_5 concentration in the control system effluent in the first phase was 64.9 mg/L. Like Figure 4A, concentration of BOD_5 decreased in the third month rather than second month. In the second phase, the average BOD_5 concentration was 56.3 mg/L, and BOD_5 concentration decreased during this phase (similar to coliform concentration). In the third phase, the average BOD_5 concentration was 78.3 mg/L; its maximum occurred in the 10th month. In this month, climatological factors such as temperature decreased at a high rate (Figure 3). It seems the worst quality of effluent was seen in this phase. According to Figure 4B, baffle and fixed media had a positive effect on decreasing BOD_5 concentration; the average BOD_5 concentrations throughout sampling were 65.5, 55.7, 50.0, and 45.8 mg/L in the control, two-, three-, and four-baffle systems, respectively.

Statistical analysis detected that P values were less than 0.0001 and 0.0005 for total coliform and BOD_5, respectively, and that indicates a significant difference for reported data.

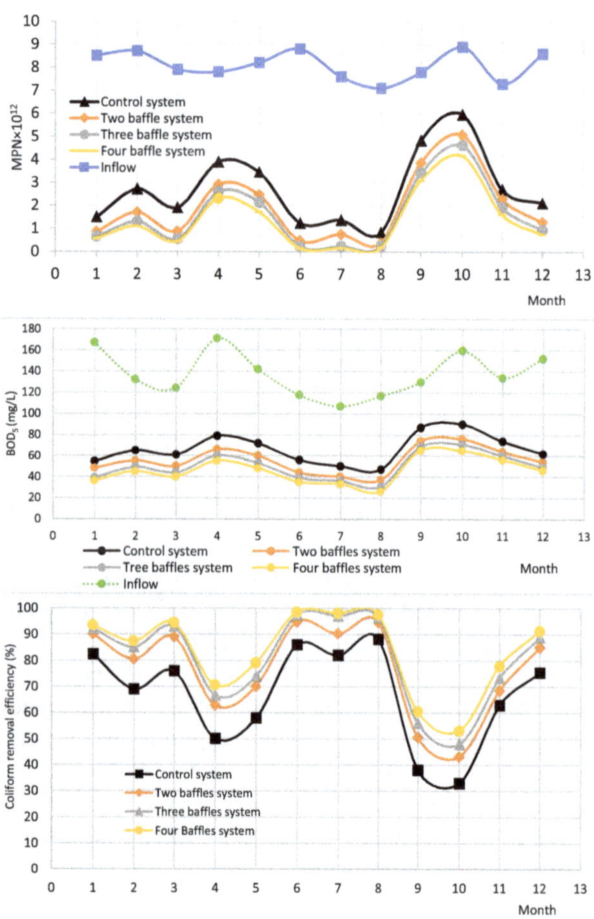

Figure 4. Effluent coliform (A), BOD_5 (B) and coliform removal efficiency (C) for systems.

Coliform removal efficiency

Figure 4C shows the efficiency of coliform removal in sampling duration. During the first phase, due to decrease in HRT, coliform removal decreased as well, except in the third month. This refers to the increased number of sunny hours and increased temperature as explained in Figure 3. Although they also increased in the fourth month, it seems it could not control the effect of decreased HRT. In the second phase duration, with increasing HRT, coliform removal increased. Although the changes in the 6th, 7th, and 8th months were slight, that indicates the increase in HRT and decrease in climatological parameters offset the effects of each other on coliform removal. As in the second phase, in the third phase duration, increased HRT resulted in increased coliform removal except in the 10th month. It seems that, according to Figure 3, a notable decrease in temperature and sunny hours caused coliform removal decreased in 10th month. Figure 4C, like Figure 4B, shows that increasing the number of baffles and the amount of fixed media improved coliform removal so that for the control, two-, three-, and four-baffle systems, the amounts were 67%, 77%, 81%, and 83%, respectively. In the baffled attached growth waste stabilization systems, factors such as filtration, adsorption, aggregate formation, and predators attributed to passive aerobic conditions and attached growth surface area may be responsible for the higher coliform removal rates observed (25,26).

Determination of coliform die-off coefficient (K_b)

Figure 5A shows K_{b20} for the first pond of all systems in the sampling duration. According to this figure, the average K_{b20} concentrations for the first pond of the control system in the first, second, and third phases were 0.11, 0.13, and 0.07 d^{-1}, respectively. This may be attributed to the climatological changes that occurred during sampling months; the warmest and coldest months belonged to the second and third phases, respectively, with average wastewater temperatures of 27°C and 19°C, respectively. This procedure happened for baffled systems too. Thus, the average of this coefficient in the 12-month period for the first pond of the control, two-, three-, and four-baffle systems was 0.11, 0.15, 0.16, and 0.18 d^{-1}, respectively. According to Figure 5B, the average K_{b20} concentrations for the second pond of the control system in the first, second, and third phases were 0.15, 0.24, and 0.23 d^{-1}, respectively. As with the first pond, increasing the number of baffles resulted in an increase in K_{b20} concentration. Concentrations in the control, 2-, 3-, and 4-baffle systems were 0.21, 0.26, 0.29, and 0.31 d^{-1}, respectively. Figure 6A shows the effect of the number of baffles on K_{b20} changes in the first pond of all systems. According to this figure, the maximum of this coefficient belongs to a flow of 1 m^3/d. It refers to the setting of the 6th month in this flow, which is the warmest month of the year and has the maximum climatological factors such as sunny hours (Figure 3). Also, the minimum concentration of K_{b20} occurred for a flow of 1.2 m^3/d. It may refers to minimum HRT that occurred by this flow, although one of the months that this flow entered the systems was fifth month that had maximum temperature and sunny hours like sixth month. So it seems the HRT (6-day) was insufficient and made it have minimum quantities of K_{b20}. In Figure 6B, the same flows have the maximum and minimum quantities of K_{b20} for second the ponds, too. According to Figure 6, adding two baffles make it grow at a notable rate, while its growth in the 3- and 4-baffle systems was at a lower rate than the 2-baffle system.

Discussion

In the current study, the average coliform removal efficiency

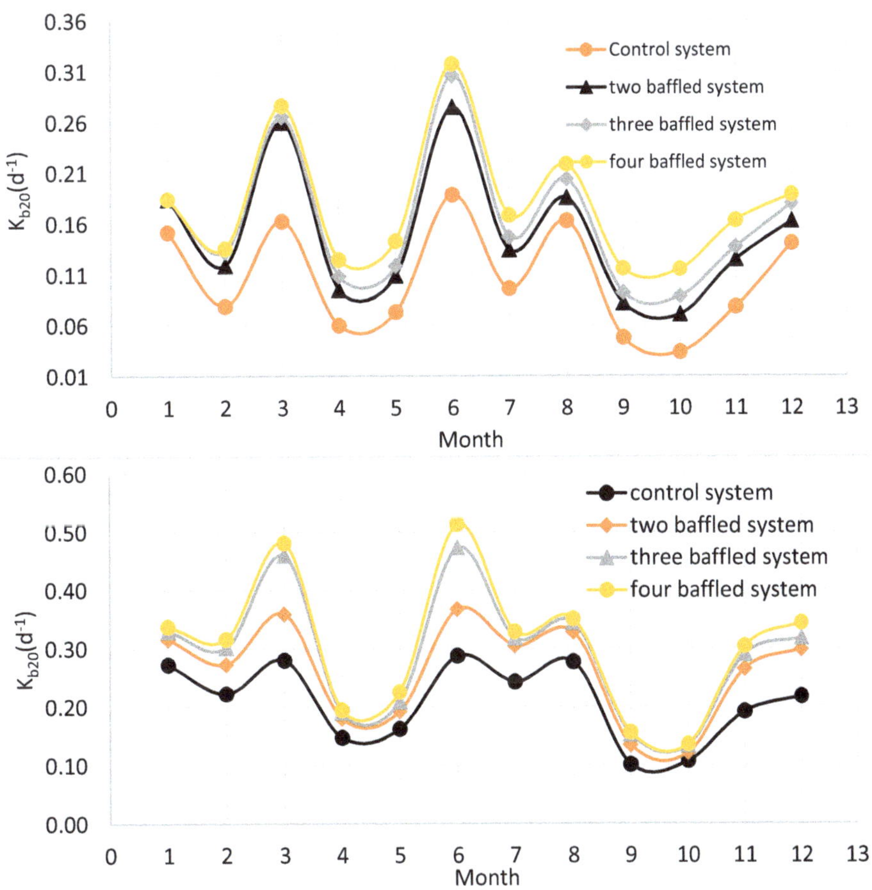

Figure 5. K_{b20} in first (A) and second (B) pond of each system in sampling duration.
Kb20: Coliform die-off coefficient in temperature of 20°C

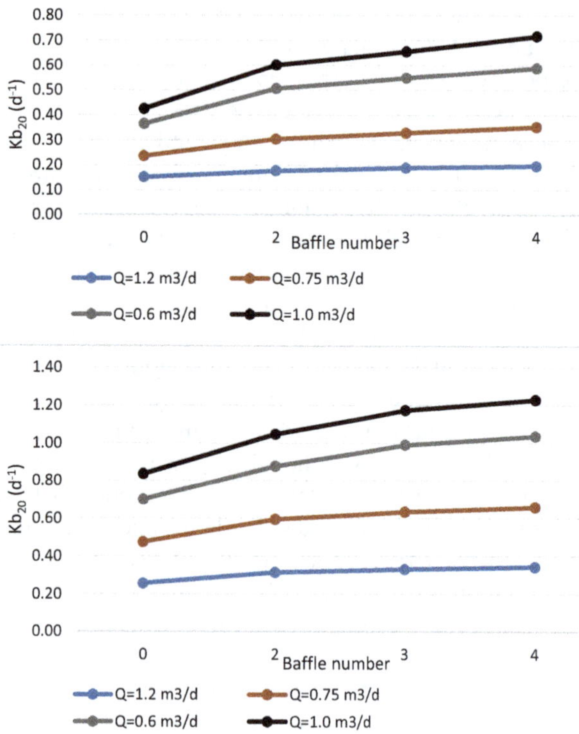

Figure 6. K_{b20} in first (A) and second (B) pond of systems in various flows and baffle numbers.
Kb20: Coliform die-off coefficient in temperature of 20˚C

in facultative lagoons was studied and determined for the control system at an HRT of 6 to 12 days; it was 67% in a 12-month period. Pearson et al determined it as 84% for facultative lagoons. This increment may be attributable to different climatological conditions. The experimental ponds were constructed at the city of Campina Grande in northeastern Brazil which has a tropical climate, and the wastewater temperature remained constant at 25˚C during a 15-month period (15).

The results of this study confirm the findings of the research conducted by Al-Sa`ed et al. In his research, the efficiency of coliform removal in a 3-baffle facultative lagoon was reported as 69%, while in the 3-baffle system of the current study was 81%. The improvement observed in the current study may be due to the higher (L/W) ratio. Furthermore, the kind of fixed media applied in the current research had more specific area than that used in the study of Al-Sa`ed et al, and that caused more attached growth area and more coliform efficiency (19).

In the current study, increasing the number of baffles and the amount of fixed media resulted in coliform removal efficiency 14.7%, 21%, and 25% greater than that of the control system. Al-Sa`ed et al reported an efficiency of 8% that is explained by the differences in geometric dimensions and kind of fixed media. Moreover, the control system in the study of Al-Sa`ed et al had 3 baffles and no fixed media (19).

The results of this study are consistent with those of a study carried out by Mattamara and Puetpaiboon in which it was reported that increasing the number of baffles from 0 to 6 made coliform removal increase to 12% in a laboratory scale facultative lagoon and an HRT of 3 days (16).

Ouali et al studied a conventional activated sludge wastewater treatment plant followed by three maturation ponds (MP1, MP2, MP3) in which 2 baffles were introduced in the first maturation pond (MP1). The results showed that, because of the well-designed baffles, the removal efficiency of *Escherichia coli* increased from 69% to 82% (20).

According to the results of the current study, the average K_{b20} concentrations of the first ponds of the control system were 0.11 d^{-1} and increased 41.6%, 54.4%, and 69% in 2-, 3-, and 4-baffle systems, respectively. For the second pond of the control system, this amount was 0.21 d^{-1}, and the 2-, 3-, and 4-baffle systems had increments of 25.1%, 40.3%, and 74% greater than the control system. Thus, the setting of two baffles and fixed media packages in the ponds resulted in a significant increase in coliform removal efficiency, while in the 3- and 4-baffle systems increased at a lower rate than the 2-baffle system. Generally, coliform die-off coefficients were higher in the baffled systems than in the control system. The range of K_{b20} obtained in this study is consistent with a study of 186 pieces of data from facultative lagoons and supplementary stabilization ponds; K_{b20} concentrations were reported as 0.1-1 d^{-1} and 0.1-0.7 d^{-1} for primary and secondary facultative lagoons, respectively (6).

Conclusion

In this study, the efficiency of baffled attached growth WSPs in the removal of pathogens from primary sedimentation effluent was evaluated in the Choneibeh wastewater treatment plant of Ahvaz. The control system could decrease total coliform of settled wastewater to an average of 67% in an HRT of 6 to 12 days. However, this system requires other supplementary ponds to decrease this index to lower than the standard limit for agricultural usage or discharge to water resources. In addition, baffled attached growth ponds had higher efficiency compared with control ponds in coliform removal. The coliform die-off coefficient (K_b) was higher in the baffled attached growth system than in the control system, and increasing the number of baffles and amount of fixed media resulted in increased K_b as well. Moreover, by reducing the depth of the second pond rather than the first resulted in increased sunlight penetration, and K_b in the second pond was determined to be more than in the first pond. Thus, putting baffles and fixed media in facultative stabilization ponds improved the efficiency of pathogen removal due to the reduction in effects of short circuiting as well as increasing the attached growth area. This method can be used to upgrade the existing WSP, and investment and maintenance costs will be reduced since no additional land is required.

Authors' contributions

NM, BA, HS, and AT conceived and designed the study. BA and HS performed the literature search and wrote the manuscript. All authors participated in data acquisition, analysis, and interpretation. All authors critically reviewed, refined, and approved the manuscript.

Competing interests

The authors declare that they have no competing interests.

Acknowledgments

This research was supported by the Water and Wastewater Company of Ahvaz and the Environment faculty of Tehran University. The authors appreciate all efforts and support.

References

1. EPA, Office of Research and Development. Principles of design and operations of wastewater treatment pond systems for plant operators, engineers, and managers. August 2011. https://www.epa.gov/nutrient-policy-data/principles-design-and-operations-wastewater-treatment-pond-systems-plant.
2. Ansa ED,Awuah E,Andoh A, Banu R, Dorgbetor WH, Lubberding HJ, et al. A review of the mechanisms of fecal coliform removal from algal and duckweed waste stabilization pond systems. Am J Environ Sci 2015; 11(1): 28.34. doi: 10.3844/ajessp.2015.28.34.
3. Owusu-Ansah E, Sampson A, Amponsah SK, Abadidoo RC, Hald T. Performance, compliance and reliability of waste stabilization pond: effluent discharge quality and environmental protection agency standards in Ghana. Research Journal of Applied Sciences, Engineering and Technology 2015; 10(11): 1293-1302.
4. Takdastan A, Mehrdadi N, Torabian A, Azimi AA, Nabi R. Investigation of excess biological sludge reduction in sequencing Bach reactor. Asian Journal of Chemistry 2009; 21(3): 2419-27.
5. Sperling MV. Waste Stabilisation Ponds. London: IWA; 2007.
6. Sperling MV. Modelling of coliform removal in 186 facultative and Maturation ponds around the world. Water Res 2005; 39(20): 5261-73. doi: 10.1016/j.watres.2005.10.016.
7. Middlebrooks EJ. Upgrading pond effluents: an overview. Water Sci Technol 1995; 31(12): 353-68. doi: 10.1016/0273-1223(95)00522-O.
8. Senzia MA, Mashauri DA, Mayo AW. Suitability of constructed wetlands and waste stabilisation ponds in wastewater treatment: nitrogen transformation and removal. Physics and Chemistry of the Earth, Parts A/B/C 2003; 28(20-27): 1117-24. doi: 10.1016/j.pce.2003.08.033
9. Juanico M. should waste stabilization ponds be designed for perfect-mixing or plug-flow? Water Sci Technol 1991; 23(7-9): 1495-502.
10. Zhao Q, Wang B. Evaluation on a pilot-scale attached growth pond system treating domestic wastewater. Water Res 1996; 30(1): 242-5.
11. Polpraset C, Agarwalla BK. Significance of biofilm activity in facultative pond design and performance. Water Sci Technol 1995; 31(12): 119-28.
12. Gullicks H, Hasan H, Das D, Moretti C, Hung YT. Biofilm fixed film systems. Water 2011; 3(3): 843-68. doi: 10.3390/w3030843.
13. Takdastan A, Azimi A, Jaafarzadeh N. Biological excess sludge reduction in municipal wastewater treatment by chlorine. Asian Journal of Chemistry 2010; 22(3): 1665-70.
14. Nielson SB, Middlebrooks EJ, Porcella DB. Effects of Baffles on the Performance of Anaerobic Waste Stabilization Ponds nUtah State University Logan; 1973. Report No: 172. http://digitalcommons.usu.edu/water_rep/172.
15. Pearson HW, Mara DD, Arridge HA. The influence of pond geometry and configuration on facultative and maturation waste stabilization pond performance and efficiency. Water Sci Technol 1995; 31(12): 129-39.
16. Mattamara S, Puetpaiboon U. Roles of baffles in waste stabilization ponds. Water Sci Technol 1997; 35(8): 275–284.
17. Abbas H, Nasr R, Seif H. Study of waste stabilization pond geometry for the waste water treatment efficiency. Ecol Eng 2006; 28(1):25-34. doi: 10.1016/j.ecoleng.2006.03.008.
18. Babu M. Effect of algal biofilm and operational conditions on nitrogen removal in wastewater stabilization ponds [dissertation]. The Netherlands: Delft University; 2011.
19. Al-Sa`ed R, Abu-Madi M, Zimmo O. Novel design concept for facultative ponds using rock filters to reclaim the effluent. J Environ Eng 2011; 137(4): 284-290. doi:10.1061/(ASCE)EE.1943-7870.0000318.
20. Ouali A, Jupsin H, Vasel JL, Marouani L, Ghrabi A. Removal improvement of bacteria (*Escherichia coli* and *enterococci*) in maturation ponds using baffles. Water Sci Technol 2012; 65(4): 589-95. doi: 10.2166/wst.2012.896.
21. Hormozi Nejad M, Takdastan A, Jaafarzadeh N, Ahmadi Mogadam M, Mengelizadeh N. Removal of orthophosphate from municipal wastewater using chemical precipitation process in Ahvaz wastewater treatment plant, Iran. Asian Journal of Chemistry 2013; 25(3): 2565-8.
22. American Public Health Association (APHA), American Water Works Association (AWWA), Water Environment Federation (WEF). Standard Methods for the Examination of water and wastewater. 22th ed. Washington, DC: American Public Health Association; 2012.
23. Short MD, Cromar NJ, Fallowfield HJ. Hydrodynamic performance of pilot-scale duckweed, algal-based, rock filter and attached-growth media reactors used for waste stabilization pond research. Ecol Eng 2010; 36(12): 1700-8. doi: 10.1016/j.ecoleng.2010.07.015.
24. IRAN Meteorological Organization products & services. Specialized products and services weather. Accessed Feb 1, 2017. http://www.irimo.ir/eng/wd/720-Products-Services.html
25. Mara DD, Johnson ML. Aerated rock filters for enhanced ammonia and fecal coliform removal from facultative pond effluents. J Environ Eng 2006; 132(4): 574-7. doi: 10.1061/(ASCE)0733-9372(2006)132:4(574).
26. Johnson M, Camargo Valero MA, Mara DD. Maturation ponds, rock filters and reedbeds in the UK: Statistical analysis of winter performance. Water Sci Technol 2007; 55(11): 135-42. doi: 10.2166/wst.2007.364.

Assessment of safety in drinking water supply system of Birjand city using World Health Organization's water safety plan

Akbar Eslami[1], Mohtasham Ghaffari[2], Behnam Barikbin[3], Farzaneh Fanaei[4*]

[1]Environmental and Occupational Hazards Control Research Center, Shahid Beheshti University of Medical Sciences, Tehran, Iran
[2]Department of Education and Health, School of Public Health, Shahid Beheshti University of Medical Sciences, Tehran, Iran
[3]Department of Environmental Health Engineering, School of Public Health, Birjand University of Medical Sciences, Birjand, Iran
[4]Department of Environmental Health Engineering, School of Public Health, Shahid Beheshti University of Medical Sciences, Tehran, Iran

Abstract
Background: The conventional method for managing drinking water quality is not a suitable preventive strategy for protecting public health. A water safety plan (WSP) presents a systematic approach to ensuring the health and quality of drinking water. This study assessed the drinking water supply system safety of Birjand city using the WHO's WSP.
Methods: This investigation employed the WSP-QA Tool and the WSP manual of the WHO and the AWI. For this purpose, software checklists were prepared and, after confirmation of content and face validity, completed based on Birjand's water and wastewater company records and interview with company's experts. Data was analyzed using WSP-QA Tool.
Results: From total possible 440 points for full use of the program and 328 points of the studied phases, a score of 190 was acquired, indicating a 43.18% coordinated implementation with WSP. System description and management procedures showed the maximum and minimum percentages of congruency with the WSP of 87.5% and 25%, respectively. Among the components of the water supply system, water resource had the most attention.
Conclusion: As regards the percentage of overall WSP-phase implementation and the low attention given to some key parameters of the water supply system, such as risk assessment and management procedures, the system currently enjoys a relatively medium level of safety. However, the high points of some phases such as system description, operational monitoring and verification enhance the system's flexibility for changing the current approach of qualitative management to the WSP.
Keywords: Drinking water, Safety, Water supply, Risk assessment, Software

***Correspondence to:**
Farzaneh Fanaei
Email: fanaei70@gmail.com

Introduction

Possessing water with a suitable quality and quantity is considered the basis for the development of sustainable health for any society (1,2). The water supply system has a wide extension requires special control considering the status and relationship of the system with the surrounding environment, the contamination potential resulting from this relationship and, in turn, threats to consumer health. Thus, this issue is of great interest to many researchers and authorities (3). Based on current estimations, one-sixth of the world's population has absolutely no access to healthy water within a distance of 1 km from their living place. Furthermore, one-fifth of the population has absolutely no suitable way to discharge waste (4). Water resources are always subject to contamination and reduction of quality because of indiscriminate exploitation, and this can adversely affect the health of society (5). For this reason, supplying adequate, quality drinking water to urban areas has always been an important challenge facing managers and decision-makers. Various parts of urban water systems are always menaced by different natural, non-natural, and functional threats which may irrecoverably affect the quantitative and qualitative reliability of the water supplied to consumers (6). Accordingly, the provision, transfer, and distribution systems of urban water are considered principal infrastructures in any society that are directly associated with the vital and primary needs of people (7). Despite all attempts to provide healthy drinking water, waterborne diseases are considered a serious health concern around the world. Digestive diseases that originate from contaminated water and deficient health measures are the reason of 2.4 million deaths and 73 million disabilities (8). Epidemic diseases resulting from unhealthy water resources affect all countries, whether developed or developing, and cause mortality and economic damage to individual people and society as a whole (9). In its newest report associated with world health statistics (WHO 2011), the WHO has introduced unsafe drinking water (along with eight other parameters) as a preventive risk factor causing disease and mortality around the world. Despite international attention to the issue of water, waterborne diseases still remain a major health concern in the world. However, around one-tenth of the load of global diseases can be prevented by improving water resources and their management (10). Drinking water safety management is changing into an integrated risk management and assessment approach. This approach encompasses all stages and processes involved in the provision of drinking water from catchment to the consumer, and it is of interest to evaluate drinking water safety in water supply systems (11). Traditionally, drinking water quality management is based on final point tests, and comprehensive attention is not paid to risk management or the assessment process. Therefore, it is not a suitable preventive strategy for protecting public health (12). Accordingly, the development of strategies for managing the quality of water resources in Iran is an important measure for policymakers in the water sector, and its most important elements will be preserving the quality of resources for drinking water from production to consumption (13). A major global transformation in the management of water resources around the world dates back to when the third qualitative guidelines of drinking water (WHO) and the charter of safe drinking water (Bonn Charter, 2004) were published in 2004. At that time, WHO and the International Water Association (IWA) both supported running a process called the water safety plan (WSP) (14,15). The best justification for this solution lay in the fact that delivering healthy water is not merely dependent on achieving a high final quality level of treated water. Rather, water supply systems may fulfill this aim while having hidden design or operation defects that could cause different incidents, including the risk of uncontrolled contaminants in the catchment, incomplete design of the treatment process, leakage in water supply systems, or improper operational methods (15). The WSP is a powerful tool for managing the quality of drinking water: a systematic approach to ensure the quality of water delivered to consumers from catchment to consumption. It also regulates step-by-step preventions for risk management regarding water contamination (16-18). This system is in charge of detecting and monitoring special risks (microbial, chemical, and physical) which have the potential to inflict damaging effects on the health of products. In this way, it is ensured that the water of interest is in line with health-based objectives. In 2009, the WHO and the IWA presented the WSP practical guidelines comprised of 11 consecutive executive steps. These guidelines state the way WSP should be developed and implemented for drinking water supply systems (1). The major steps of this plan are summarized in Table 1. Uganda was the first country in Africa to use the WSP with the aim of fulfilling its national standards without employing further treatment stages (19). Since 2008, Vietnam has been employing the WSP for its urban water supply systems and has cited the advantages of improved drinking water quality, increased consumer satisfaction, decreased number of cases of waterborne disease, ensured continual provision of water, and the purposefulness of investments (20). The WSP workshop in Berlin in 2014 also introduced this plan as an effective method for risk management in the 21th century. It stated that, even in the case of high correspondence with the European Union Drinking Water Directive (EU-DWD), the WSP is a systematic method and useful managerial tool to guarantee continuous delivery of healthy drinking water to consumers (21). In 2005, 10 Latin American countries emphasized the implementation of the WSP of interest to the WHO. Following them, the Pan American Health Organization (PAHO), the Centers for Disease Control and Prevention (CDC), and the Environmental Protection Agency (EPA) employed the WSP in Jamaica in 2006 (22). In 2010, the WHO and the IWA presented the WSP quality assurance (WSA-QA) tool software to evaluate stages of the WSP. It seems that the novelty of this software prevented its application in qualitative management of drinking water across different countries of the world from becoming prevalent, even in countries which have the experience of evaluation and application of the WSP. This plan is indeed a tool which can detect the strong and weak points in the implementation of every single stage of WSP so as to consolidate the benefits gained from its implementation. One of the characteristics of this software is its capability to be used as a guide where WSP has not been comprehensively implemented or is in the primary stages of implementation. In the present study, this feature has been taken into consideration (14,23).

Characteristics of the studied area

This descriptive-analytical study was conducted on the water supply system of Birjand city, located in southern Khorasan province, Iran in 2015. Birjand city, with an area of 5400 km^2 covering a geographical position from 58°4″0′ to 59°40″0′ E and from 32°40″0′ to 33°40″0′ N, is the capital city of southern Khorasan located in eastern Iran (24-26) (Figure 1). According to the Studies of Water and Wastewater Company of Birjand, this city is fully dependent for its drinking water on groundwater accessed through 32 deep wells with a capacity of around 800 L/s, located in different plains, including Sarbisheh, Marak, Aliabad, and Rekat. Water is supplied through transfer lines 150 km long and three pumping stations,

Table 1. Description of executive stages of water safety plan in water supply systems (1)

Stage	Stage description
Development of WSP team	An expert qualified team with sufficient experience and expertise for the extraction, treatment, and distribution of water and familiarity with the risks that can be posed to water safety from catchment to consumption is developed.
Water supply system description	The water supply system is investigated across the four stages of catchment, treatment plant, distribution system, and consumption point, which allows for the adequate management and assessment of risks with the aim of ensuring documentation of raw water quality.
Hazard identification and risk assessment	Hazards (physical, chemical, and biological) and hazardous events in water catchment, treatment plant, reservoirs, distribution system, and consumption points are identified; risk is assessed based on documentation available in the water catchment, treatment plant, reservoirs, distribution system, and consumption points.
Control measures and validation	Concurrent with hazard identification and risk assessment, the WSP team should deal with potential and existing control criteria. This can be done by field observations, the properties of the produced water, or data monitoring. The scope of risk reduction by each control criterion will specify the extent of its effectiveness.
Development of improvement plan	If an important risk associated with water safety is detected in the previous stage and it is proven that the current controls are not effective or are essentially absent, then a short-, mid-, and/or long-term improvement plan should be implemented.
Operational monitoring	Operational monitoring includes consecutive and planned observations and measurements to prove continuity of effective performance of the control criteria. Monitoring control points is essential to supporting risk management.
Verification the efficiency of WSP	Verification through activities such as internal and external audits on operational activities and investigation of satisfaction of consumers ensure that the design and operation of the system have the potential for continuous delivery of water with a special quality to fulfill health-oriented objectives.
Management procedures	Management includes standard operating procedures which are essential components of the WSP under typical operational conditions and at the incidence of accidents. The procedures should be developed by experienced employees and, if necessary, they should be updated with consideration given the implementation of an improvement plan and reviews of accidents.
Development of supporting programs	Supporting programs are activities that deal with supporting issues such as the development of knowledge and human skills, commitment to the WSP, and enhancement of the ability of management systems to deliver safe water.
Review of WSP following an incident	To ensure that new risks are regularly reviewed, the WSP should be reevaluated after any emergency or accident which threatens the provision and distribution of safe water.
Periodic review of WSP	The WSP group should deal with the development of a general, periodical review of the plan and be informed about gained experiences and new procedures.

interstitial pressure tanks, and collection reservoirs. After passing through the treatment plant, the water then enters five water storage reservoirs with a capacity of 35 000 m^3. Following chlorination, it is finally delivered to citizens through a water supply system 560 km long with 68 000 branches. By the time it reaches the consumption point, 26% of the water extracted from these resources is not considered as water or water loss leaving the consumption cycle (27). The current study analyzed the qualitative control process and provision of drinking water in the water supply system of Birjand city based on the WSP guidelines (WHO, 2009) using the WSP-QA TOOL. It was expected that this investigation would highlight the necessity of changing the current approach by identifying preventable and weak points in the water quality management in Birjand city. It could also highlight the role of the WSP in improving drinking-water quality and greatly assist in structuring decision-making in urban water supply systems by considering potential risks.

Methods

In this study, the WSP presented by the WHO and the IWA and the WSP-QA Tool were used to evaluate the safety of the water supply system in Birjand, Iran. The software can be used across a wide range of conditions from water suppliers who are at the beginning of WSP implementation to those who have used the WSP. When the tool is employed in the primary stages of WSP development, it functions as a guide and evaluates the manner of plan implementation. For data collection, software checklists consisting of 85 questions or 110 question items were prepared in the form of 12 tables by individual phases of the WSP implementation. To complete the questionnaires and record the information on these checklists, the background and information registered in an urban water and wastewater and regional water affairs company in Birjand and local observations were used. The obtained information and data were fed into the software both quantitatively and qualitatively. The answers to questions related to general information associated with water suppliers and general information associated with each water supply system was recorded quantitatively in the software. However, the answers required for the questions corresponding to wasp's executive steps were introduced into the software based on a scoring system in line with the instructions. In the

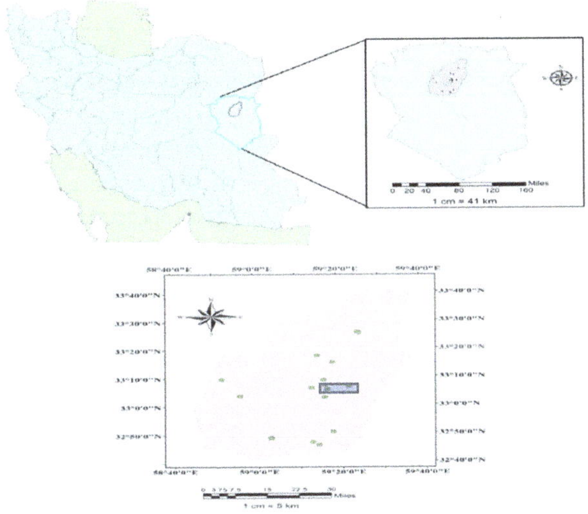

Figure 1. Geographical Position of the Studied Region (26).

system, scoring is based on the manner of implementation of every stage ranging from 0 to 4, in which every stage can range from "not started" to "done completely." Furthermore, in addition to the mentioned scores, some questions could be marked "not applicable." The possible raw score, the score acquired in this investigation, and the extent of progression in implementation in terms of percentage for each phase were calculated based on predetermined and certain relations in the software. Following completion of the evaluation process, the results were presented in the form of graphs and tables which will help water suppliers in easily identifying sites in which attempts to improve the system, points of the system that require more resources for development, and regions in which progress has occurred should be targeted. Thus, the honesty in the responses of water suppliers as well as their critical evaluation regarding the developed advances are very important (1).

Measuring the validity of the translation of the software questions

In this study, to measure the validity of the translation of the software questions, the translation-back translation technique was employed. For this purpose, the English version of the software questionnaire was first translated into Persian by two experts in Persian and English and a specialist in environmental health engineering (translation stage). In the next stage, during a meeting with the research team, the two translated texts were compared with each other and evaluated in terms of quality. In some cases, to select the most suitable translation for statements, the initial translations were combined and composed in order to reach a consensus regarding a common Persian version. Then, the single, final translated version was given to two other translators who had not seen the initial English version to translate back into English from Persian (back translate stage). This was done to gain greater confidence in the accuracy of the questionnaire translation. In another meeting, the two versions translated into English were compared with each other in terms of quality and equalized. Finally, the two translations were changed into a single English version. With another meeting of the research team, the final English translation was compared with the original English text of the questionnaire, and conceptual equality (the consideration of original questionnaire conceptions) and the overall quality of the translation were examined. Eventually, following application of required changes and modifications according to the comments of translators and experts, the final Persian version was prepared based on the final translation of the questionnaire.

The stage of validation of the instrument

For scientific validation of questionnaires, scales, and checklists designed in routine ways, face validity and content validity methods are usually used along with an expert panel (28). Following preparation of the final version of the questionnaire used in this study, its scientific validity, face validity, and content validity were investigated using validation methods and an expert panel. In this study, the instrument designed by the research team was given to eight experts experienced in environmental health engineering. Based on the findings from the validation stage, the instrument was approved by all members of the expert panel in terms of correspondence of its content with valid and scientific texts. Some suggestions and modifying comments were given, and after investigation by the research team, the required changes and modifications were applied to the instrument; eventually, an instrument with an acceptable and approvable face and content validity was prepared.

Results

The following tables and diagrams present the results of evaluation of the extent of general implementation of WSP phases in this investigation, which can help water suppliers understand their general performance.

In Table 2, the options in the first column represent the stages of implementation and step-by-step development of a complete WSP, where the application and extent of progression of implementation of each stage according to WSP in the drinking water supply management system was evaluated through responses to the questions on the relevant checklists. After performing the required calculations using the software, the percentage of implementation corresponding with the WSP was calculated for each stage as an acquired score (% implemented), which can be observed in the last column of the table.

In this investigation, the major components of a water supply system, i.e. resource, treatment, distribution, and final point of use, were also analyzed using the software. The results related to the percentage of implementation of

Table 2. Results of general evaluation of WSP phases by WSP-QA Tool for drinking water supply system in Birjand city in 2015

Overall progress with WSPs in Birjand city			
WSP implementation stages	**No. of questions**	**Total possible point**	**Score (implemented %)**
WSP team	5	20	12/20 (60.00%)
System description	2	8	7/8 (87.50%)
Hazard identification and risk assessment	7	100	52/100 (52.00%)
Control measures and validation	5	68	44/68 (64.71%)
Improvement plan	3	48	-*
Operational monitoring	4	64	43/64 (67.19%)
Verification	8	32	23/32 (71.88%)
Management procedures	3	36	9/36 (25.00%)
Supporting programs	2	8	-*
Review of the WSP	5	56	-*
Total	44	440	190/440 (43.18%)

*The evaluation of this stage necessitates full implementation of WSP.

each evaluated parameter are presented as follows:

Figure 2 indicates that, among the investigated phases, the processes associated with identification of stakeholders claimed the highest score, and the phase associated with management procedures and risk assessment had the lowest score in drinking water resources in Birjand city.

Figure 3 demonstrates that in the water treatment plant, as with the results obtained for the resources, identification of stakeholders had the highest score, while risk detection and other examined processes had a 34% and 25% correspondence with the WSP, respectively.

Figure 4 suggests that among the phases examined across the water distribution system, in addition to identification of stakeholders, the phase related to operational monitoring also gained a high score (88%). However, the process associated with risk assessment at this stage gained no score. Based on the results shown in Figure 5, it can be seen that at the point of use, as with the distribution system, the phase associated with risk assessment indicates the minimum percentage of congruent implementation with the WSP, while the processes related to operational monitoring reveal the maximum percentage of congruent implementation with the WSP. The results obtained from diagrams indicated that application of the WSP-QA Tool as a support for WSPs has well highlighted the strong and weak points as well as the areas and opportunities requiring promotion in the major components of the water supply system in Birjand city.

In an investigation of the advances achieved through implementation of the WSP in Korea in 2013, identification

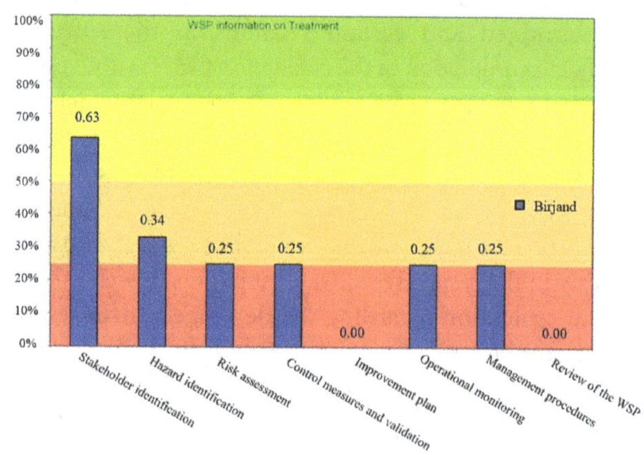

Figure 3. Results of WSP information for drinking water treatment plant in Birjand City in 2015.

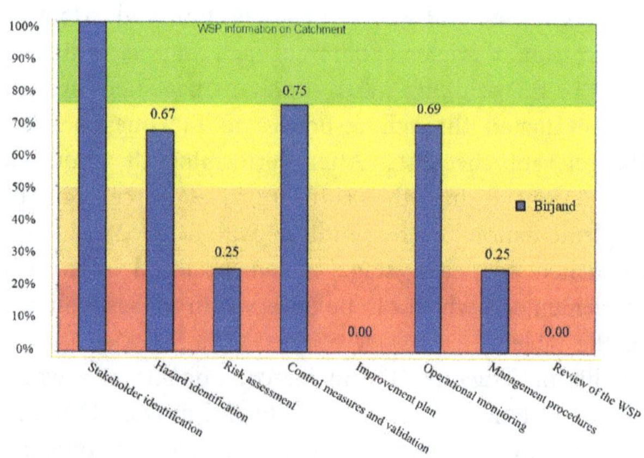

Figure 2. Results of WSP information for drinking water resources in Birjand City in 2015.

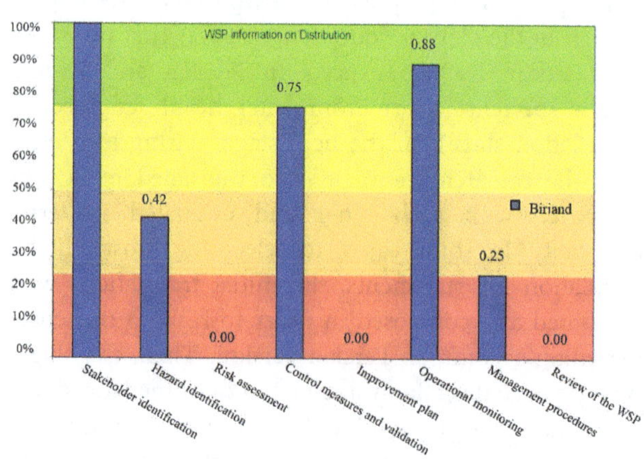

Figure 4. Results of WSP information for drinking water distribution system in Birjand city in 2015.

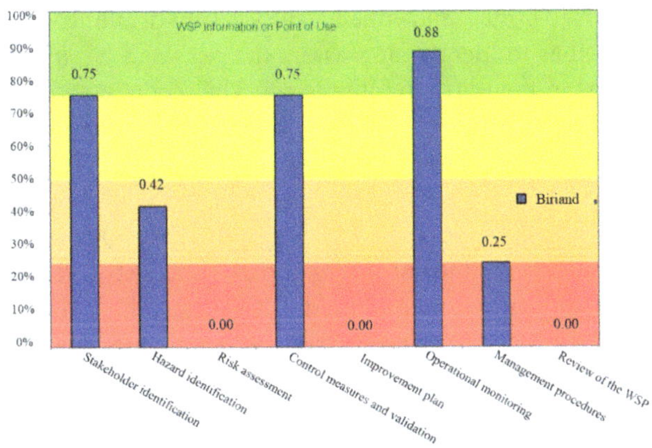

Figure 5. Results of WSP information for drinking water consumption point in Birjand city in 2015 .

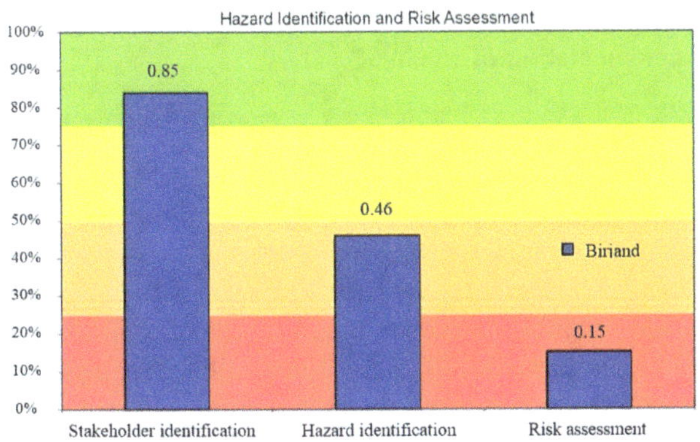

Figure 6. Extent of WSP implementation in the risk assessment and risk detection stage for the drinking water supply system in Birjand city in 2015.

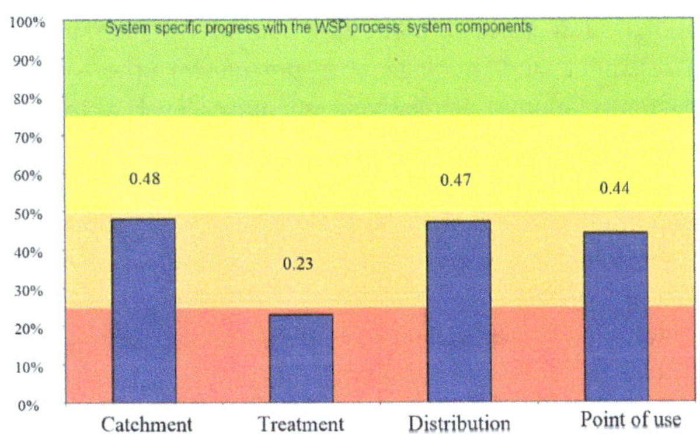

Figure 7. Extent of overall progression of implementation of WSP stages across major components of the drinking water supply system in Birjand city in 2015.

and assessment of risk was defined as a key parameter in the WSP (29). This stage can be evaluated through the three parameters of identification of stakeholders, hazard identification, and risk assessment. In this investigation, according to Figure 6 and Table 2, this phase gained 52 out of 100 raw points, representing 52% executive progression in line with the WSP. At this stage, the risk assessment parameter indicates the minimum percentage of implementation (15%). According to Figure 7, the output results associated with the extent of progression of implementation of each WSP stage in terms of major components of the water supply system in Birjand indicated that among the major components of this system, water resources enjoyed the maximum attention, whereas the water treatment plant attracted the least attention of the water supplier system. A study which evaluated drinking water safety in Khoy city in 2010 found that the distribution system enjoyed the maximum attention of the water supplier organization (14).

Discussion

In this investigation, WSP-QA Tool along with the WSP manual presented by WHO and IWA were used to evaluate the stages of implementation of the WSP in the water supply system of Birjand city. The results shown in Table 2 indicate that from the 440 points for complete application of the plan and 328 points associated with the investigated phases, 190 points were obtained and a 43.18% congruence with WSP implementation was observed. The system description stage (7 out of the total 8 points) showed the maximum percentage of congruent implementation with the WSP (87.5%), whereas the phase related to management procedures had the lowest score (9 out of the total 36 points) and claimed the minimum percentage of implementation (25%). These results reveal that, although management procedures are an important part of the WSP, in the current approach, they are not of great interest to the company. These procedures include standard operating procedures (SOPs) and corrective activities (at the time of incident). Unexpected accidents and deviations in water supply systems may occur when there is no corrective activity at a certain site. Therefore, these procedures should be developed by experienced employees and updated and documented if necessary. In 2014, Aghaei evaluated the safety of the water supply system in Ardebil city and observed 21% congruent implementation with the WSP. In this study, the system description stage also claimed the maximum percentage of congruent implementation with the WSP, whereas the phase related to management procedures claimed the minimum percentage (10). These results are in line with those of the current research. The potential of the software is such that evaluation of some stages of implementation of the WSP (including WSP review, improvement plan, and supporting programs) necessitates complete implementation of the plan in water supply system management; as WSP was not implemented completely in Birjand city, the questions and scores related to phases that could not be evaluated were not included in the final analysis. This is also justifiable considering the feature of the software in independent analysis of the data of each phase. Evaluation of the application of the eight parameters mentioned in Figures 2 to 5, which is somehow in line with corresponding phases in the WSP, for different components of water supply system (catchment, treatment, distribution, and point of use) indicated that the processes associated with the identification of stakeholders across different parts (resources, treatment plant, and distribution system) as well as the process related to operational monitoring at the final point of consumption gained the maximum score. It is notable that, although system description indicated the maximum percentage of congruent implementation with the WSP, the processes associated with risk assessment across all four parts of the water supply system (especially distribution system and the final point of use with a score of zero) were among the processes that gained the lowest scores. This suggests a lack of concentration of the water supplier system on the risk assessment process. The best way to achieve this target, according to the WSP manual, is to draw a simple table with the systematic registration of all hazardous events and relevant risks along with an estimation of their risk level. The risk assessment process can be based on a quantitative, semi quantitative (through a risk assessment matrix), or a simple qualitative solution (based on the judgment of the WSP expert panel). A WSP workshop in Berlin in 2014 also introduced this plan as an effective method for risk management in the 21st century. It further stated that, even in cases of high correspondence with the drinking water directive of the European Union (EU-DWD), the WSP is a systematic method and suitable management instrument to ensure continuous delivery of healthy drinking water to consumers (21). In a study by Gholami et al on the water supply system in Khoy city, identification and risk assessment stage showed 12% congruent implementation with the WSP (14). The results of the study by Aghaei on the water supply system in Ardebil city indicated that, considering the low percentage of general application of different WSP phases of and regarding the lack of attention to some key parameters including risk assessment and catchment management by the water supplier system, the current system does not enjoy the necessary safety, and the governing controlling approach does not have the efficiency required for integrated management of a healthy drinking water provision system (10). In this study, analysis of the results obtained from the diagrams revealed that, considering the percentage of general application of different WSP phases, the operational concentration of the supplier system is mainly on the verification stages and system description, such that the processes associated with risk assessment and management procedures gained a low score across all four main parts of the water supply system. Therefore, the major disadvantage of the current approach governing the qualitative management of drinking water in Birjand city, i.e. dependence on end-product experiments. This is in accordance with the results obtained by Gholami et al in their evaluation of drinking water safety in Khoy city (14). According to the report of progression of development of the WSP in Korea in 2013, the objectives in developing this plan is to protect and enhance the safety of drinking water as well as to identify and assess the risks and hazardous events in the water supply system. The report stated that the main characteristic of WSP implementation in this country is the presentation of a water safety index (WSI) to detect the safety level of drinking water and to identify the weak points in the water supply system. It is highly useful to gain a quick understanding of the water safety level in a water supply system and to detect the vulnerable points of the system (29). The most effective method for achieving sustainable confidence about the safety of a drinking water supply system is the application and management of a comprehensive risk assessment approach, such that it covers all the stages involved in water supply from catchment to the consumer. Such solutions are called WSP (30). The WSP quality assurance tool helps meet this need and also supports WSP teams in the development and implementation of the WSP (31).

Conclusion

Overall, the results indicate that the aims of using the WSP-QA Tool as a support for WSPs, i.e. the identification of the water supply system's weak points and evaluation of the percentage of implementation in line with each stage of the WSP in qualitative management of the drinking water in Birjand city, have been fulfilled completely. Not only did these results differentiate between areas and opportunities requiring promotion, but it also determined the inadequate efficiency of the traditional approach (relying on final point test), both of which were among the objectives of this study. Considering the percentage of the overall application of WSP phases and the low score of some key parameters such as risk assessment and management procedures, the system currently enjoys a relatively medium level of safety. However, considering the high scores of some phases, such as system description, operational monitoring, and validation, the greater attention given the parameters associated with hazard identification and risk assessment, and the application of proper management procedures which include evaluation of the system from catchment to consumer, control measures for determining risks, a plan for monitoring control measures, corrective measures if the monitoring results show deviation from critical limits in a critical control point, and communication and support plans including standard operational procedures in a documented way, make it possible to enhance the system's flexibility to change the current approach of qualitative management into the WSP, eventually increasing the drinking water safety level in the water distribution system. Drinking water should enjoy both a desirable chemical quality and a bacteriological quality. The special chemical and microbiological quality of groundwater resources for the water supply in Birjand city has caused the qualitative indices of the water to digress from desirable limits due to the geological structure of the region, high thickness of alluvial sediments on which the city is settled, and continual and progressive withdrawal from these resources because of population growth. The WHO guidelines recommend the WSP as the most effective tool for ensuring safety in a drinking water supply. As a powerful tool for the management of drinking water safety, the WSP presents a systematic approach to ensure the quality and safety of water distributed to consumers from catchment to consumption point. The aim of the WSP is to ensure the quality of drinking water based on risk management, which emphasizes the prevention from contamination of drinking water resources, water treatment for decreasing or removing contamination to achieve standards, prevention of water re-contamination during storage, distribution, and consumption. Considering the effects of the WSP on the quality of water from catchment to consumption, the implementation of this plan as the most effective tool for ensuring safety in the supply of drinking water is recommended to water supplier organizations.

Acknowledgments
This paper is adapted from part of the results of a thesis entitled "Assessment of Safety in Drinking Water Supply System of Birjand city by Using World Health Organization's WSP, Based on Risk Management Approach" and research plan No 6449 at the Health Faculty of Shahid Beheshti University of Medical Sciences.

Competing interests
The authors confirm that there are no competing interests.

Authors' contributions
All authors of this study contributed equally and participated in the collection, analysis, and interpretation of the data and manuscript writing.

References
1. Bartram J, Corrales L, Davison A, Deere D, Drury D, Gordon B, et al. Water safety plan manual: step-by-step risk management for drinking-water suppliers. Geneva: World Health Organization; 2009.
2. World Health Organization. Guidelines For Drinking-Water Quality. 3rd ed. Geneva: WHO; 2004.
3. Hrudey SE, Hrudey EJ, Pollard SJ. Risk management for assuring safe drinking water. Environ Int 2006; 32(8): 948-57. doi: 10.1016/j.envint.2006.06.004.
4. World Health Organization, UNICEF. Global Water Supply and Sanitation Assessment 2000 Report. Geneva, Switzerland: WHO; 2000.
5. Summerill C, Pollard SJ, Smith JA. The role of organizational culture and leadership in water safety plan implementation for improved risk management. Sci Total Environ 2010; 408(20): 4319-27. doi: 10.1016/j.scitotenv.2010.06.043.
6. Roozbahani A, Zahraie B, Tabesh M. Water Quantity and Quality Risk Assessment of Urban Water Supply Systems with Consideration of Uncertainties. Water and Wastewater 2013; 24(4): 2-14. [In Persian].
7. Nijhawan A, Jain P, Sargaonkar A, Labhasetwar PK. Implementation of water safety plan for a large-piped water supply system. Environ Monit Assess 2014; 186(9): 5547-60. doi: 10.1007/s10661-014-3802-x.
8. World Health Organization. Guidelines for drinking-water quality. Vol. 2, Health criteria and other supporting information: addendum. 2nd ed. Geneva: WHO; 1998.
9. Gunnarsdottir MJ, Gardarsson SM, Elliott M, Sigmundsdottir G, Bartram J. Benefits of Water Safety Plans: Microbiology, Compliance, and Public Health. Environ Sci Technol 2012; 46(14): 7782-9. doi: 10.1021/es300372h.
10. Aghaei M. Assessment Of safety in Drinking Water Supply System of Ardabil City By Using World Health Organization's Water Safety Plan, Based On Risk Management Approach [dissertaion]. Tehran: Tehran University of Medical Sciences; 2014. [In Persian].
11. Sun F, Chen J, Tong Q, Zeng S. Integrated risk assessment and screening analysis of drinking water safety of a conventional water supply system. Water Sci Technol 2007; 56(6): 47-56. doi: 10.2166/wst.2007.583.
12. Yazdanbakhsh AR, Manshoori M, Nabizadeh R, Jahed GR, Fallahzade RA. Guidelines of Water Safety Plan Based on Hazard Analysis and Critical Control Points (HACCP) System.Tehran: Avay Ghalam; 2009. [In Persian].
13. Eslami A, Barikbin B, Ghaffari M, Fanaei F. Assessment of Water Safety Plan (WSP) Implementation and Risk Management in Sarayan City. Zanko Journal of Medical Sciences 2017; 18(57): 81-95.
14. Gholami M, Bodaghi S, Babaloo E. Assessment of water safety plan in Khoy city based on step by step risk management new approach. Proceedings of the 14th National Conference on environmental health; 2011; Yazd, Iran.
15. Omar YY, Parker A, Smith JA, Pollard SJ. Risk management for drinking water safety in low and middle income countries - cultural influences on water safety plan (WSP) implementation in urban water utilities. Sci Total Environ 2017; 576: 895-906. doi: 10.1016/j.scitotenv.2016.10.131.
16. World Health Organization. A road map to support country-level implementation of water safety plans. Genova: WHO; 2010.
17. Tavasolifar A, Bina B, Amin Mohammad M, Ebrahimi A, Jalali M. Designing hazard analysis and critical control point system in drinking water supply system of Isfahan, Iran from Resource to Endpoint. J Health Syst Res 2011; 7(6): 1102-13. [In Persian].
18. Oluwasanya GO, Carter RC. Water safety planning for small water supply systems: the framework and control measures. Water Science and Technology: Water Supply 2017; 17(6): 1524-33. doi: 10.2166/ws.2017.057.
19. Davison A, Howard G, Stevens M, Callan P, Fewtrell L, Deere D, et al. Water safety plans: managing drinking-water quality from catchment to consumer. Geneva: World Health Organization; 2005.
20. World Health Organization. The report on the

implementation of the water safety plan, In pilot models and in water utilities participating in WSP – Phase 1. Viet Nam: WHO; 2010.
21. World Health Organization, International Water Association. European Strategic Workshop on Water Safety Planning Berlin. Germany: Key Outcomes; 2014.
22. EEM. Water Safety Plan, Spanish Town Water Supply. St. Catherine. Jamaica: EEM; 2007.
23. World Health Organization, International Water Association. Water safety plan quality assurance tool. Geneva: WHO; 2013.
24. Hajipoor M, Moodi M, Sharafi H, Eshaghi Nasab M, Mehrani M. Capabilities, potentials and investment opportunities of South Khorasan. [cited 2017 Aug 2]. Available from: http://investin-khj.ir/ckfinder/userfiles/images/Ebooks/farsi.pdf.
25. Introducing the city of Birjand. [cited 2016 Apr 16]; Avalable from: http://sk-shora.ir/index.php.
26. Khosravi R, Eslami H, Almodaresi SA, Heidari M, Fallahzadeh RA, Taghavi M, et al. Use of geographic information system and water quality index to assess groundwater quality for drinking purpose in Birjand City, Iran. Desalination and Water Treatment 2017; 67: 74-83. doi: 10.5004/dwt.2017.20458.
27. Pourtaghi ZS, Pourghasemi HR. GIS-based groundwater spring potential assessment and mapping in the Birjand Township, southern Khorasan Province, Iran. Hydrogeology Journal 2014; 22(3): 643-62.
28. Ghaffari M, Niknami Sh, Kazemnejad A, Mirzae E, Ghofranipour F. Designing and validating 10 conceptual scales to prevent HIV among adolescents. Journal of Kermanshah University of Medical Sciences 2007; 11(1): 38-50. [In Persian].
29. Korea Water Resources Corporation. The Development of Water Safety Plans in Korea. Daejeon: K-Water; 2013.
30. Manshouri M, Shaghaghi G. Water Safety Plan Manual Step-by-step risk management For drinking-water suppliers. Tehran: Ministry of Health, Treatment and Medical Training; 2009. [In Persian].
31. Gholami M, Boodaghi S. Water Safety Plan Quality Assurance Tool (User Manual). Tehran: Institute for Environmental Research, Tehran University of Medical Science; 2013. [In Persian].

Optimizing potassium ferrate for textile wastewater treatment by RSM

Maryam Moradnia[1], Masoud Panahifard[1], Kavoos Dindarlo[2], Hamzeh Ali Jamali[3*]

[1]MSc Student in Environmental Health Engineering, Health Faculty, Qazvin University of Medical Science, Qazvin, Iran
[2]Assistant Professor of Environmental Health Engineering, Health Faculty, Hormozgan University of Medical Science, Hormozgan, Iran
[3]Assistant Professor of Environmental Health Engineering, Health Faculty, Qazvin University of Medical Science, Qazvin, Iran

Abstract
Background: Application of potassium ferrate is a chemical oxidation approach used for water and wastewater treatment. The aim of this study is to apply central composite design (CCD) and response surface methodology (RSM) to optimize potassium ferrate consumption in the treatment of wastewater from carpet industries.
Methods: Samples in this experimental study were collected from wastewater, originating from a carpet factory. Wastewater sampling was carried out monthly for a period of two seasons. Ferrate oxidation experiments were conducted by means of a conventional jar-test apparatus. The time and speed for mixing were set with an automatic controller. Parameters of study were measured based on given methodologies in Standard method for examining water and wastewater. CCD and RSM were applied to optimize the operating variables including potassium ferrate dosage and pH.
Results: Results showed that potassium ferrate concentration (A), pH (B), their interactions (AB) and quadratic effects (A^2 and B^2) were significant in the removal of COD, turbidity, color and TSS from carpet industries effluents. At an optimum point (COD: 160 mg/L of potassium ferrate and pH 4, turbidity: 165 mg/L of potassium ferrate and pH 4, color and TSS: pH 4.5 and 150 mg/L of potassium ferrate) removal efficiencies for COD, turbidity, color and TSS were 86, 86, 87 and 89%, respectively.
Conclusion: Potassium ferrate has a significant impact on pollutants decomposition and the removal of color from wastewater produced in carpet industries. This process can be employed for the pretreatment or post treatment of wastewaters containing refractory organic pollutants. CCD and RSM are suitable tools for experimental design.
Keywords: Advanced treatment, carpet wastewater, Statistical experiment design, Response Surface Methodology.

***Correspondence to:**
Hamzeh Ali Jamali
Email: jamalisadraei@yahoo.com

Introduction

A large amount of wastewater produced in textile industries emerges from color containing units with a wide range of contaminants that must be treated before being discharged into the environment. Wastewater originating from color-containing effluents embodies a high COD, large fluctuations, high pH and temperature (1). Moreover, in many cases, these effluents are toxic and lead to ecosystem damage by producing some hazardous by-products. Application of potassium ferrate as an advanced approach is appropriate for the treatment of such types of wastewater (2,3).

Potassium ferrate (K_2FeO_4) is a ferrous salt (iron salt) containing an uncommon ion with oxidation valence +6. The ion $(FeO_4)^{-2}$ is a strong oxidant and its oxidizing potential in acidic and alkaline conditions is 2.2 and 0.72 V respectively. As potassium ferrate is reduced to Fe^{3+} ions or ferric hydroxide during oxidation process, it is capable of conducting coagulation. As a result, potassium ferrate has two functions in water and wastewater treatment: oxidation and coagulation. Hence, it is recommended for the removal of environmental contaminants. This chemical compound is a green chemical (4,5) and its reaction in aquatic environment is shown in Eq. (1).

$$FeO_4^{2-} + 8H^+ + 3e^- \rightarrow Fe^{3+} + 4H_2O \quad (1)$$

Ferrate (VI) has been successfully utilized in water and wastewater treatment (6). Currently, some researches has been carried out regarding the removal of pharmaceutical contaminants and micro pollutants utilizing ferrate (VI) and the results showed the effective impact of this compound in decomposition of such contaminants in aquatic environment.

The limitations of classical methods include high time consumption due to numerous experiments and high cost of consumables that can be omitted by statistical experimental designs. Furthermore, in this approach, it is impossible to examine the interactions among variables (7). Response surface methodology (RSM) is an appropriate method that can resolve these limitations. In this study, RSM was employed to optimize variables influencing the elimination of chemical oxygen demand (COD), color, turbidity and TSS from wastewater. RSM was derived from statistical and mathematical technique. It can be used to study the effect of several factors at three or five level as well as their influence on each other (8). In this study, central composite design (CCD) was utilized for experimental design. It is an effective design that does not need a greater number of experimental runs (9). Moreover, experimental designs such as CCD model provides valuable data on direct, pair wise interaction, and curvilinear variable effects (7). This study uses CCD and RSM to optimize potassium ferrate consumption in the treatment of wastewater from textile industry.

Methods

Experimental setup and procedure

Samples were collected from a carpet factory's effluents. The colors utilized in this industry were of azo and anthraquinone types. Sampling was carried out monthly and during summer and fall. The samples were transferred to the laboratory of Qazvin University of Medical Sciences in less than 48 hours at 4°C. The conductivity and pH of the samples were measured with a Multi WTW 340i instrument (Germany). Total organic carbon (TOC) content of the samples was determined employing an ANATEC TOC analyzer (model ANATEC II Australia). Total solids (TS), total suspended solids (TSS), total volatile solids (TVS), total dissolved solids (TDS), turbidity, alkalinity, acidity and COD were determined based on standard methods for the examination of water and wastewater (10).

Ferrate oxidation experiments were conducted by means of a conventional jar-test (zag-shimi, Iran) apparatus with impellers equipped with six of 1.5 cm × 4.5 cm rectangular blades. The time and speed for mixing were set with an automatic controller. Jar testing steps include rapid mix with the mixing speed of 200 rpm for one minute, slow mix with the mixing speed of 40 rpm for 20 minutes and settlement for 60 minutes. Potassium ferrate (97%) was provided by Sigma-Aldrich Corporation (USA). Other chemicals were of analytical grade and purchased from Merck (Germany). In order to adjust pH, 0.1M sulfuric acid and 0.05 L of sodium hydroxide solution was used. A total of 30 preliminary experiments were carried out to determine pH range and the amount of ferrate needed, based on literature reviews (3). The active ferrate solution (1 g/L) was prepared by adding the solid form of K_2FeO_4 to 0.005 M Na_2HPO_4 solution and 0.001M borate buffer at pH 9.

Experimental design and model development

In this study, determination of optimum conditions (pH and potassium ferrate consumption) for the removal of COD, color, turbidity and TSS was carried out employing CCD/RSM procedures. Design Expert software (Stat-Ease Inc., Minneapolis, USA) was applied for designing, statistical modeling and optimization. Model statistics computed were F-value, Lack of fit, coefficient of determination (R^2), and coefficient of variation (CV). Generally, a good model must be significant based on *P* value and F value as opposed to the lack of fit (insignificant).

The independent variables were concentration of potassium ferrate (mg/L) and pH denoted by A and B respectively. Furthermore, the removal efficiencies of COD, turbidity, color and TSS were considered as response variables that are denoted by Y_1, Y_2, Y_3 and Y_4, respectively. Performance of the procedure was evaluated by assessing the efficiency of COD, turbidity, color and TSS removals. The variables were coded according to Eq. (2):

$$x_i = \frac{X_i - X_0}{\Delta x}\quad i=1,2,3,...k \tag{2}$$

Where x_i is the coded value of the independent variable, x_0 is the value of x_i at the center point and Δx is the difference of variable values. The maximum, central and minimum values (+1, 0,-1) of each variable was specified based on CCD. In this study, CCD included 2^k fractional factorial points (k=2), 2k axial points and 5 center points. Thus, 13 experiments including 4 factorial points, and 4 axial points were carried out. In order to evaluate the pure error, 5 additional experiments were carried out at the center point. Following the analysis of experimental data, optimal conditions were graphically determined. Determination of optimum conditions for several parameters was carried out by overlaying of plots called graphical optimization. Graphical optimization reveals the area of feasible response values in the factor space and the regions that do fit the optimization criteria would be shaded (7,8).

To confirm the results obtained from experimental data, two additional experiments were carried out in optimum conditions and the results were compared. System behavior was described by Eq. (3) which is an experimental quadratic polynomial model.

$$Y = \beta 0 + \sum_{i=1}^{k} \beta i.xi + \sum_{i=1}^{k} \beta ii.xi^2 + \sum_{ii \le j}^{k} \sum_{j}^{k} \beta ij.xi.xj + ... + e \tag{3}$$

Where *Y*, *i*, *j*, *β*, e and *k* represent the response variable, linear constant, quadratic constant, regression constant, random error as well as the number of parameters, respectively.

The minimum and maximum range of variables was investigated along with the full experimental plan, with respect to their values in actual and coded form as listed in Table 1.

Results

The results of qualitative analysis of the wastewater are shown in Table 2.

Table 1. Independent variables and limit level for response surface study

Variables	Unit	Coded symbols	Levels				
			-1	-0.5	0	+0.5	+1
Doses of potassium ferrate	mg/L	A	50	75	150	175	200
pH	-	B	3	3.75	4.5	5.25	6

Based on the 13 experimental results, the CCD for the experimental variables is shown in Table 3.

According to the obtained results, the highest removal efficiencies for COD, turbidity, color and TSS were 90%, 87%, 89.6% and 91%, respectively.

The results obtained from analysis of variance (ANOVA) with regards to the removal of COD, turbidity, color and TSS utilizing potassium ferrate are shown in Table 4. As shown in Table 4, all the terms in the statistical quadratic model were significant ($P\leq 0.0001$). Lack of fit test for COD, turbidity, color and TSS removal ($P=0.4560$, 0.7661, 0.4735, 0.7152, respectively) were not significant. The variables R^2 (coefficient of determination), $A.R^2$ (adjusted R^2), A.P (adequate precision) and C.V (coefficient of variance) are shown in Table 4. R^2 and $A.R^2$ were found to be 0.96, 0.99, 0.99 and 0.98, 0.94, 0.97, 0.98 and 0.97 for the removal of COD, turbidity, color and TSS respectively. The obtained values of AP and C.V for the mentioned parameter were 19.21, 30.13, 30.32 and 22.3, 3.4, 2.05, 3.4 and 3.83 respectively.

The ANOVA results for each response (Table 4) indicated that the linear of two independent variables (A and B),

Table 2. Characteristics of raw wastewater from the carpet factory

Parameters	Mean ± SD
pH	6.54 ± 1.2
EC (μS/Cm)	950 ± 20
TOC (mg/L)	700 ± 30
COD (mg/L)	1600 ± 50
TSS (mg/L)	280 ± 20
TDS (mg/L)	900 ± 30
TVS (mg/L)	850 ± 40
Turbidity (NTU)	554 ± 15
Alkalinity (mg/L as $CaCo_3$)	180 ± 30
Acidity (mg/L as CaCo3)	40 ± 3.5

Table 3. CCD for the experimental variables

No.	Run	pH	Potassium ferrate (mg/L)	Removal efficiency (%)			
				COD	Turbidity	Color	TSS
1	4	6	200	75	75	55	50.4
2	10	4.5	150	85	84	83.4	90.1
3	1	3	50	75	71	49	44
4	11	4.5	150	90	87	85.7	90
5	12	4.5	150	87	87	88	83
6	13	4.5	150	83	83	89.6	91
7	2	6	50	50	52	35	45
8	8	4.5	175	83.9	85	85.5	85.2
9	9	4.5	150	85	84	88.4	87.4
10	3	3	200	65	67.5	54.3	46.3
11	5	3.75	150	80	82	78.5	86
12	6	5.25	150	82	82	80	77.8
13	7	4.5	75	81.5	80	77.8	78

Table 4. ANOVA results for the removal of COD, turbidity, color and TSS using potassium ferrate

Source	Sum of squares	df	Mean squares	F-value	*P* value
COD removal (%)					
Model	1330.18	5	266.04	36.91	<0.0001
A	43.56	1	43.56	6.04	0.0436
B	58.32	1	58.32	8.09	0.0249
AB	306.25	1	306.25	42.49	0.0003
A^2	41.08	1	41.08	5.7	0.0483
B^2	9.31	1	9.31	1.29	0.2931
Residual	50.45	7	7.21		
Lack of fit	22.45	3	7.48	1.07	0.4560
Pure error	28	4	7		
Turbidity removal (%)					
Model	1167.81	5	233.56	90.24	<0.0001
A	29.39	1	29.39	11.35	0.0119
B	107.56	1	107.56	41.56	0.0004
AB	175.56	1	175.56	67.83	<0.0001
A^2	25.57	1	25.57	9.88	0.0163
B^2	16.55	1	16.55	6.4	0.0393
Residual	18.21	7	2.59		
Lack of fit	4.12	3	1.37	0.39	0.7661
Pure error	14	4	3.5		
Color removal (%)					
Model	3916.35	5	783.27	127.7	<0.0001
A	35	1	35	5.71	0.0483
B	188.83	1	188.83	30.79	0.0009
AB	54.02	1	54.02	8.81	<0.0209
A^2	138.64	1	138.64	22.6	0.0021
B^2	49.47	1	49.47	8.07	0.0250
Residual	42.93	7	6.13		
Lack of fit	18.57	3	6.19	1.02	0.4735
Pure error	24.37	4	6.09		
TSS removal (%)					
Model	3717.15	5	743.43	90.55	<0.0001
A	54.43	1	54.43	6.63	0.0368
B	179.24	1	179.24	21.83	0.0023
AB	49	1	49	5.97	<0.0446
A^2	70.6	1	70.6	8.6	0.0219
B^2	97.69	1	97.69	11.9	0.0107
Residual	57.47	7	8.21		
Lack of fit	15.15	3	5.05	0.48	0.7152
Pure error	42.32	4	10.58		
Other statistical parameters					
	R^2	Adj. R^2	Mean ± SD	AP	CV
COD	0.96	0.94	78.65±2.68	19.21	3.4
Turbidity	0.99	0.97	78.4±1.61	30.13	2.05
Color	0.99	0.98	73.1±2.48	30.32	3.4
TSS	0.98	0.97	74.78±2.87	22.3	3.83

quadratic effects (A^2 and B^2) and their interactions (AB) were significant ($P \leq 0.0001$) in the removal of COD, turbidity, color and TSS. As presented in Table 4, the Fisher coefficient (F-value) for the removal of COD, turbidity, color and TSS were significant.

The final quadratic equations for the removal of the study parameters were obtained as shown in Eq. (4) to (7).

COD removal (%) $Y_1 = 85.2 - 3.11A + 3.6B + 8.75AB - 12.98A2 - 6.18B2$ (4)

Turbidity removal (%) $Y_2 = 84.82 - 2.56A + 4.9B + 6.62AB - 10.24A2 - 8.24B2$ (5)

Color removal (%) $Y_3 = 86.27 - 2.79A + 6.48B + 3.67AB - 23.84A2 + 14.24B2$ (6)

TSS removal (%) $Y_4 = 87.6 - 3.48A + 6.31B\ 3.5 - AB - 17.01A2 - 20.01B2$ (7)

The model equations were employed to generate 3D contour plots. The 3D contour plots for the effective factors in eliminating COD, turbidity, color and TSS are plotted in Figure 1. As shown in the plots, all responses have an optimum removal point. With respect to Figure 1A and 1B, the optimum pH for the removal of COD and turbidity was 4, while it was 4.5 for color and TSS removal (Figure 1C and 1D). Also, the optimum dose of potassium ferrate for the removal of COD and turbidity were obtained as 160 and 165 mg/L, respectively, while it was 150 mg/L for color and TSS removal. In this condition, removal efficiencies for COD, turbidity, color and TSS were 86%, 86%, 87% and 89%, respectively.

By overlaying four 3D plots, the optimum removal efficiencies for COD, turbidity, color and TSS were obtained as 84%, 85%, 85% and 87%, respectively. In this condition, the dose of potassium ferrate and pH value was obtained as 160 mg/L and 4.3 respectively after overlaying four 3D plots (Figure 2).

Based on the results obtained from the final quadratic model, to confirm the reliability of the model, two additional laboratory experiments were carried out utilizing the optimum conditions suggested by the Design-Expert software as shown in Table 5.

The optimum conditions were obtained at pH 4.3 and 160 mg/L potassium ferrate concentration. At optimum conditions, COD, turbidity, color and TSS efficiency were obtained 84.5, 85, 85.3 and 88.3% respectively.

Discussion

The quality of quadratic polynomial model fit was expressed by R^2 and Adj.R^2. R^2 should be close to 1, as this

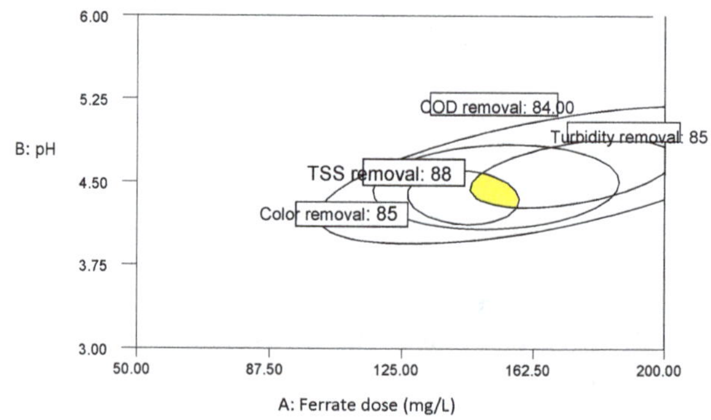

Figure 2. The overlaying plots for the removal of COD, turbidity, color and TSS

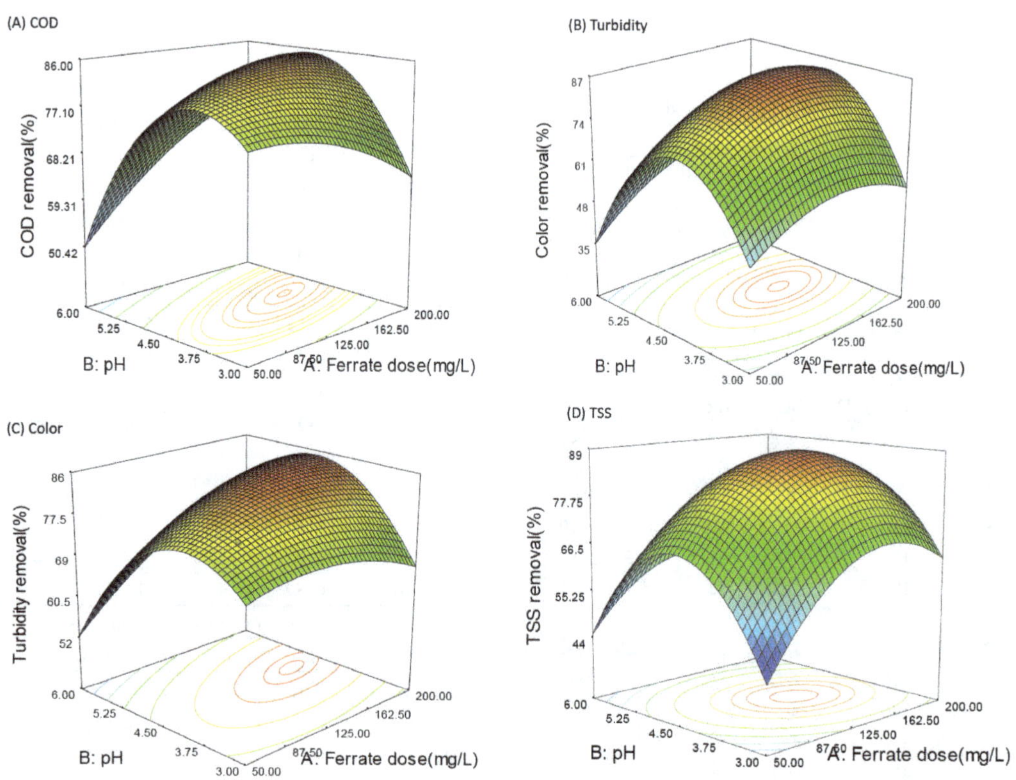

Figure 1. The 3D contour plots for the removal of COD, turbidity, color and TSS by potassium ferrate.

Table 5. Verification of experimental results at optimum conditions

Optimum condition (pH 4.3 and ferrate dose 160 mg/L)	COD removal efficiency (%)	Turbidity removal efficiency (%)	Color removal efficiency (%)	TSS efficiency (%)
Experimental results	83%	84.1%	84.5%	88%,
Model response	84.5%	85%	85.3%	88.3%
Error	1.5	0.9	0.7	0.3
Standard deviation	0.7	0.43	0.3	0.2

would show better correlation between the predicted and experimental values. A high R^2 coefficient ensures a satisfactory adjustment of the quadratic model to the experimental data. As a result, an optimal agreement with the adjusted R^2 is important (8,9). In this research, in all cases, R^2 coefficient was high. The high values of R^2 suggest the high agreement of the obtained experimental data, with the data evaluated by the model. So, it can be said that the quadratic model has a good fitness for the data. Adequate precision is a measure of the range in predicted response relative to its associated error, in other words, a signal-to-noise ratio. Its desired value is 4 or more (11,12). In this study, AP values for the removal of COD, turbidity, color and TSS were greater than 4 indicating the existence of adequate signal and the high capability of this model in predicting the results. Coefficient of variance (CV) estimates the ratio of the standard error to the mean value of the observed responses (as a percentage) and also describes the reproducibility of models. Normally, we can consider a model to be reproducible when its CV is not greater than 10% (9). CV was within the acceptable range of 10% which proves the reproducibility of the model and high precision in the results. All of these terms indicated a good fitting when there is high correlation between the observed and predicted values. The lack of fit (LOF) F-test describes the variation of data around the fitted model. If the model does not fit the data well, the lack of fit test will be significant (8). Lack of fit test for COD, turbidity, color and TSS responses were not significant, thus it shows a good fitting of the data to the model.

The response surface plots were approximately symmetrical in shape with circular contours. All response plots showed clear peaks, indicating that optimal conditions for the maximum responses depend on the dose potassium ferrate and pH in the design space. Initial concentration of ferrate and initial pH value has a significant influence in the oxidative ability of the ferrate ion. With respect to Figure 1, the interaction effects of ferrate dose and pH on responses, demonstrate concentrically closed curves. The centers represent the optimum conditions, indicating that experimental design has been well carried out. Furthermore, the experimental results which were obtained from the optimum conditions, showed the precision of the resulting data and models.

Previous studies have reported that potassium ferrate can also eliminate a range of metals (e.g., Fe^{2+}, Fe^{3+}, Mn^{2+}, and Cu^{2+}) and toxic heavy metals (e.g., Pb^{2+}, Cd^{2+}, Cr^{3+}, and Hg^{2+}) to a low level at a dose range of 10–100 mg/L as K_2FeO_4 (13, 14).

The results of a study carried out on the application of 70 mg/L potassium ferrate for the treatment of color-contained textile wastewater, indicated that high removal efficiencies of COD, color and turbidity were 91%, 50% and 93%, respectively. Except the color removal efficiency, other variables were consistent with current study (15). This can be due to different characteristics of textile wastewater.

Some researchers employed encapsulated potassium ferrate (VI) for the removal of DNBP (dinitro butylphenol). The optimal conditions regarding the degradation of DNBP were pH 6.5 and encapsulated concentration of 1.2 g/L. They also reported that encapsulated potassium ferrate (VI) was very effective in COD reduction as well as the decolorization of real wastewater from the manufacturing of DNBP. In addition, they stated that the removal of DNBP increased with encapsulated Fe (VI) concentration. The DNBP removal increased with corresponding efficiency from 68% to 93% when the encapsulated Fe (VI) dosage increases from 0.4 to 1.2 g/L (16). High consumption of ferrate for the degradation of DNBP may be as a result of complex structure of this compound.

The results of a study showed the highest removal efficiencies for Ibuprofen and Ciprofloxacin by potassium ferrate in pH 4 and 5-6, respectively (17). Moreover, another research reported complete removal of flurbiprofen from aqueous solutions by ferrate in pH 4 (18), which were in agreement with our study.

The results of a research carried out on the removal of bisphenol compounds by ferrate (VI) indicated that when pH increases from 3 to 5, bisphenol removal efficiency by ferrate (VI) would increase from 85.1% to 95.2% (19). Our results were in line with those of the mentioned study.

A study carried out in relation to the removal of turbidity and COD in the batch reactor with ferrate dose of 15 mg/L for secondary wastewater treatment, showed a significant improvement with respect to the removal of suspended solids (70.8%) and turbidity (78%) (20). The low consumption of potassium ferrate when compared to the findings of our study may be as a result of low concentration of organic matters in secondary effluent.

A report obtained on the comparative performance of potassium ferrate (VI) and ferric sulphate for the removal of COD, color and turbidity in sewage treatment, showed that potassium ferrate (VI) could reduce COD and organic contaminants in comparison with ferric sulphate for the same doses utilized. Furthermore, the volume of sludge produced by potassium ferrate (VI) was much less than ferric sulphate, which ought to make subsequent treatment of sludge easier.

Conclusion

Potassium ferrate is an effective compound for the treatment of wastewater from carpet factories. However, the limitations of the ferrate application are related to the instability of its solutions along with the complicated preparation and storage of this compound.

CCD of RSM can be utilized to determine the significant variables and optimum condition for textile wastewater treatment with respect to COD, TSS, turbidity, and color removal. Potassium ferrate concentration (A), pH (B), their interactions (AB) and quadratic effects (A^2 and B^2) were significant in the removal of COD, turbidity, color and TSS from carpet industries effluents. The statistical approach has been confirmed to be a powerful tool in studying textile wastewater treatment utilizing potassium ferrate.

Acknowledgments

The authors would like to appreciate the technical support provided by Qazvin University of Medical Sciences.

Competing interests

The authors declare that they have no competing interests.

Authors' contributions

All authors contributed equally to data collection, and article approval.

References

1. Szpyrkowicz L, Juzzolino C, Kaul SN. A comparative study on oxidation of disperse dyes by electrochemical process, ozone, hypochlorite and Fenton reagent. Water Res 2001; 35(9): 2129-36. doi: 10.1016/S0043-1354(00)00487-5.
2. Jaafarzadeh Haghighifard NA, Jorfi S, Ahmadi M, Mirali S, Kujlu R. Treatment of mature landfill leachate by chemical precipitation and Fenton advanced oxidation process. Environmental Health Engineering and Management Journal 2016; 3(1): 35-40.
3. Dindarloo K, Jamali HA, Lakbala P, Mahmoodi H, Kazemi F. Feasibility of electrochemical oxidation process for treatment of saline wastewater. Environmental Health Engineering and Management Journal 2015; 2(3): 129-34.
4. Jiang JQ, Panagoulopoulos A, Bauer M, Pearce P. The application of potassium ferrate for sewage treatment. J Environ Manag 2006; 79(2): 215-20. doi: 10.1016/j.jenvman.2005.06.009.
5. Jiang J. Research progress in the use of ferrate (VI) for the environmental remediation. J Hazard Mater 2007; 146(3): 617-23. doi: 10.1016/j.jhazmat.2007.04.075.
6. Jiang JQ, Wang S, Panagoulopoulos A. The role of potassium ferrate (VI) in the inactivation of Escherichia coli and in the reduction of COD for water remediation. Desalination 2007; 210(1): 266-73. doi: 10.1016/j.desal.2006.05.051.
7. Mohajeri S, Aziz HA, Isa MH, Zahed MA, Adlan MN. Statistical optimization of process parameters for landfill leachate treatment using electro-Fenton technique. J Hazard Mater 2010; 176(1-3): 749-58. doi: 10.1016/j.jhazmat.2009.11.099.
8. Montgomery Douglas C. Design and Analysis of Experiments. 8th ed. John Wiley & Sons; 2008.
9. Ahmed AL, Ismail S, Bhatia S. Optimization of coagulation-flocculation process for palm oil mill effluent using response surface methodology. J Environ Sci Tech 2005; 39(8): 2828-34. doi: 10.1021/es0498080.
10. APHA. Standard methods for the examination of water and wastewater 21 th ed Washington American Public Health Association; 2005.
11. Wu Y, Zhou S, Qin F, Ye X, Zheng K. Modeling physical and oxidative removal properties of Fenton process for treatment of landfill leachate using response surface methodology (RSM). J Hazard Mater 2010; 180(1-3): 456-65. doi: 10.1016/j.jhazmat.2010.04.052.
12. Arslan-Alaton I, Akin A, Olmez-Hanci T. An optimization and modeling approach for H2O2/UV-C oxidation of a commercial non-ionic textile surfactant using central composite design. J Chem Technol Biotechnol 2010; 85(4): 493-501. doi: 10.1002/jctb.2315.
13. Murmann RK, Robinson PR. Experiments utilizing FeO_4^{2-} for purifying water. Water Res 1974; 8(8): 543-7. doi: 10.1016/0043-1354(74)90062-1.
14. Bartzatt R, Cano M, Johnson L, Nagel D. Removal of toxic metals and nonmetals from contaminated water. J Toxicol Environ Health A 1992; 35(4): 205-10.
15. Ciabatti I, Tognotti F, Lombardi L. Treatment and reuse of dyeing effluents by potassium ferrate. Desalination 2010; 250(1): 222-8. doi: 10.1016/j.desal.2009.06.019.
16. Wang HL, Liu SQ, Zhang XY. Preparation and application of sustained release microcapsules of potassium ferrate (VI) for dinitro butyl phenol (DNBP) wastewater treatment. J Hazard Mater 2009; 169(1-3): 448-53. doi: 10.1016/j.jhazmat.2009.03.130.
17. Jiang JQ, Zhou Z, Patibandla S, Shu X. Pharmaceutical removal from wastewater by ferrate (VI) and preliminary effluent toxicity assessments by the zebrafish embryo model. Microchem J 2013; 110: 239-45. doi: 10.1016/j.microc.2013.04.002.
18. Barışçıa S, Ulu F, Sillanpää M, Dimoglo A. Evaluation of flurbiprofen removal from aqueous solution by electrosynthesized ferrate (VI) ion and electrocoagulation process. Chem Eng J 2015; 262: 1218-25. doi: 10.1016/j.cej.2014.10.083.
19. Han Q, Wang H, Dong W, Liu T, Yin Y, Fan H. Degradation of bisphenol A by ferrate (VI) oxidation: Kinetics, products and toxicity assessment. Chem Eng J 2015; 262: 34-40. doi: 10.1016/j.cej.2014.09.071.
20. Farooq S, Bari A. Tertiary treatment with ferrate and ozone. J Environ Eng 1986; 112(2): 301-10. doi: 10.1061/(ASCE)0733-9372(1986)112:2(301).

23

Slaughterhouse wastewater treatment by combined anaerobic baffled reactor and anaerobic filter: study of OLR and HRT optimization in ABR/AF reactors

Zabihollah Yousefi[1]*, Mohammad Behbodi [1], Reza Ali Mohammadpour [2]

[1]Department of Environmental Health Engineering, School of Public Health, Mazandaran University of Medical Sciences, Sari, Iran
[2]Department of Biostatistics, School of Public Health, Mazandaran University of Medical Sciences, Sari, Iran

Abstract
Background: Slaughterhouse wastewater (SWW) is hardly treated due to the large amount of organic matter, nutrients and suspended solids. These materials are naturally decomposed through biological processes, and then environmental pollution, transmission of pathogens and problems become smelled. Conventional purification methods require high investment costs, high energy consumption and expert workforce. Therefore, the efficient and sufficient treatment of SWW with low cost, efficient construction and operation is important.
Methods: A combined anaerobic system consisting of three pilot-scale anaerobic baffled reactors (ABRs) in the first stage and three anaerobic filters (AFs) were used to treat SWW. The ABR reactors arranged at three hydraulic retention times (HRTs) of 12, 18 and 24 hours and organic loading rates (OLRs) of 4, 7 and 10 kg/m^3/d. The OLR applied for ABR reactors, was 0.5 to 1.55 kg COD/m^3/d.
Results: Evaluation of the ABR reactor indicated that this reactor at OLR of 7 and 10 kg COD/m^3/d and HRT of 18 hours, had removal efficiency of 83.29% and 85.79%, respectively. AF reactor, at OLR of 0.981, 0.576 and 0.561 kg COD/m^3/d and HRT of 36 hours, had removal efficiency of 79.39%, 74.09% and 63.14%, respectively.
Conclusion: The optimum HRT and OLR were 24 hours and 7 kg COD/m^3/d and 36 hours and 1 kg COD/m^3/d in ABR and AF reactor, respectively.
Keywords: Waste water, Slaughterhouse, Anaerobic treatment

***Correspondence to:**
Zabihollah Yousefi
Email: zyousefi2004@gmail.com

Introduction

The slaughter industry demands large amounts of water for its processes include killing, hide or hair removal, carcass washing and cleanup operations and generates substantial volumes of wastewater (1). Slaughterhouse wastewater (SWW) contains high levels of organics (blood, fecal matter, urine, paunch and undigested food), pathogenic and non-pathogenic viruses, bacteria, nitrogen and phosphorus from organic materials (2,3). Several studies have described the common characteristics of SWW and reported that raw wastewater have biochemical oxygen demand (BOD = 834 to 16680 mg/L), chemical oxygen demand (COD = 1790 to 27800 mg/L), total phosphorus (TP = 34 to 720), total nitrogen (TN = 106 to 550 mg/L), total kjeldahl nitrogen (TKN = 90 to 525 mg/L), pH 6.7 to 8 and electrical conductivity 1.99 to 9.14 mS (4-11). SWW as a source of environmental, water and food contamination causes serious environmental problems by direct discharge into water body (deoxygenation of rivers and contamination of groundwater). So, this wastewater must be treated before final discharge (12). Similar to municipal wastewater plant, SWW treatment methods fall into three main categories include primary, secondary, and tertiary treatment by various methods such as physical (flow equalization, dissolved air flotation), chemical (coagulation–flocculation) and biological (aerobic, anaerobic) systems (1,13,14). Biological treatment is usually the most cost-effective technology for SWW. Among several biological methods, the anaerobic process is an attractive alternative for slaughterhouse wastewater treatment (SWWT) plants due to low construction and operation cost, high organic loading rate (OLR), low sludge production, simple design, high solids retention time, high stability to organic shocks and potential to produce biogas. In anaerobic systems, microorganisms break down biodegradable materials in the absence of oxygen and produce methane, carbon dioxide and hydrogen. Depending on the mechanism used to achieve biomass detention, these reactors can be classified into three major groups include the fixed film, suspended growth and hybrid reactors (1,13,14). Despite these advantages, some drawbacks of anaerobic treatment such as long retention time and instability due to hydraulic and organic shock loads lead to the development of the high-rate reactors that separate hydraulic retention time (HRT) from solid retention time (SRT). Such separation which allows the slow-growing bacteria to remain within the reactor independent of the wastewater flow and allows a higher volumetric load rate, significantly increased reactor removal efficiencies (11,15). Typical high rate anaerobic reactor used for SWW include upflow anaerobic sludge blanket (UASB), anaerobic baffled reactor (ABR), anaerobic filter (AF), anaerobic sequencing batch reactor (SBR), anaerobic attached film expanded bed reactor (AAFEB) and anaerobic rotating biological contactor (RBC) (1,13,14). Anaerobic treatment alone is not sufficient to achieve a high degree of treatment and acceptable discharge standard in terms of BOD, COD and nutrients (N, P). One of the technological solutions being considered is an ABR combined with AF (AF). ABR is one of these high-rate anaerobic systems developed by McCarty and co-workers at Stanford University. The ABR has been described as a series of UASBs, in which hanging and standing baffles divide the reactor into different compartments and flow zones (16,17). In each compartment, high concentrations of biomass that tend to settle and rise with horizontal flow stream and gas production, cause the reactor work at a relatively high SRT and low HRT and provide higher treatment rate, higher resilience to organic and hydraulic shock loads, longer biomass retention times and lower sludge yields than other anaerobic treatment systems (17,18). Moreover, the reactor ability to separate acidogenesis and methanogenesis phase and develop different bacterial groups under more favorable conditions allowing the reactor to behave as a two-phase system and enhancing reactor stability without the associated control problems and high costs (19,20). Boopathy and Tilche successfully treated high-strength molasses wastewater using a hybrid ABR. At OLR of 20 kg COD/m^3/d, soluble COD removal efficiencies over 70% were achieved (18). Gopala Krishna et al reported COD and BOD removal efficiencies more than 90% at 8 and 10 hours HRT and OLRs of 1.5 to 1.2 kg COD/m^3/d (21). Bodkhe reported removal efficiencies of 84% for COD and 87% for ABR in treating domestic wastewater at an HRT of 6 hours (22). Cao and Mehrvar used the combined ABR and UV/H_2O_2 processes to treat a synthetic SWW, and reported treatment efficiencies of 89.9% for TOC, 97.7% for COD and 96.6% for BOD$_5$. They also found that combined processes had higher treatment efficiency than those of individual processes (23). Yousefi et al investigated the Landfill leachate treatment by modified ABR/AF at different OLRs and HRTs. They reported maximum COD removal efficiency in 10.72 kg COD/m^3/d OLR and 3-day HRT and average removal efficiency varied from 39% to 96%, 39% to 58% and 4% to 16% for COD, BOD$_5$ and orthophosphate, respectively (24). Different studies reported the efficiency of AF for a wide range of OLR and HRT (25-30).

In this study, combined ABR and AF was used to treat SWW at a pilot scale. Among different parameters, OLR and HRT are the most important parameters in anaerobic processes that must be optimally controlled. The aim of this study was to investigate the optimum OLR and HRT of two high-rate reactors (ABR and AF) in anaerobic treatment of high-strength SWW.

Materials and Methods

Slaughterhouse wastewater

Raw wastewater used in this basic and applied study was collected three times in a week from an industrial slaughterhouse in Sari, Iran. The raw wastewater was screened to separate coarse particles larger than 3 mm and stored in a refrigerator at 4-8°C before use. Analysis of the wastewater was carried out several times and the average composition is displayed in Table 1. By changing blood and water ratio, different OLRs and COD concentrations were provided for the reactors.

Experimental set-up

In this study, a three-pilot-scale system consisted of ABR and an AF was designed and built up in Mazandaran University of Medical Sciences wastewater treatment plant. Schematic representations of the reactors used in the experiments are shown in Figure 1.

Anaerobic baffled reactor

Three ABR reactors (A_1, A_2 and A_3) with effective volume of 250, 200 and 130 L were used, respectively. These reactors were right-angle shaped and entirely made up by PE (Population Equivalent) consisted of 4 equal-volume compartments. Each compartment was further divided into two small chambers by a 45° slanted-edge baffle leading to downflow and upflow of the wastewater.

Table 1. Characteristics of the slaughterhouse wastewater

Parameters	Range	Number of samples
COD (mg/L)	2000-10000	23
TKN (mg/L)	42-227	23
TP (mg/L)	23-129	23
pH	7.15-7.68	27
Temperature (°C)	15-30	27

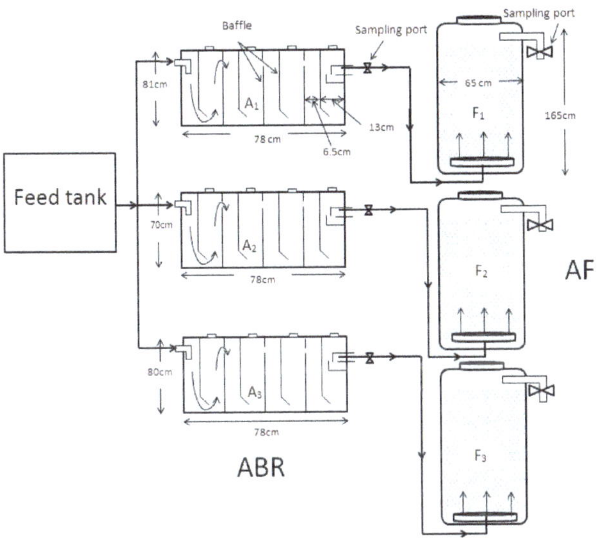

Figure 1. Schematic diagrams of the experimental set-up for the combined processes of ABR and AF reactor.

This structure provided an effective mixing and contact time between the biomass and wastewater within each compartment. The wastewater sampling ports were located at the inlet and outlet zone of each reactor.

Anaerobic filter

Three identical polyethylene (PE) cylinders (65 cm inner diameter, 165 cm height) with 300 L effective volume were used as AF (F_1, F_2 and F_3) in this study. The wastewater entered each reactor from the bottom (under the bed support media) through a T-inlet connected to the ABR reactor outlet. Two-thirds of the reactor (100 cm) was filled with polyvinyl chloride (PVC) corrugated tubes (3 cm height and 1.5 cm in diameter) that were employed as random supporting materials. The reactor packing medium provided a specific surface area of 250 m^2/m^{-3} and a porosity of 93% for biomass attachment, thus, the working volume of AF reactors was 280 L. At the outlet of the AF reactor, one sampling port was allowed to take grab samples. All calculations of HRTs and OLRs were based on the reactors effective volumes and the wastewater daily flow rates.

Inoculum

The inoculum used for reactors start-up was a mixture of 200 L digested sewage sludge obtained from Sari municipal wastewater treatment plant (WWTP) and 80 L SWW taken from the equalization basin of Mazandaran slaughterhouse WWTP. The mixture was screened to remove coarse materials and then transferred to ABR reactors (30% of total volume) and AF (10% total volume). After three days, 100 l diluted SWW with 1000 mg COD/L were added to ABR and these reactors effluent used as AF reactors influent. During the acclimatization period (153 days), raw wastewater concentration gradually increased up to 4000, 3000, 2000 mg COD/l in ABR reactors A_1, A_2 and A_3, respectively. As ABR and AF were fully acclimatized, the combined processes were operated continuously for a total of 152 days at various HRTs (12-24 hours) and OLRs under ambient temperature (15.2–35.5°C). In order to obtain a steady-state condition after changing the OLRs and HRTs from one to another, the systems were allowed to run for three HRTs before taking any sample for measurement.

Analytical methods

The following parameters were analyzed twice a week during the 6-month study period: COD, TKN, TP, total solids (TS) and total suspended solids (TSS) according to standard methods for the examination of water and wastewater (20th edition). The COD was determined by the closed reflux method and direct reading spectrophotometer (DR/2800, HACH, USA). TKN was measured using the macro-Kjeldahl procedure and titrimetric method for ammonia nitrogen determination, TP was measured by the persulfate digestion and stannous chloride method, TSS and TS were measured by the gravimetric method, pH was determined electrometrically and nitrate was determined by following the procedure described by USEPA (method 352.1).

Data were analyzed using analytical statistics (analysis of variance, ratio, etc) by SPSS and Excel software.

Results

Effects of HRT and OLR in ABR

The effect of changing HRTs and OLRs on the effluent characteristics was analyzed to determine the relationship between both parameters and optimize condition for systems performance in terms of removal efficiencies, reactors stability etc.

Figure 2 shows the evaluation of COD removal efficiency for three HRTs (12, 18 and 24 hours) and three OLRs (4, 7 and 10 kg COD/m^3/d).

Results showed that when OLR increased from 4, 7 to 10 kg COD/m^3/d, the COD removal efficiency had an increasing trend and reached maximum COD removal

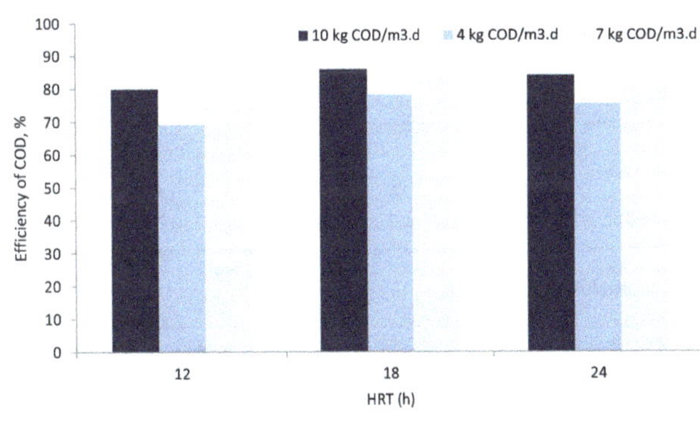

Figure 2. Average COD removal efficiency at different HRTs and OLRs in ABR treating raw SWW.

efficiency of 83.29% and 85.72% at an OLR of 7 and 10 kg COD/m³/d, respectively (Figure 2).
According to Table 2, OLR 7 kg/m³/d and HRT 18 h was the optimum condition for ABR and the highest removal efficiency and stability were obtained in this condition.

Effects of HRT and OLR in AF performance
In this section, different OLRs and HRTs were investigated in AF to determine the effects of the process parameters (HRT and OLR) on the reactor performance and stability and also to develop an optimum situation for pollutant removal using a continuous AF reactor. For this reason, three AF reactors (F_1, F_2 and F_3) were used at different HRTs (18, 24 and 36 hours) and OLRs (0.4-1.55 kg COD/m³/d). After reactors start-up, the initial HRT was set at 36 hours and then decreased to 24 and 18 hours (first phase), then it was set at 18 and increased to 24 and 36 hours (second phase), and finally, it was set at 36 hours and then decreased to 24 and 18 hours (third phase). Table 3 shows different runs performed and the HRT and OLR tested for each reactor in this study. Removal efficiency at each HRT and OLR under steady-state operations is presented in Figures 3-5. The experimental analyses were intermitted because of the adaptation of the reactor to the OLR changes. In this study, 27 measurements were performed for all reactors.
As shown in Figures 3-5, at HRT 36 hours, the highest removal efficiency was observed in the reactors F1, F2 and F3, respectively. The F_1 reactor at an OLR of 1 kg COD/m³/d showed the highest removal efficiency of 68.83%, 78.46% and 79.39% at HRT of 18, 24 and 36 hours,

Table 2. Summary of COD performance of three ABRs at different OLRs and HRTs in treating SWW

Reactor number	OLR (kg COD/m³/d)	HRT (h)	Percent removal
A_1	4	12	68.96
A_2	4	18	78.13
A_3	4	24	75.3
A_1	7	12	74.35
A_2	7	18	83.29
A_3	7	24	81.34
A_1	10	12	79.76
A_2	10	18	85.72
A_3	10	24	83.91

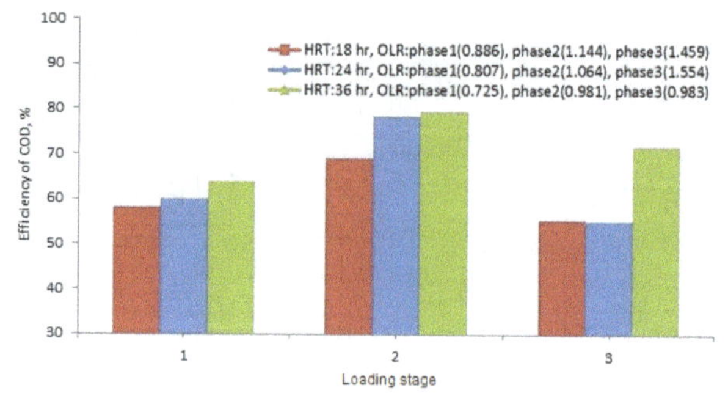

Figure 3. Summary of COD performance of AF_1 at different OLRs and HRTs in treating ABR_1 effluent.

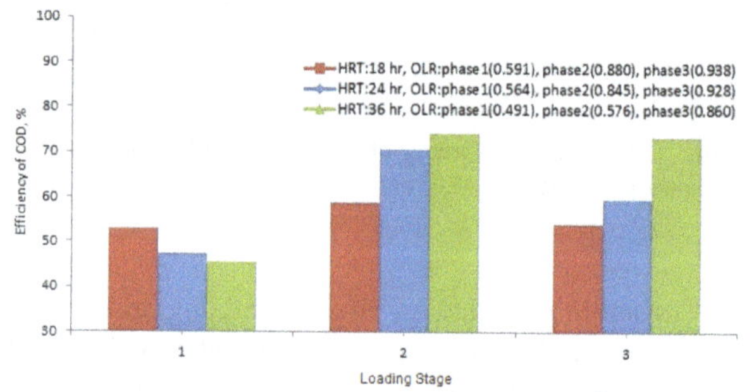

Figure 4. Summary of COD performance of AF_2 at different OLRs and HRTs in treating ABR_2 effluent.

respectively. F_2 showed removal efficiency of 59.2%, 70.97% and 74.09% at OLR of 0.760 kg COD/m³/d and HRT of 18, 24 and 36 hours, respectively. And F_3 showed COD removal efficiency of 52.6%, 58.5% and 63.14% at OLR of 0.764 kg COD/m³/d and HRT of 18, 24 and 36 hours, respectively.

Discussion

According to previous studies, the increase of OLR and HRT had significant effects on the effluent characteristics, thus, the COD was used as one of the most important parameters to compare the performance of the reactors and monitor the effect of these parameters throughout the study (20-24,29). These studies showed that COD concentrations in the influent (raw wastewater) ranged between 2000 and 10000 mg/L for ABR and 190 and 970 mg/L for AF.
The effects of OLRs were studied in a continuous system by varying the influent COD concentrations; it was revealed

Table 3. Details of operation conditions of AF used in the treatment of SWW in treatment studies

Parameters		Phase 1			Phase 2			Phase 3		
		Run 1	Run 2	Run 3	Run 4	Run 5	Run 6	Run 7	Run 8	Run 9
HRT (h)		36	24	18	18	24	36	36	24	18
OLR (kg COD/m³.d)	F_1	0.725	0.807	0.886	1.144	1.064	0.981	0.983	1.554	1.459
	F_2	0.491	0.564	0.591	0.880	0.845	0.576	0.860	0.928	0.938
	F_3	0.417	0.584	0.574	0.906	0.826	0562	0.846	0.788	0.874

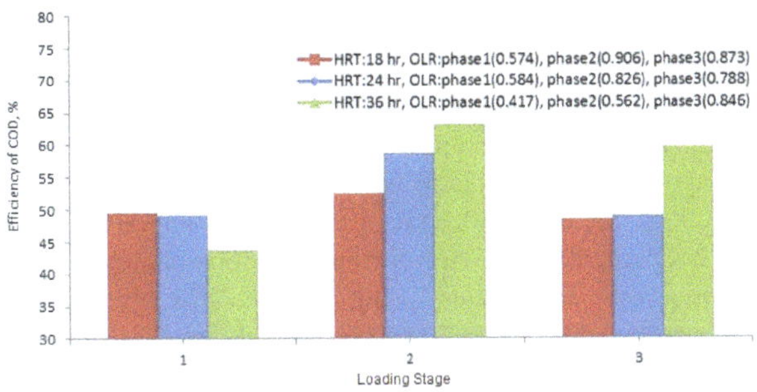

Figure 5. Summary of COD performance of AF_3 at different OLRs and HRTs in treating ABR_3 effluent

that by increasing the OLR, the effluent COD was also increased and a significant increase in the effluent COD was observed at the highest OLR (10 kg/m^3/d) (Figure 2). The results showed that COD removal efficiency was significantly increased with increase of the HRT from 12 to 18 and 24 hours at a constant OLR, however, with increase of the HRT from 18 to 24 hours, a slight increase in the system removal efficiency was observed. The maximum COD removal efficiency in the ABR processes was observed at 18 hours HRT, in this situation, the COD removal efficiencies were 78.13%, 83.29% and 85.72% at 4, 7 and 10 kg/m^3/d OLR, respectively, which is consistent with the results of the previous studies (14,26,29). At the lowest HRT (12 hours), the effluent COD removal rate was reduced to 68.96%, 74.35% and 79.76% at OLR 4, 7 and 10 kg/m^3/d, respectively, which was attributed to the incomplete anaerobic digestion and an onwards slight sludge wash-out.

According to Table 2, at a fixed OLR, the treatment efficiency did not depend on the HRT when it was longer than 18 h. But when a considerably lower HRT (12 hours) and a high OLR (10 kg/m^3/d) were applied, removal efficiency decreased to 68.96%. However, with higher HRT (18-24 hours), an OLR of 10 kg COD/m^3/d did not show such destabilization, which shows the great influence of high HRT over process stability. Despite an increase in the COD removal efficiencies, by increasing OLR from 7 to 10 kg, the variance explained by this variable was 2.77 and 6.8 in the optimum HRT (18 hours), respectively.

As shown in Figures 3-5, the removal of COD was not significantly declined when the OLR increased except at HRT 24 hours by 23%, 11% and 9% reduction in F_1, F_2 and F_3, respectively. The results indicate that the increase of HRT from 18 to 36 hours significantly improved the COD removal in AF reactor because the longer HRT allowed an effective conversion of the substrate and COD removal was relatively high at all OLRs.
The results of this research are not consistent with some research (21-23). The reason for this discrepancy is that the loading conditions for the published results were not the same and also the researchers have announced the results in lower loading. However, they are consistent with the findings of some studies (24-30), and the higher efficiency for more powerful sewage in this study was obtained, as with other studies (9,11,15,24).

Conclusion

In summary, the combined ABR and AF appears to be a potentially useful reactor for use in SWW treatment. It was concluded that the ABR processes exhibited higher treatment ability in the optimal conditions for the COD reduction. The optimum HRT in ABR was determined as 18 hours; at this situation, the impacts of different OLRs on the removal efficiency were investigated. The optimum OLR was 7 kg COD/m^3/d for the ABR in this study. The results showed that the increase of HRT from 18 to 36 hours significantly improved the COD removal in AF because longer HRT allowed an effective conversion of the substrates and COD removal was relatively high at all OLRs. The optimum HRT was determined as 36 h and the optimal OLR was 1, 0.760 and 0.764 kg COD/m^3/d in F_1, F_2 and F_3 reactors, respectively.

Acknowledgements

This study was supported by Sari Public Health School and Vice Chancellor for Research and Technology in Mazandaran University of Medical Sciences. Also, this paper extracted from an MSc thesis and research project that approved by Vice Chancellor for Research and Technology in Mazandaran University of Medical Sciences (code: 952). The authors would like to gratitude Masoumali Movahedi for his sincere cooperation in this research.

Competing interests

The authors declare that they have no competing interests.

Authors' contribution

All authors contributed and participated in the data collection, analysis, and interpretation. All authors reviewed, refined, and approved the manuscript.

References

1. United States Environmental Protection Agency. EPA-821-R-04-011. Technical Development Document for the Final Effluent Limitations Guidelines and Standards for the Meat and Poultry Products Point Source Category (40 CFR 432). Washington, DC: EPA; 2004.
2. Bibbal D, Kerouredan M, Loukiadis E, Scheutz F, Oswald E, Brugere H. Slaughterhouse effluent discharges into rivers not responsible for environmental occurrence of enteroaggregative Escherichia coli. Vet Microbiol 2014; 168(2-4): 451-4. doi: 10.1016/j.vetmic.2013.11.042.

3. Jensen PD, Sullivan T, Carney C, Batstone DJ. Analysis of the potential to recover energy and nutrient resources from cattle slaughterhouses in Australia by employing anaerobic digestion. Appl Energy 2014; 136: 23-31. doi: 10.1016/j.apenergy.2014.09.009.
4. Rajakumar R, Meenambal T, Banu JR, Yeom IT. Treatment of poultry slaughterhouse wastewater in upflow anaerobic filter under low upflow velocity. Int J Environ Sci Technol 2011; 8(1): 149-58. doi: 10.1007/bf03326204.
5. Keskes S, Hmaied F, Gannoun H, Bouallagui H, Godon JJ, Hamdi M. Performance of a submerged membrane bioreactor for the aerobic treatment of abattoir wastewater. Bioresour Technol 2012; 103(1): 28-34. doi: 10.1016/j.biortech.2011.09.063.
6. Jensen PD, Yap SD, Boyle-Gotla A, Janoschka J, Carney C, Pidou M, et al. Anaerobic membrane bioreactors enable high rate treatment of slaughterhouse wastewater. Biochem Eng J 2015; 97: 132-41. doi: 10.1016/j.bej.2015.02.009.
7. Del Nery V, Damianovic MH, Pozzi E, de Nardi IR, Caldas VEA, Pires EC. Long-term performance and operational strategies of a poultry slaughterhouse waste stabilization pond system in a tropical climate. Resources Conservation and Recycling 2013; 71: 7-14. doi: 10.1016/j.resconrec.2012.11.006.
8. Budiyono, Widiasa IN, Johari S, Sunarso. Study on slaughterhouse wastes potency and characteristic for biogas production. Int J Waste Resour 2011; 1(2): 4-7. doi: 10.4172/2252-5211.1000102.
9. Bazrafshan E, Kord Mostafapour F, Farzadkia M, Ownagh KA, Mahvi AH. Slaughterhouse wastewater treatment by combined chemical coagulation and electrocoagulation process. PLoS One 2012; 7(6): e40108. doi: 10.1371/journal.pone.0040108.
10. Bayar S, Yildiz Y, Yilmaz A, Koparal AS. The effect of initial pH on treatment of poultry slaughterhouse wastewater by electrocoagulation method. Desalination Water Treat 2014; 52(16-18): 3047-53. doi: 10.1080/19443994.2013.800268.
11. Sunder GC, Satyanarayan S. Efficient treatment of slaughter house wastewater by anaerobic hybrid reactor packed with special floating media. International Journal of Chemical and Physical Sciences 2013; 2: 73-81.
12. Hossaini H, Fatehizadeh A, Yousefi N, Reshadat S, Rajabi Gilan N, Ghasemi SR, et al. Application of enhanced softening process in slaughterhouse wastewater treatment. Indian J Chem Technol 2013; 20(3): 217-21.
13. Johns MR. Developments in wastewater treatment in the meat processing industry: a review. Bioresour Technol 1995; 54(3): 203-16. doi: 10.1016/0960-8524(95)00140-9.
14. Mittal GS. Treatment of wastewater from abattoirs before land application--a review. Bioresour Technol 2006; 97(9): 1119-35. doi: 10.1016/j.biortech.2004.11.021.
15. Nguyen H, Turgeon S, Matte J. The Anaerobic Baffled Reactor: A study of the wastewater treatment process using the anaerobic baffled reactor. Massachusetts: Worcester Polytechnic Institute; 2010.
16. Shakeri M, Vossoughi M. Coupled anaerobic baffled reactor (ABR)/activated sludge treatment of synthetic wastewater with high concentration of sulfate and COD. International Journal of Engineering 2003; 16(1): 11-20.
17. Foxon KM, Pillay S, Lalbahadur T, Rodda N, Holder F, Buckley CA. The anaerobic baffled reactor (ABR): an appropriate technology for on-site sanitation. Water SA 2004; 30(5): 44-50. doi: 10.4314/wsa.v30i5.5165.
18. Boopathy R, Tilche A. Anaerobic digestion of high strength molasses wastewater using hybrid anaerobic baffled reactor. Water Res 1991; 25(7): 785-90. doi: 10.1016/0043-1354(91)90157-L.
19. Barber WP, Stuckey DC. The use of the anaerobic baffled reactor (ABR) for wastewater treatment: a review. Water Res 1999; 33(7): 1559-78. doi: 10.1016/S0043-1354(98)00371-6.
20. Liu RR, Tian Q, Yang B, Chen JH. Hybrid anaerobic baffled reactor for treatment of desizing wastewater. Int J Environ Sci Technol 2010; 7(1): 111-8. doi: 10.1007/bf03326122.
21. Gopala Krishna GV, Kumar P, Kumar P. Treatment of low strength complex wastewater using an anaerobic baffled reactor (ABR). Bioresour Technol 2008; 99(17): 8193-200. doi: 10.1016/j.biortech.2008.03.016.
22. Bodkhe SY. A modified anaerobic baffled reactor for municipal wastewater treatment. J Environ Manage 2009; 90(8): 2488-93. doi: 10.1016/j.jenvman.2009.01.007.
23. Cao W, Mehrvar M. Slaughterhouse wastewater treatment by combined anaerobic baffled reactor and UV/H2O2 processes. Chem Eng Res Des 2011; 89(7): 1136-43. doi: 10.1016/j.cherd.2010.12.001.
24. Yousefi Z, Yazdani Cherati J, Movahedi M, Kariminejad F. Effect of Organic Loading Rate on the Performance of Anaerobic Process in Treatment of Pulp and Paper Mill Effluents. Journal of Mazandaran University of Medical Sciences 2015; 25(131): 136-50. [In Persian]
25. Rajakumar R, Meenambal T. Comparative Study on start-up performance of HUASB and AF reactors treating poultry slaughterhouse wastewater. Int J Environ Res 2008; 2(4): 401-10. doi: 10.22059/ijer.2010.221.
26. Tritt WP. The anaerobic treatment of slaughterhouse wastewater in fixed-bed reactors. Bioresour Technol 1992; 41(3): 201-7. doi: 10.1016/0960-8524(92)90002-F.
27. Omil F, Garrido JM, Arrojo B, Mendez R. Anaerobic filter reactor performance for the treatment of complex dairy wastewater at industrial scale. Water Res 2003; 37(17): 4099-108. doi: 10.1016/s0043-1354(03)00346-4.
28. del Pozo R, Diez V, Beltrán S. Anaerobic pre-treatment of slaughterhouse wastewater using fixed-film reactors. Bioresour Technol 2000; 71(2): 143-9. doi: 10.1016/S0960-8524(99)90065-2.
29. Kobayashi HA, Stenstrom MK, Mah RA. Treatment of low strength domestic wastewater using the anaerobic filter. Water Res 1983; 17(8): 903-9. doi: 10.1016/0043-1354(83)90164-1.
30. Bodik I, Herdova B, Drtil M. The use of upflow anaerobic filter and AnSBR for wastewater treatment at ambient temperature. Water Res 2002; 36(4): 1084-8. doi: 10.1016/s0043-1354(01)00308-6.

Use of bauxite from active Iranian mines for the removal of fluoride from drinking water

Mohammad Malakootian[1,2*], Marzie Javdan[2], Farnaz Iranmanesh[1]

[1]Environmental Health Engineering Research Center, Kerman University of Medical Sciences, Kerman, Iran
[2]Department of Environmental Health, School of Public Health, Kerman University of Medical Sciences, Kerman, Iran

Abstract
Background: Fluoride plays an important role in bone and dentin mineralization; however, excess fluoride intake is harmful to mankind.
Methods: This study evaluated the performance of bauxite from active Iranian mines in removing fluoride from drinking water. The effects of pH, contact time, adsorbent dose, and fluoride concentration on defluoridation and removal efficiency were determined. Kinetics and adsorption isotherms were studied. Fluoride levels were measured using SPADNS. Data analysis was performed using SPSS16.
Results: Bauxite from the Jajarm mine had the lowest adsorbency (20 g/L) and required the shortest contact time (90 minutes) to reach equilibrium compared with the ore from bauxite mines evaluated in another study which had greater efficiency rates in removing fluoride from drinking water (58.15%). The fluoride removal efficiency rates of the other bauxite mines were as follows: Mendon > Sadrabad > Khidabas > Khezri > Shahbalaghi > Tash > Biglar. Bauxite from Shomal-e Yazd, Hasanabad, and Shahid Nilchian mines could not achieve the required efficiency to remove fluoride from drinking water without initial preparation and modification. The removal efficiency rates of actual samples were much lower than the synthetic samples because of confounding factors.
Conclusion: As a result of the low cost and abundant availability of bauxite and the fact that its use does not require a particular expertise or sophisticated technology, the removal efficiency of this adsorbent can be increased to desirable levels through the use of corrective methods such as heating, acidifying, particle crushing, or the mixing of two or more removal systems.
Keywords: Fluoride, Iran, drinking water, kinetics

***Correspondence to:**
Mohammad Malakootian
Email:
m.malakootian@yahoo.com

Introduction

Fluoride occurs naturally in mineral deposits and natural water systems (1). Since groundwater is a major source of human exposure to fluoride (2,3), determining the amount of fluoride present in drinking water is an important factor in human health. The treatment of groundwater resources is often the only disinfection performed on drinking water; fluoridation and defluoridation rarely occur (4).

Fluoride has been classified as a water pollutant by the World Health Organization (WHO). The presence of fluoride in drinking water, dependent upon concentration (5) and duration of continuous absorption (2), can be either helpful or harmful to humans. Maintaining a fluoride concentration of about 1 mg/L can prevent skeletal and dental problems, especially in children (2,6). However, the intake of excessive fluoride can cause serious concerns for public health (7).

Fluorosis is one of the most important public health issues around the world (5). Dental and skeletal fluorosis is irreversible, and there is no treatment for it yet. The only remedy is prevention, which is possible by maintaining safe limits of fluoride intake. Following the WHO guidelines, many countries have reported 1.5 mg/L as the maximum contaminant level. Many epidemiologists have indicated in their reports, however, that there is the possibility of adverse effects with long-term exposure to fluoride, even where consumption is at the permissible limit (8,9). Global estimates suggest that over 70 million people suffer from fluorosis (10). Fluoride concentrations in the drinking water of large parts of the world are higher than allowed by the WHO guidelines (11). It is estimated that more than 200 million people worldwide use drinking water that has a fluoride concentration higher than set standards (2). The presence of high levels of fluoride in groundwater is a global problem and is found in various locations in Africa, Asia, and America (9).

Mesdaghinia et al studied the monitoring of fluoride levels in the groundwater of Iran (4). Their study showed that in various regions of Iran, especially Bushehr province and parts of Hormozgan, Khorasan-e Razavi, Fars, and Yazd provinces, fluoride levels in groundwater were higher than standard levels (4). A study conducted in 2012 by Haghighat et al showed that fluoride concentrations in the drinking water of Dashtestan and Borazjan from Bushehr port functions, Hormozgan (Bandar Abbas, Bandare Lenghe, Qeshm), Yazd (Ardakan), Hamedan, Tehran, Kerman (Shahrbabak, Kuhbanan), Khorasan-e Razavi, South Khorasan, Semnan, Zabol, and Zahedan were higher than set standards (11).

The effects of excess fluoride on teeth and bones have been documented (12). Research on the removal of fluoride from water in order to prevent dental and skeletal fluorosis is improving around the world.

Methods such as coagulation, chemical precipitation, surface adsorption, electrochemical and membrane processes, and ion exchange are used to remove excess fluoride from water (13). The high costs and complexity of application of many of these methods prohibit their use in developing countries and low-income areas (14). There is a growing tendency to use local materials and low-cost methods for the removal of fluoride from water (15). Removing fluoride from water by surface adsorption is known as the ideal method in small communities. In recent years, the use of low-cost adsorbents to remove fluoride from water has been investigated. Hydroxyapatite, calcite, fluorspar and quartz (16), fly ash (17), red mud (18), and bentonite (19) are some such adsorbents. Bauxite is a native and available mineral consisting of aluminum hydroxides with minor and variable amounts of silica, iron oxide, titanium oxide, and alumina silicate (20). Many studies have investigated using bauxite to remove pollutants from water, such as Bhakat et al in India (21), Baral et al in India (22), and Debasish et al in Korea (23) which used bauxite to remove arsenic and chromium (VI) and chromium, respectively. In Tanzania, Thole used bauxite to remove fluoride from water (24). Sajidu et al In 2008 in India, used raw bauxite for the removal of fluoride from water (25). Atasoy et al in Turkey (26), Lavecchia et al in Italy (20), and Malakootian et al in Iran used bauxite to remove fluoride from water (3). Some areas of Iran have high fluoride concentrations; however, prior to this study, the efficiency of bauxite from Iranian mines in fluoride removal had not been studied. Thus, the current study evaluated the efficiency of activated bauxite from ore mined in Iran in removing fluoride from aqueous solutions.

Methods

This experimental study was conducted on synthetic and actual samples in the Environmental Health Engineering Research Center, Kerman University of Medical Sciences, from May to November of 2013. Actual samples were collected from drinking water from the Kuhbanan region in northwestern Kerman province (31°24′37″N, 56°16′57″E) (Figure 1).

Previous studies reported that fluoride concentrations in the water resources of this region ranged from 2.3-5.4 mg/L (27). First, the chemical quality of actual samples was determined. Bauxite, used as adsorbent in this study, was prepared from bauxite mines that were active in Iran until 2011 (Jajarm mine located in north Khorasan; Shomal-e Yazd and Sadrabad mines located in Yazd; Shahblaghi mine located in Damavand; Hassanabad and Biglar mines located in Qazvin; Mendon mine located in Kohgeloye and Bovair Ahmad; Shahid Nilchian mine in Chaharmahal and Bakhtiari; Khezri and Khidbas mines in South Khorasan; and Tash mine in Semnan province). Initially, stones were crushed, ground, and sieved to obtain <250 μm size (mesh 60). The resultant bauxite powder was washed with deionized water, dried in an oven at 200°C in an open-air chamber muffle furnace for 2 hours, and cooled at room temperature. Samples were stored in a desiccator. X-ray fraction (XRF) and x-ray diffraction (XRD) were used to determine the chemical composition and the phases in the bauxite powder samples. XRD and XRF analyses were done using the Philips X'PERT device in Iran Mineral Processing Research Center. Stock solution was prepared by dissolving 0.221 mg sodium fluoride (NaF) in 1000 mL deionized water (1 mL = 100 μg/F), and experimental solutions of the required concentrations were prepared from this solution. Fluoride was measured using the SPADNS method and spectrophotometry (wavelength = 570 nm). All analysis methods were taken from the *"Standard Methods for Examination of Water and Wastewater"*, 20th edition (28). Absorption tests were used to determine equilibrium time, adsorption kinetics, and optimal conditions (pH, contact time, fluoride concentration, and amount of adsorbent), and the adsorption isotherm models were studied. Absorption experiments were used to determine equilibrium time, synthetic absorption, and optimum conditions and to study the adsorption isotherm models. The operational parameters were pH values (3–34); initial

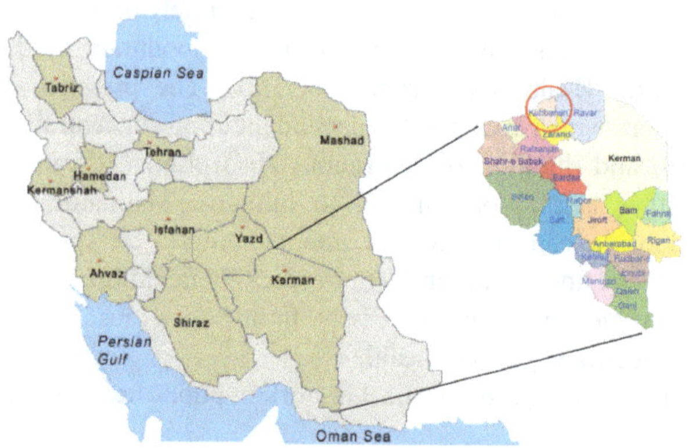

Figure 1. Geographical map of Kuhbanan region in northwestern Kerman province (31°24′37″N, 56°16′57″E).

fluoride concentrations (2, 4, 6 mg/L); dosages of bauxite (2, 5, 34, 15, 20, 25, 30 g/L); contact times (5-240 minutes). Tests were then performed under optimal conditions on actual samples, and removal efficiency was measured. The pH value of the samples was set to 0.1*N* by H_2SO_4 and NaOH. The pH was measured by a microprocessor pH meter (model 211 Hanna). Deionized water was used in the preparation of all solutions and in all analyses. Removal efficiency (E%) and adsorption capacity (q_e mg/L) were calculated using equations 1 and 2:

$$E = [(C_i - C_e)/C_i] \times 100 \quad (1)$$

$$q_e = [(C_i - C_e)V] / m \quad (2)$$

where C_i and C_e are the initial and final concentrations of fluoride in solution (mg/L), respectively, V is the volume of solution (L), and m is mass of the adsorbent (g).

Langmuir and Freundlich isotherm models were investigated to study the adsorption isotherms. Data analysis was performed using SPSS software version 16.

Results

Composite sampling was done 3 times from the Kuhbanan water resource. Analysis of the samples was expressed as the average of three replicates. The commonly occurring cations and anions present in the Kuhbanan drinking water sample were found in mg/L as NO_3^- (5.6), SO_4^{2-} (139), Cl^- (126), Total solid (711), Ca^{2+} (83), K^+ (1.9), Mg (49), Na^+ (95), F^- (2.3–2.7), Fe^{3+} (<0.05), Al^{3+} (<0.01), pH (6.9–7.66), and temperature (25–27°C). Sajidu et al examined the effects of bauxite on water quality after fluoride removal in the villages of southern Malawi in order to measure the amount of fluoride and fluoride removal by bauxite. In this study, a slight increase in aluminium and silicon contents and the pH of the water after bauxite was used; however, the amounts were less than the WHO standards (15). On this basis, there was no need to measure Al and Fe contents.

The names of each mine and the chemical analyses of ore from each using XRF and XRD are presented in Table 1.

The greatest constituents in all mines were Al_2O_3, Fe_2O_3, SiO_2, and TiO_2. The main phases of most mines were diaspore, anatase and boehmite.

The maximum fluoride removal efficiency rate achieved using bauxite from the studied mines and the optimal amount of bauxite, contact time, and pH for both the synthetic and actual water samples from Kuhbanan are given in Table 2.

The results regarding the pseudo first order and pseudo second order kinetics for initial fluoride concentrations (2, 4, 6 mg) are presented in Table 3.

The Langmuir and Freundlich isotherm parameters for the adsorption of fluoride using 5 g of activated bauxite from ore mined in Iran, pH = 7, and initial fluoride concentrations of 2, 4, and 6 mg/L are given in Table 4.

Discussion

The results showed that the ore from all active bauxite mines in Iran (with the exception of Nilchian, Hasanabad, and Shomal-e Yazd mines) had the required qualities for fluoride removal from drinking water samples and could achieve much lower concentrations than the WHO guidelines (from 2.74 mg to less than 1.5 mg) without any preparation or modification. The bauxite particles from Jajarm, Tash, Mendon, and Sadrabad mines proved more efficient than that from other mines in adsorbing fluoride from drinking water. The differences in performance are associated with several factors, such as the composition of the adsorbent.

Table 1 shows that the ore from all studied bauxite mines have aluminium, silicon, titanium, and iron oxide (the four main elements of bauxite). The fluoride removal efficiency is correlated with the percentage of AL_2O_3 and Fe_2O_3 compounds. Percentages of silicon oxide (SiO_2) and titanium oxide (TiO_2) in each of the 11 studied mines were higher than the percentages reported in other studies. Gomoro et al in Ethiopia used 3 lateritic soils including red gullale soil (RGS), yellow gullale soil (YGS), and ambo soil

Table 1. Chemical analysis of Iranian active bauxite mines by XRF and XRD

Name of mine	XRD pattern of the bauxite ore mines	Percentage of chemical compounds of Bauxite by XRF							
	The main minerals of bauxite	P_2O_5	K_2O	MgO	CaO	Fe_2O_3	Al_2O_3	TiO_2	SiO_2
Tash	Diaspore Kaolinite Chamosite	0.05	0.52	0.48	0.66	20.60	49.31	3.98	10.39
Nilchian	Diaspor-Boehemite-Anatase-Kaolinit	0.05	0.12	0.13	0.42	1.07	53.90	2.78	30.51
Mendon	Boehemite-Hematie-Kailinite	0.06	0.05	0.17	0.29	19.51	53.64	2.48	12.43
Jajarm	Diaspore- Anatase -Chamosite	0.09	0.22	0.69	1.53	21.06	51.05	5.56	9.82
Biglar	Dioaspore-Anatase-Hematite-Chlorite	0.16	0.65	0.12	0.44	38.50	24.96	6.18	11.87
Khidbas	Boehemite-Hematite-Kaolinite	0.07	0.25	0.33	0.44	20.97	40.04	2.43	24.00
Sadrabad	Diaspore-Hematite-Cholorite-Anatas	0.10	0.03	2.35	0.52	24.77	45.77	5.87	10.20
Hassanabad	Diaspore-Anatase	0.12	0.99	0.04	0.45	1.89	59.90	11.60	9.31
Shahbalaghy	Kaolinite-Chamosite-Anatase	0.09	<0.03	0.67	0.29	39.80	21.85	3.79	16.97
Kaezri	Boehemite-Kaolinite	0.08	0.20	0.19	0.30	17.67	44.36	3.76	13.28
Shomal-e-Yazd	Diaspore-Hematite-Dolomite-Quartz	0.09	1.00	3.35	6.62	19.93	34.38	4.74	17.27

Table 2. Maximum efficiency of fluoride removal by using bauxite of studied mines with optimal amounts of bauxite, contact time and pH for synthetic and water samples of Kuhbanan

Entry	Name of mine	Contact time (min)	pH	Fluoride removal from Kuhbanan drinking water (2.74 mg/L)	Efficiency of fluoride removal from synthetic samples (2 mg/L)	Adsorbent dose (g)
1	Tash	120	7	61.67	70.39	25
2	Nilchian	180	6	30.19	60.16	35
3	Mendon	120	7	56.64	85.75	20
4	Jajarm	90	7	58.15	93.12	20
5	Biglar	180	6	49.31	66.47	25
6	Khidbas	180	6	52.69	80.41	30
7	Sadrabad	180	7	53.21	84.69	25
8	Hassanabad	180	6	36.05	62.72	30
9	Shahbalaghy	180	6	46.05	77.17	25
10	Kaezri	180	7	50.87	82.91	35
11	Shomal-e-Yazd	180	7	36.68	64.74	30

Table 3. The results of investigation of pseudo first and pseudo second order of kinetics for initial fluoride concentrations (2, 4, 6 mg)

Entry	Name of mine	Fluoride concentrations mg/L	First class synthetic			Second class synthetic		
			K_1	q_e	R^2	K_2	q_e	R^2
1	Tash	2	0.023	0.562	0.976	0.148	0.303	0.985
		4	0.023	0.707	0.956	0.062	0.501	0.967
		6	0.016	0.664	0.959	0.086	0.494	0.979
2	Nilchian	2	0.013	0.540	0.975	0.292	0.053	0.911
		4	0.018	0.652	0.998	0.419	0.047	0.933
		6	0.018	0.648	0.950	0.466	0.049	0.933
3	Mendon	2	0.023	0.553	0.942	0.218	0.347	0.991
		4	0.027	0.762	0.988	0.116	0.613	0.988
		6	0.034	0.810	0.940	0.156	0.667	0.989
4	Jajarm	2	0.018	0.567	0.958	0.138	0.339	0.984
		4	0.032	0.748	0.982	0.090	0.543	0.983
		6	0.018	0.816	0.872	0.025	0.754	0.906
5	Biglar	2	0.009	0.484	0.950	0.089	0.254	0.929
		4	0.011	0.634	0.979	0.043	0.447	0.919
		6	0.006	0.710	0.984	0.027	0.580	0.887
6	Khidbas	2	0.016	0.549	0.962	0.063	0.320	0.940
		4	0.013	0.752	0.973	0.025	0.683	0.842
		6	0.013	0.816	0.968	0.024	0.803	0.866
7	Sadrabad	2	0.009	0.538	0.936	0.078	0.331	0.939
		4	0.011	0.692	0.975	0.043	0.559	0.948
		6	0.013	0.800	0.987	0.035	0.734	0.954
8	Hassanabad	2	0.009	0.478	0.975	0.063	0.248	0.883
		4	0.009	0.648	0.985	0.025	0.484	0.851
		6	0.006	0.653	0.965	0.029	0.501	0.878
9	Shahbalaghy	2	0.011	0.502	0.886	0.222	0.296	0.991
		4	0.011	0.653	0.909	0.105	0.563	0.990
		6	0.020	0.669	0.949	0.114	0.656	0.996
10	Khezri	2	0.011	0.521	0.977	0.062	0.292	0.940
		4	0.009	0.661	0.982	0.034	0.509	0.928
		6	0.009	0.714	0.950	0.029	0.623	0.932
11	Shomal-e-Yazd	2	0.004	0.511	0.780	0.146	0.223	0.949
		4	0.004	0.644	0.784	0.154	0.352	0.989
		6	0.004	0.747	0.888	0.042	0.494	0.891

Table 4. Langmuir and Freundlich isotherm (5 g/L bauxite, pH = 7, initial fluoride concentrations 2, 4 and 6 mg)

Entry	Name of mine	Freundlich isotherm			Langmuir isotherm		
		n	k_f	R^2	q_m	k_L	R^2
1	Tash	3.484	0.616	0.884	0.504	2.319	0.991
2	Nilchian	2.906	0.540	0.980	0.470	1.090	0.999
3	Mendon	3.225	0.732	0.908	0.712	2.900	0.997
4	Jajarm	2.801	0.677	0.994	0.718	1.500	0.998
5	Biglar	1.766	0.526	0.981	0.750	0.438	0.995
6	Khidbas	1.795	0.623	0.915	0.899	0.645	0.955
7	Sadrabad	1.992	0.513	0.998	0.581	0.603	0.982
8	Hassanabad	1.992	0.513	0.917	0.581	0.603	0.968
9	Shahbalaghy	1.988	0.664	0.873	0.859	0.926	0.959
10	Kaezri	1.751	0.563	0.977	0.841	0.476	0.992
11	Shomal-e-Yazd	2.016	0.521	0.993	0.663	0.485	0.964

(AS) to remove fluoride from drinking water (9). Under oxidative conditions and as a result of the weathering of rocks, this soil loses silica. The main component of this soil depends on lateritic phase. The last step is the formation of bauxite. Maximum fluoride removal using 2 g/L of RGS soil was 32.32%. The findings of a more complex fluoride formation with RGS soil could be due to the presence of high levels of alumina (Al_2O_3) (9).

In a study conducted in Tanzania in 2009, Peter used two types of kaolinite and bauxite-rich soil to remove fluoride (29). His results showed that soil rich in bauxite had a higher adsorption capacity than soil rich in kaolinite. He identified the chemical and mineral composition of soils as reasons for different removal rates. Soil chemical analyses revealed that soil rich in bauxite had a large amount of aluminum oxide compared to soil rich in kaolinite. Peter introduced aluminum oxide as a good adsorbent for fluoride removal due to the Al and F ions reaction (29). XRF analyses of ore from all bauxite mines achieved results similar to those of Altundoğan and Tümen who used bauxite in phosphate removal (30), Ayoob et al who used arsenic in bauxite removal (31), Sujana et al (8), Sajidu et al (25), and Malakootian et al (32) who used bauxite in fluoride removal. However, there are differences in the percentages of major bauxite compositions among all studies. Gomoro et al indicated that changes in the various formation stages of bauxite are reasons for different percentages of bauxite ore compounds in various locations (9). According to the results of the presented XRD patterns, all mines included the same minerals, such as bauxite, diaspore (AlOOH), anatase (TiO_2), and quartz (SiO_2). Diaspore is from the aluminum hydroxide family, and this can be the main factor in removing fluoride with bauxite. High amounts of anatase and kaolinite in some mines can cause decreasing fluoride removal efficiency. Peter attributed the low fluoride removal efficiency using kaolinite to the high content of SiO_2 (50.60%) (29). Gomoro et al found that anatase and quartz rarely have a good adsorption capacity for fluoride removal (9).

With the ore from all bauxite mines in Iran, with a fixed amount of adsorbent (5 g/L), initial concentrations of fluoride (2, 4, and 6 mg/L), and over time, the removal efficiency increased to the maximum extent. However, it did not increase for a longer time. After equilibrium time, the repulsive forces between the pollutant on the adsorbent and contaminants in the solution prevented contact between the pollutants and the adsorbent. The shortest time to reach equilibrium was 90 minutes, achieved using bauxite from the Jajarm mine; the longest time to reach equilibrium was 180 minutes, achieved using bauxite from Khidbas, Khezri, Sadrabad, Shomal-e-Yazd, Biglar, Hasanabad, Nilchian, Mendon, Tash, and Shahbalaghi mines. The equilibrium time for bauxite from Tash mine was 120 minutes. In all cases, equilibrium time was independent of the initial concentration of fluoride. Peter used soils rich in bauxite and kaolinite to remove fluoride from water with high concentrations of fluoride. He achieved a range of 90 to 180 minutes as the maximum time to achieve efficiency of fluoride removal by both soils. Peter expressed that saturation of the ion-exchange positions explained achieving equilibrium over time (29). Sujana et al used bauxite to remove fluoride. Equilibrium time was calculated at 120 minutes for concentrations of 4 and 12 mg-F (8). Atasoy et al used raw and modified bauxite in fluoride removal and reached equilibrium after 3 hr. They found that the removal rate was fast at first, because all the absorption positions were empty and free fluoride ions were available (26). Evaluations of the effect of time on removal efficiency in the mentioned studies revealed results similar to those of the present study.

One factor that has a significant impact on the adsorption process is pH. Basically, bauxite is a mixture of iron, aluminum, silica, titanium oxide, and hydroxide. When bauxite is hydrated at neutral pH, a net positive or negative charge will be formed which is able to adsorb the fluoride anion. According to the results for bauxite from all 11 mines, the highest fluoride removal efficiency was observed at neutral pH, and the lowest fluoride removal efficiency was observed at acidic and alkaline pHs. The reduction in removal efficiency at pH <5 could be attributed to the formation of poor hydrofluoric acid (HF). The reduction in efficiency in alkaline pHs could be due to competition between hydroxide ions (OH^-) and fluoride ions (F^-) in the absorption process. Shrivastava and Vani reviewed fluoride removal methods and found that the maximum fluoride absorption by alumina takes place in pH 5-7. They justified reduction efficiency at pHs less than 5 and greater than 7 with these two reasons: 1) At pH <5, activated alumina is insoluble in water, and that leads to the loss of surface absorption; and 2) at pH >7, silicate and hydroxide have a stronger bond with the positions of absorption toward fluoride ions (33). These results concur with those of Sujana et al (pH = 6.4) (8), Malakootian et el (pH = 7) (34-36), and Sajidu et al (pH = 7) (15). At acidic and alkaline pHs, they showed the same trend.

Initial fluoride concentration is the most effective factor for fluoride ion transfer from the liquid phase to a solid phase adsorbent (37). The results showed that removal efficiency rates are higher in low concentrations of fluoride. By using fixed amounts of all bauxite samples, increasing the fluoride concentration from 2 to 6 mg/L increased the absorption capacity, but decreased the removal efficiency. A study by Atasoy et al showed that increasing the fluoride concentration from 1 to 10 mg/L reduced the removal efficiency from 66% to 36% (26). Lv et al used double-modified hydroxides with polycarbonate, aluminum, and manganese in fluoride removal. By increasing the fluoride concentration from 20 to 200 mg/L (in a fixed amount of adsorbent), the residual fluoride was changed from 0.4 to 50 mg/L. This result showed that the fluoride removal efficiency increased when the initial concentration of fluoride was decreased. The results of the mentioned studies were consistent with the results of the current study (38).

The results showed that by increasing amount of bauxite in all studied samples, the removal efficiency was increased, but the absorption capacity was decreased. Efficiency was almost constant with the addition of higher absorption. The maximum adsorption efficiency was 46.58%, obtained using 2 g/L of bauxite from the Sadrabad mine in 6 mg//L of fluoride concentration. In similar conditions, when bauxite was increased to 25 g/L, the adsorption efficiency was 75.31%. Increasing the amount of adsorbent made more active positions available for fluoride ions, but the removal efficiency did not show any dramatic increase for values higher than 25 gr/L; in this amount only a limited amount of free fluoride ions remained. A study was conducted in Iran in 2011 by Malakootian et al to remove fluoride using regenerated spent bleaching earth. In this study, increasing the amount of adsorbent from 10 to 20 g/L (concentration of fluoride = 2.5 mg/L) increased removal efficiency from 66.41% to 74.01% (34). In India, Mohapatra et al used mineral oxides to remove fluoride. Similar behavior was observed with increased amounts of adsorbent (39).

In this study, fluoride removal using bauxite from the Sadrabad, Shahid Nilchian, Biglar, Khidbas, Hassanabad, and Khezri mines and using pseudo first order kinetics with a correlation coefficient greater than 0.950 showed better results than pseudo second order kinetics. The results of this study were consistent with those of other similar studies, such as Mohapatra et al (39), Das et al (40), and Sujana et al (8).

Fluoride removal using bauxite from the Shomal-e-Yazd, Tash, Mendon, Jajarm, and Shahbalaghi mines and using pseudo second order kinetics with a correlation coefficient greater than 0.984 showed better results than the pseudo first order kinetics. Thole (24) and Tor et al (41) found similar results. Analysis of the data regarding fluoride absorption on bauxite particles by isotherm showed that using bauxite from the Jajarm, Khidbas, Khezri, Shahid Nilchian, Mendon, Hassanabad, Biglar, Shahbalaghi, and Tash mines with a correlation coefficient greater than 0.955, fluoride absorption obeyed the Langmuir adsorption isotherm. Sujana et al, Malakootian et al, and Das et al used bauxite as a fluoride adsorbent, and their experimental data was better described by the Langmuir absorption model (8,34,40). These mentioned studies are consistent with this study.

Fluoride absorption by bauxite from the Sadrabad and Shomal-e Yazd mines with a correlation coefficient greater than 0.993 obeyed the Freundlich adsorption isotherm. Lavecchia et al (20) and Wang et al (42) obtained similar results. In this study, 20 g/L of bauxite from the Jajarm and Mendon mines, 25 g/L of bauxite from the Sadrabad, Biglar, Shahbalaghi and Tash mines, 30 g/L of bauxite from the Hassanabad, Khidbas, and Shomal-e Yazd mines, and 35 g/L of bauxite from the Khezri and Shahid Nilchian mines was added to the drinking water of Kuhbanan as the actual sample. Under these conditions and with no change to the pH of the water, fluoride concentrations changed from 2.74 mg/L to 1.15, 1.14, 1.28, 1.42, 1.48, 1.48, 1.75, 1.30, 1.74, 1.35, and 1.91 mg/L, respectively. Generally, under the same conditions, the removal efficiency is less in actual samples than in synthetic samples. The presence of interfering ions in the actual samples (e.g., nitrate, sulfate, chloride) compete with fluoride to adsorb the iron and aluminum of the bauxite, resulting in a decrease in the removal efficiency of the actual sample. Sujana et al used bauxite to remove fluoride from groundwater (8). At pH = 6, increasing the sulfate, nitrate, and phosphate concentrations (0-20 mg/L) in solution caused a reduction in fluoride removal, and increasing the number of carbonate ions had a negligible effect on absorption (43). The absorption rate was lower in actual samples with normal pH values than in synthetic samples. In addition to interfering anions, water pH can cause a reduction. Das et al reported water pH as a reason for the reduction in absorption efficiency using activated titanium rich bauxite in the groundwater samples (40). According to the results, bauxite from the mines of Jajarm, Mendon, Sadrabad, Khezri, Khidbas, Shahbalaghi, Tash, and Biglar performed as needed to reach the WHO guidelines (<1.5 mg/L) to remove fluoride from actual (Kuhbanan drinking water with 2.74 mg/L fluoride concentration) and synthetic samples (2, 4, and 6 mg/L concentrations of fluoride) with no initial preparation and as a raw material; however, the residual concentration of fluoride in the drinking water samples after using bauxite from the Hassanabad, Shahid Nilchian, and Shomal-e Yazd mines was greater than the WHO guidelines (>1.5 mg/L). The highest removal efficiency obtained using bauxite was seen at the Jajarm mine and equaled 93.12% for synthetic samples and 58.15% for actual samples from Kuhbanan. The use of the following methods is suggested to increase the fluoride removal efficiency, it is suggested that raw bauxite be improved by milling it to nanometer-sized particles, a high sorbent be used in actual samples, or the contact time with adsorbent be increased, and bauxite be modified using acid or heat.

Conclusion

Bauxite from the Jajarm mine had a higher removal efficiency (58.15%) than the other studied mines. The removal efficiency rates of the other mines were as follows: Jajarm > Mendon > Sadrabad > Khidabas > Khezri > Shahbolaghi > Tash > Biglar. Bauxite from the Shomal-e Yazd, Hasanabad, and Shahid Nilchian mines were not sufficiently efficient without initial preparation and modification. Although the removal efficiency in actual samples is much lower than that in synthetic samples, its low cost, abundant availability, and no need for expertise and sophisticated technology are beneficial. The removal efficiency of this adsorbent can be increased to the desirable level using corrective methods such as heating, acidifying, particle crushing, and the mixing of two or more removal systems.

Acknowledgements

This study extracted from a master's thesis conducted at the Environmental Health Engineering Research Center. It was sponsored by the Vice-Chancellor for Research and Technology of Kerman University of Medical Sciences. The author would like to express his gratitude for all the assistance provided by this University and to Iran's Mineral Processing Research Center which cooperated on the XRD and XRF analyses of bauxite ore mines and to all others involved in the realization of this study.

Competing interests

The author declares that he has no competing interests.

Authors' contributions

The author contributed and is involved in the suggestion of the problem, design of experiments, data collection, and article approval.

References

1. Chen N, Zhang Z, Feng C, Sugiura N, Li M, Chen R. Fluoride removal from water by granular ceramic adsorption. J Colloid Interface Sci 2010; 348(2): 579-84. doi: 10.1016/j.jcis.2010.04.048.
2. Garmes H, Persin F, Sandeaux J, Pourcelly G, Mountadar M. Defluoridation of groundwater by a hybrid process combining adsorption and Donnan dialysis. Desalination 2002; 145(1): 287-91. doi: 10.1016/S0011-9164(02)00424-1.
3. Malakootian M, Javdan M, Iranmanesh F. Fluorid removal study from aqueous solutions using Jajarm bauxite: Case study on Koohbanan water. Fluoride 2015; 48(2): 113-122.
4. Mesdaghinia A, Vaghefi KA, Montazeri A, Mohebbi MR, Saeedi R. Monitoring of fluoride in groundwater resources of Iran. Bull Environ Contam Toxicol 2010; 84(4): 432-7. doi: 10.1007/s00128-010-9950-y.
5. Nie Y, Hu C, Kong C. Enhanced fluoride adsorption using Al (III) modified calcium hydroxyapatite. J Hazard Mater 2012; 233-234: 194-9. doi: 10.1016/j.jhazmat.2012.07.020.
6. Thole B, Mtalo F, Masamba W. Groundwater Defluoridation with Raw Bauxite, Gypsum, Magnesite, and Their Composites. Clean Soil Air Water 2012; 40(11): 1222-8. doi: 10.1002/clen.201100111.
7. Nagendra Rao CR. Fluoride and Environment-a Review; 2003 Dec 15-17; Department of Geography, University of Madras and Faculty of Environmental Studies, York University, Chennai; 2003.
8. Sujana MG, Anand S. Fluoride removal studies from contaminated ground water by using bauxite. Desalination 2011; 267(2-3): 222-7. doi: 10.1016/j.desal.2010.09.030.
9. Ayoob S, Gupta AK. Fluoride in Drinking Water: A Review on the Status and Stress Effects. Crit Rev Environ Sci Technol 2006; 36(6): 433-87.
10. Malakootian M, Fatehizadeh A, Yousefi N, Ahmadian M, Moosazadeh M. Fluoride removal using regenerated spent bleaching earth (RSBE) from groundwater: case study on Kuhbonan water. Desalination. 2011; 277(1): 244-9. doi: 10.1016/j.desal.2011.04.033.
11. Haghighat GA, Dehghani MH, Nasseri S, Mahvi AH, Rastkari N. Comparison of carbon nonotubes and activated alumina efficiencies in fluoride removal from drinking water. Indian J Sci Technol 2012; 5(S3): 2432-5.
12. Malakootian M, Moosazadeh M, Yousefi N, Fatehizadeh A. Flouride removal from aqueous solution by pumice casestudy on Kuhbonan water. Afr J Environ Sci Tech 2011; 5(4): 299-306. doi: 10.5897/AJEST10.308.
13. Dou X, Mohan D, Pittman CU, Yang S. Remediating fluoride from water using hydrous zirconium oxide. Chem Eng J 2012; 198-199: 236-45. doi: 10.1016/j.cej.2012.05.084.
14. Chidambaram S, Ramanathan AL, Vasudevan S. Fluoride

removal studies in water using natural materials: technical note. Water SA 2003; 29(3): 339-44. doi: 10.4314/wsa.v29i3.4936.

15. Sajidu S, Kayira C, Masamba W, Mwatseteza J. Defluoridation of Groundwater Using Raw Bauxite: Rural Domestic Defluoridation Technology. Environment and Natural Resources Research 2012; 2(3): 1-9. doi: 10.5539/enrr.v2n3p1.
16. Fan X, Parker DJ, Smith MD. Adsorption kinetics of fluoride on low cost materials. Water Res 2003; 37(20): 4929-37. doi: 10.1016/j.watres.2003.08.014.
17. Xu X, Li Q, Cui H, Pang J, Sun L, An H, et al. Adsorption of fluoride from aqueous solution on magnesia-loaded fly ash cenospheres. Desalination 2011; 272(1): 233-9. doi: 10.1016/j.desal.2011.01.028.
18. Çengeloğlu Y, Kır E, Ersöz M. Removal of fluoride from aqueous solution by using red mud. Sep Purif Technol 2002; 28(1): 81-86. doi: 10.1016/S1383-5866(02)00016-3.
19. Ma Y, Shi F, Zheng X, Ma J, Gao C. Removal of fluoride from aqueous solution using granular acid-treated bentonite (GHB): batch and column studies. J Hazard Mater 2011; 185(2-3): 1073-80. doi: 10.1016/j.jhazmat.2010.10.016.
20. Lavecchia R, Medici F, Piga L, Rinaldi G, Zuorro A. Fluoride Removal from Water by Adsorption on a High Alumina Content Bauxite. Chem Eng Trans 2012; 26: 225-30. doi: 10.3303/CET1226038.
21. Bhakat PB, Gupta AK, Ayoob S, Kundu S. Investigations on arsenic(V) removal by modified calcined bauxite. Colloids Surf A Physicochem Eng Asp 2006; 281(1-3): 237-45. doi: 10.1016/j.colsurfa.2006.02.045.
22. Baral SS, Das SN, Rath P, Chaudhury GR. Chromium(VI) removal by calcined bauxite. Biochem Eng J 2007; 34(1): 69-75. doi: 10.1016/j.bej.2006.11.019.
23. Debasish M, Debaraj M, Kyung Ho P. A laboratory scale study on arsenic(V) removal from aqueous medium using calcined bauxite ore. J Environ Sci 2008; 20(6): 683-9. doi: 10.1016/S1001-0742(08)62113-0.
24. Thole B. Defluoridation kinetics of 200°C calcined bauxite, gypsum, and magnesite and breakthrough characteristics of their composite filter. J Fluor Chem 2011; 132(8): 529-35. doi: 10.1016/j.jfluchem.2011.05.016.
25. Sajidu SMI, Masamba WRL, Thole B, Mwatseteza JF. Groundwater fluoride levels in villages of Southern Malawi and removal studies using bauxite. Int J Phys Sci 2008; 3(1): 1-11.
26. Atasoy AD, Yesilnacar MI, Sahin MO. Removal of fluoride from contaminated ground water using raw and modified bauxite. Bull Environ Contam Toxicol 2013; 91(5): 595-9. doi: 10.1007/s00128-013-1099-z.
27. Sadeghkasmaei L, Gorgin N, Fekrikohbanani M. Survey the amount of fluoride in Kohbanan (Kerman—Iran) soil and water and them effect on dental health. The First International Conference on Water Crisis; March 10-12, 2013; Zabol University, Zabol; 2013.
28. American Public Health Association. Standard Methods For the Examination of Water and Wastewater. Washington, DC: American Public Health Association; 1998.
29. Peter KH. Defluoridation of high fluoride waters from natural water sources by using soils rich in bauxite and kaolinite. J Eng Appl Sci 2009; 4(4): 240-6.
30. Altundoğan HS, Tümen F. Removal of phosphates from aqueous solutions by using bauxite. I: Effect of pH on the adsorption of various phosphates. J Chem Technol Biotechnol 2002; 77(1): 77-85. doi: 10.1002/jctb.525.
31. Ayoob S, Gupta AK, Bhakat PB. Analysis of breakthrough developments and modeling of fixed bed adsorption system for As(V) removal from water by modified calcined bauxite (MCB). Sep Purif Technol 2007; 52(3): 430-8. doi: 10.1016/j.seppur.2006.05.021.
32. Malakootian M, Javdan M, Iranmanesh F, Yaghmaeian K. Evaluation of active bauxite mine (Tash) in the removal of fluoride from drinking water: a case study on Kuhbonan water. Journal of Sabzevar University of Medical Sciences 2016; 23(3): 540-9. doi: 10.21859/sums-2303540. [In Persian].
33. Shrivastava BK, Vani A. Comparative study of Defluoridation technologies in India. Asian Journal of Experimental Sciences 2009; 23(1): 269-74.
34. Malakootian M, Fatehizadeh A, Yousefi N, Ahmadian M, Moosazadeh M. Fluoride removal using Regenerated Spent Bleaching Earth (RSBE) from groundwater: Case study on Kuhbonan water. Desalination 2011; 277(1-3): 244-9. doi: 10.1016/j.desal.2011.04.033.
35. Malakootian M, Javdan M, Iranmanesh F. Fluoride Removal Study from Aqueous Solutions using Bauxite Activated Mines in Yazd Province (Case Study: Kuhbanan Water). Journal of Community Health Research 2014; 3(2): 103-14.
36. Malakootian M, Fatehizadeh A, Yousefi N. Evaluating the effectiveness of Modified pumice in fluoride Removal from Water. Asian J chem 2011; 23(8): 3691-4.
37. Gulnaz O, Saygideger S, Kusvuran E. Study of Cu(II) biosorption by dried activated sludge: effect of physico-chemical environment and kinetics study. J Hazard Mater 2005; 120(1-3): 193-200. doi: 10.1016/j.jhazmat.2005.01.003.
38. Lv L, He J, Wei M, Evans DG, Duan X. Factors influencing the removal of fluoride from aqueous solution by calcined Mg-Al-CO3 layered double hydroxides. J Hazard Mater 2006; 133(1-3): 119-28. doi: 10.1016/j.jhazmat.2005.10.012.
39. Mohapatra D, Mishra D, Mishra SP, Chaudhury GR, Das RP. Use of oxide minerals to abate fluoride from water. J Colloid Interface Sci 2004; 275(2): 355-9. doi: 10.1016/j.jcis.2004.02.051.
40. Das N, Pattanaik P, Das R. Defluoridation of drinking water using activated titanium rich bauxite. J Colloid Interface Sci 2005; 292(1): 1-10. doi: 10.1016/j.jcis.2005.06.045.
41. Tor A, Danaoglu N, Arslan G, Cengeloglu Y. Removal of fluoride from water by using granular red mud: Batch and column studies. J Hazard Mater 2009; 164(1): 271-8. doi: 10.1016/j.jhazmat.2008.08.011.
42. Wang Y, Chen N, Wei W, Cui J, Wei Z. Enhanced adsorption of fluoride from aqueous solution onto nanosized hydroxyapatite by low-molecular-weight organic acids. Desalination 2011; 276(1-3): 161-8. doi: 10.1016/j.desal.2011.03.033.
43. Malakootian M, Javdan M, Iranmanesh F. Performance evaluation of the removal of fluoride from drinking water using mendon bauxite mine: case study on Kuhbonan Water, Kerman, Iran. Journal of Health System Research 2016; 12(1): 14-20. [In Persian].

25

Investigating the efficiency of single-walled and multi-walled carbon nanotubes in removal of penicillin G from aqueous solutions

Soheila Chavoshan[1], Maryam Khodadadi[2*], Negin Nasseh[3], Ayat Hossein Panahi[4], Aliyeh Hosseinnejad[1]

[1]Student Research Committee, Birjand University of Medical Sciences, Birjand, Iran
[2]Social Determinants of Health Research Center, Birjand University of Medical Science, Birjand, Iran
[3]Yong Research and Elites Club, Science and Research Branch Islamic Azad University, Tehran, Iran
[4]Office of Research and Technology, Birjand University of Medical Sciences, South Khorasan, Iran

Abstract
Background: Drugs, especially antibiotics, are one of the serious problems of modern life and the main pollution sources of the environment, especially in the last decade, which are harmful to human health and environment. The aim of this study was to investigate the removal of penicillin G from aqueous solutions using single-walled and multi-walled carbon nanotubes.
Methods: In this study, the effect of different parameters including pH (3, 5, 7, 9, and 11), initial concentration of pollutant (50, 100, 150, and 200 mg/l), absorbent dose (0.25, 0.5, 0.75, and 1 g/L), mixing speed (0, 100, 200, and 300 rpm), and temperature (10, 15, 25, 35, 45°C) were investigated. The Langmuir, Freundlich, Temkin, BET, Dubinin-Radushkevich isotherms and adsorption kinetics of the first- and second-order equations were determined.
Results: The results showed that the efficiency of single-walled and multi-walled carbon nanotubes in the removal of penicillin G was 68.25% and 56.37%, respectively, and adsorption capacity of the nanotubes was 141 mg/g and 119 mg/g at initial concentration of 50 mg/l and pH=5 with adsorption dose of 0.8 g/L for 105 minutes at 300 rpm and temperature of 10°C from aqueous solutions. Also, it was revealed that the adsorption process had the highest correlation with the Langmuir model and second-order kinetics, and the maximum adsorption capacity based on Langmuir model was 373.80 mg/g.
Conclusion: According to the results, it was found that single-walled and multi-walled carbon nanotubes can be used as effective absorbents in the removal of penicillin G from aqueous solutions.
Keywords: Single-walled carbon, Multi-walled carbon, Adsorption, Penicillin G, Antibiotics

***Correspondence to:**
Maryam Khodadadi
Email:
maryam.khodadadi@gmail.com

Introduction

Drugs are considered as highly important compounds and an integral part of modern life, which are used for treatment of human and animal diseases. Today, the presence of drugs in the environment is one of the important and considerable issues in the world. Recently, some of developed countries such as the United States, England, Germany, and Italy have started to examine the negative effects of these pollutants on the environment. Drugs essentially have very strong biological activity, which affects the living organisms (1). In addition, they are highly resistant to biological degradation and are not degraded under normal conditions but require specific reactions under special conditions. Finally, drug compounds enter water resources, and affect the environment (1). Among different drug compounds, antibiotics are the most important group of drugs because these compounds are extensively used for the treatment of various diseases (2). The studies have shown that annually 100-200 tons of antibiotics are consumed in the world (3,4). These substances enter the environment through various ways such as pharmaceutical industry wastewater, hospital sewage and fecal and urine of humans and animals (5). Penicillin is the common group of antibiotics that is applied to treat the infections of ear, nose, throat, urinary tract, and skin due to gram-positive bacteria. The mechanism of this antibiotic is degradation the wall cell of bacterial preventing the formation of peptidoglycan (6). In recent years, the bacterial resistance has increased and researchers believe that the increased resistance is due to the excessive consumption of antibiotics (7). The problem is the bacterial resistance against antibiotics despite their low concentrations (7). Moreover, the antibiotic-resistant genes can enter drinking water sources, thereby, causing antibiotic-resistant diseases (8). Recent studies have shown that antibiotics can be found in surface waters and sewage treatment plants (3,4,9). The results show that these drugs cannot be completely removed during biological purification and enter receiving waters, therefore, the removal of antibiotics from wastewater before discharge to receiving water systems is highly important (10). Different methods are used for the removal of antibiotics from aquatic solutions (water and sewage) such as adsorption (11), biological processes (12,13), and oxidations processes (14-17). Biological methods have low removal efficiency for antibiotics but oxidation processes result in degradation of antibiotics although they make borderline materials dangerous. Adsorption is an effective method for the removal of pollutants from water and wastewater even at very low concentrations (below 1 mg/L). The adsorption is a simple and feasible method with low operating cost in comparison with other methods for the separation of pollutants from aqueous solutions (18). Therefore, many researchers have focused on the optimization of adsorption process and finding new absorbents with high adsorption capacity and low cost (18). Despite the widespread use of activated carbon in the adsorption process, the production and recovering problems have encouraged researchers to find new recoverable adsorbents with high adsorption capacity (11). In this regard, nanotechnology has received considerable attention. The carbon nanotubes are graphite sheets that are wrapped as cylindrical tubes and divided into single-walled (SWCNTs) and multi-walled (MWCNTs) groups based on the number of layers in their structure (19-21). The unique properties of carbon nanotubes include high specific surface and permeability, good mechanical and thermal resistance, electrical stability, and reusability. Also, the carbon nanotubes have shown excellent adsorption properties for various organic compounds and mineral ions (21-23).

In a study conducted by Ncibi and Sillanpaa, the removal of antibiotic drugs from aqueous solutions was investigated using single-, double-, and multi-walled carbon nanotubes and the results showed that the highest adsorption capacity for oxytetracycline and ciprofloxacin was 554 and 724 mg/g, respectively (24). In another study conducted by Kim et al, the adsorption of antibiotics on single-walled and multi-walled carbon nanotubes was investigated and it was reported that the adsorption generally followed the order SWCNT > MWCNT. The relatively low adsorption on MWCNT was probably due to its lower specific surface area than other carbon materials (25). The other study conducted by Peng et al, on the adsorption of ofloxacin using carbon nanotubes showed that cationic and anionic ofloxacin had much higher solubility than zwitterionic ofloxacin and with increase of methanol fraction, both OFL solubility and sorption decreased (26).

The aim of this study was to evaluate the efficiency of single-walled and multi-walled carbon nanotubes in the removal of penicillin G from aqueous solutions.

Materials and Methods

In the present study, penicillin G, single-walled and multi-walled nanotubes (manufactured by Noyan TebPadideh, Western Tehran), HCL and NaOH 0.1 N (manufactured by Merck Germany) specifications related to penicillin G were used (Table 1).

In order to study the shape, average diameter, surface details, determine functional groups on the absorbent surface and analyze the structure of carbon nanotubes, scanning electron microscope (SEM) (SIGMA VP-500 manufactured by ZEISS, Germany), X-ray diffraction (XRD) (Pert x, Pro model, manufactured by PANalytical Corporation) and Fourier-transform infrared spectroscopy (FT-IR) by AVATAR 370 were used. The adsorption test was carried out in a continuous adsorption system using penicillin G as an absorbed material and carbon nanotubes as adsorbents. The stock solution was prepared by dissolving 1 g penicillin G in deionized water, and then, the required concentrations were made by diluting the stock solution. The parameters studied in the removal process include pH (3, 5, 7, 9, and 11), initial concentration (50, 100, 150, and 200 mg/L), adsorbent dose (0.25, 0.5, 0.75, and 1 g/L), mixing speed (0, 100, 200, and 300 rpm), and temperature (10, 15, 25, 35, and 45°C). For measurement, first, the calibration curve was drawn with different concentrations of penicillin G, where R^2 was 0.99, then, the residual concentration of penicillin G was measured using a spectrophotometer at wavelength of 283 nm (wavelength of maximum adsorption) (27). To ensure the accuracy of results, each step of the experiment was repeated twice and the average results were reported. The adsorption isotherms were evaluated using Langmuir, Freundlich, Temkin, BET and Dubinin-Radushkevich models, and the adsorption kinetics was investigated using the first-order and second-order equations. The results of adsorption and its isotherms were analyzed using

Table 1. Physical and chemical properties of penicillin G

Penicillin G	Shape	Brand	Molecular Weight (g/mol)	Formula
	Crystal and white powder	BenPen	356. 37	$C_{16}H_{18}N_2O_4S$

Langmuir and Freundlich models. The linear equations of these two models are shown in equations (1) and (2), respectively.

$C_e/q_e = 1/bq_m + C_e/q_m$ (1)

$Lnq_0 = LnK_f + 1/n\ LnC_0$ (2)

In these equations, C_e and q_e are the equilibrium concentration in the soluble (mg/L) and solid (mg/L) phase, respectively, *qm* is maximum adsorption capacity (mg/g), and *k*, *n*, and *b* are model constants.

In the Temkin adsorption isotherm model, the positive or negative values of b_T indicate the exothermic or endothermic adsorption process.

$q_e = (RT/b_T)Ln(A_TC_e)$ (3)

In the equation (3), A_T, b_T, *R*, and *T* are bond constants, which indicate the maximum bond energy ($L.g^{-1}$), Temkin constant (J/mol), gas constant, and absolute temperature, respectively.

$q_e = (q_{max} K_bC_e)/(C_s-C_e)[1+(K_b-1)(C_e/C_s)]$ (4)

In the equation (4), C_S, K_b, and q_{max} are saturation concentration of the soluble phase (mg/L), model constant, and absorbed content per adsorbent unit, respectively. The Dobinin-Radushkevich model is often used to determine the nature and characteristics of the adsorption process and free energy.

qe = exp (-k2) E=R.T.ln(1+1/ce) E= 1/√2B (5)

In the equation (5), *E*, *B*, *R*, and *T* are Polanyi adsorption potential ($KJ^2.mol^{-2}$), free energy of adsorption ($mol^2.kj^{-2}$), gas constant ($J.mol^{-1}.k^{-}$), and temperature (k), respectively (25).

Chemical kinetics indicates the speed of chemical reactions. In this study, penicillin G adsorption rate constant with single-walled and multi-walled carbon nanotubes was checked using pseudo-first-order and pseudo-second-order kinetics models, and the most appropriate model was determined. The pseudo-first-order and pseudo-second-order kinetic models are represented in equations (6) and (7), respectively.

$Ln(q_e-q_t)=Lnq_e- K_1t$ 6

$t/q_t=1/k_2q_e^2 + t/q_e$ 7

In equations (6) and (7), q_e, q_t, k_1, and k_2 are the equilibrium concentration of adsorption phase (mg/g), antibiotic concentration at time of *T* (mg/g), rate constants of pseudo-first-order and pseudo-second-order equations, respectively (28).

Results

The properties of adsorbents were investigated using specific analysis techniques, as following.

FT-IR analysis

To determine the functional groups on the adsorbent surface after the adsorption process of pollutant, FT-IR spectrum was used and the results are shown in Figure 1. In the FT-IR spectrometry of single-walled carbon nanotubes after penicillin G adsorption, the peaks in range of 3432.59 cm^{-1} are related to vibrating modes of O-H, the

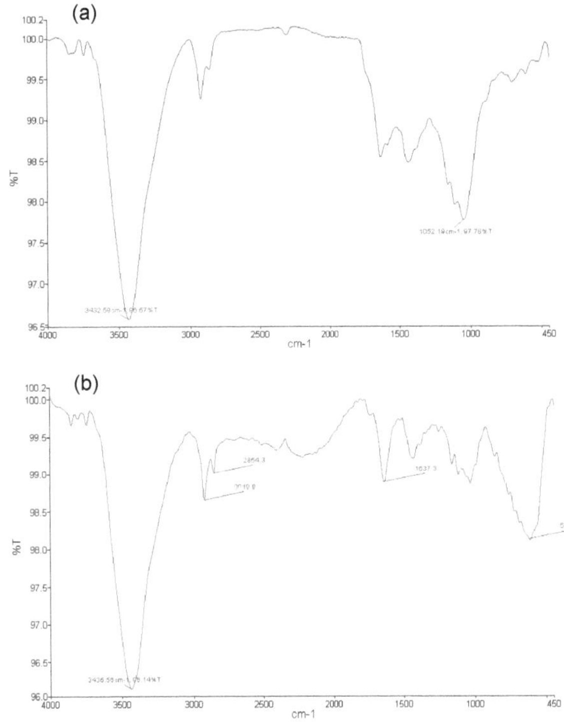

Figure 1. FT-IR analysis of single-walled carbon nanotubes (a) and multi-walled carbon nanotubes (b) after the adsorption process.

peaks in range of 1052.19 cm^{-1} are related to tensile C-O groups in the composition, and the peaks in range of 2900, 1600, and 2300 cm^{-1} are respectively related to tensile bonds C-H, C=O, and S-H. The adsorption peaks below 100 cm^{-1} are related to tensile peaks of C-S groups. And in FT-IR spectroscopy of the multi-walled carbon nanotubes after penicillin G adsorption, the peaks in the range of 3435.5 cm^{-1} are related to vibrating modes of O-H, the peaks in range of 1037.3 cm^{-1} are related to tensile C-O groups in the composition, and the peaks in range of 2919.6, 1637.3 and 2300 cm^{-1} are respectively related to tensile bonds C-H, C=O, and S-H. The adsorption peaks below 1000 cm^{-1} are related to tensile bonds of the C-S groups, and all of these variations in the FT-IR spectrum approve the functional groups of penicillin G adsorbed on the structure of single-walled and multi-walled nanotubes. These variations in the FT-IR spectrum show that penicillin G is located on the surface of single-walled and multi-walled carbon nanotubes.

XRD analysis

The XRD spectrum of single-walled carbon nanotubes before adsorption of penicillin G (a), and multi-walled carbon nanotubes before adsorption of penicillin G (b), single-walled carbon nanotubes after adsorption of penicillin G (c), and multi-walled carbon nanotubes after adsorption of penicillin G (d) is presented in Figure 2a-d. The crystal structure of single-walled and multi-walled carbon nanotubes before and after adsorption of the pollutant is presented in Figure 2a-d. XRD patterns show

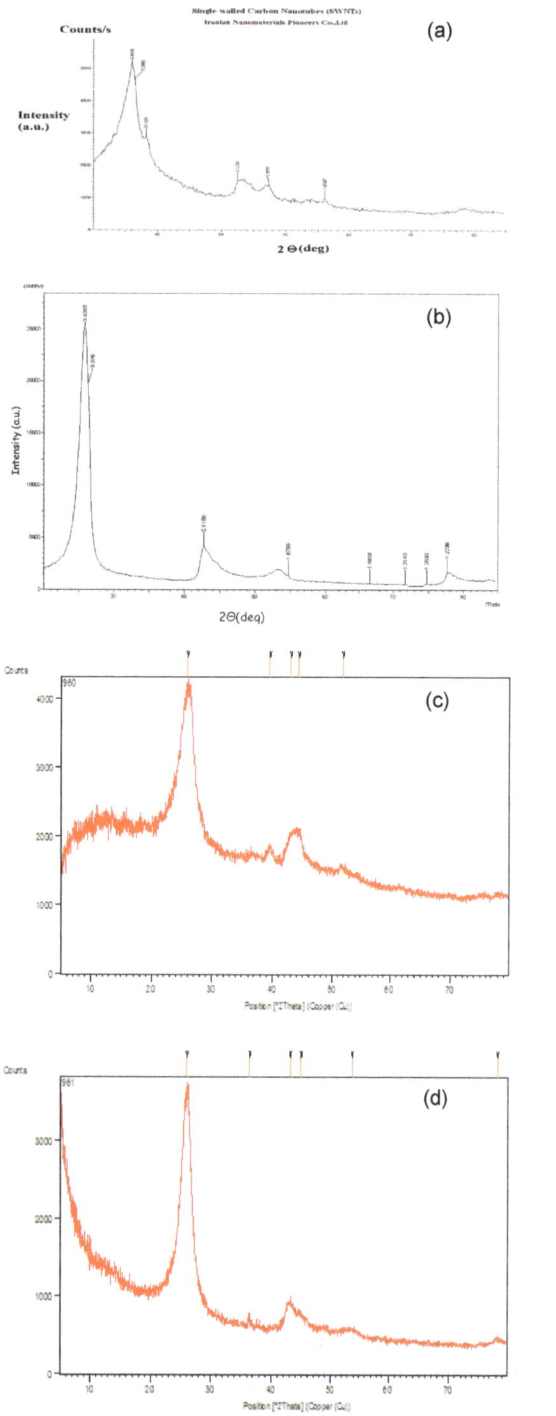

Figure 2. XRD images of single-walled carbon nanotubes (a) and multi-walled carbon nanotubes (b) before the adsorption process, single-walled carbon nanotubes (c), and multi-walled carbon nanotubes (d) after the adsorption process.

certain peaks for SWCNT before penicillin G adsorption at angles 2, 16, 18, 24, 43, 48, and 56 **θ**, which is in accordance with the JCPDS database. XRD patterns show certain peaks for SWCNT after penicillin G adsorption at angles 2, 26.021, 43.1, and 44.4 **θ**, and for MWCNT at angles 2, 25.98, and 43.12 **θ**, which made few changes in the adsorption structure compared to the peaks before the adsorption process, which is in accordance with the JCPDS database. The shape and size of carbon nanotubes were determined by SEM.

SEM images

Figure 3 shows SEM images of (a) single-walled carbon nanotubes powder before adsorption process and (b) single-walled carbon nanotubes after adsorption process (c) SEM images of multi-walled carbon nanotubes powder before adsorption process and (d) multi-walled carbon nanotubes after adsorption process. SEM is widely used to determine the morphology, shape, and size of particles in micro- and nano-scale. These images indicate the nanotube string structures. SEM images of these nanoparticles can be seen and size of single-walled carbon nanotube before adsorption was in range of 1-3 nm, and diameter of these nanotubes after penicillin G adsorption increased to 42.95-89.32 nm, and diameter of multi-walled carbon nanotubes is in the range of 20-30 nm, and diameter of the multi-walled nanotubes after penicillin G adsorption increased to 31.26-63.94 nm. As shown in Figure 3a and Figure 3d, the diameter of the nanotubes increased after adsorption process; the adsorption images also indicate that after the adsorption process, most of the sites and pores of the studied adsorbents were filled with penicillin G in the spherical and string forms in line with the surface of the nanotubes.

Effect of pH on removal efficiency of penicillin G using single-walled and multi-walled carbon nanotubes

Figure 4 shows the effect of pH on the removal of penicillin G by single-walled carbon nanotubes (a) and multi-walled carbon nanotubes (b). Based on the results of the experiments on single-walled and multi-walled carbon nanotubes as adsorbents, the increase of pH from 3 to 9 has a positive effect on the removal of penicillin G, and

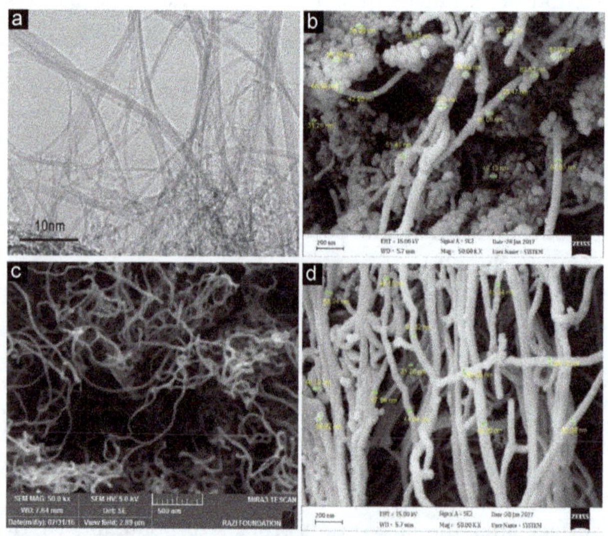

Figure 3. SEM images of single-walled carbon nanotubes powder before adsorption process (a), single-walled carbon nanotubes after adsorption process (b), multi-walled carbon nanotubes powder before adsorption process (c), and multi-walled carbon nanotubes after adsorption process (d).

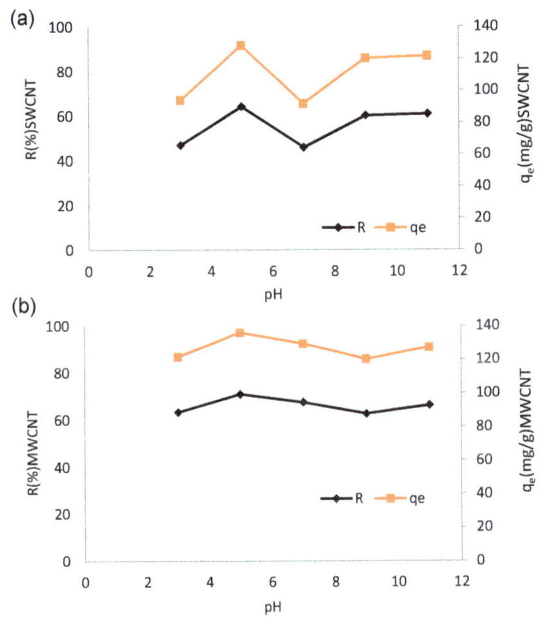

Figure 4. Effect of pH on the removal of penicillin G using single-walled carbon nanotubes (a) and multi-walled carbon nanotubes (b).

by increasing pH to above 9, the efficiency of the removal process of penicillin G was kept constant. SWCNTS and MWCNTs had the highest removal efficiency at pH=5 (64.13% and 70.83%, respectively). According to the results, pH 5 was chosen as the optimum pH.

Effect of adsorbent dose on the removal efficiency of penicillin G using single-walled and multi-walled carbon nanotubes

Effect of adsorbent dose on the removal of penicillin G using single-walled carbon nanotubes powder (a) and multi-walled carbon nanotubes powder is shown in Figure 5. According to the figure, an increase in the carbon nanotubes dosage has a significant effect on the removal of penicillin G by the carbon nanotubes. Although by increasing carbon nanotubes from 0.25 g/L to 1 g/L, the efficiency augmented, but 0.8 g/L of carbon nanotube had the highest removal efficiency (nearly 80%), so the optimum dosage was chosen 0.8 g/L and used in the following steps.

Effect of contact time and concentration on the removal efficiency of penicillin G using single-walled and multi-walled carbon nanotubes

Figure 6 indicates the effect of initial concentration and contact time on the removal of penicillin G using single-walled carbon nanotubes powder (a) and multi-walled carbon nanotubes powder (b). By increasing the contact time, the removal efficiency also increased. The optimum contact time was 105 minutes for SWCNT and MWCNT, the quantity of which is shown in Figure 6a and b. As the initial concentration of penicillin increased, the removal efficiency was reduced but the adsorption capacity was increased, which can be due to the filling active adsorption

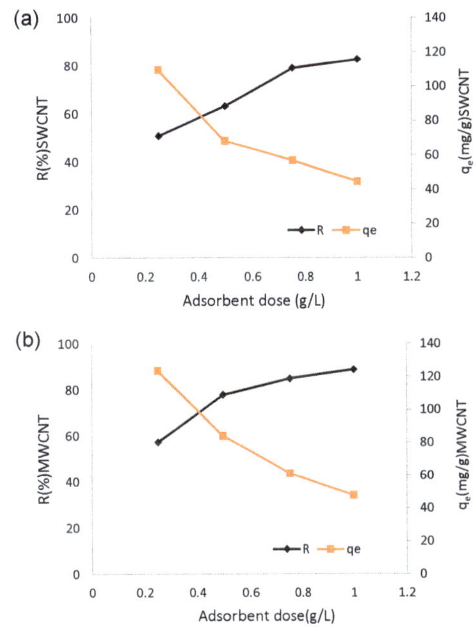

Figure 5. Effect of adsorbent dose on the removal of penicillin G using single-walled carbon nanotubes powder (a) and multi-walled carbon nanotubes powder (b).

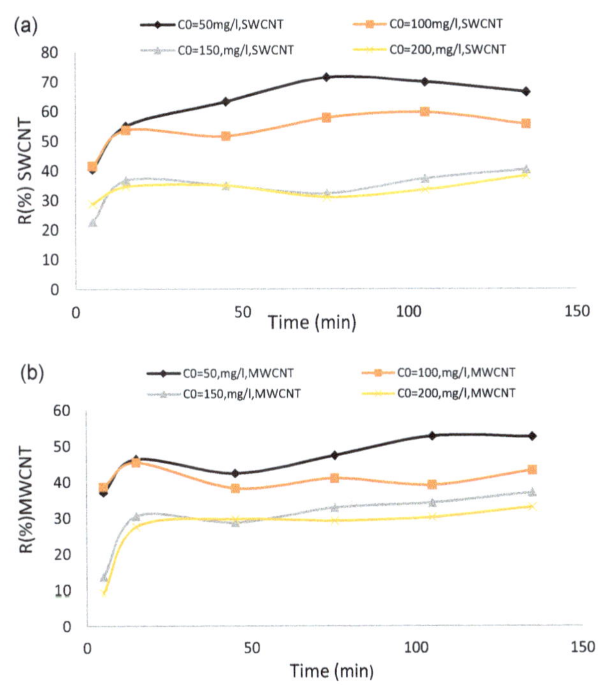

Figure 6. Effect of initial concentration and contact time on the removal of penicillin G using single-walled carbon nanotubes powder (a) and multi-walled carbon nanotubes powder (b).

sites of nanotubes, and as the concentration of penicillin G increased, adsorption process was carried out gradually from surface of the nanotubes. According to the figure, the highest adsorption at initial concentration 50 mg/L and contact time 70 and 105 minutes for single-walled and multi-walled carbon nanotubes was respectively achieved, and in this condition, the removal efficiency was 71% and 53% for single-walled and multi-walled carbon nanotubes, respectively.

Effect of mixing speed on the removal efficiency of penicillin G using single-walled and multi-walled carbon nanotubes

Figure 7 represents the effect of stirring on the removal of penicillin G using single-walled carbon nanotubes (a) and multi-walled carbon nanotubes (b). According to the figure, mixing speed had a direct effect on the removal efficiency. By increasing the mixing speed of the nanotubes and aqueous phase of penicillin G and the contact surface between the two phases, the removal efficiency also increased. As the mixing speed increased, the removal of penicillin G was increased, and the optimum speed of 300 rpm was considered for the above mentioned adsorbents.

Adsorption isotherms

In the adsorption process of pollutants on different adsorbents, determination of adsorption isotherm is one of the most important characteristics that should be considered. In fact, the isotherm parameters provide important information for design and modeling of the adsorption process. The adsorption isotherms are often used to explain the adsorption of materials on the adsorbent. In this study, Langmuir, Freundlich, BET, Temkin and Dubinin-Radushkevich isotherms were examined for SWCNT and MWCNT nanotubes after the adsorption of penicillin G, and the coefficients and constants of each one are shown in Table 2 and 3. As shown in these tables, the studied adsorption process followed the Langmuir isotherm model.

Adsorption kinetics

Table 4 shows kinetic parameters for the adsorption of penicillin G onto single-walled and multi-walled carbon nanotubes. Chemical kinetics indicates the chemical reaction rate. In this study, penicillin G adsorption rate constant with single-walled and multi-walled carbon nanotubes was compatible with pseudo-first-order and pseudo-second-order kinetic models, and the most

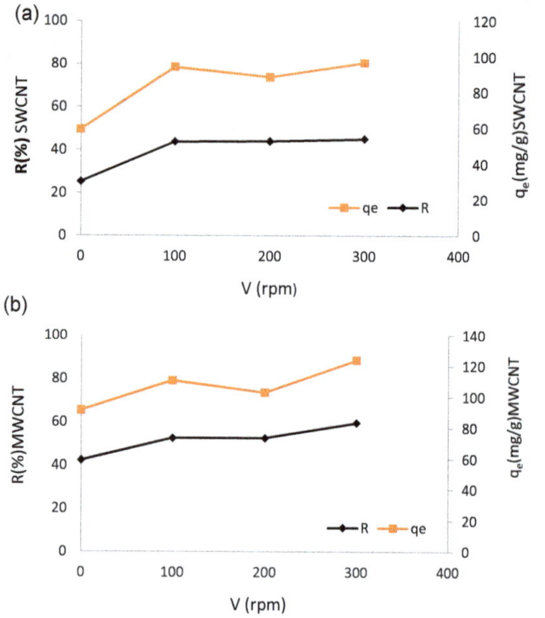

Figure 7. Effect of mixing rate on the removal of penicillin G using single-walled carbon nanotubes (a) and multi-walled carbon nanotubes (b).

suitable model was determined.

Table 4 shows the kinetic parameters and their correlation coefficients. As shown in this table, penicillin G adsorption by SWCNT and MWCNT was better matched with pseudo-second-order model.

Adsorption thermodynamic

The efficiency of temperature and thermodynamics of pollutant adsorption by the adsorbents considered in this study is shown in Table 5. After calculating the constant of equilibrium of thermodynamic for various temperatures and the free energy released during the adsorption process of penicillin G, the diagram of LnKd versus 1/T was plotted. As presented in this table, values of (ΔH) and (ΔS) were positive and the values of (ΔG) were negative. Negative variation of Gibbs free energy (Δg) represents possibility of the adsorption process. Positive value of (ΔH) indicates that the adsorption process is naturally endothermic and the capacity of adsorption increases with

Table 2. Isotherm parameters for various common adsorption isotherms for the adsorption of penicillin G on single-walled carbon nanotubes

		q_{max}, mg/g	q_{max}, mg/g	R_L	R^2
Langmuir constants	L1	337.80	0.041	0.50	0.97
	L2	368.25662	0.0418653		
Freundlich constants		k_f, mg/g	n	1/n	R^2
		56.21	2.74	0.36	0.90
BET constants	X_m	A	(A-1)/(A.Xm)	1/A.X_m	R^2
	133	1.0	0.0075	0.0003	0.790
Temkin constants		B	b_T	A_T, L/mg	R^2
		77.93	31.79	0.46	0.90
Dubinin- Radushkevich constants		q_m, mg/g	E, kJ/mole	B, mole^2/kJ^2	R^2
		290.08682	0.1318402	2.877E-05	0.94

Table 3. Isotherm parameters for various common adsorption isotherms for the adsorption of penicillin G on multi-walled carbon nanotubes

		q_{max}, mg/g	q_{max}, mg/g	R_L	R^2
Langmuir constants	L1	550.54	0.009	0.50	0.97
	L2	713.06901	0.0064013		
Freundlich constants		kf, mg/g	n	1/n	R^2
		10.47	1.41	0.71	0.97
BET constants	X_m	A	(A-1)/(A.Xm)	1/A.Xm	R^2
	-37	1.0	-0.0271	0.0218	0.433
Temkin constants		B	bT	AT, L/mg	R^2
		132.14	18.75	0.08	0.93
		10.47	1.41	0.71	0.97

Table 4. Kinetic parameters for the adsorption of penicillin G on single-walled and multi-walled carbon nanotubes

Concentration (mg/L)	Pseudo-first-order Kinetic Model			Pseudo-second-order Kinetic Model		
	MWCNT	SWCNT		MWCNT	SWCNT	
50	0.0066	0.0482	K_1	0.0052	0.0015	K_2
	12.0697	64.5216	q_e	108.1081	156.25	q_e
	0.156	0.991	R^2	0.994	0.988	R^2

increase of temperature. Positive value of (ΔS) indicates the tendency of absorbent to absorb the liquid and make some structural changes in the adsorbate and adsorbent. Positive value of (ΔS) suggests the adsorbate tendency to be adsorbed in the solution and some structural changes in. Positive entropy value of (ΔS) indicates that the freedom degree of the solid-liquid transition state was increased during adsorption.

Discussion

The pH of solution is an important parameter that influences the adsorption of pollutant using adsorbent. The results of the experiments on single-walled and multi-walled carbon nanotube adsorbents showed that with an increase in pH from 3 to 5, the uptake of penicillin G also increased, but by increasing pH above 5, the rate of penicillin G adsorption decreased abruptly. Penicillin G have a pka=2.75 with a weak moocarboxylic acid (29). Therefore, penicillin G is an acidic compound that could be dissociated at pH>pka (30). At higher pH>pka, penicillin G ionization and OH^- ions increase, resulting in the electrostatic repulsion between the OH^- and negative charge on the adsorbents (SWCNT and MWCNT) surface (30,31).

According to above description, at lower pH, the removal efficiency of penicillin G was notably higher. In similar studies (29,30), pH 5 was selected as the optimum pH. By increasing the contact time from 5 to 105 minutes, the removal efficiency was increased, so, the removal efficiency was constant. Therefore, the optimum contact time for SWCNT and MWCNT was 105 minutes. Also, by increasing the initial concentration of penicillin G, the removal efficiency was reduced but the adsorption capacity was increased It can be concluded that in the early stages of the reaction, the effect of contact time and initial concentration on the adsorption of antibiotics penicillin G using single-wall and multi nanotubes was slowly developed, so that the equilibrium time was observed at 105 minutes. Secondly, the highest antibiotic adsorption was seen at lower concentrations. In the adsorption process, since the removal of pollutant is done through adsorption on the surface of adsorbent, therefore, specified ion exchange sites can be seen on the adsorbent surface that bind to pollutants. This is done when the amount of pollutant remains constant around the adsorbent. As the concentration of pollutant increases in the solution, the thrust force will be generated by the concentration gradient that will lead pollutants into the adsorbent internal pores, and due to the gradual filling of the sites, by increasing the concentration of penicillin G, the desorption process occurs on the surface of nanotube. According to the study by Gao et al, an increase in ionic concentration of the solution led to a decrease in the adsorption of tetracycline by graphene oxide (32). The adsorption rate at initial concentration of tetracycline (166.67 mg/L) and NaCl (100 Mmol/L) was above 50%. This decrease was not considerable in the tetracycline adsorption at low concentrations. Another study by Ji et al showed that the increased ionic concentration of the solution on tetracycline adsorption onto multi-walled carbon nanotubes decreased the reaction efficiency (33).

Table 5. Thermodynamic parameters for the adsorption of penicillin G onto single-walled and multi-walled carbon nanotubes

Temp	ΔG	ΔS	ΔH
10	-149.65	0.105	1.05
15	-187.065		
25	-274.34		
35	-331.73		

The effect of ionic concentration on the adsorption on single-walled carbon nanotubes was more than the multi-walled ones, however, it was unclear in the graphite due to the lack of surface required (34). Chang et al used palygorskite to absorb tetracycline. They reported that the equilibrium time to absorb tetracycline on palygorskite was about 2 hours (35). In the other study on the removal of tetracycline from synthetic solutions using raw LECA by Noori Sepehr et al, it was shown that the highest removal efficiency occurred at 180 minutes (36). In this study, the penicillin G adsorption on single-walled and multi-walled carbon nanotubes was tested as a function of adsorbents in dosages of 0.25-1 g/L and at constant conditions (initial concentration= 50 mg/L, pH=11, and contact time= 60 minutes). The effect of adsorbent dosage on the removal of penicillin G on single-walled carbon nanotubes powder (a) and multi-walled carbon nanotubes powder has been shown in Figure 5. According to this figure, removal efficiency increased with increase of adsorbent dose. Physically, it can be described by the aggregation of exchangeable binding sites of the adsorbents used for HA molecules. In a study by Ncibi and Sillanpaa (24) on the removal of antibiotics from aqueous solutions using single-, double- and multi-walled carbon nanotubes and the study by Peng et al (37) on the adsorption of ofloxacin and norfloxacin on carbon nanotubes, similar results were obtained. In this study, since the adsorption capacity was not changed much with an increase in the adsorbent dose, and the lowest amount of adsorbent with the highest removal efficiency was chosen as the optimal dose, 0.25 g/L of absorbent was used in the next steps. In this study, stirring the solution had a direct effect on the removal efficiency. Since the mixing speed of the nanotubes and aqueous phase of penicillin G increased the contact surface between two phases, therefore, the removal efficiency was also increased. By increasing the stirring speed, the removal efficiency of penicillin G was increased and the speed of 300 rpm was considered as the optimum speed for these absorbents. However, by increasing temperature, the removal efficiency of penicillin G by nanotubes was reduced. The Langmuir and Freundlich isotherm models are widely used to study the adsorption process. The Langmuir model is based on the single-layer adsorption in few equal sites. In this model, it is assumed that the adsorption occurs in sites with the same energy on the absorbent surface, while in the Freundlich isotherm, the soluble matter is absorbed on uneven surfaces in several layers. In the present study, five isotherm models including Langmuir, Freundlich, BET, Temkin, and Dubinin-Radushkevich were studied, and it was revealed that the highest removal efficiency belonged to Langmuir, Dubinin-Radushkevich, Temkin Freundlich, and BET models, respectively. According to Table 2, the maximum penicillin G adsorption (G q_{max}) by single-walled nanotubes was calculated as 373 mg/g. The results of this study indicated better compatibility with Langmuir isotherm model, which represented the adsorption of a penicillin G layer on single-walled carbon nanotubes and homogeneous absorbent surface. For the multi-walled nanotubes, the isotherm models were also investigated, and it was revealed that the highest removal efficiency belonged to Langmuir, Freundlich, Dubinin-Radushkevich, Temkin, and BET models. According to Table 3, the highest penicillin G adsorption (G q_{max}) by multi-walled nanotubes was 550.50 mg/g. The results of this study showed better compatibility with Langmuir isotherm model. In the study of Zhang et al, the tetracycline adsorption by multi-walled carbon nanotubes had better compatibility with Langmuir isotherm (28). The adsorption kinetics describes important information about the adsorption mechanism, uptake rate of carbon nanotubes and contact time of adsorption process (38). In a pseudo-first-order kinetic model, it is assumed that the rate of changes in removal of penicillin G per time is directly proportional to the variations in saturation concentration and removal of adsorbent per time (39).

The pseudo-second-order model assumes that two reactions (parallel or serial) are effective in adsorption of penicillin G on the carbon nanotubes; the first one is fast and rapidly reaches equilibrium, while the second one is slow and continues for longer period of time (40). In this study, the laboratory data showed that the penicillin G adsorption follows the second-order equation. The second-order reactions proceed at a speed proportional to the second exponent of the primary material (40). This result shows that the penicillin G adsorption mechanism on carbon nanotubes can be a physical adsorption. In the research by Samadi et al, use of multi-walled carbon nanotubes in the removal of amoxicillin from aqueous solution was investigated and the results showed that the adsorption equilibrium data were compatible with Langmuir isotherm (R2=0.9108) and adsorption kinetics had better compatibility with the second-order kinetic equation (41).

In a research by Naghizadeh et al, the efficiency of single-walled and multi-walled carbon nanotubes in the removal of arsenic from aqueous solutions was compared and the results showed that the adsorption capacity of single-walled and multi-walled carbon nanotubes is 148 and 95 mg/g, respectively. In this study, the laboratory data were analyzed using Freundlich and Langmuir isotherm models, and the equilibrium data showed that the Langmuir isotherm model is the best compatible model to describe the behavior of arsenic adsorption on carbon nanotubes (42). In the present study, the adsorption capacity of single-walled and multi-walled carbon nanotubes was 141 and 119 mg/g, respectively, and the results showed that single-walled and multi-walled carbon nanotubes can be used as effective absorbents in the removal of penicillin G from aqueous solutions.

Conclusion

The results showed that the removal efficiency of penicillin G by single-walled and multi-walled nanotubes in the optimal conditions (initial concentration=50 mg/L, pH=5, adsorption dose= 0.8 g/L, contact time= 105 minutes, mixing speed= 300 rpm and temperature= 10°C) was 68.25% and 56.37%, respectively. Adsorption capacity of the single-walled and multi-walled carbon nanotubes was 141 mg/g and 119 mg/g, respectively. According to the results, it was found that single-walled and multi-walled carbon nanotubes can be used as an effective adsorbent in the removal of penicillin G from aqueous solutions.

Acknowledgements

The authors would like to appreciate the Vice Chancellor for Research and Technology of Birjand University of Medical Sciences for funding this study with project code (1164) and ethics code (Ir.bums.1395.33).

Competing interests

The authors have declared that they have no conflicts of interest.

Authors' contribution

All authors contributed in data collection, analysis, and interpretation. All authors reviewed, refined, and approved the manuscript.

References

1. McConnell MM, Truelstrup Hansen L, Jamieson RC, Neudorf KD, Yost CK, Tong A. Removal of antibiotic resistance genes in two tertiary level municipal wastewater treatment plants. Sci Total Environ 2018; 643: 292-300. doi: 10.1016/j.scitotenv.2018.06.212.
2. Jeong J, Song W, Cooper WJ, Jung J, Greaves J. Degradation of tetracycline antibiotics: mechanisms and kinetic studies for advanced oxidation/reduction processes. Chemosphere 2010; 78(5): 533-40. doi: 10.1016/j.chemosphere.2009.11.024.
3. Cho JY. Evaluation of degradation of antibiotic tetracycline in pig manure by electron beam irradiation. Bull Environ Contam Toxicol 2010; 84(4): 450-3. doi: 10.1007/s00128-010-9967-2.
4. Dimitrakopoulou D, Rethemiotaki I, Frontistis Z, Xekoukoulotakis NP, Venieri D, Mantzavinos D. Degradation, mineralization and antibiotic inactivation of amoxicillin by UV-A/TiO(2) photocatalysis. J Environ Manage 2012; 98: 168-74. doi: 10.1016/j.jenvman.2012.01.010.
5. Elmolla ES, Chaudhuri M. The feasibility of using combined TiO2 photocatalysis-SBR process for antibiotic wastewater treatment. Desalination 2011; 272(1-3): 218-24. doi: 10.1016/j.desal.2011.01.020.
6. Peterson JW, Petrasky LJ, Seymour MD, Burkhart RS, Schuiling AB. Adsorption and breakdown of penicillin antibiotic in the presence of titanium oxide nanoparticles in water. Chemosphere 2012; 87(8): 911-7. doi: 10.1016/j.chemosphere.2012.01.044.
7. Smalla K, Cook K, Djordjevic SP, Klumper U, Gillings M. Environmental dimensions of antibiotic resistance: assessment of basic science gaps. FEMS Microbiol Ecol 2018; 94(12). doi: 10.1093/femsec/fiy195.
8. Zheng W, Chen K, Zhu J, Ji L. A novel process for erythromycin separation from fermentation broth by resin adsorption-aqueous crystallization. Sep Purif Technol 2013; 116: 398-404. doi: 10.1016/j.seppur.2013.06.019.
9. Kummerer K. Drugs in the environment: emission of drugs, diagnostic aids and disinfectants into wastewater by hospitals in relation to other sources--a review. Chemosphere 2001; 45(6-7): 957-69.
10. Akmehmet Balcioglu I, Ötker M. Treatment of pharmaceutical wastewater containing antibiotics by O3 and O3/H2O2 processes. Chemosphere 2003; 50(1): 85-95. doi: 10.1016/S0045-6535(02)00534-9.
11. Crisafully R, Milhome MA, Cavalcante RM, Silveira ER, De Keukeleire D, Nascimento RF. Removal of some polycyclic aromatic hydrocarbons from petrochemical wastewater using low-cost adsorbents of natural origin. Bioresour Technol 2008; 99(10): 4515-9. doi: 10.1016/j.biortech.2007.08.041.
12. Elmolla ES, Chaudhuri M. Comparison of different advanced oxidation processes for treatment of antibiotic aqueous solution. Desalination 2010; 256(1-3): 43-7. doi: 10.1016/j.desal.2010.02.019.
13. Kummerer K. Antibiotics in the aquatic environment--a review--part I. Chemosphere 2009; 75(4): 417-34. doi: 10.1016/j.chemosphere.2008.11.086.
14. Daghrir R, Drogui P, Ka I, El Khakani MA. Photoelectrocatalytic degradation of chlortetracycline using Ti/TiO2 nanostructured electrodes deposited by means of a Pulsed Laser Deposition process. J Hazard Mater 2012; 199-200: 15-24. doi: 10.1016/j.jhazmat.2011.10.022.
15. Gad-Allah TA, Ali ME, Badawy MI. Photocatalytic oxidation of ciprofloxacin under simulated sunlight. J Hazard Mater 2011; 186(1): 751-5. doi: 10.1016/j.jhazmat.2010.11.066.
16. Hapeshi E, Achilleos A, Vasquez MI, Michael C, Xekoukoulotakis NP, Mantzavinos D, et al. Drugs degrading photocatalytically: kinetics and mechanisms of ofloxacin and atenolol removal on titania suspensions. Water Res 2010; 44(6): 1737-46. doi: 10.1016/j.watres.2009.11.044.
17. Xiong P, Hu J. Degradation of acetaminophen by UVA/LED/TiO2 process. Sep Purif Technol 2012; 91: 89-95. doi: 10.1016/j.seppur.2011.11.012.
18. Putra EK, Pranowo R, Sunarso J, Indraswati N, Ismadji S. Performance of activated carbon and bentonite for adsorption of amoxicillin from wastewater: mechanisms, isotherms and kinetics. Water Res 2009; 43(9): 2419-30. doi: 10.1016/j.watres.2009.02.039.
19. Ntim SA, Mitra S. Adsorption of arsenic on multiwall carbon nanotube-zirconia nanohybrid for potential drinking water purification. J Colloid Interface Sci 2012; 375(1): 154-9. doi: 10.1016/j.jcis.2012.01.063.
20. Upadhyayula VK, Deng S, Mitchell MC, Smith GB. Application of carbon nanotube technology for removal of

contaminants in drinking water: a review. Sci Total Environ 2009; 408(1): 1-13. doi: 10.1016/j.scitotenv.2009.09.027.
21. Malakootian M, Mahdizadeh H. Performance evaluation of multi-walled carbon nanotubes oxidized with a mixture of H2SO4/HNO3 in removal of 4-chlorophenol from aqueous solutions. Journal of Ilam University of Medical Sciences 2017; 25(1): 80-91. doi: 10.29252/sjimu.25.1.80. [In Persian].
22. Sheng GD, Shao DD, Ren XM, Wang XQ, Li JX, Chen YX, et al. Kinetics and thermodynamics of adsorption of ionizable aromatic compounds from aqueous solutions by as-prepared and oxidized multiwalled carbon nanotubes. J Hazard Mater 2010; 178(1-3): 505-16. doi: 10.1016/j.jhazmat.2010.01.110.
23. Trojanowicz M. Analytical applications of carbon nanotubes: a review. TrAC Trends Anal Chem 2006; 25(5): 480-9. doi: 10.1016/j.trac.2005.11.008.
24. Ncibi MC, Sillanpaa M. Optimized removal of antibiotic drugs from aqueous solutions using single, double and multi-walled carbon nanotubes. J Hazard Mater 2015; 298: 102-10. doi: 10.1016/j.jhazmat.2015.05.025.
25. Kim H, Hwang YS, Sharma VK. Adsorption of antibiotics and iopromide onto single-walled and multi-walled carbon nanotubes. Chem Eng J 2014; 255: 23-7. doi: 10.1016/j.cej.2014.06.035.
26. Peng H, Pan B, Wu M, Liu R, Zhang D, Wu D, et al. Adsorption of ofloxacin on carbon nanotubes: solubility, pH and cosolvent effects. J Hazard Mater 2012; 211-212: 342-8. doi: 10.1016/j.jhazmat.2011.12.063.
27. Pouretedal HR, Sadegh N. Effective removal of amoxicillin, cephalexin, tetracycline and penicillin G from aqueous solutions using activated carbon nanoparticles prepared from vine wood. Journal of Water Process Engineering 2014; 1: 64-73. doi: 10.1016/j.jwpe.2014.03.006.
28. Zhang L, Song X, Liu X, Yang L, Pan F, Lv J. Studies on the removal of tetracycline by multi-walled carbon nanotubes. Chem Eng J 2011; 178: 26-33. doi: 10.1016/j.cej.2011.09.127.
29. Malakootian M, Balarak D, Mahdavi Y, Sadeghi S, Amirmahani N. Removal of antibiotics from wastewater by *Azolla filiculoides*: kinetic and equilibrium studies. International Journal of Analytical, Pharmaceutical and Biomedical Sciences 2015; 4(7): 105-13.
30. Balarak D, Kord Mostafapour F, Joghataei A. Experimental and kinetic studies on penicillin G adsorption by *Lemna minor*. Br J Pharm Res 2016; 9(5): 1-10. doi: 10.9734/BJPR/2016/22820.
31. Rahardjo AK, Susanto MJ, Kurniawan A, Indraswati N, Ismadji S. Modified Ponorogo bentonite for the removal of ampicillin from wastewater. J Hazard Mater 2011; 190(1-3): 1001-8. doi: 10.1016/j.jhazmat.2011.04.052.
32. Gao Y, Li Y, Zhang L, Huang H, Hu J, Shah SM, et al. Adsorption and removal of tetracycline antibiotics from aqueous solution by graphene oxide. J Colloid Interface Sci 2012; 368(1): 540-6. doi: 10.1016/j.jcis.2011.11.015.
33. Ji L, Shao Y, Xu Z, Zheng S, Zhu D. Adsorption of monoaromatic compounds and pharmaceutical antibiotics on carbon nanotubes activated by KOH etching. Environ Sci Technol 2010; 44(16): 6429-36. doi: 10.1021/es1014828.
34. Ji L, Chen W, Bi J, Zheng S, Xu Z, Zhu D, et al. Adsorption of tetracycline on single-walled and multi-walled carbon nanotubes as affected by aqueous solution chemistry. Environ Toxicol Chem 2010; 29(12): 2713-9. doi: 10.1002/etc.350.
35. Chang PH, Li Z, Yu TL, Munkhbayer S, Kuo TH, Hung YC, et al. Sorptive removal of tetracycline from water by palygorskite. J Hazard Mater 2009; 165(1-3): 148-55. doi: 10.1016/j.jhazmat.2008.09.113.
36. Noori Sepehr M, Mohebi S, Abdlollahi Vahed S, Zarrabi M. Removal of tetracycline from synthetic solution by natural LECA. Journal of Environmental Health Engineering 2014; 1(4): 301-11. doi: 10.18869/acadpub.jehe.1.4.301. [In Persian].
37. Peng H, Pan B, Wu M, Liu Y, Zhang D, Xing B. Adsorption of ofloxacin and norfloxacin on carbon nanotubes: hydrophobicity- and structure-controlled process. J Hazard Mater 2012; 233-234: 89-96. doi: 10.1016/j.jhazmat.2012.06.058.
38. Zhu HY, Jiang R, Xiao L, Zeng GM. Preparation, characterization, adsorption kinetics and thermodynamics of novel magnetic chitosan enwrapping nanosized gamma-Fe2O3 and multi-walled carbon nanotubes with enhanced adsorption properties for methyl orange. Bioresour Technol 2010; 101(14): 5063-9. doi: 10.1016/j.biortech.2010.01.107.
39. Nandi BK, Goswami A, Purkait MK. Adsorption characteristics of brilliant green dye on kaolin. J Hazard Mater 2009; 161(1): 387-95. doi: 10.1016/j.jhazmat.2008.03.110.
40. Zazouli MA, Yousefi Z, Taghavi M, Akbari-adergani B, Yazdani Cherati J. Cadmium removal from aqueous solutions using L-cysteine functionalized single-walled carbon nanotubes. Journal of Mazandaran University of Medical Sciences 2013; 23(98): 37-47. [In Persian].
41. Samadi MT, Shokoohi R, Araghchian M, Tarlani Azar M. Amoxicillin removal from aquatic solutions using multi-walled carbon nanotubes. Journal of Mazandaran University of Medical Sciences 2014; 24(117): 103-15. [In Persian].
42. Naghizadeh A, Derakhshani E, Javid A. Comparison between single and multi wall carbon nanotubes in removal of arsenic from aqueous solution. Journal of Health 2014; 5(1): 36-44.

Phosphate removal from aqueous solutions using polyaniline/ $Ni_{0.5}Zn_{0.5}Fe_2O_4$ magnetic nanocomposite

Mohammad Hossein Tarmahi[1], **Farid Moeinpour**[2*]

[1]MSc in Water and Wastewater Engineering, Department of Chemistry, College of Science, Bandar Abbas Branch, Islamic Azad University, Bandar Abbas, Iran
[2]Assistant Professor of Chemistry, Department of Chemistry, College of Science, Bandar Abbas Branch, Islamic Azad University, Bandar Abbas, Iran

Abstract
Background: Phosphorus is an indispensable element for the growth of animals and plants. There are several environmental problems related to phosphate; therefore, the technical and economic methods of removing phosphate are of great importance. This study evaluated the efficiency of polyaniline/ $Ni_{0.5}Zn_{0.5}Fe_2O_4$ magnetic nanocomposite in removing phosphate from aqueous environments.
Methods: The adsorbent was characterized by several methods, including X-ray diffraction (XRD), scanning electron microscopy (SEM), vibrating sample magnetometer (VSM), and Fourier transform infrared (FT-IR) spectroscopy. Then, the potential of the adsorbent**to** adsorb phosphate was investigated. The effects of the parameters of contact time (5-60 minutes), pH (3-9), adsorbent dosage (0.05-0.6 g), and initial phosphate concentration (2-100 mg/L) on the phosphate removal yield were studied. All phosphate ion concentrations were measured using the ammonium molybdate spectrophotometric method.
Results: The results showed that a time of 30 minutes, pH of 5, and adsorbent dose of 0.4 g were the optimum conditions for phosphate removal through adsorption. Increasing the initial concentration of phosphate from 2 to 100 mg/L decreased the removal efficiency from 90.3% to 32%. The experimental data was fitted well with the Freundlich isotherm model (R^2 = 0.997).
Conclusion: Polyaniline/$Ni_{0.5}Zn_{0.5}Fe_2O_4$ magnetic nanocomposite removes phosphate from aqueous solutions with a simple and environmentally benign procedure. The maximum adsorption capacity based on Langmuir isotherm (R^2 = 0.931) is 85.4 mg/g. This magnetic nanocomposite is applicable in managing water resource pollution caused by phosphate ions.
Keywords: Adsorption, Phosphates, Nanocomposites, Polyaniline

***Correspondence to:**
Farid Moeinpour
Email: fmoeinpour52@iauba.ac.ir

Introduction

Adsorption has been established as an important and economically practical treatment technology for removing phosphate ions from water and wastewater. Activated carbon is the most common adsorbent used for this purpose. Despite the abundance of applications for activated carbon, its uses are sometimes limited by its high cost and loss during reformation (1-4). Therefore, researchers are seeking new low-cost substitute adsorbents for water pollution control, especially where cost plays an important role. Many efforts have been made to develop other adsorbents that are effective and inexpensive. They can be produced from a wide variety of raw materials which are abundant and have high carbon and low inorganic content. Because of the low cost and high accessibility of these materials it is not essential to have complex regeneration processes; inexpensive adsorption methods have attracted the attention of many researchers. Often, the adsorption capabilities of such adsorbents are not great; therefore, studies of more and more new adsorbents are still being developed. Some inexpensive adsorbents used to remove phosphate ions have been studied, such as fly ash (5), red mud (6), aluminum hydroxide (7), iron oxide (8), zirconium oxide (9), manganese dioxide (10), modified clinoptilolite (11) and horizontal roughing filters (12).

In recent years, economic problems have encouraged the creation of inexpensive, efficient alternative methods of wastewater treatment. Use of magnetic adsorbents is one of the most efficient, technical, and economic methods of treating wastewater. These adsorbents have magnetic properties and, by using an external magnetic field, can be easily separated from solutions. In magnetic separation, the usual high costs of separation, e.g., centrifugation and filtration, are not incurred (13). Extensive research has been done into the magnetization of materials such as chitosan (14), silica (15), polymer (16), and activated carbon (17) for water contaminant removal. The use of this property in nanoparticles is remarkable because of their high specific surface area and adsorption capacities (18). Nickel-zinc ferrites have drawn noticeable consideration from researchers because of their remarkable magnetic properties, large permeability, and very high electrical resistivity (19). They have an extensive list of potential applications in such areas as high-density information storage devices, microwave devices, transformer cores, magnetic fluids (20), etc. This study investigated the capability of $Ni_{0.5}Zn_{0.5}Fe_2O_4$ magnetic nanoparticles that were surface-modified with polyaniline (PANI) and used a polyaromatic amine as an effective adsorbent for removing phosphate from aqueous solutions (Figure 1). PANI, a familiar conducting polymer, is one of the most conceivably useful conducting polymers. It has received significant consideration by many researchers (21-23). PANI may be readily synthesized chemically or electrochemically from acidic aqueous solutions (24,25). The chemical structure of PANI is shown in Figure 2.

Methods

Materials

Anhydrous potassium dihydrogen orthophosphate (KH_2PO_4) was purchased from Merck. A 1000 ppm stock solution of the KH_2PO_4 was provided in deionized water. All solutions used in this study were made by consecutive diluting with double distilled water. All chemicals [$Fe(NO_3)_3.9H_2O$, $Zn(NO_3)_2.6H_2O$ and $Ni(NO_3)_2.6H_2O$, NaOH and HNO_3] were of analytical grade and obtained from Sigma-Aldrich. X-ray diffraction analysis (XRD) was carried out using a PAN analytical X'Pert Pro X-ray diffractometer. Surface morphology and particle size were investigated with a Hitachi S-4800 scanning electron microscopy (SEM) instrument. Infrared (IR) spectra of the samples were obtained with a Bruker model 470 Fourier transform infrared (FT-IR) spectrometer. Concentrations of phosphate ion solutions were analyzed using the ammonium molybdate spectrophotometric method.

Synthesis of $Ni_{0.5}Zn_{0.5}Fe_2O_4$/PANI

$Ni_{0.5}Zn_{0.5}Fe_2O_4$ nanoparticles were initially prepared using stoichiometric ratios of metal nitrates and freshly extracted egg-white (26). The metal nitrates [$Fe(NO_3)_3.9H_2O$, $Zn(NO_3)_2.6H_2O$ and $Ni(NO_3)_2.6H_2O$] were dissolved together in a minimum amount of double-distilled water to obtain a clear solution. Sixty milliliters of extracted egg-white was dissolved in 40 mL of double-distilled water by vigorous stirring and was added to the nitrate mixture at ambient temperature. After constant stirring for 30 minutes, the resultant sol-gel was evaporated at 353 K until dry precursor was obtained. The dried precursors were ground and calcined in a muffle furnace at 823 K for 2 hours. The PANI–$Ni_{0.5}Zn_{0.5}Fe_2O_4$ composite was prepared by precipitating PANI on the surface of pre-synthesized $Ni_{0.5}Zn_{0.5}Fe_2O_4$ nanoparticles. In this method, 50 mL of freshly prepared reaction mixture (0.1 M aniline, 0.125 M ammonium peroxydisulfate in 0.5 M nitric acid) was added to 1.6 g of $Ni_{0.5}Zn_{0.5}Fe_2O_4$ at room temperature (27). The mixture was stirred during the polymerization of aniline, which was completed within 1 hour. Next, the $Ni_{0.5}Zn_{0.5}Fe_2O_4$/PANI nanocomposite was magnetically separated, washed with 0.5 M nitric acid and with acetone, and finally dried at 333 K in a vacuum oven for 6 hours.

Batch adsorption experiments

Phosphate adsorption ions onto $Ni_{0.5}Zn_{0.5}Fe_2O_4$/PANI was studied in aqueous phosphate solutions under diverse practical conditions (pHs = 3–9) at 298 K and 5 mg/L phosphate ion solution. 0.10 g $Ni_{0.5}Zn_{0.5}Fe_2O_4$/PANI was added to 50 mL of phosphate solution. After that, the resultant mixture was shaken in a shaker at 1000 rpm. The initial pHs of the phosphate solutions were balanced in the range of 3 to 9 with 0.1 mol/L HNO_3 or 0.1 mol/L NaOH solutions by a pH meter. When equilibrium was reached, the solutions were centrifuged and the $Ni_{0.5}Zn_{0.5}Fe_2O_4$/PANI was magnetically removed. The concentration of phosphate ions in the supernatant was measured spectrophotometrically using the ammonium molybdate method. The effects of various parameters (contact time, initial phosphate concentration, pH, $Ni_{0.5}Zn_{0.5}Fe_2O_4$/

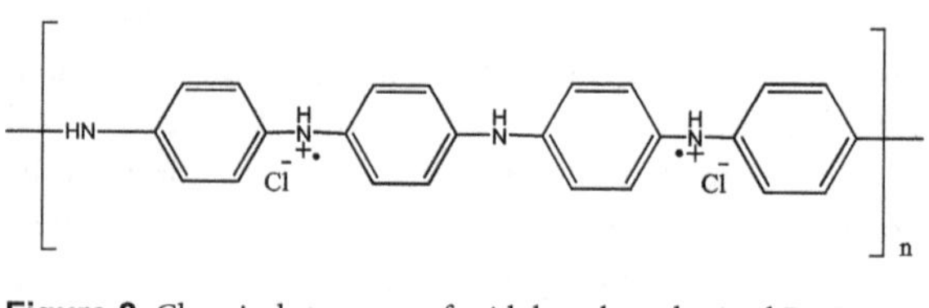

Figure 2. Chemical structure of acid doped synthesized PANI.

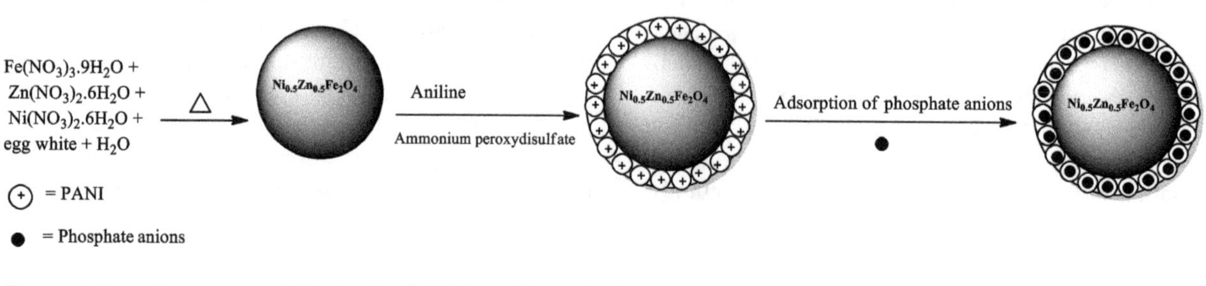

Figure 1. Formation process of $Ni_{0.5}Zn_{0.5}Fe_2O_4$/PANI and its application as a phosphate adsorbent.

PANI dose, and temperature) on phosphate removal were studied.

Phosphate removal (%R) was computed using Eq. (1):

$$\%R = \frac{C_0 - C_t}{C_0} \times 100 \quad (1)$$

where C_0 and C_t (mg/L) are initial phosphate concentration and time *t*, respectively.

The adsorption capacity of the adsorbent (Q_e) was computed by Eq. (2):

$$Q_e = \frac{(C_0 - C_e)V}{m} \quad (2)$$

where C_0 and C_e are the initial and equilibrium concentrations of the phosphate (mg/L), respectively, *m* is the mass of the adsorbent (g), and *V* is the volume of phosphate solution (L).

Adsorption Isotherms

For adsorption equilibrium studies, 0.40 g of $Ni_{0.5}Zn_{0.5}Fe_2O_4$/PANI was added to 50 mL of different concentrations of phosphate solutions (2-100 mg/L and pH = 5) at 298 K. These solutions were shaken at 1000 rpm for 30 minutes. After adsorption, the concentrations of the phosphate solutions were analyzed on a UV-Vis spectrophotometer using the ammonium molybdate method.

Results

In this section, the results of this study are presented in the form of diagrams and tables. $Ni_{0.5}Zn_{0.5}Fe_2O_4$/PANI nanocrystallites were specified by FT-IR spectroscopy (Figure 3), XRD (Figure 4), SEM (Figure 5), and VSM (Figure 6). Figure 7 shows the different contact times, removal efficiency of the remaining phosphate, and the adsorption equilibrium time. As shown in Figure 7, removal efficiency increased in the time range of 0 to 30 minutes, became fixed at 30 minutes and partly increased at 60 minutes. Thus, the adsorption equilibrium time was determined to be 30 minutes in this test.

The effect of solution pH (from 3 to 9) on phosphate removal was studied at 298 K. Figure 8 shows the removal efficiency of phosphate at the equilibrium time and at 4 different pHs (3, 5, 7, and 9), equaling, respectively, 78%, 82%, 77.5%, and 56%. Maximum adsorption occurred at around pH 5.0, and this value was therefore selected for all adsorption experiments in this study.

The effect of $Ni_{0.5}Zn_{0.5}Fe_2O_4$/PANI dosage on phosphate adsorption was investigated with various adsorbent doses (from 0.05 to 0.60 g). The results are exhibited in Figure 8. The greatest adsorption of phosphate was about 89.0%, achieved when applying a $Ni_{0.5}Zn_{0.5}Fe_2O_4$/PANI dosage of 0.40 g in 50 mL of 5 mg/L phosphate solution.

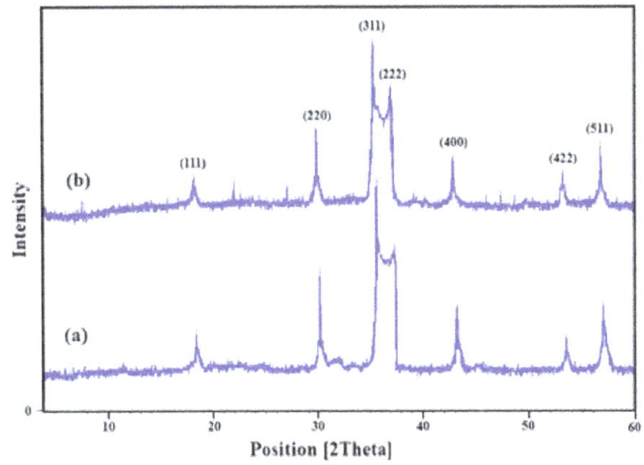

Figure 4. XRD patterns of $Ni_{0.5}Zn_{0.5}Fe_2O_4$. (a) Synthesized $Ni_{0.5}Zn_{0.5}Fe_2O_4$; (b) Standard $Ni_{0.5}Zn_{0.5}Fe_2O_4$ (JCPDS 08-0234).

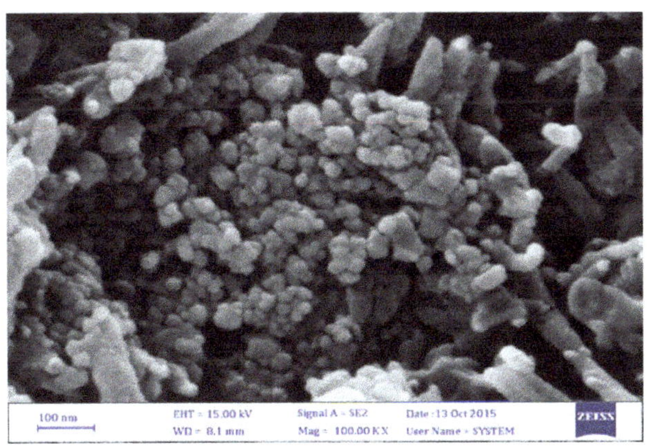

Figure 5. SEM image of $Ni_{0.5}Zn_{0.5}Fe_2O_4$ /PANI nanocomposite.

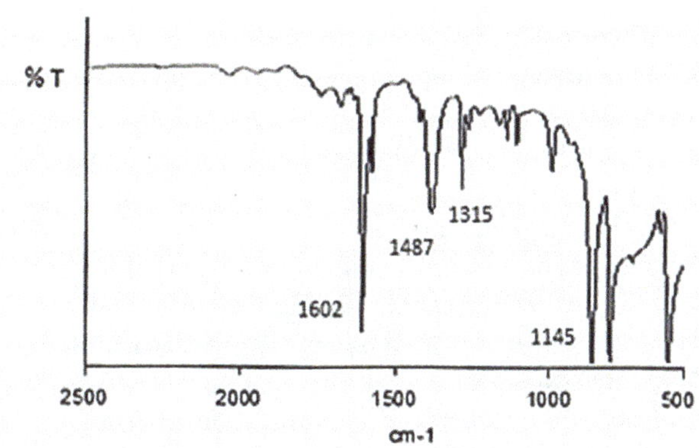

Figure 3. FT-IR spectrum of $Ni_{0.5}Zn_{0.5}Fe_2O_4$/PANI.

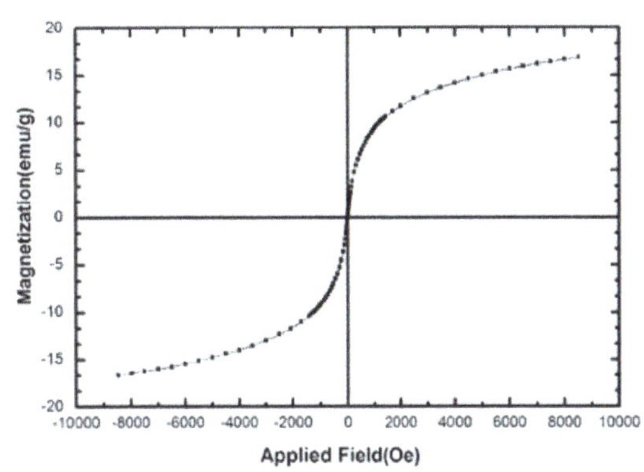

Figure 6. VSM curve of $Ni_{0.5}Zn_{0.5}Fe_2O_4$/PANI at room temperature.

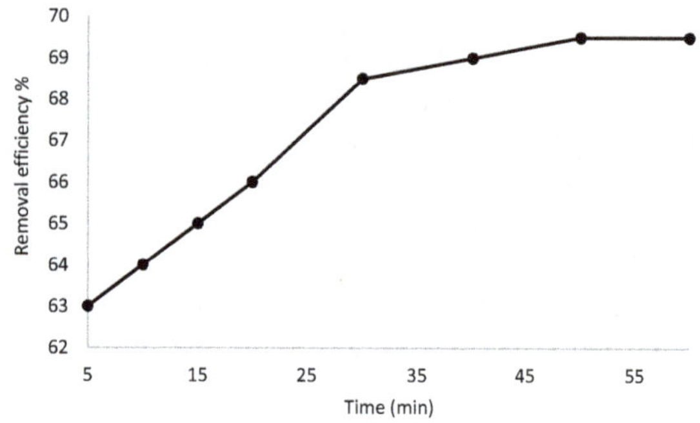

Figure 7. Effect of contact time.

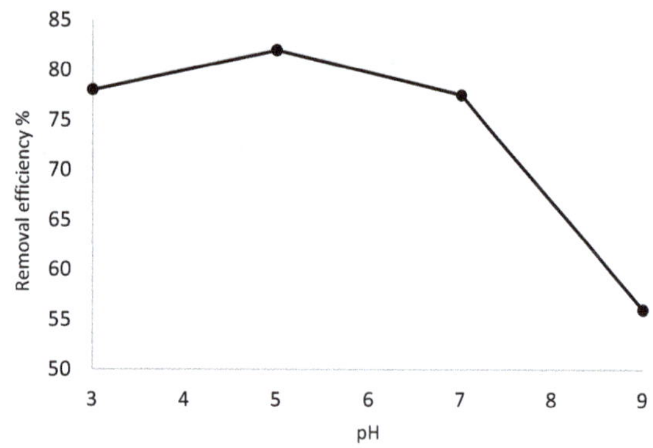

Figure 8. Effect of pH on phosphate removal.

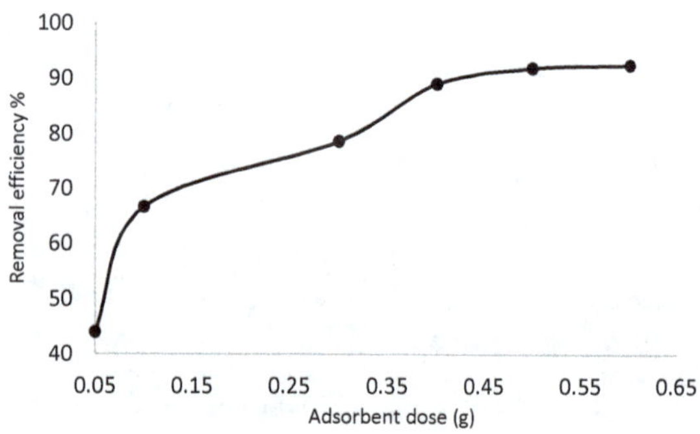

Figure 9. Effect of adsorbent dosage on phosphate removal.

Batch adsorption experiments were performed at different initial phosphate concentrations (2, 5, 10, 30, 50, and 100 mg/L) while other experimental parameters were kept constant. Figure 9 shows that the percentage of phosphate removal (%R) decreased when the initial concentration was increased, indicating that the adsorption of phosphate onto $Ni_{0.5}Zn_{0.5}Fe_2O_4$/PANI is highly related to initial phosphate concentration.

The adsorption isotherms of phosphate by the adsorbent are shown in Table 1. As can be seen, $R^2 = 0.931$ in the Langmuir model, $R^2 = 0.997$ in the Freundlich model, and $R^2 = 0.687$ in the Dubinin–Radushkevich model.

Discussion

In this study, the SEM technique was used to perform morphological analyses and characterize the size and shape

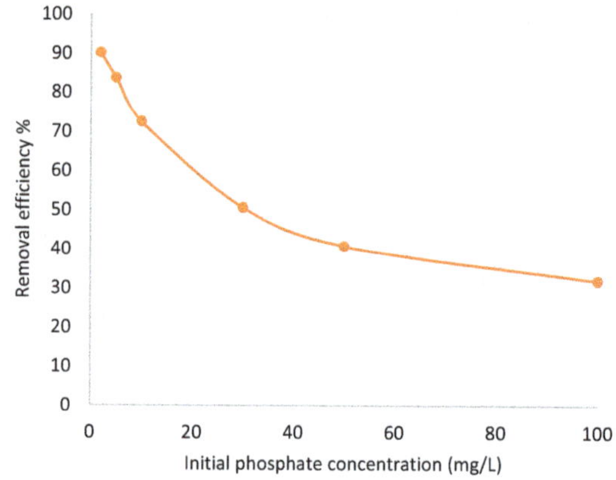

Figure 10. Effect of initial phosphate concentration on removal efficiency of phosphate.

of the resultant $Ni_{0.5}Zn_{0.5}Fe_2O_4$ /PANI nanocomposites. XRD was used to estimate the morphologic type of $Ni_{0.5}Zn_{0.5}Fe_2O_4$/PANI. Functional group analyses of $Ni_{0.5}Zn_{0.5}Fe_2O_4$/PANI were performed using the FT-IR technique. The magnetic properties of the $Ni_{0.5}Zn_{0.5}Fe_2O_4$/PANI were evaluated with a VSM.

$Ni_{0.5}Zn_{0.5}Fe_2O_4$ nanocrystallites were prepared according to the procedure reported by Gabal et al (26). Nano $Ni_{0.5}Zn_{0.5}Fe_2O_4$/PANI were characterized by FT-IR (Figure 3). In the FT-IR spectrum of $Ni_{0.5}Zn_{0.5}Fe_2O_4$ /PANI, many of the $Ni_{0.5}Zn_{0.5}Fe_2O_4$ and PANI bands, with some having a slight shift, are distinct, which shows that PANI well coated the $Ni_{0.5}Zn_{0.5}Fe_2O_4$. The peaks in the range of 500-1000 cm^{-1} were caused by the Fe_3O_4 framework, which is in accordance with the iron oxide spectrum (28). The peaks at 1602 and 1487 cm^{-1} were attributed to the characteristic C = C stretching of the quinoid and benzenoid rings; the peaks at 1315 cm^{-1} were assigned to the C – N stretching of the benzenoid ring; and the broad peak at 1145 cm^{-1} were related to the vibration mode of N=Q=N (Q refers to the quinonic-type rings) (29,30). This apparently shows that the $Ni_{0.5}Zn_{0.5}Fe_2O_4$ magnetic core was coated with PANI and a core/shell nanostructure was formed. To verify the formation of Ni-Zn ferrite in the prepared magnetic nanoparticles, the XRD pattern of the sample was investigated. The XRD patterns (Figure 4) showed that $Ni_{0.5}Zn_{0.5}Fe_2O_4$ nanoparticles have a spinel framework with all the main peaks compatible with the $Ni_{0.5}Zn_{0.5}Fe_2O_4$ standard pattern (JCPDS 08-0234). The adsorbent particle size was investigated using SEM. The SEM photograph of the sample (Figure 5) showed that the average size of $Ni_{0.5}Zn_{0.5}Fe_2O_4$/PANI is slightly less than 100 nm. The magnetic properties of the $Ni_{0.5}Zn_{0.5}Fe_2O_4$ /PANI were evaluated using a VSM. As shown in Figure 6, the saturation magnetic moments of the $Ni_{0.5}Zn_{0.5}Fe_2O_4$/PANI reached about 17 emu/g. It showed superparamagnetic behavior that would enable easy recovery of the adsorbent from solution under an applied magnetic field.

The time efficacy on the phosphate removal was studied with an initial phosphate concentration of 5 mg/L. As can be seen in Figure 7, the percentage of adsorptions increased with increases in contact time. The least removal occurred in 5 minutes (63.0%), and maximum adsorption was 69.0% at the time of 30 minutes. The process showed a fast phosphate adsorption. The adsorption level, however, reached a stable stage as time was increased, because all accessible positions were occupied.

The phosphate adsorption clearly increased with decreases in pH. When the pH was low, the positive charge on the surface of PANI was concentrated, leading to a static electricity force between positive PANI surfaces and negative phosphate ions. Then the mechanism of phosphate ion removal may be related to the electrostatic force as well as the interaction of ion-exchange (31). At a low pH, the phosphate anions can be present in solution as HPO_4^{-2} or $H_2PO_4^-$ as follows:

$PO_4^{-3} + H^+ \rightarrow HPO_4^{-2}$

$HPO_4^{-2} + H^+ \rightarrow H_2PO_4^-$

PANI in the doped state (having releasable dopants such as Cl^-) can be conveniently exchanged with anions like phosphate which exist in anionic form in aqueous solutions. At pH values higher than 3, a de-doping process took place in the polymer (PANI), and at pH > 5 the desorption of phosphate becomes the significant process, lowering the amount of phosphate removed (32).

Figure 9 indicates that, when the amount of $Ni_{0.5}Zn_{0.5}Fe_2O_4$/PANI was increased, the percentage of adsorption also increased and approached satiation stage, where increases in $Ni_{0.5}Zn_{0.5}Fe_2O_4$/PANI extent did not alter phosphate removal. An increment in removal level with $Ni_{0.5}Zn_{0.5}Fe_2O_4$/PANI quantity could be ascribed to the enhanced space and accessibility of further adsorption sites.

Batch adsorption experiments were performed at different initial phosphate concentrations. As shown in Figure 10, these observations can be explained by the fact that increasing the initial phosphate concentration made more phosphate ions available, while the amount of active sites on the adsorbent remained constant which led to a lower removal efficiency percent.

The adsorption isotherm is important for determining the adsorption behavior of an adsorbent. The experimental data corresponded with the Langmuir, Freundlich, and Dubinin–Radushkevich models as shown in Table 1.

In the Dubinin–Radushkevich model, for E < 8 kJ mol^{-1}, the adsorption process might be performed physically, while chemical adsorption occurs when E > 8 kJ mol^{-1} (33). All parameters are listed in Table 1. Looking at Table 1 in which the Langmuir, Freundlich, and D–R isotherm constants for the adsorption of phosphate are summarized, it can be derived from R^2 that the Freundlich model matched the experimental data better than either the Langmuir or D–R models. Moreover, it is clear that the adsorption of phosphate by $Ni_{0.5}Zn_{0.5}Fe_2O_4$/PANI may be explained as a physical adsorption process, for the value of E is 2.236 kJ.

Adsorption capacity is a significant parameter which determines the performance of an adsorbent. Table 2 compares the maximum adsorption capacity of $Ni_{0.5}Zn_{0.5}Fe_2O_4$/PANI for phosphate adsorption with that of other adsorbents in the literature.

Table 1. Langmuir, Freundlich, D–R isotherm constants for the adsorption of phosphate ions onto $Ni_{0.5}Zn_{0.5}Fe_2O_4$/PANI

	Langmuir		
q_m (mg g^{-1})	K_L	R_L	R^2
85.40	0.0096	0.51	0.931
	Freundlich		
1/n	K_F		R^2
0.47	1.07		0.997
	Dubinin–Radushkevich (D–R)		
q_m (mg g^{-1})	β (mol^2 kJ^{-2})	R^2	E (kJ mol^{-1})
3.57	1×10^{-7}	0.687	2.236

Table 2. Maximum adsorption capacity of different adsorbents for phosphate removal

Adsorbents	q_m (mg/g)	References
Modified fly ash	12.69	(34)
Magnetic biochar	1.24	(35)
Magnetic iron oxide	5.03	(36)
$ZnCl_2$-activated carbon	5.1	(37)
Hydroxy-aluminum pillared bentonite	12.7	(38)
Metal-loaded orange waste	14	(39)
Ground burnt patties	0.41	(40)
Zeolite	2.15	(41)
$Ni_{0.5}Zn_{0.5}Fe_2O_4$/PANI	85.40	Present study

Conclusion

$Ni_{0.5}Zn_{0.5}Fe_2O_4$/PANI magnetic nanoparticles were used in the adsorption of phosphate ions from aqueous solutions. The study demonstrated that the adsorption of phosphate increased with decreases in pH and was enhanced with increments in the extent of $Ni_{0.5}Zn_{0.5}Fe_2O_4$/PANI. Maximum phosphate adsorption was achieved at pH 5 with a maximum adsorption capacity of 85.40 mg/g at 298 K. The adsorption isotherm fit the Freundlich model well. The manufactured $Ni_{0.5}Zn_{0.5}Fe_2O_4$/PANI nanoparticles could be spread in the solution well and simply collected by a magnet. The water treatment described here is efficient using $Ni_{0.5}Zn_{0.5}Fe_2O_4$/PANI. The current results provide a practical way for water treatment and the removal of phosphate ions.

Acknowledgments

Financial support from the Islamic Azad University-Bandar Abbas Branch (No. 11450515932016) is gratefully acknowledged.

Competing interests

The authors declare that they have no competing interests.

Authors' contributions

All authors contributed equally and were involved in the study design, data collection, and article approval.

References

1. Can MY, Yildiz E. Phosphate removal from water by fly ash: Factorial experimental design. J Hazard Mater 2008; 135(1-3): 165-70. doi: 10.1016/j.jhazmat.2005.11.036.
2. Tillotson S. Phosphate removal: an alternative to chemical dosing. Filtr Separat 2006; 43(5): 10-2. doi: 10.1016/S0015-1882(06)70884-9.
3. İrdemez Ş, Yildiz YŞ, Tosunoğlu V. Optimization of phosphate removal from wastewater by electrocoagulation with aluminum plate electrodes. Sep Purif Technol 2006; 52(2): 394-401. doi: 10.1016/j.seppur.2006.05.020.
4. Karageorgiou K, Paschalis M, Anastassakis GN. Removal of phosphate species from solution by adsorption onto calcite used as natural adsorbent. J Hazard Mater 2007; 139(3): 447-52. doi: 10.1016/j.jhazmat.2006.02.038.
5. Pengthamkeerati P, Satapanajaru T, Chularuengoaksorn P. Chemical modification of coal fly ash for the removal of phosphate from aqueous solution. Fuel 2008; 87(12): 2469-76. doi: 10.1016/j.fuel.2008.03.013.
6. Huang W, Wang S, Zhu Z, Li L, Yao X, Rudolph V, et al. Phosphate removal from wastewater using red mud. J Hazard Mater 2008; 158(1): 35-42. doi: 10.1016/j.jhazmat.2008.01.061.
7. Guan XH, Chen GH, Shang C. Adsorption behavior of condensed phosphate on aluminum hydroxide. J Environ Sci 2007; 19(3): 312-18. doi: 10.1016/S1001-0742(07)60051-5.
8. Persson P, Nilsson N, Sjöberg S. Structure and bonding of orthophosphate ions at the iron oxide–aqueous interface. J Colloid Interface Sci 1996; 177(1): 263-75. doi: 10.1006/jcis.1996.0030.
9. Chitrakar R, Tezuka S, Sonod A, Sakane K, Ooi K, Hirotsu T. Selective adsorption of phosphate from seawater and wastewater by amorphous zirconium hydroxide. J Colloid Interface Sci 2006; 297(2): 426-33. doi: 10.1016/j.jcis.2005.11.011.
10. Mustafa S. Zaman MI, Khan S. pH effect on phosphate sorption by crystalline MnO_2. J Colloid Interface Sci 2006; 301(2): 370-75. doi: 10.1016/j.jcis.2006.05.020.
11. Malakootian M, Yousefi N, Jaafarzadeh Haghighifard N. Kinetics modeling and isotherms for adsorption of phosphate from aqueous solution by modified clinoptilolite. Water and Wastewater 2011; 22(4): 21-9. [In Persian].
12. Khezri SM, Majidi G, Jafari Mansoorian H, Ansari M, Atabi F, Tohidi Mogaddam T, et al. Efficiency of horizontal roughing filter in removing nitrate, phosphate and chemical oxygen demand from effluent of waste stabilization pond. Environ Health Eng Manag J 2015; 2(2): 87-92.
13. Shen Y, Tang J, Nie Z, Ren Y, Wang Y, Zuo L. Preparation and application of magnetic Fe_3O_4 nanoparticles for wastewater purification. Sep Purif Technol 2009; 68(3): 312-19. doi: 10.1016/j.seppur.2009.05.020.
14. Cho DW, Jeon BH, Chon CM, Kim Y, W. Schwartz F, Lee ES, et al. A novel chitosan/clay/magnetite composite for adsorption of Cu (II) and As (V). Chem Eng J 2012; 200-202: 654-62. doi: 10.1016/j.cej.2012.06.126.
15. Shi G. Synthesis of SiO_2/Fe_3O_4 nanomaterial and its application as cataluminescence gas sensor material for ether. Sensors and Actuators B: Chemical 2012; 171-172: 699-04. doi: 10.1016/j.snb.2012.05.059.
16. Dallas P, Georgakilas V, Niarchos D, Komninou P, Kehagias T, Petridis D. Synthesis, characterization and thermal properties of polymer/magnetite nanocomposites. Nanotechnology 2006; 17(8): 2046.
17. Mohan D, Sarswat A, Singh VK, Alexandre-Franco M, Pittman C. Development of magnetic activated carbon from almond shells for trinitrophenol removal from water. Chem Eng J 2011; 172(2-3): 1111-25. doi: 10.1016/j.cej.2011.06.054.
18. Nitayaphat W, Jintakosol T. Removal of silver (I) from aqueous solutions by chitosan/bamboo charcoal composite beads. J Clean Prod 2015; 87(1): 850-55. doi: 10.1016/j.jclepro.2014.10.003.
19. Sharma S, Verma K, Chaubey U, Singh V, Mehta BR. Influence of Zn substitution on structural, microstructural and dielectric properties of nanocrystalline nickel ferrites. Mat Sci Eng B 2010; 167(3): 187-92. doi: 10.1016/j.mseb.2010.02.015.
20. Virden A, O'Grady K. Structure and magnetic properties of NiZn ferrite nanoparticles. J Magn Magn Mater 2005; 290-291(Pt 2): 868-70. doi: 10.1016/j.jmmm.2004.11.398.
21. Zeng XR, Ko TM. Structures and properties of chemically reduced polyanilines. Polymer 1998; 39(5): 1187-95. doi: 10.1016/S0032-3861(97)00381-9.
22. Kang ET, Neoh KG, Tan KL. Polyaniline: a polymer with many interesting intrinsic redox states. Prog Polym Sci 1998; 23(2): 277-324.
23. Negi YS, Adhyapak P. Development in polyaniline conducting polymers. J Macromol Sci Polymer 2002; 42(1): 35-53. doi: 10.1081/MC-120003094.
24. Cao Y, Andreatta A, Heegert AJ, Smith P. Influence of chemical polymerization conditions on the properties of polyaniline. Polymer 1989; 30(12): 2305-11.doi: 10.1016/0032-3861(89)90266-8.
25. Chiang JC, MacDiarmid AG. 'Polyaniline': protonic acid doping of the emeraldine form to the metallic regime. Synthetic Met 1986; 13(1-3): 193-205. doi: 10.1016/0379-6779(86)90070-6.
26. Gabal MA, El-Shishtawy RM, Al Angari YM. Structural and magnetic properties of nano-crystalline Ni–Zn ferrites synthesized using egg-white precursor. J Magn Magn Mater 2012; 324(14): 2258-64. doi: 10.1016/j.jmmm.2012.02.112.
27. Stejskal J, Trchová M, Brodinová J, Fedorova SV, Prokes J, Zemek J. Coating of zinc ferrite particles with a conducting polymer, polyaniline. J Colloid Interface Sci 2006; 298(1): 87-93. doi: 10.1016/j.jcis.2005.12.034.
28. Pol VG, Daemen LL, Vogel S, Chertkov G. Solvent-free fabrication of ferromagnetic Fe_3O_4 octahedral. Ind Eng Chem Res 2009; 49(2): 920-24. doi: 10.1021/ie9011062.
29. Khan JA, Qasim M, Singh BR, Singh S, Shoeb M, Khan W, et al. Synthesis and characterization of structural, optical, thermal and dielectric properties of polyaniline/$CoFe_2O_4$ nanocomposites with special reference to photocatalytic activity. Spectrochim Acta A Mol Biomol Spectrosc 2013; 109: 313-21. doi: 10.1016/j.saa.2013.03.011.
30. Stejskal J, Gilbert RG. Polyaniline, preparation of a conducting polymer (IUPAC technical report). Pure Appl Chem 2002; 74(5): 857-67. doi: 10.1351/pac200274050857.

31. Long F, Gong JL, Zeng GM, Chen L, Wang XY, Deng JH, et al. Removal of phosphate from aqueous solution by magnetic Fe–Zr binary oxide. Chem Eng J 2011; 171(2): 448-55. doi: 10.1016/j.cej.2011.03.102.
32. Ansari R. Application of polyaniline and its composites for adsorption/recovery of chromium (VI) from aqueous solutions. Acta Chim Slov 2006; 53: 88-94.
33. Tan Y, Chen M, Hao Y. High efficient removal of Pb (II) by amino-functionalized Fe_3O_4 magnetic nano-particles. Chem Eng J 2012; 191: 104-11. doi: 10.1016/j.cej.2012.02.075.
34. Xu K, Tao H, Deng T. Removal of phosphate from coating wastewater using magnetic Fe-Cu bimetal oxide modified fly ash. Journal of Water Reuse and Desalination 2016; 6(3) 430-6. doi: 10.2166/wrd.2015.105.
35. Chen B, Chen Z, Lv S. A novel magnetic biochar efficiently sorbs organic pollutants and phosphate. Bioresour Technol 2011; 102(2): 716-23. doi: 10.1016/j.biortech.2010.08.067.
36. Yoon SY, Lee CG, Park JA, Kim JH, Kim SB, Lee SH, et al. Kinetic, equilibrium and thermodynamic studies for phosphate adsorption to magnetic iron oxide nanoparticles. Chem Eng J 2014; 236: 341-47. doi: 10.1016/j.cej.2013.09.053.
37. Namasivayam C, Sangeetha D. Equilibrium and kinetic studies of adsorption of phosphate onto $ZnCl_2$ activated coir pith carbon. J Colloid Interface Sci 2004; 280(2): 359-65. doi: 10.1016/j.jcis.2004.08.015.
38. Yan LG, Xu YY, Yu HQ, Xin XD, Wei Q, Du B. Adsorption of phosphate from aqueous solution by hydroxy-aluminum, hydroxy-iron and hydroxy-iron–aluminum pillared bentonites. J Hazard Mater 2010; 179(1-3): 244-50. doi: 10.1016/j.jhazmat.2010.02.086.
39. Biswas BK, Inoue K, Ghimire KN, Harada H, Ohto K, Kawakita H. Removal and recovery of phosphorus from water by means of adsorption onto orange waste gel loaded with zirconium. Bioresour Technol 2008; 99(18): 8685-90. doi: 10.1016/j.biortech.2008.04.015.
40. Rout PR, Bhunia P, Dash RR. Modeling isotherms, kinetics and understanding the mechanism of phosphate adsorption onto a solid waste: ground burnt patties. J Environ Chem Eng 2014; 2(3): 1331-42. doi: 10.1016/j.jece.2014.04.017.
41. Sakadevan K, Bavor HJ. Phosphate adsorption characteristics of soils, slags and zeolite to be used as substrates in constructed wetland systems. Water Res 1998; 32(2): 393-99. doi: 10.1016/S0043-1354(97)00271-6.

Dental waste characterization in the city of Ilam in 2014

Farzad Kazemi[1], Zabihollah Yousefi[2*], Reza Ali Mohammadpour[3]

[1]MSc, Department of Environmental Health Engineering, Student Research Committee, Faculty of Health, Mazandaran University of Medical Sciences, Sari, Iran
[2]Professor, Department of Environmental Health Engineering, Faculty of Health, Mazandaran University of Medical Sciences, Sari, Iran
[3]Associate Professor, Department of Biostatistics, Faculty of Health, Mazandaran University of Medical Sciences, Sari, Iran

Abstract

Background: Dental wastes are one of the environmental issues due to toxic and pathogenic agents such as pathological wastes, pharmaceutical and chemical etc have particular sensitivity. The aim of this study was to determine the dental waste management and related factors in the city of Ilam.

Methods: In this cross-sectional study, the studied community was all the sixteen dental clinics in Ilam. Five samples of each clinic per week (Saturday, Sunday and Wednesday) were selected. Thereafter waste sample was manually separated into 36 components and were weighed using a laboratory scale with an accuracy of 0.01 g. Each component was weighed five times and the mean value obtained for each component was considered. Production per capita was calculated for each person. Data were analyzed using descriptive statistics, SPSS and Excel software.

Results: The per capita percent for infectious waste section was 51%. The average of infectious waste is 201.13 g. The per capita percent for chemical, pharmaceutical waste section was 36% with an average of 142.48 g. The per capita percent for toxicity section was 13% in the dental clinics with the weighted average of 48.78 g. According to the results of the checklist, further dental clinics have been poorly managed.

Conclusion: According to the presence of various materials and different components with different characteristics in the dental wastes, the optimal management of this type of wastes should be carried out based on the specific characteristics, which include programs to reduce waste production, segregation, recycle and reuse.

Keywords: Waste management, dental waste, infectious waste, toxic waste

***Correspondence to:**
Zabihollah Yousefi
Email: zyousefi2004@gmail.com

Introduction

The collection, transportation and disposal of waste is considered as an important and complex issue due to development of cities, increase in the number and density of population, industrialization and increase in per capita waste generation which need a regular system to implement the rules. The suitable collection and disposal of these wastes have specific importance due to their direct and indirect risks on human, animal, plants and environment (1).

On the other hand, the rapid population growth and increasing healthcare needs of human societies has led to develop and increase in the various clinics providing health services such as hospitals, clinics, public and private laboratories in the world which increases the production of medical wastes (2,3). Based on Iranian Waste Management Act, all of the infectious and hazardous wastes produced by hospitals, health clinics, laboratories, dental clinics and other similar clinics are called medical wastes (4). Medical waste is assumed to be a major problem in the management of municipal solid waste in developing countries. When these wastes are mixed with municipal waste, they can create dangerous condition for the environment alongside those who are in relation with these materials (5).

Solid wastes have different qualitative and quantitative characteristics depending on the type and source of production. The awareness of the type, source and specifications of quantity and quality is important for better planning for waste management (6). Dental clinics are considered as one of the wastes production clinics (7). Dental wastes in most countries are classified as medical waste. Medical wastes are divided into two major groups namely; common and special waste. Common wastes are similar to municipal waste in terms of quality and they should be transported in accordance with the laws related to municipal solid waste. Special waste includes infectious, radioactive and chemical waste and each have certain management method (8). It is important that these types of wastes should not be mixed with municipal waste due to the spreading of chemical and biological materials which contain dangerous and pathogenic agent that ultimately lead to health and environmental risks (9). The solid or semi-solid material is one of the hazardous waste in the dental clinics that have the ability to create risk (10-12). A small quantity of hazardous waste is produced in dental clinics. Some of these hazardous wastes are recyclable. Hazardous dental wastes include amalgam fillings, x-ray film and lead foil, residues of drugs and unused chemicals (13). It is also possible that the dental clinics participate in air pollution through the water spray system as well as by blood and infected tissues. The air pollution can be in the form of aerosol or mist (14). Mercury as a heavy metal included in amalgam can lead to environmental contamination. Improper disposal of amalgam waste in dental clinics and centers can lead to environmental exposure to mercury which can have nephrotoxic and neurotoxic effects. Gastrointestinal, respiratory, immune and renal systems can be affected by mercury in amalgam waste.

The solid waste management, especially hazardous wastes, is based on the prevention and reduction of waste production, the methods which are eco-friendly and are able to ensure the health of staff working in these clinics. Lack of attention on waste management has brought dangers and irreparable damages for the staff and patients. The most important work in the field of dental waste management is to prevent the mixing of all the components, because the components of dental wastes have different characteristics and the management method should be carried out based on these characteristics (15). Thus, proper and hygienic monitoring and management on how to collect, store, transport and dispose hazardous wastes is important in order to protect the health of patients and practitioners (dentists, nurses and medical and non-medical staff) as well as referrals to dental clinics. In this study, dental waste management and the related factors were discussed in the city of Ilam in 2014. Moreover, this study has investigated the description of current situation and has carried out a detailed study on dental wastes for future planning and correct optimized management.

Methods

In this study, the studied population was all the sixteen dental clinics of Ilam (five public and 11 private) which were selected according to a list provided by medical deputy which were still active during the study period and had the necessary cooperation in the field of data collection. Each sampling was carried out at the end of working day of clinics and five samples were obtained from each clinic in three days (Saturday, Monday and Wednesday) per week. The sampling was carried out in the weeks when there were no closure and for days between Saturdays to Wednesdays. This cross-sectional study was carried out from December 2014 to February 2015. Prior to sampling, the dentists and their assistants were notified about this study. Based on World Health Organization guideline for medical waste management in developing countries (World Health Organization, 2001), a questionnaire with 54 questions were designed. Interviews with dentists and also field consideration were carried out to collect data. The collected samples were analyzed after transferring to an appropriate location. The special color-coded plastic containers were used to separate and collect each type of waste from different parts of the clinics. The common waste were collected and separated in orange plastic containers while the infectious waste with sharp objects and chemical and pharmaceutical waste were separated in safety box and brown containers, respectively and were weighed. The canvas gloves, boots, overalls and special pants for the analysis of waste were used in this study and vaccination against diseases such as tetanus, Hepatitis b and c were carried out for collecting and weighing workers. The samples should be physically analyzed up to 3 h after sampling. The accuracy of digital laboratory carriage scales (model EK120A) utilized in this study was 0.1 g. At first, the waste samples were manually separated into 36 components and were weighed. Each component was weighed five times, and the average value was calculated for each component. The obtained value represents the amount of various components of waste produced at the end of the working day of the clinic. The number of patients per working day was determined in each clinic. The per capita wastes generated were determined according to the number of patients per working day and total amount of waste produced. The collected data were analyzed using Excel software. Moreover, the per capita waste generation of each patient in any working day was obtained by dividing the amount of daily waste produced by the number patients referred in working day. In the next stage, the average daily production of various components of waste in each office was achieved by calculating the average of obtained values from 5 days sampling in each clinic.

A check list containing 45 questions were used in order to evaluate the management method of each clinic. Each of the questions had a point between 0 and 2. These obtained points were multiplied in given weight related to that points to calculate the certain number which can be assessed as follows: Good: 130-172, Average: 87-129, Weak: 44-86, Unacceptable: 0-43.

Results

Based on the points obtained from the checklist, the highest poor management in the office is related to lack of attention to eliminate the risk of infectious waste at the clinic. Other poor management factors can be expressed as follows: lack of dental amalgam recycling, fixer and developer drugs, x-ray films and its lead coatings, mercury leakage and improper storage location of wastes. Based on the results of the checklist, most of the dental clinics are faced with management deficiencies.

Table 1 shows the different components of infectious waste produced in dental clinics based on services provided (restoration, dental extraction, removal of calculus, endodontic therapy, orthodontic and implant and surgery). The greatest waste among all services provided in latex and plastic gloves was related to suture needles and its thread except for surgery which has the maximum amount in this case. The highest amount per capita for each patient was 64.43 and 44.15 g for dental extraction,

Table 1. Infectious waste components in dental clinics on the basis of services provided (g)

Components of infectious wastes in dental clinics	Restoration	Dental extraction, removal of calculus	Endodontic therapy	Orthodontic and Prosthesis	Surgery
Paper napkins stained with blood	10.78	150.1	18.1	8.58	18.63
Paper napkins contaminated with saliva	8.54	84.7	7.3	18.6	11.62
Gauze contaminated with blood		353.8	23.56	53.4	164.42
Gauze contaminated with saliva		234.65	44.32	64.32	58.54
Cotton contaminated with blood	34.43		28.23		
Cotton contaminated with saliva	17.43	65.5	17.86		
Dental roll contaminated with blood	7.4	75.43	8.65	15.43	
Dental roll contaminated with saliva	13.1	123.32	14.2	24.12	
Nylon gloves	54.22	362.67	64.43	73.92	64.31
Latex gloves	68.85	593.21	98.15	113.65	118.63
Plastic syringes		267.43	33.43	55.1	96.5
Suction head	14.42	109.91	9.73	17.83	23.1
Extracted teeth		312.95			
Absolang	11.43	67.43	8.48	24.26	14.54
Dental crowns		64.86			
Matrix tape	6		1.14	0.94	
Air spray coverage	2.21	6.72	2.37	5.94	
Orthodontic wire				14.41	
The injection and washing needle	11.71	77.37	8.21	6.77	13.97
Suture needles and its thread					252.82
Gutta-percha			33.43		
Surgical Blade				43.52	61.87
Disposable plastic cover	26.79	78.51	19.95	21.9	43.82
Total	287.31	3028.56	441.54	562.69	942.77
The number of patients	19	47	10	16	30
Per capita of patient	15.12	64.43	44.15	35.16	31.42

removal of calculus and endodontic therapy, respectively. The least amount was related to restoration which was equal to 15.12 g.

Table 2 shows various components of toxic dental waste of the dental clinics based on different services (restoration, dental extraction, removal of calculus, endodontic therapy, orthodontic and implant and surgery). The roll contaminated with mercury is the only waste generated in orthodontics and dental prosthesis and x-ray films and its lead coatings are related to surgery which is the higher waste in dental extraction, removal of calculus, restoration, and endodontic therapy too. The dental amalgam and material contaminated with dental amalgam has the highest amount in restoration, dental extraction, removal of calculus, and endodontic therapy.

The highest amounts per capita for each patient were related to endodontic therapy (24.87 g) and restoration (11.58 g). The least amount was obtained in orthodontics and prosthetics (1.02 g).

Table 3 presents the various components of the chemical, pharmaceutical waste in dental clinics based on the services provided. Except for orthodontics and prosthetics that has more than 96% of waste allocated to the sector, the Karpool constitutes all of the waste in the remaining cases.

The highest amounts per capita for each patient were related to orthodontics and Prosthesis (132.67 g) and endodontic therapy (2.43 g). The least amount was observed in restoration (0.647 g).

Figure 1 indicates the average percentage and quantity of dental wastes components in Ilam city. The highest values were related to infection (62%), chemical-pharmaceutical (27%) and toxic (11%).

Figure 2 illustrates the per capita infectious waste, chemical-pharmaceutical, the total toxic waste generated in the dental clinics for each person. The percent per capita for infectious waste was 51%. The average of per capita infectious waste was 201.13 g.

The percent per capita chemical-pharmaceutical for dental clinics was 36%. Average per capita chemical-pharmaceutical was 142.48 g.

Percent per capita for toxic in the office was 13% with an average weight of 48.78 g.

Figure 3 shows the total waste produced in dental clinics based on services (restoration, dental extraction and removal of calculus, endodontic therapy, orthodontic and implant and surgery). As can be seen, the highest waste produced is related to dental extraction, removal of calculus but the least was related to endodontic therapy and restoration.

Discussion

In this study, the lack of mercury and amalgam recycling was found to be as one of the main problems related to poor management in all dental clinics and the units of these clinics, which is in accordance (based on checklist) with other studies carried out in different cities of Iran such as Sari (16), Qaemshahr, Hamadan (17) as well as re-

Table 2. Various components of toxic waste in dental office based on the services provided (g)

Toxic waste	Restoration	Dental extraction, Removal of calculus	Endodontic therapy	Orthodontic and implant	Surgery
Paper napkins stained with amalgam	47.1	41.23	9.2		
Gauze contaminated with amalgam	70.73	97.84	17.43		
Cotton contaminated with amalgam	6.14	16.23	4.64		
Dental roll contaminated with amalgam	9.25	24.28	5.16	16.43	
Radiography film coating with lead foil	58.71	63.61	179.93		77.83
Radiography film	1.2	3.45	14.87		4.96
Amalgam particles	26.97	167.87	17.54		
Total	220.1	414.51	248.77	16.43	82.79
The number of patients	19	47	10	16	30
Per capita of patient	11.58	8.81	24.87	1.02	2.75

Table 3. Various components of the chemical, pharmaceutical waste dental office on the basis of services provided (g)

Pharmaceutical waste	Restoration	Dental extraction, Removal of calculus	Endodontic therapy	Orthodontic and implant	Surgery
Molding dough				2100.96	
Karpool	12.31	44.18	24.3	21.91	23.27
Total	12.31	44.18	24.3	2122.87	23.27
The number of patients	19	47	10	16	30
Per capita of patient	0.647	0.94	2.43	132.67	0.77

search carried out in Palestine (18), nevertheless, the several studies carried out in Sweden (19), the United States (20) and India (21,22) indicated the recycling of amalgam. In this study, the dental clinics did not carry out any safety measure on infectious waste. Treasure and Treasure (23) conducted a study in New Zealand with the questionnaire. The results from 767 filled questionnaires showed that 56.4% of dentists wrapped the swab contaminated with blood on the paper and threw them in the trash. A 24.4% of contaminated waste and sharp materials are discharged in household waste collection containers (23). One of the main causes of improved management in developed countries is the awareness of dentists on the policy of waste management (22). Germany has also mandated that all dentistry should have a dental amalgam separation and removal with efficiency of at least 95% (24). The management level and recycling in dental clinics and other clinics can be improved by adopting laws, training classes and providing conditions. In a study carried out at the University of Sydney, Australia in 1997 to evaluate the general condition of dental solid waste, it was determined that occupational health and safety equipment and containers for the collection of sharp instruments should be considered for this area. A 91% of all infectious waste was constituted with items such as gloves, masks and protective coatings form. The radiographic processing was disposed to sewer in 3 cases of the 10 (25). The fixer liquid along with developer solution are commonly utilized for radiology in the mouth and teeth where the fixer liquid is classified as hazardous material due to its high concentration of silver and it should not be directly discharged into the sewer system or the trash. Silver recycling is the best way to manage it (15); while, it was observed that this method was used for only few units of the clinics in this study to manage the fixer. The study used only 6.5% of dental clinics in the city of Surrey due to their recycling. Barafrashtehpour et al

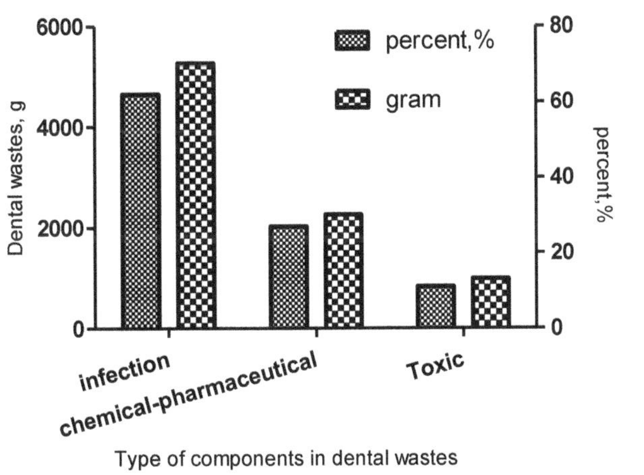

Figure 1. Average percentage and gram of dental wastes components in Ilam city.

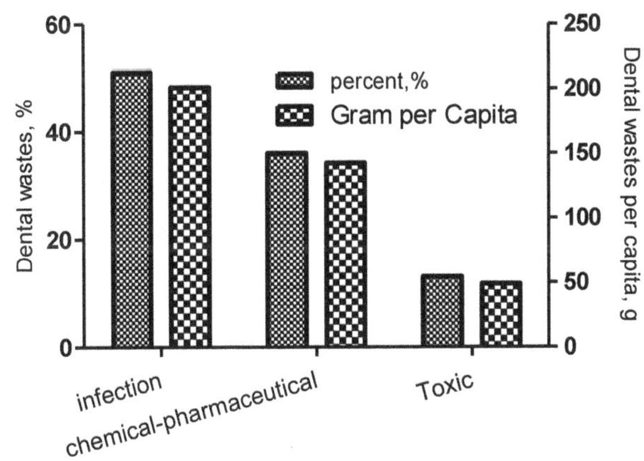

Figure 2. Average percentage and gram of dental wastes components per capita in Ilam city.

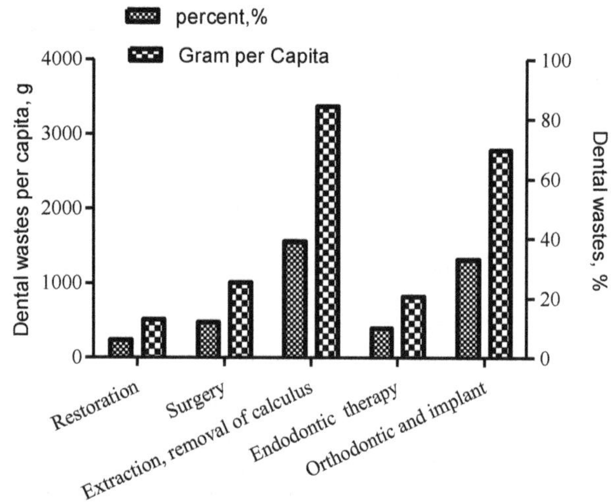

Figure 3. Waste production at the clinic based on different services (g and percent).

has found that the recycling was only performed in 5.6% of clinics (15). The results of another study have also revealed that there was no plan to reduce production in 64% of dental wastes. The waste segregation has performed in 83% office. A 90% of dental amalgam was directly released into the sewer system. The fixer was discharged into toilet and sewer system in 90% of the dental clinics. The use of safety box was the sharp objects management in 95% dental clinics and other clinics (15). The highest per capita amount of infectious waste for each patient was related to dental extraction, removal of calculus and endodontic therapy. The least amount was observed for dental restoration (Table 1). The percentage per capita for infectious waste was 51%. The average per capita infectious waste was 13/201 g which is the maximum amount of waste among the three studied groups of waste (Figure 2) which is consistent with the study carried out in Shiraz (26) and Ardebil (27). Thus it can be concluded that managerial models utilized for dental waste management in Iran and other countries can be applied in Ilam due to the similarity of the type and amount of waste produced in the dental setting of Ilam and other places. Nevertheless, it should be noted that the composition and amount of dental waste varied from one office to another office at different time. As earlier mentioned, the optimized management of dental wastes is not possible only by relying on a principle, but its management should be implemented with a purposeful plan and correctly modeled from the successful management used in the world (28). For example, the replacement of amalgam with safer substances can be considered as a good option to control amalgam and its risks. Today, in countries that have effective measures in this regard, the resin material is used instead of amalgam (29). Moreover, a significant portion of the waste such as paper and plastic coatings used tools, graph paper, disposable gowns of patient and others which have not come in contact with blood and infectious material and have not been contaminated can be separated from other infectious waste and be disposed with other materials such as paper, envelopes, food cover and food debris (26).

The composition and amount of solid waste in four dentistry laboratories were studied in Xanthium, Greece in 2002. These four laboratories have produced 75% of total dental waste in the state. Sampling was carried out in a period of 2 months and was divided into 3 major categories: (*a*) infectious and potentially infectious waste, (*b*) non-infectious toxic waste and (*c*) household type solid waste. The maximum amount of wastes produced in these laboratories was related to infectious and potentially infectious waste (74%) (29), which is in line with the results of our study.

In another study which was carried out in Qazvin to investigate the quantitative and qualitative effect of dental waste, the results showed that 12%, 1%, 36% and 51% of total waste produced was related to contaminated instruments, patient's tissue, sharp instrument and consumable items, respectively but these amount was 13%, 8%, 13% and 66% for contaminated instruments, patient's tissue, sharp instrument and consumable items, respectively (30) which are similar to the results of our study presented in Tables 1-3.

The average per capita for each person was 382.72 ± 9.67 g and more than 50% were infectious wastes. The study carried out in Greece by Kizlary et al (29) showed that the amount of infectious and semi-household waste was 94.7% and 3.3%, respectively. Moreover, the rate of dental waste production was 513 g per person per day while the rate of infectious waste generation was about 486 g per person per day. The results of this study were also in accordance with that of our study.

In present study, two other groups of dental wastes such as chemical-pharmaceutical waste and toxic waste constitute 7.7% of total waste production. These two parts contain toxic elements and chemicals and some of its elements such as amalgam particles and lead foil coated radiographic films are recyclable. Generally, we can say that these two parts of the dental waste should be collected and separated according to their characteristics and sometimes considered as hazardous material. The results of this study are in agreement with the results of the study of Nabizadeh et al (17) where the amount of household, infectious, chemical-pharmaceutical and toxic waste was 91.14%, 2.14%, 6.7% and 0.02%, respectively.

A study carried out in Shiraz, Southern Iran indicated that 89.1% of dental offices and clinics disposed their infectious waste with domestic waste. Standard method for the disposal of sharp objects was obtained as 60% by these centers and zero percent for their pharmaceutical and x-ray wastes. Amalgam recycling and lead foil pockets waste manufactured by these centers were less than 10% (31).

An investigation carried out on the characteristics and management of dental wastes in Urmia, Iran, showed that 31 components were manually separated and weighted. The results showed that domestic, potentially infectious, toxic, chemical and pharmaceutical wastes were 35.46%, 34.24%, 11.83% and 5.56%, respectively (32). In dental centers in Esfahan city, the amount of infectious, non-infectious and domestic wastes were obtained as 45.07%, 12.15% and 42.78%, respectively (33).

Conclusion

Finally, it can be suggested that the training of dentists for activities related to the reduction, separation and recycling in the dental clinics can play important role in developing the dental waste management.

Nevertheless, providing guideline and criteria for collection, transportation and separating disposal of any waste sector dental medications (like household waste, potentially infectious, chemical medicinal and poisonous) seems to be necessary. Undoubtedly, the continuous and correct monitoring of the above actions is necessary to obtain best results. Due to the implementation of problems, it is recommended that the dental wastes should be classified into four groups including semi-household waste, infectious, sharp and chemical objects and be separated at site. Separation and source reduction and separate disposal of wastes can be best way to control dental wastes.

Acknowledgments

The authors are grateful to the committee of student research of Mazandaran University of Medical Sciences for their financial support. This article was extracted from the MS Project (Research project code 162 in Mazandaran University of Medical Sciences) of F. Kazemi.

Competing interests

We affirm that this article is the original work of the authors and have no conflict of interest to declare.

Authors' contributions

All authors were involved in all stages of the article. On behalf of the co-authors, the corresponding author bears full responsibility for this submission.

References

1. Center for Disease Control and Prevention, Healthcare Infection Control Practices Advisory Committee (HICPAC). Guidelines for Environmental Infection Control in Healthcare Facilities. 2001. Available from: http://www.cdc.gov/hicpac/pdf/guidelines/eic_in_HCF_03.pdf.
2. Tchobanoglous G, Theisen H, Vigil SA. Integrated Solid Waste Management: Engineering Principles and Management Issues. New York: McGraw-Hill; 1996.
3. Omrani G, Alavi N. Hospital Solid Waste. Tehran: Tehran University Publication; 2008. [In Persian].
4. Department of Environmental Iran.Waste Management ACT. Departmentof Environmental Publications; 2004.
5. Tsakona M, Anagnostopoulou E, Gidarakos E. Hospital waste management and toxicity evaluation: a case study. Waste Manag 2007; 27(7): 912-20.
6. Birpinar ME, Bilgili MS, Erdoğan T. Medical waste management in Turkey: a case study of Istanbul. Waste Manag 2009; 29(1): 445-8.
7. Wong KV, Kashyap R. Medical waste characterization. J Environ Health 1994; 57(1): 19-25.
8. Green AE. Medical Waste Inciration and Pollution Prevention. New York: Springer; 1992.
9. Verma LK, Mani S, Sinha N, Rana S. Biomedical waste management in nursing homes and smaller hospitals in Delhi. Waste Manag 2008; 28(12): 2723-34.
10. Arenholt-Bindslev D. Environmental aspect of dental filling materials. Eur J Oral Sci 1998; 106(2 Pt 2): 713-20.
11. La Grega MD. Hazardous waste management. 2nd ed. New York: McGraw Hill; 2001.
12. Fan PL, McGill SL. How much waste do dentists generate? J Calif Dent Assoc 1989; 17(12): 39-40.
13. Trip L. Canada-wide standards: a pollution prevention program for dental amalgam waste. J Can Dent Assoc 2001; 67(5): 270-3.
14. ADA Council on Scientific Affairs and ADA Council on Dental Practice. Infection control recommendations for the dental office and the dental laboratory. J Am Dent Assoc 1996; 127(5): 672-80.
15. Barafrashtehpour M, Rezayi S, Alinejad A, Sadat A. Evaluation of dental wastes management In Yasouj. Thirteenth Congress of Iran Environmental Health; Kerman, Iran; 2010. [In Persian].
16. Zazoli MA, Barafrashteh Pour M, Mahdavi Y, Rostamali E. Qualitative and quantitative analysis and Dental waste management in the city of Sari, Iran. Seventh Congress of Epidemiology; Yasouj, Iran; 2013. [In Persian].
17. Nabizadeh R, Kulivand A, Jonidi Jafari A, Younesian M, Omrani GH, Evaluation of dental solid waste in Hamedan. Journal of Dental Medicine 2009; 22(1): 66-73. [In Persian].
18. Darwish RO, Al-Khatib IA. Evaluation of dental waste management in two cities in Palestine. East Mediterr Health J 2006; 12(2): 217-22.
19. Ogden GR, Bahrami M, Sivarajasingam V, Phillips G. Dental students' knowledge and compliance in cross infection control procedures at a UK dental hospital. Oral Dis 1997; 3(1): 25-30.
20. Jamie S. Reducting dental mercury discharge in Missoula, Montana: collaborative opportunites. Master of Science Environmental Studies, California; 2001.
21. Sudhakar V, Chandrashekar J. Dental health care waste disposal among private dental practices in Bangalore city, India. Int Dent J 2008; 58(1): 51-4.
22. Sushma M, Bhat S, Shetty S, Babu S. Biomedical dental waste management and awareness of waste management policy among private dental practitioners in Mangalore city, India. Tanzania Dental Journal 2010; 16(2): 39-43.
23. Treasure ET, Treasure P. An investigation of the disposal of hazardous wastes from New Zealand dental practices. Community Dent Oral Epidemiol 1997; 25(4): 328-31.
24. Koulivand A, Nbizadeh R, Joneidy A, Younesian M, Omrany G. Quantity and quality analysis and management of solid waste produced in dentistry laboratories and practical dentist office in Hamedan, 2007. Iranian Journal of Health and Environment 2009; 2(1): 36-45. [In Persian].
25. Cannata S, Bek M, Baker P, Fett M. Infection control and contaminated waste disposal practices in southern Sydney area health service dental clinics. Aust Dent J 1997; 42(3): 199-202.
26. Ghanbarian M, Majlesi M, Samaei MR. Analysis of Solid Waste Products Disposed by Dental Clinics in Shiraz. Journal of Research in Dental Sciences 2013; 10(4): 246-51. [In Persian].

27. Sotoudeh M, Dindra S, Merrikhi H, Talebi M, Ebrahimi S. Qualitative and quantitative analysis and Dental waste management in Ardabil city. The Third Student Congress of Ardabil University of Medical Sciences; Ardabil, Iran; 2011. [In Persian].
28. Ozbek M, Sanin FD. Astudy of dental solid waste produced in a school of dentistry in turkey. Waste Manag 2004; 24(4): 339-45.
29. Kizlary E, Losifidis N, Voudrias E, Panagiotakopoulos D. Composition and production rate of dental solid waste in Xanthi, Greece. Waste Manag 2005; 25(6): 582-91.
30. Nafez A, Karbord A, Sharifi M, Jabari R, Abotorabi M. Qualitative and quantitativ of dental wastes in Qazvin. Twelveth Congress of Iran Environmental Health; Tehran, Iran; 2009. [In Persian].
31. Danaei M, Karimzadeh P, Momeni M, Palenik CJ, Nayebi M, Keshavarzi V, et al. The management of dental waste in dental offices and clinics in Shiraz, Southern Iran. Int J Occup Environ Med 2014; 5(1): 18-23.
32. Koolivand A, Gholami-Borujeni F, Nourmoradi H. Investigation on the characteristics and management of dental waste in Urmia, Iran. J Mater Cycles Waste Manag 2015; 17(3): 553-9.
33. Nafez AH, Ebrahimi A, Hajian Nejad M, Bina B. Quantity and composition of produced dental solid waste in Isfahan, 2011. Int J Environ Health Eng 2014; 3: 17.

Organoleptic and palatability properties of drinking water sources and its health implications in Ethiopia

Sisay Derso Mengesha[1*], Abel Weldetinsae[1], Kirubel Tesfaye[1], Girum Taye[2]

[1]Environmental Health Research Team, Ethiopian Public Health Institute, Addis Ababa, Ethiopia
[2]Health System Research Directorate, Ethiopian Public Health Institute, Addis Ababa, Ethiopia

Abstract

Background: This retrospective study aimed to investigate the physicochemical properties of drinking water sources in Ethiopia and compare the water quality with the health-based target. For this purpose, the water quality database of Ethiopian Public Health Institute (EPHI) from 2010 to 2016 was used.

Methods: The concentration and other properties of the water samples were analyzed according to the Standard Methods of Water and Wastewater analysis. Quality control and quality assurance were applied in all stages following our laboratory standard operation procedures (SOPs).

Results: The concentration of the selected parameters varied based on the type of water sources. The mean concentration of turbidity was higher in spring water (21.3 NTU) compared to tap (12.6 NTU) and well (3.9 NTU) water sources. The mean concentration of total dissolved solids (TDS), electrical conductivity (EC), sodium (Na^+), and sulfate (SO_4^{-2}) was found to be higher in spring water sources than tap and well water sources. Comparably, the concentration of hardness, calcium, and magnesium was found to be higher in well water sources than spring and tap water sources. The bivariate analysis indicated that out of 845 analyzed water samples, more than 50% of the samples from Oromia region had turbidity, pH, TDS, hardness, Ca^{++}, K^+, and Na^+ within an acceptable limit. In addition, the logistic regression analysis showed that water quality parameters were strongly associated with the type of water sources and regional administration at $P < 0.05$.

Conclusion: More than 80% of the samples analyzed from drinking water sources were in agreement with WHO guidelines and national standards. However, the remaining 20% specifically, pH (25%), calcium (20%), hardness (18.1%), TDS (15.5%), and turbidity (13.3%) analyzed from improved water sources did not comply with these recommendations. Due to objectionable or unpleasant taste, people may force to look for alternative unprotected water sources that lead to health concerns.

Keywords: Drinking water, Water quality, Water sources, Taste, Physicochemical properties, Retrospective study, Ethiopia, Logistic models

***Correspondence to:**
Sisay Derso Mengesha
Email: sisdres23@yahoo.com

Introduction

Drinking water quality parameters are often the most important tools for measuring access to improved water sources. The drinking water with an acceptable quality shows the water safety in terms of its physical, chemical and bacteriological parameters (1). Some of the attributes, including color, turbidity, odor, hardness, etc, are substantially influenced by the acceptability of drinking water (2). Consumers' perceptions of quality also have a great value in their drinking water safety. Consumers can have different opinion towards aesthetic values of drinking water quality based on taste, odor, and appearance. Having a good knowledge of the factors that influence public perception can help improve water management, consumer services, the acceptance of water reuse and risk communication, among other areas (3). Therefore, consumer perceptions and aesthetic criteria need to be considered in assessment of drinking water supplies even if they may not adversely affect human health (1).

In developing water supply units, regulations and standards, both aesthetic criteria and health-related guideline values should be given a prior attention. This is attributed to the fact that one complements the other (4). In assessing the quality of drinking water, consumers rely principally upon their senses. A water source with poor microbial and physicochemical properties could hamper the aesthetic value of the water (5). Any complication on the aesthetic value of the water could also make the consumer to have a less preference for consumption of the developed water source, otherwise, it would lead consumers to consume acceptable but potentially unsafe alternative water sources (6).

According to the previous studies, environmental pollution, mainly of water sources, in underdeveloped countries have been suffering the impact of pollution due to disordered economic growth associated with the exploitation of natural resources (7). Similarly, stone mining around water sources in Ethiopia is very common. The quality of any water sources is subjected to both natural influences and human activities. In the absence of anthropogenic impact, water quality would be determined by natural origins, including weathering, natural leaching of organic matter and nutrients, and hydrological factors (8,9). According to Ethiopian Ministry of Health National Drinking Water Quality Monitoring and Surveillance Strategy, possible sources of pollutants to drinking water sources that threat the public health, include open field defecation, animal wastes, economic-related activities (agricultural, industrial and businesses) and even wastes from residential areas as well as transportation systems (10).

In the rural and urban areas of Ethiopia, spring water, surface water (streams, ponds, and rivers), groundwater and rainwaters are the main sources of drinking water. According to Ethiopian Demographic and Health Survey 2016, 97% of the urban households and 57% of rural households have access to an improved source of drinking water. The most common source of improved drinking water sources in urban households is water piped into the household's residence (63%), water piped into a public tap/ standpipe (13%), and water piped to a neighbor (12%). By disparity, rural households obtain their drinking water majorly from public taps/standpipes (19%), protected springs (14%), and tube wells and boreholes (13%) (11). Water in a distribution system could not have a similar condition across the distribution system until it reaches the distribution system. A constituent of water can be altered and also violate the health-related guideline value. To ascertain this, continuous quality monitoring of the water source at a different time and space will help to assure the desired goal. In this regard, the Environmental Health Laboratory of Ethiopian Public Health Institute (EPHI) is a spearhead in assisting the monitoring activities in the country. Data from the continuous water monitoring service were used in this retrospective study. The prime objective was to fill the void in the availability of studies that focused on organoleptic and palatability properties of drinking water sources in Ethiopia.

Materials and Methods

Country description

Ethiopia found in the Horn of Africa and located between 30N and 150N of the equator and 330E and 480E longitudes. The country is surrounded by Sudan to the west, Kenya to the south, Somalia to the southeast, Djibouti to the east, and Eritrea to the north and northeast. The total area of the country is about 1 127 127 km². In this country, there is great geographical variation, with a topography ranging from 4550 m above sea level to 110 m below sea level (12). Such variations have its own influences on spatial and temporal distributions of water sources availability in the country.

As indicated in Figure 1, Ethiopia is composed of nine administrative regions (Afar, Amhara, Benshangul-Gumuz, Gambella, Harari, Oromia, Somali, Tigray and the Southern Nations, Nationalities and Peoples Region (SNNPR)) and two city administration (Addis Ababa and Dire Dawa). Each regional state is subdivided into zones, woredas (districts), and Kebele.

In each woreda, there is a Water, Mining, and Energy Office which is responsible for water supply development and management and ensuring quality along with Woreda Water, Sanitation, and Hygiene Committee (WASHCO). The WASHCO always is organized from Health Office, Education Office, Woman Affair Office and Finance and Development Office. In addition, to ensure the water supply system through regular maintenance and follow up, each water supply scheme has its own WASHCO that is gathered from the community.

Sample collection and analysis

The water samples were collected by the environmental health professionals using plastic bottles. The concentration and other properties of the water samples were analyzed according to the *Standard Methods of Water and Wastewater analysis* (13). Quality control and quality assurance were applied in all stages of sample preparation, preservation and analysis according to the National Environmental Public Health laboratory's standard operation procedures (SOPs).

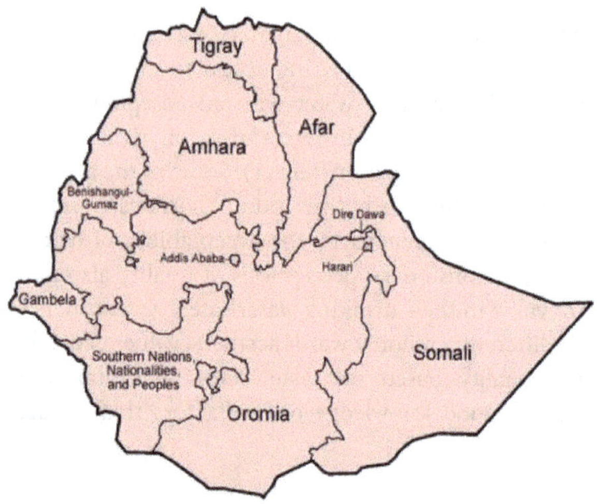

Figure 1. Map of Ethiopia with the nine regional states and two city administrations.

Data analysis

This retrospective study of physicochemical properties was conducted using the water quality database of EPHI. The physicochemical water quality data during seven years (2010 to 2016) were used. The data were stored in an Excel worksheet and transferred to SPSS version 20 for further statistical analysis. The physicochemical data of a total of 845 water samples from different sources (395 well, 242 spring, and 208 tap) were analyzed. The concentration physicochemical properties of the water samples were compared with Ethiopian drinking water standards and World Health Organization (WHO) guidelines (6,14). In addition to descriptive analysis, bivariate (chi-square test) and multivariate (logistic regression analysis) analyses were also conducted to determine the association of water quality parameters with each type of water source and regional administration.

Results

Descriptive physicochemical characteristics of water samples

Water samples examined during 2010 to 2016 in our laboratory were used for this study. Amongst all water samples, 46.7% were well water, 28.6% were spring water

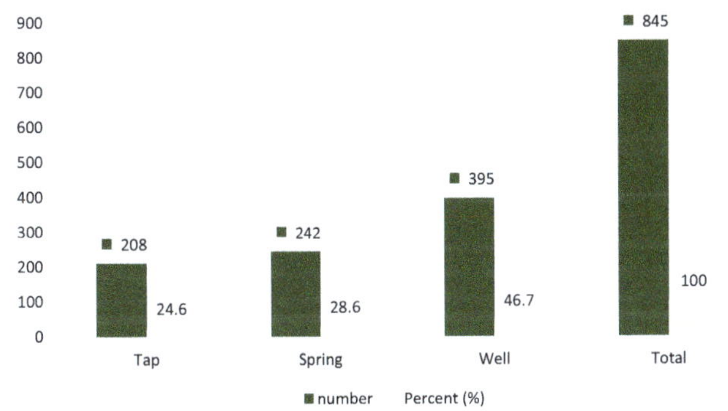

Figure 2. The number and proportion of water source types considered for retrospective analysis, brought to EPHI laboratory during 2010-2016.

and 24.6% were tap water as shown in Figure 2.

The mean, minimum and maximum concentrations of the selected physicochemical parameters along with the type of water sources at the national level are presented in Table 1. The concentration of the selected palatability and aesthetic parameters varied with the type of water sources. The mean concentration of turbidity in Nephelometer Turbidity Unit (NTU) was higher in spring water samples (21.3 NTU) compared to tap (12.6 NTU) and well water samples (3.9 NTU). The maximum turbidity level was also recorded in spring water samples (919 NTU). In addition, the mean concentration of total dissolved solids (TDS), electrical conductivity (EC), sodium (Na^+), and sulfate (SO_4^{-2}) was found to be higher in spring water samples than tap and well water samples. Comparably, the concentration of hardness, calcium, and magnesium was found to be higher in well water samples than spring and tap water samples.

The number or frequency and proportion of palatability parameters in drinking water samples are summarized in Table 2. More than 90% of magnesium, potassium, sodium, chloride, and sulfate were within the ranges of the Ethiopian standards and WHO guidelines. About 12% and 13% of the water samples had a pH value less than 6.5 and above 8.5, respectively. In addition, approximately, 13% of the water samples reportedly had a turbidity level above 5 NTU.

Bivariate analysis

The status of palatability of drinking water in different water sources in Ethiopia is presented in Table 3. The results show that the palatability parameters within the recommended limits of the national standards and WHO guidelines were significantly associated with the type of water sources ($P=0.001$).

The number of water samples that were in the acceptable level was higher in well water sources for turbidity (354), pH (309), calcium (314), potassium (373), sodium (354),

Table 1. Mean distribution of water quality parameters by the type of water sources, retrospective data during 2010-2016

Water source type	Parameter	Turbidity (NTU)	TDS at 105°C	EC	pH	Hardness	Na^+	K^+	Ca^{2+}	Mg^{2+}	Cl^-	SO_4^{2-}
Spring	Number	242	242	242	242	242	242	242	242	242	242	242
	Mean	21.25	854.33	1291.75	7.16	200.79	228.11	16.88	49.10	17.37	108.37	77.40
	Minimum	0	0	3.6	0	0	0	0	0	0	0	0
	Maximum	919	43326	63600	10.8	2450	17375	920	721.44	189.7	9497.06	2216.8
Tap	Number	208	208	208	208	208	208	208	208	208	208	208
	Mean	12.62	500.16	664.22	7.11	219.33	32.81	22.77	51.52	20.65	12.85	44.78
	Minimum	0	1.4	2.1	2.05	0	0	0	0	0	0	0
	Maximum	472	3420	7300	9	1500	353.6	288.75	292.58	189.7	92	1199.4
Well	Number	395	395	395	395	395	395	395	395	395	395	395
	Mean	3.92	571.07	766.71	7.43	267.37	96.94	7.02	70.22	126.57	74.92	91.06
	Minimum	0	0	0	3	0	0.014	0	0	0	0	0
	Maximum	201	7664	10650	9.41	3000	3000	462	801.6	41830	2500	2216

Table 2. Percentage of water quality parameters in drinking water sources, retrospective data during 2010-2016

Parameter	Range	N	%
Turbidity, NTU	0-5	733	86.7
	>5	112	13.3
pH	<6.5	102	12.1
	6.5-8.5	626	74.1
	>8.5	117	13.8
EC, µS/cm	0-1000	144	17
	>1000	701	83
TDS, mg/L	0-1000	716	84.7
	>1000	129	15.3
Hardness, mg/L	0-300	692	81.9
	>300	153	18.1
Calcium, mg/L	0-75	674	79.8
	>75	171	20.2
Magnesium, mg/L	0-50	766	90.7
	>50	79	9.3
Potassium, mg/L	0-1.5	773	91.5
	>1.5	72	8.5
Sodium, mg/L	0-200	774	91.6
	>200	71	8.4
Chloride, mg/L	0-250	812	96.1
	>250	33	3.9
Sulfate, mg/L	0-250	772	91.4
	>250	73	8.6

and chloride (369) compared with spring and tap water sources (Table 3).

Table 4 presents the status of palatability of drinking water sources in different regions of Ethiopia. The proportion of palatability parameters that complied with the recommended limits of the national standards and WHO guidelines was significantly associated with regions ($P = 0.001$).

Among a total of 845 water samples, more than 50% of turbidity, pH, TDS, total hardness, calcium, potassium, and sodium were in the range of acceptable ranges in Oromia. Next, the water samples of Addis Ababa were within the tolerable limit of both national standards and WHO guidelines compared to those of the rest regions in all parameters.

Multivariate analysis of the selected physicochemical parameters of water samples

Table 5-8 present adjusted odds ratios ($P < 0.001$) that were obtained from multivariate logistic regression model by considering the variable as acceptable and unacceptable limit of standards as outcome variables. The results show

Table 3. Distribution of water quality parameters in drinking water sources by the type of water sources, retrospective data during 2010-2016

Parameters	Ranges	Water Sources			Total	Chi-square	*P* value
		Spring	Tap	Well			
Turbidity, NTU	0-5	183	196	354	733		
	>5	59	12	41	112	39.028	0.001
	Total	242	208	395	845		
pH	<6.5	45	33	24	102		
	6.5-8.5	146	171	309	626	63.188	0.001
	>8.5	51	4	62	117		
	Total	242	208	395	845		
EC, µS	0-1000	65	23	56	144		
	>1000	177	185	339	701	24.061	0.001
	Total	242	208	395	845		
Total dissolved solid, mg/L	0-1000	206	173	337	716		
	>1000	36	35	58	129	0.524	0.770
	Total	242	208	395	845		
Hardness, mg/L	0-300	207	167	318	692		
	>300	35	41	77	153	3.041	0.219
	Total	242	208	395	845		
Calcium, mg/L	0-75	205	155	314	674		
	>75	37	53	81	171	7.231	0.027
	Total	242	208	395	845		
Magnesium, mg/L	0-50	224	182	360	766		
	>50	18	26	35	79	3.591	0.166
	Total	242	208	395	845		
Potassium, mg/L	0-1.5	218	182	373	773		
	>1.5	24	26	22	72	9.244	0.01
	Total	242	208	395	845		
Sodium, mg/L	0-200	214	206	354	774		
	>200	28	2	41	71	20.125	0.001
	Total	242	208	395	845		
Chloride, mg/L	0-250	235	208	369	812		
	>250	7	0	26	33	16.657	0.001
	Total	242	208	395	845		
Sulfate, mg/L	0-250	223	192	357	772		
	>250	19	16	38	73	0.908	0.635
	Total	242	208	395	845		

Table 4. Distribution of water quality parameters in drinking water sources by regional state, retrospective data during 2010-2016

Parameter	Region									$\chi 2$	*P* value
	A. A	Amhara	DD	Gambella	Oromia	SNNPR	Somalia	Tigray	A. A		
Turbidity, NTU	0-5	25.8%	7.5%	1.8%	1.8%	51.4%	3.8%	5.5%	2.5%		
	>5	2.7%	0.0%	0.0%	0.0%	27.7%	0.0%	0.0%	69.6%	442.981	0.001
	Total	192	55	13	13	408	28	40	96		
pH	<6.5	10.8%	4.9%	2.9%	2.9%	64.7%	2.9%	1.0%	9.8%		
	6.5-8.5	28.6%	7.3%	1.6%	1.6%	50.2%	4.0%	5.9%	0.8%	486.688	0.001
	>8.5	1.7%	3.4%	0.0%	0.0%	23.9%	0.0%	1.7%	69.2%		
	Total	192	55	13	13	408	28	40	96		
EC, μS/cm	0-1000	17.4%	2.8%	0.7%	0.7%	11.1%	0.0%	0.7%	66.7%		
	>1000	23.8%	7.3%	1.7%	1.7%	55.9%	4.0%	5.6%	0.0%	536.260	0.001
	Total	192	55	13	13	408	28	40	96		
TDS, mg/L	0-1000	26.4%	6.7%	1.3%	1.3%	47.8%	3.6%	1.4%	11.6%		
	>1000	2.3%	5.4%	3.1%	3.1%	51.2%	1.6%	23.3%	10.1%	145.223	0.001
	Total	192	55	13	13	408	28	40	96		
Hardness, mg/L	0-300	27.2%	6.8%	1.0%	1.0%	45.5%	3.8%	1.4%	13.3%		
	>300	2.6%	5.2%	3.9%	3.9%	60.8%	1.3%	19.6%	2.6%	155.722	0.001
	Total	192	55	13	13	408	28	40	96		
Calcium, mg/L	0-75	26.0%	6.8%	1.0%	1.0%	45.8%	3.9%	1.6%	13.8%		
	>75	9.9%	5.3%	3.5%	3.5%	57.9%	1.2%	17.0%	1.8%	118.919	0.001
	Total	192	55	13	13	408	28	40	96		
Magnesium, mg/L	0-50	25.1%	6.4%	1.3%	1.3%	48.2%	3.4%	2.6%	11.7%		
	>50	0.0%	7.6%	3.8%	3.8%	49.4%	2.5%	25.3%	7.6%	105.004	0.001
	Total	192	55	13	13	408	28	40	96		
Potassium, mg/L	0-1.5	24.8%	7.1%	1.7%	1.7%	47.7%	3.6%	5.2%	8.2%		
	>1.5	0.0%	0.0%	0.0%	0.0%	54.2%	0.0%	0.0%	45.8%	114.652	0.001
	Total	192	55	13	13	408	28	40	96		
Sodium, mg/L	0-200	24.8%	6.6%	1.4%	1.4%	47.3%	3.4%	2.7%	12.4%		
	>200	0.0%	5.6%	2.8%	2.8%	59.2%	2.8%	26.8%	0.0%	109.558	0.001
	Total	192	55	13	13	408	28	40	96		
Chloride, mg/L	0-250	23.6%	6.8%	1.4%	1.4%	49.3%	3.2%	3.2%	11.2%		
	>250	0.0%	0.0%	6.1%	6.1%	24.2%	6.1%	42.4%	15.2%	127.551	0.001
	Total	192	55	13	13	408	28	40	96		
Sulfate, mg/L	0-250	24.9%	6.7%	1.4%	1.4%	48.3%	3.4%	2.1%	11.8%		
	>250	0.0%	4.1%	2.7%	2.7%	47.9%	2.7%	32.9%	6.8%	155.564	0.001
	Total	192	55	13	13	408	28	40	96		

that the type of water sources (spring, tap, and well) and region were both statistically significant at a 5% level of significance ($P<0.5$). According to this association, the safety of drinking water samples was determined. As a result, the above-mentioned palatability parameters were considered in the multivariate logistic regression model as explanatory variables.

The odd ratios shown in these tables were used to compare whether the probability of an event is the same for the two groups. In this study, it helped to test the following hypotheses.

H0: there is no association between the type of water supply sources and their special distribution with palatability of drinking water when $P>0.05$.

H1: there is an association between the type of water supply sources and their special distribution with palatability of drinking water when $P<0.05$.

The *P* values for hardness, TDS, and calcium by region were less than 5% level of significance, indicating that there was an association between region and these parameters in drinking water. As a result, the odds of having water supplies within the acceptable limit of hardness, TDS, and calcium were less likely in drinking water supplies analyzed from Amhara, Dire Dawa, Gambella, Oromia, SNNPR, Somalia, and Tigray compared to those from Addis Ababa (Tables 5-7). In other way, Addis Ababa water supply sources were more likely to be in acceptable limit of these parameters compared to all regions of Ethiopia. At a 5% level of significance, almost all of the *P* values in analyzed water supply sources and palatability parameters were less than 0.05, indicating that there was an association between the type of water supply sources (spring, tap, and well) and palatability parameters (hardness, calcium, magnesium, TDS, sodium, potassium, and sulfate) (Table 8). As a result, the odds of having acceptable limits of hardness, calcium, magnesium, TDS, sodium, potassium, and sulfate were more likely in spring water sources compared to tap and well water sources except for sodium (1.520), which had good quality in well water sources compared to spring and tap water sources (Table 8).

At a 5% level of significance, there was no association between the type of water sources and region with turbidity, pH, and EC.

Table 5. Logistic regression of hardness by region, retrospective data during 2010-2016

Region	Adjusted odds ratio	*P* value
Addis Ababa	1	0.001
Amhara	0.079	0.001
Dire Dawa	0.015	0.001
Gambella	0.015	0.001
Oromia	0.051	0.001
SNNPR	0.143	0.03
Somalia	0.004	0.001
Tigray	0.269	0.07

Table 6. Logistic regression of TDS by region, retrospective data during 2010-2016

Region	Adjusted odds ratio	*P* value
Addis Ababa	1	0.001
Amhara	0.063	0.001
Dire Dawa	0.021	0.001
Gambella	0.021	0.001
Oromia	0.054	0.001
SNNPR	0.100	0.01
Somalia	0.002	0.001
Tigray	0.051	0.001

Table 7. Logistic regression of calcium by region, retrospective data during 2010-2016

Region	Adjusted odds ratio	*P* value
Addis Ababa	1	0.001
Amhara	0.313	0.01
Dire Dawa	0.071	0.001
Gambella	0.071	0.001
Oromia	0.214	0.001
SNNPR	0.656	0.5
Somalia	0.020	0.001
Tigray	1.659	0.4

Discussion

The mean concentration of the selected palatability parameters in this study varied according to the type of water sources. Most of the parameters including turbidity, TDS, EC, Na+, Cl-, and SO_4^{-2} were found to be higher in spring water samples compared to well and tap water samples. Mostly spring water is subjected to anthropogenic-induced factors due to its subsurface nature compared to groundwater that originates below the water table. Especially parameters like turbidity and chloride are signs of physical degradation of water sources by human and animal pressures along with open defecation and overgrazing near to the water sources. Turbidity is mostly related to rain. Studies conducted in Ethiopia reported high levels of turbidity in rainy season compared to the dry season (15-17).

According to Ethiopian national standards, pH of the water should be between 6.5 and 8.5 to avoid such a bad water taste. This pH range has also been recommended by WHO guidelines for drinking water quality. However, about 25% of water samples (12% and 14% below and above the recommended boundaries, respectively) in this study didn't comply with this recommendation (Table 2), which is consistent with a similar study conducted by Alemu et al (18). So, a pH value below 6 leads to toxic nature and a pH greater than 9 turns the water taste to a bitter taste (19). As a result, it would affect the user's perceptions and may make them to change their drinking water sources.

Table 8. Logistic regression of palatability parameters in drinking water sources, retrospective data during 2010-2016

Parameter	Water sources	Adjusted odds ratio	*P* value
Hardness	Spring	1	0.001
	Tap	0.350	0.001
	Well	0.814	0.4
Calcium	Spring	1	0.001
	Tap	0.348	0.001
	Well	0.851	0.05
Magnesium	Spring	1	0.001
	Tap	0.179	0.001
	Well	0.852	0.6
TDS	Spring	1	0.001
	Tap	0.338	0.001
	Well	1.113	0.6
Sodium	Spring	1	0.01
	Tap	9.387	0.001
	Well	1.520	0.1
Potassium	Spring	1	0.001
	Tap	0.144	0.001
	Well	0.998	0.9
Sulfate	Spring	1	0.01
	Tap	0.373	0.01
	Well	0.897	0.7

On the other hand, parameters like hardness, calcium, and magnesium were found to be higher in well water samples than tap and spring water samples. Nationally, calcium concentration (20%) in the water samples was not found in agreement with Ethiopian national standards. In addition, the level of hardness in the water samples that did not meet the national standards was estimated about 18.1% (Table 2). Since calcium and magnesium are well known to occur naturally in groundwater because of its channel through mineral deposits and rock layers that result in its total hardness (20), the results of the present study are in agreement with such facts. Another study conducted by Sisay et al in the south western part of Ethiopia indicated a similar scenario of hardness levels comparing with the type of water sources (21). Health-based guideline value for hardness is not proposed yet, but it has economic implications due to detergent consumption and scale formation during boiling. Even though the public acceptance level for hardness varies from community to community and person to person, hardness above 500 mg/L is classified as unacceptable (22). This unacceptable taste may make the public to look for other unprotected water sources, which consequently may threat the public health.

In addition, when the concentration of calcium exceeds the acceptable limit or national standards (greater than 75 mg/L), it will have its own effect on water quality. The taste threshold for the calcium ion is in the range of 100–300 mg/L as set by WHO but it depends on the other anions in the water (23). Due to the high content of calcium and carbonates as the most dissolved ions in hard water, hardness is usually expressed as calcium carbonate ($CaCO_3$). However, calcium is one of the essential elements that is required by human beings particularly for infants and under five years. It is crucial for bone formation and development. The basis for skeletal well-being is founded so early in life, and osteoporosis prevention begins by optimizing obtains in bone mineral throughout childhood and adolescence (19,24).

In the present study, a chi-square test and row percentage were used to determine the significant association of drinking water palatability with the type of water sources and administrative regions. Since the chi-square can provide information on the significance of any observed differences and determine the exact categories account for any differences found (25). Based on this analysis, the type of water sources has a direct relationship with water quality. The acceptable level of turbidity (354), pH (309), calcium (314), potassium (373), sodium (354), and chloride (369) was found to be higher in well water samples than in tap and spring water samples ($P = 0.001$) (Table 3). In addition, the region has a direct association with quality of water supply in the country ($P = 0.001$). Among a total of 845 water samples, more than 50% of turbidity, pH, TDS, total hardness, calcium, potassium, and sodium were in the range of acceptable limits in Oromia. After Oromia, Addis Ababa water samples were within the acceptable limits of recommended standards than the rest administrative regions in all parameters.

Furthermore, a logistic regression analysis was conducted to determine the significance level of water quality parameters with spatial distribution by region and type of water sources. In a developing country, water sources vulnerability has typically been assessed using largely qualitative methods and expressed as relative measures of risk (26). The Ethiopian scenario was not different and water quality data have usually summarized in a descriptive manner. However, determining the likelihood that different water sources contain elevated concentrations of contaminants can help water resources managers protect drinking water supplies. Based on this analysis, *P* values for hardness, TDS, and calcium by region were less than 5% level of significance ($P < 0.05$) indicating that there was an association between region and these parameters in drinking water. It is furtherly interpreted as the odds of having water supplies within an acceptable limit of hardness, TDS, and calcium were less likely in drinking water supplies in Amhara, Dire Dawa, Gambella, Oromia, SNNPR, Somalia, and Tigray compared to water sources in Addis Ababa, an administration city (Table 5-7). In addition, at $P < 0.05$, almost all the *P* values in analyzed water supply sources and palatability parameters were less than 0.05, it shows that there was an association between the type of water sources (spring, tap, and well) and palatability parameters (hardness, calcium, magnesium, TDS, sodium, potassium, and sulfate) (Table 8). This implies that the odds of having acceptable limit of hardness, calcium, magnesium, TDS, sodium, potassium, and sulfate were more likely in spring water sources compared to tap and well water sources except for sodium (1.520), which has a good quality in well water sources compared to spring and tap water sources (Table 8). But, at 5% level of significance, there was no association between the type of water supply sources and region with turbidity, pH, and EC.

In general, this study showed that more than 80% of all parameters analyzed from improved water sources (spring, tap, and well) were found within the national standards and WHO guidelines. These findings are in agreement with the previous study summarized from the same water quality database (18). However, the remaining 20% of water quality parameters collected from improved water sources which failed to meet both national standards and WHO guidelines along with above 35% of overall Ethiopian population relied on unprotected water sources reported by Ethiopian Demographic and Health Survey (11), that could threat the public health from drinking water supply. Unaccepted palatability parameters have equal influences and threat water supply sources even if they have not a track record on health concerns. Due to bad taste of drinking water, people will force to get alternative water sources that may be unprotected sources especially in rural area and dry places where water is not easily accessible. Other studies showed that high concentration of organoleptic parameters including turbidity, and salts of calcium, sodium, magnesium, and sulfate are the main influencing factors for objectionable water taste and influence community acceptability (27,28).

Conclusion

This retrospective study investigated the water quality status of improved water sources at a national level through cost-effective approaches i.e. analyzing water quality databases. The descriptive analysis showed that the mean concentration of the selected parameters in this study varied according to the type of water sources. Parameters like turbidity, TDS, EC, Na^+, Cl^-, and SO_4^{-2} were found to be higher in spring water samples compared to well and tap water samples. Contrarily, hardness, calcium, and magnesium were found to be higher in well water samples than tap and spring water samples.

Based on bivariate analysis, at *P* value of 0.000, the palatability parameters within the recommended limits of the national standards and WHO guidelines were significantly associated with the type of water sources and administrative regions. Among a total of 845 water samples, more than 50% of turbidity, pH, TDS, total hardness, calcium, potassium, and sodium were in the range of acceptable limits in Oromia. Next, Addis Ababa water samples were within the acceptable limits of recommended standards than the rest administrative regions in all parameters. The logistic regression analysis indicated that water quality was strongly associated with special distribution and type of water sources. This implies that the odds of having the acceptable limits of hardness, calcium, magnesium, TDS, sodium, potassium, and sulfate were more likely in spring water sources and Addis Ababa (an administration city) compared to tap and well water supply sources and other regions.

In general, more than 80% of all parameters analyzed from improved water sources were in agreement with WHO guidelines and national standards. However, the remaining 20% specifically, pH (25%), calcium (20%), hardness (18.1%), TDS (15.5%), and turbidity (13.3%) analyzed from improved water sources did not comply with these recommendations. Due to unpleasant taste, people will force to look alternative unprotected water sources that lead to health concerns. This calls to enhance the awareness and knowledge of the community on such water quality and water authority should put into consideration these water quality parameters during the development of water supply schemes.

Acknowledgments

The authors would like to thank the EPHI for water quality databaseaccess.

Competing interests

The authors declared that they have no conflicts of interest.

Author contribution

SD has participated in data analysis and manuscript preparation. AW and KT have participated in the manuscript preparation. And GT has participated in data analysis.

References

1. World Health Organization (WHO). Guidelines for Drinking-Water Quality: Recommendations. Geneva: WHO; 2004.
2. Addisie MB. Assessment of drinking water quality and determinants of household potable water consumption in Simada District, Ethiopia. New York: Cornell University; 2012.
3. Doria Mde F, Pidgeon N, Hunter PR. Perceptions of drinking water quality and risk and its effect on behaviour: a cross-national study. Sci Total Environ 2009; 407(21): 5455-64.
4. Bartram J, Cotruvo J, Exner M, Fricker C, Glasmacher A. Heterotrophic Plate Counts and Drinking-water Safety. London: IWA Publishing; 2003.
5. Gadgil A. Drinking water in developing countries. Annual Review of Environment and Resources 1998; 23: 253-86. doi: 10.1146/annurev.energy.23.1.253.
6. World Health Organization (WHO). Guideline for Drinking Water Quality. Geneva: WHO; 2011.
7. Kazi TG, Arain MB, Jamali MK, Jalbani N, Afridi HI, Sarfraz RA, et al. Assessment of water quality of polluted lake using multivariate statistical techniques: a case study. Ecotoxicol Environ Saf 2009; 72(2): 301-9. doi: 10.1016/j.ecoenv.2008.02.024.
8. Ritter L, Solomon K, Sibley P, Hall K, Keen P, Mattu G, et al. Sources, pathways, and relative risks of contaminants in surface water and groundwater: a perspective prepared for the Walkerton inquiry. J Toxicol Environ Health A 2002; 65(1): 1-142. doi: 10.1080/152873902753338572.
9. Khatri N, Tyagi S. Influences of natural and anthropogenic factors on surface and groundwater quality in rural and urban areas. Front Life Sci 2015; 8(1): 23-39. doi: 10.1080/21553769.2014.933716.
10. Ministry of Health. National Drinking Water Quality Monitoring and Surveillance Strategy. Addis Ababa: A Federal Democracy of Ethiopian Ministry of Health; 2011.
11. Central Statistical Agency (CSA) [Ethiopia], International Coach Federation (ICF). Ethiopia Demographic and Health Survey. Addis Ababa, Rockville: CSA, ICF; 2017.
12. Segele ZT, Leslie LM, Tarhule AA. Sensitivity of horn of Africa rainfall to regional sea surface temperature forcing. Climate 2015; 3(2): 365-90.
13. American Public Health Association (APHA). Standard Methods for the Examination of Water and Wastewater. Washington, DC, USA: APHA; 2005.
14. Ethiopian Standards Agency (ESA). Compulsory ethiopian standard: Drinking water Specification. ESA; 2013. ICS: 13.060.20.
15. Tadesse D, Desta A, Geyid A, Girma W, Fisseha S, Schmoll O. Rapid assessment of drinking-water quality in the Federal Democratic Republic of Ethiopia: country report of the pilot project implementation in 2004-2005. Geneva: WHO; 2010.
16. Sisay T, Beyene A, Alemayehu E. Assessment of drinking water quality and treatment plant efficiency in southwest Ethiopia. J Environ Sci 2017; 3(3): 208-12.
17. Yasin M, Ketema T, Bacha K. Physico-chemical and bacteriological quality of drinking water of different sources, Jimma zone, Southwest Ethiopia. BMC Res Notes 2015; 8: 541. doi: 10.1186/s13104-015-1376-5.
18. Alemu ZA, Teklu KT, Alemayehu TA, Balcha KH, Mengesha SD. Physicochemical quality of drinking water sources in Ethiopia and its health impact: a retrospective study. Environ Syst Res 2015; 4(1): 22. doi: 10.1186/s40068-015-0049-7.
19. Mohsin M, Safdar S, Asghar F, Jamal F. Assessment of drinking water quality and its impact on residents health in Bahawalpur city. Int J Humanit Soc Sci 2013; 3(15): 114-28.
20. Karavoltsos S, Sakellari A, Mihopoulos N, Dassenakis M, Scoullos MJ. Evaluation of the quality of drinking water in

regions of Greece. Desalination 2008; 224(1-3): 317-29. doi: 10.1016/j.desal.2007.06.013.
21. Sisay T, Beyene A, Alemayehu E. Spatiotemporal variability of drinking water quality and the associated health risks in southwestern towns of Ethiopia. Environ Monit Assess 2017; 189(11): 569. doi: 10.1007/s10661-017-6277-8.
22. Asadullah A, Nisa K, Khan SI. Physico-chemical properties of drinking water available in educational institutes of Karachi city. Science Technology and Development. 2013; 32(1): 28-33.
23. Bernard E, Ayeni N. Physicochemical analysis of groundwater samples of Bichi local government area of Kano State of Nigeria. World Environment 2012; 2(6): 116-9. doi: 10.5923/j.env.20120206.02.
24. Bachrach LK. Acquisition of optimal bone mass in childhood and adolescence. Trends Endocrinol Metab 2001; 12(1): 22-8. doi: 10.1016/S1043-2760(00)00336-2.
25. McHugh ML. The chi-square test of independence. Biochem Med (Zagreb) 2013; 23(2): 143-9.
26. Tesoriero AJ, Inkpen EL, Voss FD. Assessing ground-water vulnerability using logistic regression. Dallas, TX: Proceedings for the Source Water Assessment and Protection 98 Conference; 1998. p. 157-165.
27. Doria Mde F, Pidgeon N, Hunter PR. Perceptions of drinking water quality and risk and its effect on behaviour: a cross-national study. Science of the Total Environment 2009; 407(21): 5455-64.
28. Zoeteman BCJ, de Grunt FE, Koster EP, Smit KGJ, Punter PH. Taste assessment of individual salts in water methodology and preliminary findings by a selected national panel. Chem Senses 1978; 3(2): 127-39. doi: 10.1093/chemse/3.2.127.

Bioaccumulation and translocation factors of petroleum hydrocarbons in *Aeluropus littoralis*

Mohammad Rafiee[1], Mahsa Jahangiri Rad[2*], Arash Afshari[3]

[1]Assistant Professor of Environmental Health Engineering, Department of Environmental Health, School of Public Health, Shahid Beheshti University of Medical Sciences, Tehran, Iran
[2]Assistant Professor of Environmental Health Engineering, Department of Environmental Health Engineering, Tehran Medical Sciences Branch, Islamic Azad University, Tehran, Iran
3. MSc Student of Environmental Engineering, Department of Environmental Engineering, Tehran Shomal Branch, Islamic Azad University, Tehran, Iran

Abstract
Background: Polycyclic aromatic hydrocarbons (PAHs), which consist of fused conjugated aromatic rings, not only are toxic to humans and other living organisms, but will also pollute groundwater. These compounds can be point source or non-point source and are one of the most widespread organic pollutants. Some of them are suspected carcinogens and are linked to other health problems. This study was carried out to assess the petroleum hydrocarbon phytoremediation potential of the *Aeluropus littoralis* species. Accumulation of PAHs in roots and upper parts of *A. littoralis* has been determined. Translocation factor (TF) was also calculated.
Methods: In this study, soil samples were taken from the vicinity of Isfahan oil refinery, and the PAHs compounds were analyzed with gas chromatography. One-meter soil columns were prepared from the control and contaminated soil. Unplanted *A. littoralis* treatments were also prepared to eliminate the effects of environmental factors on the reduction of oil-based contaminants. Seventeen weeks after planting, soil columns were sampled at 25, 50, 75, and 100 cm depths, and the concentration of petroleum hydrocarbons and number of oil-degrading bacteria were counted. Moreover, the roots and shoots of *A. littoralis* were separated and weighed.
Results: Results indicated that *A. littoralis* reduced the concentration of petroleum hydrocarbons to a significantly higher extent than the control treatment. Increasing depth was associated with improved petroleum hydrocarbon concentrations and decreased number of oil-degrading bacteria. Mostly, the bioaccumulation factor (BAF) of PAHs was <1, which shows *A. littoralis* could be an excluder of PAHs. The results also showed that the TF of PAHs was less than one, and hence, *A. littoralis* could be considered as an accumulator of PAHs.
Conclusion: It can be concluded that this species is suitable for use in the phytoremediation of PAHS contamination. For further confirmation, an evaluation under field conditions should be undertaken.
Keywords: Biodegradation, Soil, Polycyclic aromatic hydrocarbons, *Aeluropus littoralis* plant

***Correspondence to:**
Mahsa Jahangiri Rad
Email: m.jahangiri@iautmu.ac.ir

Introduction

Petroleum is a compound consisting of hundreds of organic and trace amounts of inorganic substances. Although each organic compound has unique physical and chemical properties, they are often divided into the paraffin, naphthenes, and aromatic hydrocarbons (1). Some of the most problematic organic components in crude oil are the polycyclic aromatic hydrocarbons (PAHs). Their bioaccumulation, airborne transportation, and persistence in the environment are of great concern since PAHs have been identified as carcinogens, mutagens, and teratogens (2). Various methods have been applied to clean up oil contaminated soil, including chemical, biological, and physical methods (3). Phytoremediation is a relatively new method that relies on the use of plants to remove a wide range of contaminants without the need to excavate the contaminant material. It has gained special attention over the past 20 years because plants have an extraordinary capacity to clean up the environment (4). Plants not only clean up contaminants, they also slow groundwater hydraulics and reduce the spread of pollutants. Each plant affects the rate of degradation diversely (5). The plant's ability to degrade the compound, the lethal dosage, rate of degradation, and type of plant should be considered when phytoremediation is applied. The type of soil with which the oil has contact also affects how well the oil is degraded. It has been shown that soils higher in organic matter actually reduce the rate of oil decay (6). There are also studies to suggest that the type of soil affects the rate of degradation. Soils containing larger pores are capable of breaking down hydrocarbons faster since the oil can percolate through the soil at a faster rate (7). Phytoremediation as applied to petroleum clean-up in salt marshes has received great attention owing to its low cost and low environmental impacts. Six ways are known as the predominant processes for contamination cleanup by phytoremediation, including rhizodegradation, phytoextraction, phytovolatilization, phytodegradation, rhizofiltration, and rhizostabilization (4). The first and most important method of petroleum hydrocarbon removal is rhizodegradation, a process which occurs in the rhizosphere (8). Research has shown that the larger the microbial population is, the faster the rate of degradation will be (9). In the rhizosphere, petroleum compounds can be degraded to simpler, less toxic molecules or mineralized entirely to water and carbon dioxide. The rhizosphere also removes harmful petroleum hydrocarbons from soil and water by promoting volatilization through transpiration and by accumulating organic compounds through lignification (10).

Various plant species have been proven capable of remediating petroleum-contaminated soils in bench-top studies, including *Vicia faba*, *Zea mays*, *Triticum aestivum* L, and *Glycine max* (11,12). *Aeluropus littoralis* from the Poaceae family is characterized by its high tolerance to soluble salts and can be grown in saline habitats. *A. littoralis* exhibits a rapid growth rate, high biomass production, and widespread root system distribution in saline and non-saline soils (13). In a research conducted by Rezvani et al, *A. littoralis* showed a high capability in cadmium and lead removal from contaminated soils (14). The present study was therefore designed to investigate the bioaccumulation and translocation characteristics of *A. littoralis* in hydrocarbon removal from contaminated soil around the Isfahan Oil Refinery (Isfahan, Iran).

Materials and methods

Selection of soil sampling

The Isfahan Oil Refinery selected for this study is located about 340 km south of Tehran. It is Iran's second-largest refinery with a production capacity of 265 000 bpd (Figure 1). The city is located on the main north–south and east–west routes crossing Iran (32°38′N51°39′E). It has an arid climate and is situated at 1590 m (5217 ft) above sea level on the eastern side of the Zagros Mountains (15). The refinery, established in 1976, receives its crude oil feedstock from the Maroon and Shadegan fields. Esfahan's products include liquid petroleum gas (LPG), gasoline, light naphtha, light jet fuel, ATK, petrochemical feedstock, kerosene, gas oil, lube oil, fuel oil, and asphalt. Soil samples were taken from land near the refinery's Sulfur Recovery Unit. Control samples were also chosen from uncontaminated nearby land. Considering the size of the area (1000 m^2), 5 locations were selected out of every 200 square meters of land. At each location, 2 kg of soil was sampled from 30 cm below the surface in 5 geographical directions; the samples were mixed to form a compound sample of about 10 kg. Finally, three subsamples were drawn from each compound sample and stored at a controlled temperature (4°C) before being analyzed.

Measuring physical and chemical properties of soil

Physical and chemical properties of soil play significant roles in the efficacy of processes reducing petroleum hydrocarbons (16). In this study, soil samples were air dried and passed through a 2-mm sieve. Soil texture (through hydrometry), pH, electrical conductivity, organic matter, total nitrogen, available phosphorus, and potassium content of each sample were determined according to EPA methods 9071 A and 3540 B (17).

Determination of total petroleum hydrocarbons

PAH$_s$ and total petroleum hydrocarbons (TPHs) in the soil were measured by Soxhlet extraction using a 1:1 (v/v) n-hexane/dichloromethane solvent (150 mL) mixture for 24 hours. The extracted compounds were then evaporated and condensed using a rotary evaporator under vacuum conditions. Column chromatography (using silica gel and alumina as absorbent) was employed to purify the samples (18,19). The concentration of PAHs in soil was determined with gas chromatography (17).

Phytoremediation procedure

Selected for the phytoremediation experiments was *A.*

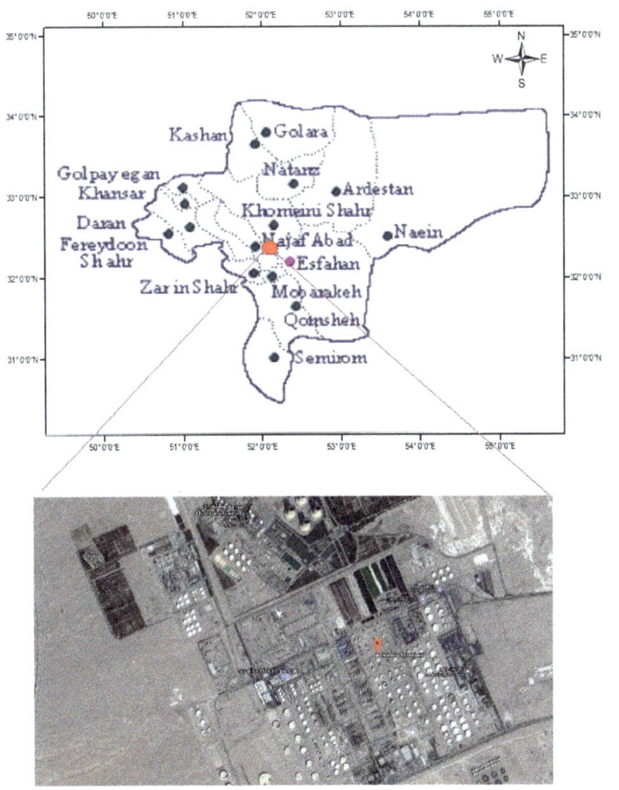

Figure 1. A map location of Isfahan province and Isfahan oil refinery that selected in this research work.

littoralis, a permanent plant regrowing from underground parts. Polyvinyl chloride (PVC) pipes (20 cm in diameter and 100 cm long) were employed. The pipes were holed at 20, 40, and 60 cm for the final sampling with 10-cm drainage at the bottom. Considering the density of soil (2.5 g/cm^3) and volume of PVC pipes, the required amount of contaminated soil for each pot was calculated. Phytoremediation experiments were conducted on 3 conditions: 1) TPH removal in the presence of *A. littoralis,* 2) treatments without plants intended to eliminate the effects of environmental factors on the reduction of oil-based contaminants for each column, and 3) controlled soil samples (without TPH contamination). The experiment pots were filled with the specified amount of either contaminated or control soil samples while 10 cm empty space was left on top. The TPH concentration in each column was measured as 50 000, 55 000, 57 000, and 60 000 ppm, respectively. Three repetitions of 4 treatments were conducted on a total of 12 soil columns. The cultured, contaminated, and control soil columns were treated with *A. Littoralis* seed placed 2 cm below the top soil (3 of them remained untreated to exclude the environmental factors on oil reduction). Thirteen weeks after planting, soil columns were sampled at depths of 20, 40, and 60 cm and concentrations of petroleum hydrocarbons were analyzed. Finally, the roots and shoots of the plants were separated, dried in an oven at 80°C for 48 hours, and weighed.

Plant analysis

Samples were taken from the plant shoots and roots and analyzed for PAH 11, 12, 13, 14, 15, 16, and 17 weeks after planting. At each stage, plant samples were washed thoroughly with distilled water and oven dried at 70°C for 24 hours for further analyses. Plant samples were then ground and digested overnight in a 5-mL acid mixture of $HClO_4$ (Merck) (60%) and HNO_3. Digestion was completed by gradually increasing the temperature from 60°C to 195°C (60°C, 3 hours; 100°C, 1 hour; 120°C, 1 hour; 150°C, 30 minutes; 175°C, 30 minutes; 195°C, 2.5 hours). After cooling, HCl (20%, 2.5 mL) was added, and the mixture was whirl-mixed and warmed to 80°C for 30 minutes. The final volume was brought to 10 mL with double-distilled water and rewarmed to 80°C for another 30 minutes. Samples were filtered through a cellulose filter (0.2 μM). The PAH_s concentrations in plants were determined using gas chromatography (17-19).

Counting oil-degrading bacteria

In order to count oil degrading bacteria, one-gram soil samples from the microplots were aseptically collected and transported in sterile containers and on ice to the laboratory. A series of dilutions (10^{-1}-10^{-8}) were prepared, and total bacterial numbers were determined using a culture medium which consisted of 990 mL sterile agar solution and $CaCl_2\_H_2O$ (0.02), $FeCl_3$(0.05), $MgSO_4\_7H_2O$(0.2), K_2HPO_4(1), NH_4NO_3(1), KH_2PO_4 (1) (pH = 7) (20).

Translocation and bioaccumulation factors

Oil translocation from shoot to root was measured by T_F which is given below (18,19):

$$T_f = \frac{C_{shoot}}{C_{root}} \quad (1)$$

where C_{shoot} and C_{root} are oil concentrations in the shoot (mg kg^{-1}) and root of plant (mg kg^{-1}), respectively. T_F>1 indicates that translocation of oil is mostly due to the shoot from the root.[7] Bioaccumulation factor (BAF) was calculated as follows:

$$BAF_f = \frac{C_{root}}{C_{soil}} \quad (2)$$

where C_{root} and C_{soil} are oil concentrations in the plant root (mg kg^{-1}) and soil (mg kg^{-1}), respectively.

BAF>1 and BAF<1 represent accumulatory and excludery properties.

Results

Physical and chemical properties of contaminated and control soils are summarized in Table 1.

The results of this experiment showed that soil contaminated with oils has a lower pH compared to the control soil. Table 2 shows the fraction concentration of each PAHs measured in nearby soil. The mean petroleum hydrocarbon concentration of 65 000 mg/kg suggested that the soil from areas near the oil refinery is extremely contaminated.

Statistical analyses of the main and interaction effects of soil and plants on shoot and root dry matter yield (Figures

Table 1. Physical and chemical properties of soil in the selected area

Characteristic	Control soil	Contaminated soil
Texture	Sandy clay loam	Sandy clay loam
pH (1:2.5)	7.5	7.1
EC (ds/m)	2	3.5
Organic matter (%)	0.5	5.2
Total nitrogen (%)	0.07	0.90
Available phosphorus (mg/kg)	36	80
Available potassium (mg/kg)	15	21

Table 2. Concentrations of the measured polycyclic aromatic hydrocarbons and total petroleum hydrocarbons in contaminated soil

PAHs	Concentration (mg/kg)
Phenantherene	31
Naphthalene	40
Fluoranthene	21
Pyrene	16
Anthracene	7.3
Benzo[k]fluoranthene	0.3
Benzo[a]pyrene	0.6
TPHs	65000

Abbreviations: TPHs, total petroleum hydrocarbons; PAHs, polycyclic aromatic hydrocarbons.

2 and 3) of *A. littoralis* revealed that plant type and soil influence dry matter yield. In fact, dry weight comparisons between the contaminated and control treatments showed reductions of about 22% and 51% in root and shoot dry matter of *A. littoralis* in petroleum hydrocarbon-containing soil, respectively.

Shoot and root weights were drastically reduced with increases in oil concentration ($P < 0.05$). In fact, dry weight comparisons between contaminated and control treatments indicated reductions of about 10% and 20% in the root and shoot dry matter of *A. littoralis* in petroleum hydrocarbon-containing soil, respectively.

The effects of depth on petroleum hydrocarbon concentrations and number of oil-degrading bacteria in *A. littoralis* treatments and control soils at three different depths (Figure 4) were determined. Apparently, the lowest concentration of petroleum hydrocarbons and greatest number of oil-degrading bacteria were observed at 0-20 cm depth. The comparison of means showed a significant difference between unplanted and planted treatments in terms of petroleum hydrocarbon concentration and number of oil-degrading bacteria ($P < 0.05$) (Table 3 and Figure 4).

The BAFs of PAH (initial concentration of 50 000 ppm) for *A. littoralis* samples in different weeks (from 1 to 7) of growth are given in Figure 5. The BAF represents the contaminant concentration in plants compared with the environment concentration (in soil) (20). TF was calculated to determine the translocation of PAHs from soil to root or from root to upper parts of plants. TF indicates the ability of plants to accumulate PAHs from soils. The calculated TF values are shown in Figure 5.

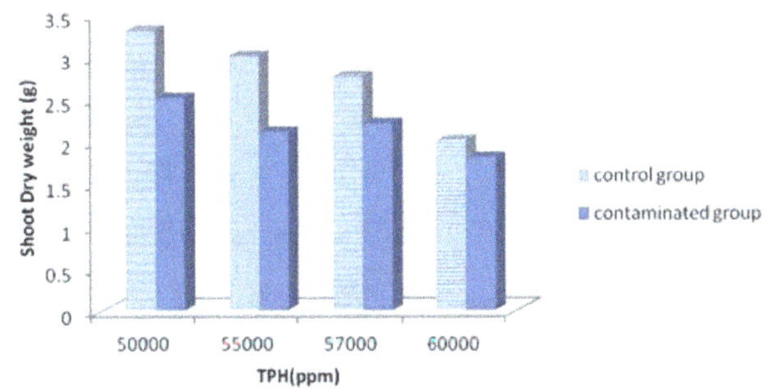

Figure 2. Shoot dry weight of *Aeluropus littoralis* in the contaminated and control treatments.

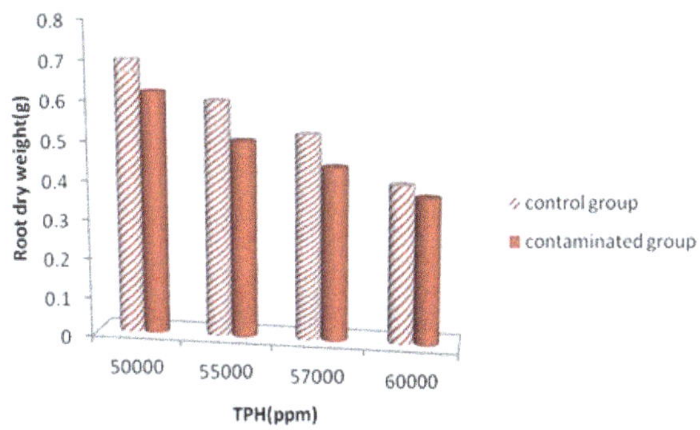

Figure 3. Root dry weight of *Aeluropus littoralis* in the contaminated and control treatments.

Table 3. Comparison of mean values of percentage reductions in petroleum hydrocarbons and number of oil-degrading bacteria

	Number of oil-degrading bacteria	Percentage reductions in petroleum hydrocarbons
A. littoralis	600 000	55
Unplanted	460 000	31

Discussion

The results of soil property analyses are consistent with those of other studies and suggest that organic acid production rooting from microorganisms' activity in soil coupled with the presence of sulfur and sulfur oxides in waste oil are responsible for pH reduction (16,21). On the question of electrical conductivity, this study found that soils rich in oil showed higher EC. A possible explanation for this might be the presence of metal cations (e.g., nickel and vanadium) in waste oil. The findings of the current study are consistent with those of other researchers who found elevated electrical conductivity

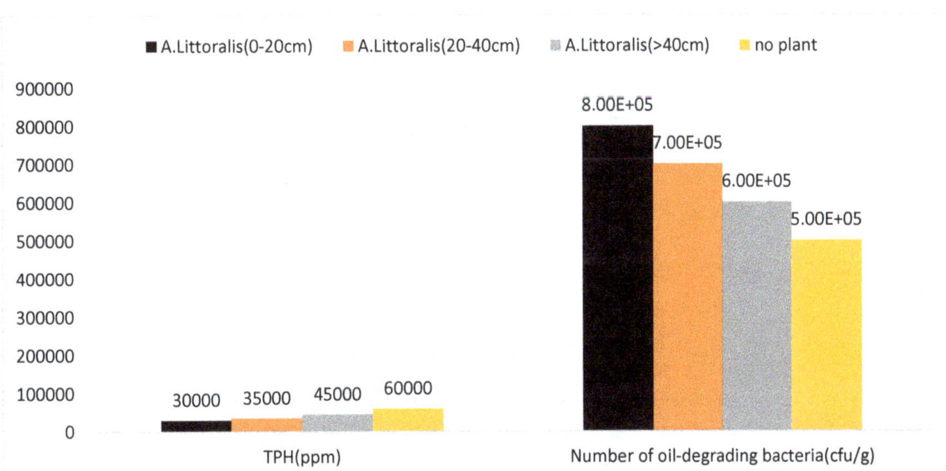

Figure 4. Changes in the concentration of total petroleum hydrocarbons and number of oil-degrading bacteria at different depths of soil.

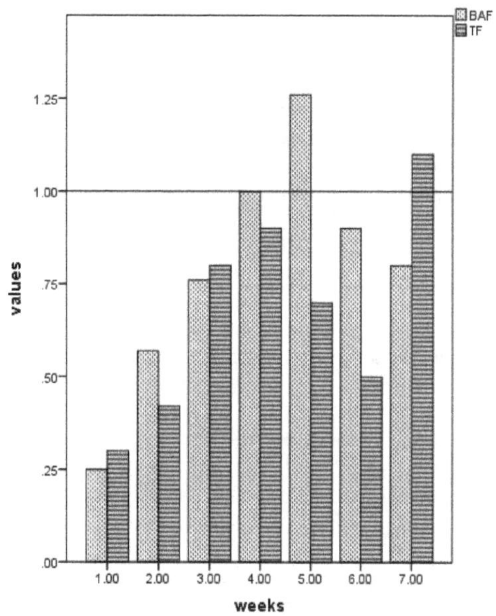

Figure 5. Root dry weight of *Aeluropus littoralis* in the contaminated and control treatments.

in soils contaminated with petroleum hydrocarbons (22). Another finding was that crude oil degradation in soil increased the organic matter percentage and organic carbon content of the contaminated soil compared to the control soil. These conditions, in turn, led to a decrease in the oxygen content of the contaminated soil and the formation of anaerobic environments which significantly restrict biodegradation. As shown in Figures 2 and 3, the dry matter weights of shoots and roots decreased as compared to control groups (23,24). The effects of depth on petroleum hydrocarbon concentrations and number of oil-degrading bacteria in treatments at three different depths were evaluated. Clearly, the lowest concentration of petroleum hydrocarbons and greatest number of oil-degrading bacteria were observed at 0-25 cm depth. Phytoremediation is responsive to a variety of organic and inorganic compounds. In fact, plants are capable of enhancing the degradation of organic pollutants by releasing nutrients and secretions in soil, transporting oxygen to the root zone, and stimulating and increasing the activity of oil-degrading microbial populations (25). It is encouraging to compare this figure with that found by Xu et al who explained that the presence of plant roots at this depth explains this finding (26,27). At 20-40 cm depths, the planted and control treatments were still different, since plant roots can reach this depth. However, higher concentrations of petroleum hydrocarbons and lower numbers of oil-degrading bacteria were detected compared to 0-20 cm depths.

The findings regarding reductions in dry matter yields support previous research which links the reduction of root and shoot dry matter yield to higher concentrations of PCB (23,24). The BAFs which exhibited concentrations of PAH in the shoots of *A. littoralis* divided by their values in soil ranged from 0.25 to 1.25, indicating the accumulation behavior of this plant. The minimum values for accumulation in root samples were observed in five of the seven weeks. BAF values higher than 1 indicate that the plant is highly efficient in removing contaminants from soil. The main tasks of the root include fixing in soil and absorbing water and dissolved compounds (25).

The root contributes to the metabolism of *A. littoralis*, feeding the whole plant, and often serves to store materials. The ability of roots to accumulate PAH is considered to be a way in which the aerial parts of the plant are protected. In the current experiment, BAF of PAH ranged between 0.25 and 1.3, suggesting a moderate performance by *A. littoralis* in the phytoremediation of PAH-contaminated soil. Translocation factors (TFs), presented in Figure 5, ranged from 0.3 to 1.1. As shown, in most cases, the PAH is uniformly distributed in root parts of *A. littoralis,* because the value of TF is less than 1. This indicates that *A. littoralis* was unable to translocate PAH from the roots to the shoots. Baker reported that plants are classified as accumulators if the heavy metal concentration ratio (shoot to root) is more than one and as excluders if this ratio is less than one (28). Substances that have a high solubility in water are not easily absorbed by plant roots (29). Due to their low solubility in water, hydrophobic chemicals are not easily absorbed in plant roots or remain around the roots. As a result they cannot be translocated into the plant (30).

Conclusion

The current study used *A. littoralis* to remediate petroleum hydrocarbon-contaminated soil from lands near the Isfahan Oil Refinery. The results showed that *A. littoralis* is capable of tolerating high concentrations of petroleum hydrocarbons and toxic materials in soil. *A. littoralis* successfully decreased the concentration of petroleum hydrocarbons in soil. Reductions in contaminants of unplanted soil might be attributed to leaching, adsorption, oxidation in exposure to light, evaporation, or biodegradation. The mechanisms by which plants escalate pollutant removal are different, comprising uptake and concentration, transformation, stabilization and rhizosphere degradation in which plants advance the growth of bacteria in the root zone. At depths of 20-40 cm and more than 40 cm where *A. littoralis* roots cannot penetrate, TPH removal decreased. The results of this study indicate that, in all 3 treatments, the maximum concentrations of petroleum hydrocarbons and minimum numbers of oil-degrading bacteria were observed at depths of 60 cm and more. The number of oil-degrading bacteria in 0-40 cm depths (with root penetration) was actually twice that in 60 cm depths (without root penetration).

Acknowledgments

This study was part of an MS project conducted in the laboratory of the Tehran Medical Sciences Branch of the Islamic Azad University. The authors extend their thanks to all the people who provided help, especially the laboratory staff.

Competing interests

The authors declare that they have no conflicts of interest.

Authors' contributions

Concept and design of experiments: AF. Performance of experiments: AF. Analysis of data: MJR. Manuscript preparation: MR and MJR.

References

1. U.S. Environmental Protection Agency (U.S. EPA) Hazard Characterization Document. Screening-Level Hazard Characterization. Washington, DC: EPA; 2011.
2. Kalf DF, Crommentuijn T, van de Plassche EJ. Environmental quality objectives for 10 polycyclic aromatic hydrocarbons (PAHs). Ecotoxicol Environ Saf 1997; 36(1): 89-97. doi: 10.1006/eesa.1996.1495.
3. United States Environmental Protection Agency (EPA). Oil Spill Response Techniques. Available from: https://www.epa.gov/emergency-response/epas-response-techniques. Accessed Nov 17, 2016.
4. Pilon-Smits E. Phytoremediation. Annu Rev Plant Biol 2005; 56: 15-39. doi: 10.1146/annurev.arplant.56.032604.144214.
5. Reilley KA, Banks MK, Schwab AP. Dissipation of Polycyclic Aromatic Hydrocarbons in the Rhizosphere. J Environ Qual 1996; 25(2): 212-9. doi: 10.2134/jeq1996.00472425002500020002x.
6. Dowty RA. Wetland restoration: the bioremediation of small-scale oil spills in fresh marsh environments and adventitious rooting response in marsh vegetation [dissertation]. Hammond, LA: Masters thesis, Southeastern Louisiana University; 1998.
7. Fayiga AQ, Ma LQ. Using phosphate rock to immobilize metals in soils and increase arsenic uptake in *Pteris vittata.* Sci Total Environ 2006; 359(1-3): 17-25. doi: 10.1016/j.scitotenv.2005.06.001.
8. Rovira AD. Plant Root Exudates. Botanical Review 1969; 35(1): 35-57. doi: 10.1007/BF02859887.
9. Günther T, Dornberger U, Fritsche W. Effects of ryegrass on biodegradation of hydrocarbons in soil. Chemosphere 1996; 33(2): 203-15. doi: 10.1016/0045-6535(96)00164-6.
10. Kathi S, Khan AB. Phytoremediation approaches to PAH contaminated soil. Indian J Sci Technol 2011; 4(1): 56-63.
11. Diab EA. Phytoremediation of oil contaminated desert soil using the rhizosphere effects of some plants. Res J Agric Biol Sci 2008; 4(6): 604-610.
12. Njoku KL, Akinola MO, Oboh BO. Phytoremediation of crude oil contaminated soil: the effect of growth of glycine max on the physicochemistryand crude oil contents of soil. Nature and Science 2009; 7(10): 79-87.
13. Shahin Kaleybar B, Nematzadeh G, Askari H, Hashemi HR, Kabirnataj S. Physiological and genetic responses of halophyte *Aeluropus* to salinity. Journal of Crop Breeding 2014;5(12):15-28.
14. Rezvani M, Zaefarian F, Miransari M, Nematzadeh GA. Uptake and translocation of cadmium and nutrients by *Aeluropus littoralis.* Arch Agron Soil Sci 2012; 58(12):1413-25. doi: 10.1080/03650340.2011.591385.
15. Assari A, Mahesh TM. Demographic comparative in heritage texture of Isfahan city. Journal of Geography and Regional Planning 2011; 4(8): 463-70.
16. Tang J, Lu X, Sun Q, Zhu W. Aging effect of petroleum hydrocarbons in soil under different attenuation conditions. Agric Ecosyst Environ 2012; 149: 109-117. doi: 10.1016/j.agee.2011.12.020.
17. US Environmental Protection Agency (US EPA). EPA Quality Manual For Environmental Programs. Washington, DC: EPA; 1994.
18. Wuana RA, Okieimen FE, Imborvungu JA. Removal of heavy metals from contaminated soil using chelating organic acids. Int J Environ Sci Technol (Tehran) 2010; **7**(3): 485-96. doi: 10.1007/BF03326158.
19. Sleegers F. Phytoremediation as green infrastructure and a landscape of experiences. Proceedings of the Annual International Conference on Soils, Sediments, Water and Energy. Available from: http://scholarworks.umass.edu/soilsproceedings/vol15/iss1/13.
20. Scragg A. Environmental Biotechnology. New York: Oxford University Press; 2005.
21. Spinelli LF, Schnaid F, Selbach PA, Bento FM, Oliveira JR. Enhancing bioremediation of diesel oil and gasoline in soil amended with an agroindustry sludge. J Air Waste Manag Assoc 2005; 55(4): 421-9. doi: 10.1080/10473289.2005.10464639.
22. Haines JR, Wrenn BA, Holder EL, Strohmeier KL, Herrington RT, Venosa AD. Measurement of hydrocarbon-degrading microbial populations by a 96-well most-probable-number procedure. J Ind Microbiol 1999; 16(1): 36-41.
23. Wrenn BA, Venosa AD. Selective enumeration of aromatic and aliphatic hydrocarbon degrading bacteria by a most-probable-number procedure. Can J Microbiol 1996; 42(3): 252-8.
24. Passatore L, Rossetti S, Juwarkar AA, Massacci A. Phytoremediation and bioremediation of polychlorinated biphenyls (PCBs): state of knowledge and research perspectives. J Hazard Mater 2014; 278: 189-202. doi: 10.1016/j.jhazmat.2014.05.051.
25. TRIPOD. Bioremediation. Available from: http://bioremediationstudy.tripod.com/id8.html. Accessed Nov 25, 2016.
26. Xu L, Teng Y, Li Z, Norton JM, Luo YM. Enhanced removal of polychlorinated biphenyls from alfalfa rhizosphere soil in a field study: the impact of a rhizobial inoculum. Sci Total Environ 2010; 408(5): 1007-13. doi: 10.1016/j.scitotenv.2009.11.031.
27. Zuberer DA. Recovery and enumeration of viable bacteria. In: Weaver RW, ed. Methods of Soil Analysis, Part 2. Microbiological and Biochemical Properties. Madison, USA: Soil Science Society of America; 1994.
28. Baker AJM. Accumulators and excluders strategies in the response of plants to heavy metals. J Plant Nutr 1981; 3(1-4): 643-54. doi: 10.1080/01904168109362867.
29. Schnoor JL, Licht LA, McCutcheon SC, Wolfe NL, Carreira LH. Phytoremediation of organic and nutrient contaminants. Environ Sci Technol 1995; 29 (7): 318-23. doi: 10.1021/es00007a747.
30. Schwab AP, Banks MK, Arunachalam M. Biodegradation of polycyclic aromatic hydrocarbons in rhizosphere soil. In: Hinchee RE, Anderson DB, Hoeppel RE, eds. Bioremediation of Recalcitrant Organics. Columbus: Battelle Press; 1995.

Evaluation of physical and chemical quality of groundwater in the villages of Qom province and zoning them by GIS

Ahmadreza Yari[1], Hossein Jafari Mansoorian[2], Gharib Majidi[3], Ghazal Yazdanpanah[4], Ali Sayfouri[3*]

[1]Assistant Professor, Research Center for Environmental Pollutants, Qom University of Medical Sciences, Qom, Iran
[2]Lecturer, Environmental Health Engineering Research Center, Department of Environmental Health, Kerman University of Medical Sciences, Kerman, Iran
[3]MSc in Environmental Health Engineering, School of Public Health, Qom University of Medical Sciences, Qom, Iran
[4]MSc of Microbiology, Environmental Health Engineering Research Center, Kerman University of Medical Sciences, Kerman, Iran

Abstract
Background: The aim of this study was to evaluate the physical and chemical quality of groundwater in the villages of Qom province during summer 2014, and to compare the results with Iran's Standard No. 1053.
Methods: This research is a descriptive sectional study. During the investigation, one step sampling with the necessary repetitions (if necessary) was performed of each well for evaluation of physical and chemical quality of groundwater in the villages of Qom province. SPSS version 16 and ARCGIS software were employed for data analysis and data zoning respectively.
Results: The average concentration of magnesium, sulfate and ammonia in wells water were higher than standard. The average concentration of fluoride was lower than the standard limit. pH and turbidity in all wells were within the standard limit but sulfate in 33.33%, ammonia in 13.33%, magnesium in 10.66%, sodium in 40%, calcium in 5.33%, nitrate in 17.33%, total dissolved solids (TDS) in 14.66% and electrical conductivity (EC) in 24% of wells were higher than the standard limit, and fluorine in 98.66% of wells was lower than the standard limit. The results of zoning in the villages of Qom province revealed that measured values at most wells were in the maximum amount in central and northeastern areas and minimum amount in southern and western areas.
Conclusion: This study indicated that most of the physicochemical parameters such as fluoride in almost all villages, and sulfates, chloride, magnesium, sodium, calcium, nitrate, TDS and EC in some of the wells did not fall within the permissible limit.
Keywords: Physical and chemical quality, groundwater, Qom, zoning, GIS

***Correspondence to:**
Ali Sayfouri
Email: lisaifouri@gmail.com

Introduction

Approximately, 65% of an organism's weight comprises water, and certainly it is a vital compound for all organisms on earth. Access to safe drinking water sources in many countries around the world is an important issue (1). According to the World Health Organization (WHO), drinking water is water that is suitable for human consumption and all home applications, and should be available in sufficient quantity and good quality for society. According to the viewpoint of the agency, the water utilized for human consumption must be free from microorganisms and chemicals that are dangerous to human health. The most important public preference attributes of public drinking water resources are that they have no turbidity and color, or unpleasant smell or taste. The absence of turbidity, color, odor or unpleasant taste of drinking water is the most important characteristics of drinking water public resources (2). Water does not exist in pure form in nature; it always contains levels of salts, suspended solids and dissolved gases. This causes the water in different regions to have different characteristics. For example, some salts in water are essential to human health, while levels of salts higher than the established standard limit would be dangerous for human health (3). The physical and chemical qualities of water in viewpoint of its acceptability for the consumer, protecting the health of consumers and for the maintenance of water network systems is necessary (4). Among the most important physical and chemical properties of drinking water are parameters like pH, electrical conductivity (EC), total dissolved solids (TDS), turbidity, alkalinity, hardness, chloride, calcium, magnesium, iron, sulfate, fluorine and phosphate (1). pH indicates the acidity or alkalinity property of water; so it depends on the consideration of H^+ or OH^- levels. pH is important in water treatment, disinfection and corrosion control. Corrosion control is dependent on the pH of water (5). Solids are one of the physical impurities of water. This parameter, involving protection from microorganisms via direct contact with disinfectants such as chlorine, can have an essential role in water pollution and may also cause an increase in the amount of disinfectant required to counteract the chemical action of the solids. Suspended materials can create tendencies for the adsorption of chemical and biological agents (4). High TDS in water can cause a salty taste and reduce the desire of consumers to utilize these waters. Drinking water with a TDS of less than 500 mg/L is considered very good water (1). One of the chemical characteristics of water is hardness (4). Hardness is caused by polyvalent metals cations. Sodium, potassium, calcium and magnesium are the most abundant type of polyvalent metal cations found in natural waters. Calcium and magnesium cause sedimentation of soap which decreases their cleaning performance. This reaction leads to the formation of layers of $CaCO_3$ and $MgOH_2$ in mainline water distribution channels and in hot water heaters. Water 'hardness' can be divided into 3 categories: slightly hard water, containing less than 50 mg/L TDS, moderately hard water, containing up to 150 mg/L TDS and very hard water, containing more than 300 mg/L TDS (6). Calcium and magnesium in excess amounts can have negative effects such as increased consumption of soap, creating stains on dishes, bleaching vegetables, and bursting boilers (4). Higher levels of sodium, potassium and magnesium compounds create taste in drinking water. Waters with large amounts of chloride are salty and waters containing large amounts of sodium sulfate and magnesium sulfate have a bitter taste. Sulfate in drinking water can substantially change the taste (7). Drinking water containing high concentrations of sulfate (200 ppm+), or ammonia levels of 100 ppm+, or calcium carbonate of over 300 ppm will cause dyspepsia in most people (8). High fluoride concentrations in drinking water can cause tooth enamel fluorosis, and low concentrations can cause tooth decay at an early age (2). Standards fluoride concentration in drinking water which provided by different organizations are different. These values in different standards are dependent on seasonal changes. The reason for this issue is a direct relationship between air temperature and water consumption. The amount of drinking water consume in warm months is higher than on cold months. Therefore, with the assumption of constant fluoride ion concentration in the water, the amount of fluoride is more in warmer months than other months (8). The fluoride concentration in drinking water should not be less than 0.5 mg/L (2). Nitrogen is a complex element. Different forms of nitrogen are found in natural waters that contain organic nitrogen, ammonia, nitrogen gas, nitrite and nitrate (9). In areas where rainfall is low and has little vegetation, groundwater pollution with nitrate is more than other areas (10). According to the Iran national standard (Standard No. 1053), the standard limits of nitrate and nitrite in drinking water are 50 mg/L in terms of NO_3^- and 3 mg/L in terms of NO_2^- respectively (11). The ill health effects of nitrate and nitrite are related to the occurrence of methemoglubinemia disease or blue baby syndrome, the production of carcinogenic nitrosamine compounds, and an increased risk of miscarriage (12). At present, many studies have been carried out on the physical and chemical quality of water in rural areas. In a study carried out in the city of Zahedan in 2008, it was found that the concentration of fluoride in the water of 6.48% of the villages were in optimal range, 87.3% were higher than optimal range, and 6.48% were lower than the optimal range (13). In Pourmoghadas' study, it was found that all groundwater of Lenjan city of Isfahan were very hard water (14). Dindarlo et al in assessing the quality of drinking water of Bandar Abbas concluded that the amount of sulfate, ammonia, sodium, total hardness, EC and TDS in groundwater sources are at the maximum standard limits (15). Sadeghi et al, in evaluating the physico-chemical indicators of drinking water in the city of Ardabil reported that sulfate in 9% of total samples, total hardness in 41% and phosphate in 71% of samples were higher than the standard limit, and fluorine is less than the minimum allowed in 57% of the samples (16). The results of investigating the chemical quality of drinking water in rural areas of the Boyer Ahmad's central part in 2009 revealed that the average concentration of chemical parameters (except fluoride and chlorine residual) are in the range of Iran's national and international standards (3).

The main purpose of reviewing water quality is to ensure the protection of public health. The aim of this study was to evaluate the physical and chemical quality of groundwater in the villages of Qom province during summer 2014 and the results were compared with the Iran's standard 1053 issue.

Methods

This research is a descriptive cross-sectional study carried out for the evaluation of physical and chemical quality of groundwater in the villages of Qom province (75 wells in 6 regions). Since there are no special guidelines for the evaluation of physico-chemical quality of water and due to the chemical water quality parameters not having significant changes in the short term; during the investigation, one step sampling with necessary repetitions (if necessary) was carried out on each well to evaluate the physical and chemical quality of groundwater in the villages of Qom province. A total number of 75 wells were evaluated. The areas studied were numbered from 1 to 6 and the studied wells were numbered from 101 to 606. The study areas included regions 1 (21 wells), 2 (23 wells), 3 (5 wells), 4 (6 wells), 5 (16 wells) and 6 (4 wells). The locations of studied wells are shown in Figure 1. Sampling was carried out in 3.5 L plastic containers. Before each sampling, bottles were washed with the well water. After recording characteristics (sampling places and sampling time), the samples were immediately sent to the laboratory to measure various parameters. Chemical and physical parameter measurements were carried out according to standard testing methods in Standard Methods for the Examination of

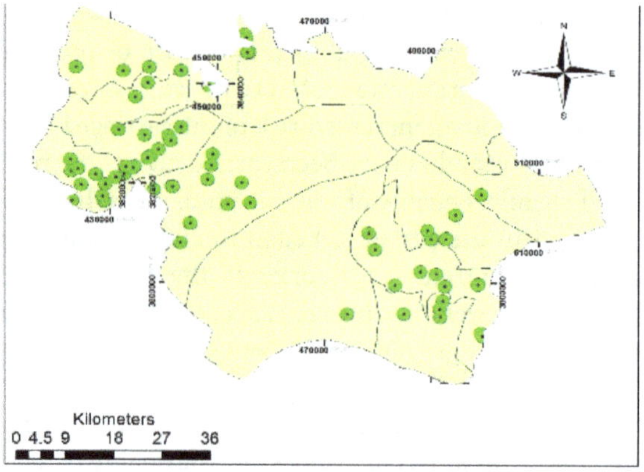

Figure 1. The location of sampled wells.

Water and Wastewater (17). SPSS version 16 and ARCGIS software were utilized for data analysis and data zoning respectively. In descriptive statistics, mean and standard deviation were calculated. Generally, experiments were carried out in two groups of test devices and titrimetric. Testing methods are listed in Table 1.

Results

According to Tables 2 and 3, the quality evaluation result of chemical and physico-chemical parameters of the quality of groundwater in the villages of Qom province at summer 2014 was determined and compared with standard 1053. The Standard No. 1053 is an instruction provided by industrial research and the Standard Institute of Iran for evaluating the quality of drinking water. The number and percentage of drinking water wells with concentration beyond Standard No. 1053 are shown in Table 4. Based on this study, the average concentration of magnesium, sulfate and ammonia in wells water were higher than Iran's Standard No. 1053. In addition, the average concentration of fluorine was lower than the standard limit. Sulfate in 33.33% of wells, ammonia in 13.33% of wells, magnesium in 10.66% of wells, sodium in 40% of wells, calcium in 5.33% of wells, nitrate in 17.33% of wells, TDS in 14.66% of wells and EC in 24% of wells were higher than the standard limit, and fluorine in 98.66% of wells was lower than the standard limit.

Figures 2 to 4 present geographical distributions of fluoride, total hardness, magnesium, calcium, electrical conductivity, total solid solution, chloride, potassium, nitrite, sulfate and total alkalinity in water tables in the villages of Qom province.

Villages

Summarized results of zoning parameters are shown in Table 5. Chemical and physical parameters of zoning results in well water in the villages of Qom province showed that the wells' content levels measured in the central and northeast regions had the higher content and the south and east had lower content.

Table 1. Testing methods

Parameter	Test method	Model	Parameter	Test method	Model
Chloride	Argenometric Method	-	Calcium	EDTA Titration Method	-
Fluoride	Spectrophotometric method	USA– Hack-DR4000	Magnesium	MgB. Calculation Method	-
pH	Electreometric Methods	USA– Hack- HQ	Sodium	Flame Emission Photometric Method	JENWAY,PFP7/C
Total solids	Total Dissolved Solids Dried at 180°C	-	Potassium	Flame Emission Photometric Method	JENWAY,PFP7/C
Electrical conductivity	Laboratory Method	Hach 50161/co150	Bicarbonate	Titration Method	-
Total hardness	EDTA Titration Method	-	Sulfate	Turbidomethric Method	HACH,2100QIS01
Total alkalinity	Titration Method	-	Turbidity	Nephlometric Method	USA– Hack-P2100
Nitrite	Spectrophotometric method	USA– Hack-DR4000	Nitrate	Spectrophotometric method	USA– Hack-DR4000

Table 2. Results of chemical examination parameters of drinking water in the villages of Qom province in summer 2014

Parameter (mg/L)	Ca	Mg	Na	K	CO_3^{2-}	HCO_3^-	NO_3^-	NO_2^-	SO_4^{2-}	Cl^-	F^-
Mean	120.72	24.06	199	1.2	0	259.19	30.28	0.012	317	187	0.55
Maximum	646.40	221.49	1920	11.5	0	528	91.52	0.03	1500	2872	1.84
Minimum	32	1.97	2	0.2	0	28.80	4.84	0.006	8	12	0.03
Standard deviation	±91.31	±28.64	±267	±1.51	0	±89.46	±19.21	±0.005	±322	±366	±0.32
Standard	250	50	200	-	-	-	50	3	400	400	0.8-1.8

Table 3. Result of chemical and physico-chemical parameters quality of groundwater in the villages of Qom province at summer 2014

Parameter (mg/L)	Turbidity (NTU)	pH	TDS (mg/L)	EC (Mmho/cm)	Total Hardness (mg/L as $CaCO_3$)	Total alkalinity (mg/Las $CaCO_3$)
Mean	0.42	7.6	1040	1598	401	213
Maximum	3.5	8.38	6971	13500	1836	440
Minimum	0.12	6.39	123	210	88	24
Standard deviation	±0.36	±0.32	±1011	±1722	±303	±73.5
Standard	5	6.5-9	1500	2000	500	-

Table 4. Number and percentage of wells' drinking water with concentration higher than the standard limit in the villages of Qom province in 2014

Parameter	SO_4^{2-} (mg/L)	Cl^- (mg/L)	F^-(mg/L)		NO_3^- (mg/L)	NO_2^- (mg/L)	Mg (mg/L)	Na (mg/L)	Ca (mg/L)	TH (mg/L)	EC (Mmoh/cm)	TDS (mg/L)	pH	Turbidity (NTU)
			Lower than the permissible limit	Higher than permissible limit										
Standard	400	400	0.8	1.8	50	3	50	200	250	500	2000	1500	6.5-9	5
No. of well	25	10	74	1	13	0	8	30	4	23	18	11	0	0
%	33.33	13.33	98.66	1.33	17.33	0	10.66	40	5.33	30.66	24	14.66	0	0

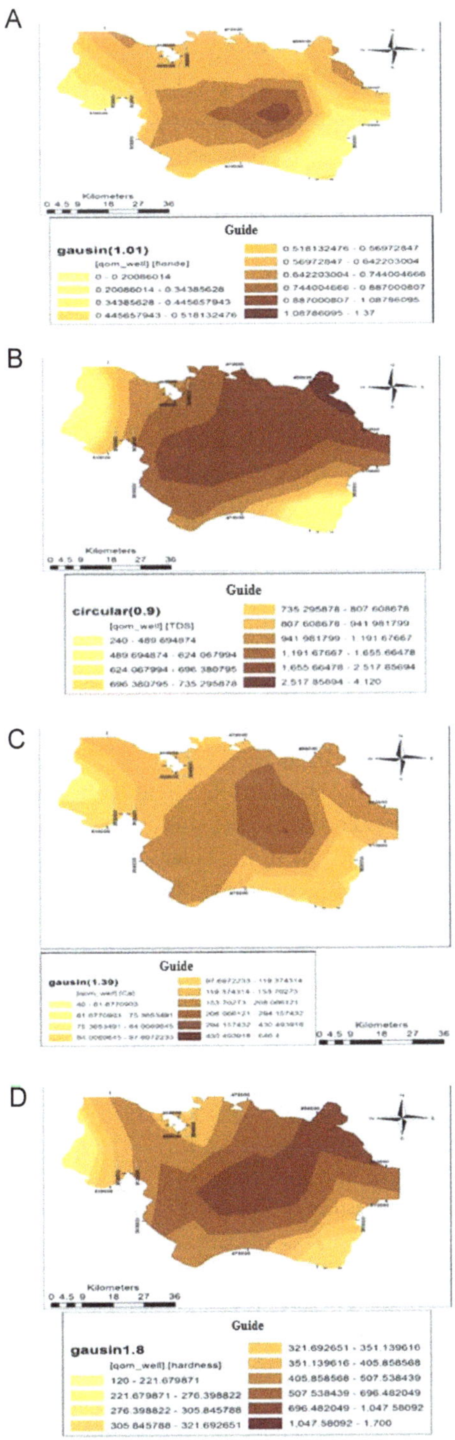

Figure 2. Geographical distribution of drinking water resources in the villages of Qom province: (A) Fluoride, (B) Water hardness, (C) dissolved solids, (D) Calcium.

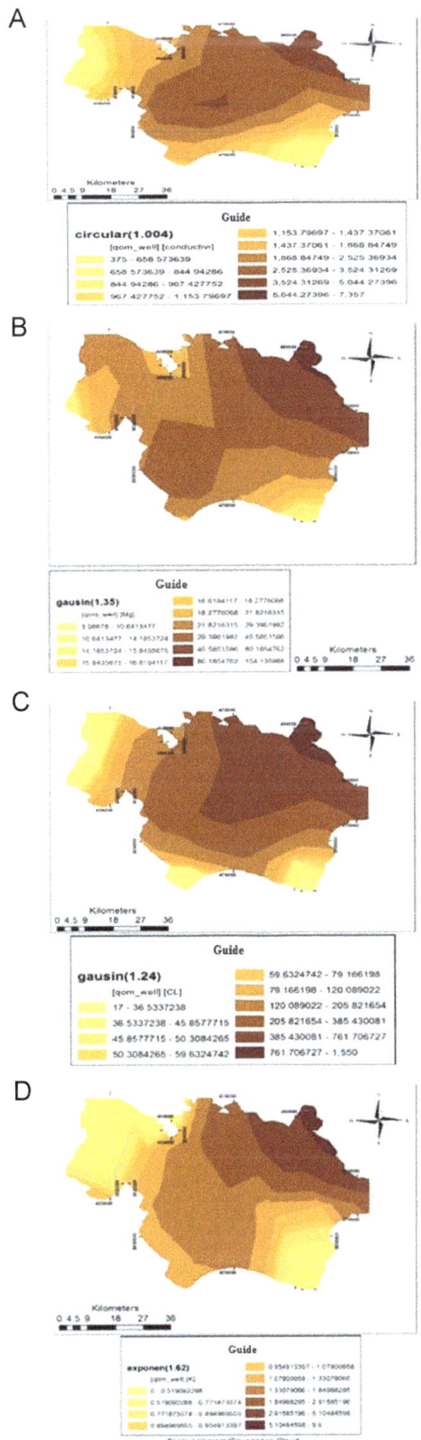

Figure 3. Geographical distribution of drinking water resources in the villages of Qom Province: (A EC, (B) Magnesium, (C) Chloride, (D) Potassium

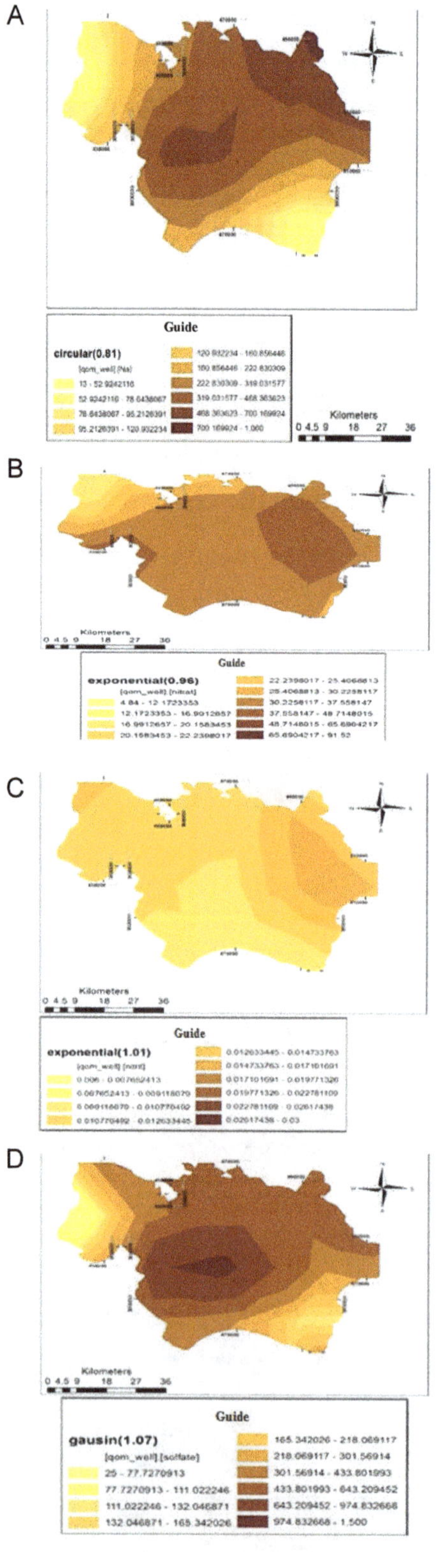

Figure 4. Geographical distribution of drinking water resources in the villages of Qom Province: (A) Sodium, (B) Nitrate, (C) Nitrite, (D) Total alkalinity

Discussion

In this study, minimum, average and maximum turbidity was 0.12, 0.42 and 3.5 NTU respectively. The authorized range of NTU by Iranian national standard is 5 NTU. In a study carried out in 2012 by Rezai et al, corrosion and scaling potential in drinking water resources in the villages of Qom province were investigated. Average turbidity in Rezai et al study was 0.78 NTU (18). In a study carried out by Fahiminia et al in 2012, the average turbidity in the water wells of the Qom city were 0.58 NTU (19). The results of this study are in line with the studies of Fahiminia et al and Rezai et al.

In this study, maximum and minimum pH was 8.38 and 6.39, respectively. Average pH was 7.6. The accepted range of pH by Iranian national standard is 7-8.5 and the maximum is 6.5-9. pH of samples when compared with national standards were in the authorized range (20). The study of Fahiminia et al demonstrated that the average pH values in Qom city groundwater is 7.4 (19). The average pH values in the study of Rezai et al was 7.27 (18). The results of this study are in line with the studies of Fahiminia et al and Rezaei et al.

Minimum and maximum chloride concentration was 12 and 2872 mg/L respectively, in groundwater resources in the villages of Qom province. Average chloride concentration was 187 mg/L. Chloride concentrations higher than 250 mg/L can make a distinguishable effect on water taste (18). Based on national standards, the maximum chloride concentration in drinking water should not be more than 400 mg/L (11). Chloride levels in 13.33% of wells sampled were higher than the national standard. The average amount of chloride in the study of Rezaei et al was 230.83 (18), which is higher than those obtained in the present study.

In this study, the variations sulfate content ranged between 8 to 1500 mg/L. In studied wells, the average sulfate was 317 mg/L. Sulfate concentrations between 300-400 mg/L can create a distinguishable taste in water. Diarrhea is often the result of high sulfate concentrations. Sulfate can also cause water hardness and foam in steam boilers (21). The Iranian national standard for sulfate concentration is 400 mg/L (11). According to the study of Rezaei et al, sulfate levels in water wells in the villages of Qom province were obtained as 309 mg/L (18) which is in line with the current study. In the present study, the amount of sulfate in 33.33% of wells was higher than the national standard. The study of Miranzadeh et al showed that the sulfate content in 5.3% of drinking water samples in the villages of Kashan city were higher than the national standard (2). In a study carried out by Rajaei et al, amounts of sulfate in drinking water samples of Birjand and Qaen plain at 33% were higher than national standards (8). The results of this study were not in line with those of Miranzadeh et al but are similar to the study of Rezaei et al (18).

EC, or EC of water, represents the ionic strength of a solution to conduct an electric current. Its unit is microsiemens per centimeter (µs/cm) or micromhos per centimeter (µmho/cm). Since ions carry electric current, the EC is directly related to TDS. Therefore, EC increases as TDS does and consequently, the corrosive effects of TDS in affected water increases. Water EC is 56 µmho/cm (9). In fact, TDS is total dissolved solids in the water; that is, the sum of the concentrations of all ions present in the water. TDS is an important parameter that could cause some

Table 5. Chemical and physical zoning parameter results

Parameter	Region with the most content	Region with the fewest content
F^-	Central	East and West
Total Hardness	Central and East	West and South
TDS	Central and East north	West and South
Ca	Central point	West and South
EC	Central and East North	West and South
Mg, Cl^-, K, Na	Central and East North	West and South
NO_3^-	East	West
NO_2^-	East	Central and West
SO_4^{2-}	Central point	West and South
Total Alkalinity	South and South West	East

adverse effects of water quality and can have a direct or indirect important effect on other quality characteristics of water. We can calculate the corrosion of metals in water and the leakage of metals (18). In this study, TDS content and EC were 123 to 6971 mg/L and 210 to 13500 µmho/cm, and the mean was 1040 mg/L and 1598 µmho/cm, respectively. The national standard range for TDS is 1500 mg/L and for EC 2000 µmho/cm (11). Water from wells with TDS of more than 1500 mg/L are non-drinkable (1). If water EC is more than 1500 µmho/cm, it can have a corrosive effect on any iron structures within a water distribution network (1). According to this issue, the amount of TDS in 14.66% of wells and the amount of EC in 24% of wells was higher than the national standard. In the study of Rezai et al, mean values of TDS in drinking water wells of villages in Qom province equal 1205 mg/L (18) which is in line with the current study.

Results of groundwater analysis in villages of Qom province have shown that the fluoride range is between 0.03-1.84 mg/L and the fluoride's mean is 0.55 mg/L, which is lower than the national standard. Based on Iranian national standards, the best of fluoride concentration in warm months is 0.7 mg/L, and in cold months is 1.2 mg/L (11). Results of the study of Rezaei et al showed that the fluoride concentration in drinking water wells in the villages of Qom province is equal to 0.6 mg/L (18). The results of this study corresponded to those of Rezai et al. In the present study, fluoride content in 98.66% of wells was below the authorized range and in 1.33% of wells was higher than authorized range. The study of Rajaei et al showed that the fluoride content in 92% of drinking water samples in the Birjand and Qaen plain is lower than the permissible limit (8) which is in line with the current study. The important issue about the chemical quality of water in the villages of Qom province is the low fluoride concentration. Most general hygiene agencies suggest the addition of fluoride in water as an effective way to prevent caries in populations with low fluoride levels in water (2). The range of sodium and potassium's content in studied wells were 2-1920 and 0.2-11.5 mg/L, respectively. Average sodium and potassium in wells were 199 and 1.2 mg/L, respectively. The national standard range for sodium is 200 mg/L (11). In 40% of studied wells, sodium was above the national standard. In a study carried out by Miranzadeh et al, sodium and potassium in drinking water in Kashan city were equal to 110.4 and 2.5 mg/L respectively (2). The average of sodium and potassium in the study of Rezaei et al were 245 and 1.76 mg/L, respectively (18). The values obtained in the above studies are higher than those obtained in the present study.

Hardness is caused by polyvalent metals cations such as calcium and magnesium (6). In this study, the concentration variations for calcium and magnesium range from 32-646.40 and 1.97-221.49 and their mean were 120.72 and 24.06 mg/L respectively. Also, the total hardness variation ranges from 88-1836 mg/L while $CaCO_3$ average content was 401 mg/L. The national standard for calcium, magnesium and total hardness is 250, 50 and 500 mg/L respectively (11). Calcium, magnesium and total hardness were higher than national standard in 5.33%, 10.66% and 30.66% of wells, respectively. A high concentration of magnesium creates an unpleasant taste in water. Magnesium can also cause diarrhea, but the most harmful effect is on pipelines transporting water and thermal installation; therefore, it is expensive for thermal and water treatment facilities (1). Sadeghi and Rohollahi in assessing the physico-chemical indicators of drinking water in the city of Ardabil reported that 41% of the samples exceeded the limit allowed (16). In the study of Rezaei et al, mean values of calcium, magnesium and total hardness in drinking water wells in the villages of Qom province was 132, 25 and 435 mg, respectively (18), which corresponded to the current study.

Based on the national standard, the standard limit of nitrate and nitrite in the water are 50 and 3 mg/L, respectively (11). The average concentrations of nitrate and nitrite in wells in the villages of Qom province were 34.73 and 0.01 mg/L, respectively. Izanloo et al examined the trend in nitrate concentrations in drinking water resources in the villages of Qom province and its zoning by GIS software (22). The results showed that the average nitrate concentration in studied wells was 23.12 mg/L between the years 2006 to 2016. This study showed more concentration of nitrate than in the study of Izanloo et al but it is still in the authorized range. In another study carried out by Izanloo et al, average nitrate in Jiroft groundwater resources was 10.5 mg/L (23). The average amount of nitrate in the present study is the higher of the two studies above; however, it is within the standard range. The study shows that in 17.33% of well samples, nitrate levels were higher than the national standard. Generally, chemical and physical parameters of zoning results in water wells in the villages of Qom province showed that the wells' content levels measured in the central and northeast regions had the higher content and the south and east had lower content. The nitrate concentration distribution map shows that the villages in Qom province have different concentrations of nitrate. Nitrate and nitrite have the highest concentration in the eastern region. Nitrate has less concentration in the western region and nitrite has less concentration in the central and eastern regions. Zoning nitrate concentration in drinking water in villages of Qom province during 2006 to 2012 has been studied with GIS software by Izanloo et al (22). Nitrate has the highest concentration in the south-east. These results are in line with the study of Izanloo et al.

Conclusion

Chemical and physical parameters are important specifications in determining the quality of drinking water, which should be of concern in ensuring the health and development of society, and some rules should be enacted to achieve these purposes. Undesirable drinking water, besides having chemical, physical and microbial problems, also has unacceptable appearance and taste, and consequently the water distribution network will not be trusted by consumers, and they will use other sources of water instead. Results of this study showed the average concentration of fluoride in most villages was lower than the standard limit. The average concentration of magnesium, sulfate and ammonia in the wells water were higher than Iran's Standard No. 1053. Also, the average concentration of fluoride was lower than the standard limit. pH and turbidity in all wells were within the standard limit, but sulfate in 33.33% of wells, ammonia in 13.33% of wells, magnesium in 10.66% of wells, sodium in 40% of wells, calcium in 5.33% of wells, nitrate in 17.33% of wells, TDS in 14.66% of wells and EC in 24% of wells were higher than the standard limit, and fluorine in 98.66% of wells was lower than the standard limit.

The results of physical and chemical parameters zoning of waters in the villages of Qom province showed that measured values at most wells were in the maximum amount in the central and northeastern areas and minimum amount in the southern and western areas.

Acknowledgments

The authors would like to acknowledge the Qom University of Medical Sciences' and the School of Public Health's laboratory staffs for their support. Special thanks go to Alfred Smith Jr. for editing this article.

Competing interests

The authors declare that they have no competing interests.

Authors' contributions

Ahmadreza Yari and Ali Sayfouri designed the study. Gharib Majidi, Hossein Jafari Mansoorian and Sharam Nazari performed the literature search and wrote the manuscript. Ghazal Yazdanpanah provided significant English translation. All authors participated in data acquisition, analysis, and interpretation. All authors critically reviewed, refined, and approved the manuscript.

References

1. Shabankareh Fard E, Hayati R, Dobaradaran S. Evaluation of physical, chemical and microbial quality of distribution network drinking water in Bushehr, Iran. Iranian South Medical Journal 2015; 17(6): 1223-35. [In Persian].
2. Miranzadeh MB, Mesdaghinia AR, Heidari M, Younesian M, Nadafi K, Mahvi AH. Investigating the chemical quality and chlorination status of drinking water in Kashan's villages. Health System Research 2010; 6(2): 889-97. [In Persian].
3. Raygan Shirazi AR, Rezaei S, Jamshidi A, Fararoei M, Sadat A, Hashemi H. Investigating the microbial and chemical quality of drinking water. Health System Research 2012; 8(3): 431-7. [In Persian].
4. Keramati H, Mahvi AH, Abdulnezhad L. The survey of physical and chemical quality of Gonabad drinking water in spring and summer of 1386. Ofogh-e-Danesh Journal 2007; 13(3): 25-32. [In Persian].
5. Samaei MR, Ebrahimy A, Ehrampoosh MH, Talebi P, Khalili MH, Morovati R. A study of the physical and chemical quality of potable water in Yazd. Journal of Toloo-e-Behdasht 2007; 6(2): 50-7. [In Persian].
6. Hammer M. Water and Wastewater Technology. 5th ed. New York: John Wiley & Sons Inc; 2004. p. 52-3.
7. Lalehzari R, Tabatabaei SH. Groundwater Quality Mapping in Shahrekord Aquifer. Journal of Environmental Studies 2010; 36(53): 55-62.
8. Rajaei Q, Mehdinejad MH, Hesari Motlagh S. A survey of chemical quality of rural drinking water of Birjand and Qaen plains, Iran. Journal of Health System Research 2011; 7(6): 737-45. [In Persian].
9. Qasim SR, Motely EM, Zhu G. Water Works Engineering: Planning, Design and Operation. Upper Saddle River, NJ: Prentice Hall; 2000. p. 45-60.
10. Malakootian M, Yaghmaian K, Tahergorabi M. The efficiency of nitrate removal in drinking water using iron nano-particle: determination of optimum conditions. Journal of Toloo-e-Behdasht 2011; 10(2):35-44. [In Persian].
11. Institute of Standards and Industrial Research of Iran. Drinking water-physical and chemical specifications. 5th ed. Available from: http://markazsalamat.behdasht.gov.ir/uploads/1053_180364.pdf.
12. Nanbakhsh H, Mohammadi A, Ebrahimi A. Investigating of nitrate and nitrite concentration of drinking water wells in villages around of the industrial park, in Urmia city. Journal of Health System Research 2011; 6: 881-8 .[In Persian].
13. Kamani H, Shah Bakhsh MH, Noormandi MH, Mirpour AA. A survey on flouride concentration in drinking water sources of Zahedan villages in 2008 in cooperation with national standards and meteorological conditions. Proceedings of the 12th National Congress on Environmental Health; 2009 November 12-14, 2009; Shahid Beheshti University of Medical Sciences, Tehran, Iran.
14. Pour Moghadas H. Astudy of ground water quality in lenjan township of Isfahan province. Journal of School of Public Health and Institute of Public Health Research 2003; 1(4): 31-40. [In Persian].
15. Dindarlo K, Alipour V, Farshidfar GR. Chemical quality of Bandar Abbas drinking water. Medical Journal Of Hormozgan University 2006; 10(1): 57-62. [In Persian].
16. Sadeghi H, Rohollahi S. Measure physical and chemical parameters of drinking water in Ardabil. Journal of Ardabil

University of Medical Sciences 2007; 7(1): 52-6. [In Persian].
17. APHA/AWWA/WEF. Standard Methods for the Examination of Water and Wastewater. 20th ed. Washington DC: American Water Work Association; 2005.
18. Rezaei Kalantari R, Yari AR, Azari A, Ahmadi E, Azari A, Tahmasbi Zade M, Gharagazlo F. Survey of corrosion and scaling potential in drinking water resources of the villages in Qom province by use of four stability indexes (with quantitative and qualitative analysis). Arch Hyg Sci 2013; 2(4): 127-34.
19. Fahiminia M, Jafari Mansoorian H, Ansari M, Saifour Mofrad A, Majidi G, Ansari Tadi R, et al. Evaluation of trends for iron and manganese concentrations in wells, reservoirs, and water distribution networks, Qom city, Iran. Environmental Health Engineering and Management Journal 2015; 2(2): 67-72.
20. Malakootian M, Ehrampoosh MH, Jafari Mansoorian H. Quality of drinking water consumed in interurban bus transportation system of Kerman in the first half of 2008. Journal of Toloo-e-Behdasht 2008; 7(1,2): 22-29.
21. Nabizadeh R, Naddafi K, Mohebbi M, Yonesian M, Mirsepasi A, Oktaie S et al. Evaluating the microbial content of the drinking water in rural areas of Tehran province. Journal of School of Public Health and Institute of Public Health Research 2008; 5(4): 63-73.
22. Izanloo H, Khezri S, Majidi G, Alesheikh A, Tashauoei H, Khazae M. A GIS survey of trends for nitrate concentration in drinking water sources, rural areas of Qom province, Iran. Journal of Sabzevar University of Medical Sciences 2015; 21(6): 1194-204. [In Persion].
23. Izanloo H, Majidi G, Nazari Sh, Maleki A, Khazae M, Tabatabaei Majd MS, et al. Survey of nitrate and nitrite concentration in Jiroft groundwater resources in 2009. Journal of Sabzevar University of Medical Sciences 2016; 22(6): 1035-43. [In Persion].

Optimization of nitrogen and phosphorus removal from meat processing wastewaters using microalgal biofilms

Soroosh Danaee, Hamideh Ofoghi*, Seyed Mohammad Heydarian, Abbas Farazmand

Biotechnology Department, Iranian Research Organization for Science and Technology, Tehran, Iran

Abstract

Background: The conventional activated sludge processes are not able to completely treat the nitrogen and phosphorus compounds of meat processing wastewaters. Microalgal biofilm, which has the advantages of easiness and inexpensiveness, can be applied as a post-treatment method.

Methods: The optimal levels of micronutrients and N/P ratio were aimed in the cultivation of *Scenedesmus quadricauda* as a biofilm formed on the two types of commercially available bed packing. For this purpose, the Factorial design was run to maximize the removal of ammonium, nitrate and phosphate concentration in the synthetic meat processing wastewater.

Results: Experimental data showed better results for the removal of biofilm formed on the sponge filter. In addition, a maximum removal efficiency of 55.6%, 47% and 48.% was respectively achieved for ammonium, nitrate and phosphate at optimal N/P ratio of 10.5 and volumetric percentage of 50% of micronutrients.

Conclusion: The results demonstrated that microalgal biofilm of *Scenedesmus* is able to remove the nutritional components of meat processing wastewaters. The process is needed to be optimized for the effective treatment.

Keywords: Meat, Biofilms, Microalgae, Wastewater, *Scenedesmus*

***Correspondence to:**
Hamideh Ofoghi
Email: ofoghi@irost.ir

Introduction

Common wastewater treatment systems, known as conventional activated sludge processes, are not sufficiently effective in complete removal of some nutrients such as nitrogen and phosphorus compounds (1-4). Additional cost of the post-treatments caused release of some enriched effluents without proper treatment which in turn led to the environmental problems (1,2,4). Biological treatment processes have been introduced as an alternative for their low cost and high efficiency. Nitrogen compounds can be converted through nitrification/ denitrification processes which have some disadvantages: additional demand of oxygen and carbon sources, N loss in gaseous form, toxic side products (NO/N_2O), etc (5). Phosphorus compounds can be accumulated by polyphosphate accumulating organisms (PAOs) which have the disadvantages of disposal of sludge residues and unstable removal efficiencies (5).

Cultivation of microalgae is rising up in the removal process of nutrient compounds with advantages of nutrient recovery, valuable biomass production, inexpensive oxygenation and heavy metals adsorption (5). However, most of the single cell microalgae are accompanied by the problems of separation in downstream processes in which 20%-30% of the total cost of the process accounts for the biomass separation. Algal immobilization techniques can be used as alternatives to reduce the process costs and separation problems (1-3).

Numerous studies have been conducted on the removal of nutrient compounds from wastewater using immobilized microalgae; in addition, the effects of various factors have been investigated in recent years (2,6-9). Optimization of the nitrogen to phosphorus (N/P) ratio in the wastewater seems to be of great importance to increase the efficiency of compounds removal. Guzzon et al used a photosynthetic biofilm, taken from a sedimentation tank, to remove phosphorus from the wastewater of Rome airport, and modified N/P ratio up to 15 by adding nitrate and phosphate to the wastewater (10). Boelee et al also assessed the removal capability of nutrient compounds from synthetic urban wastewater in post-treatment process using microalgal biofilms. Although micronutrients of WC medium were used in this study, but the effects of using these compounds in the microalgal treatment were not investigated (2). The effect of nutritional components on the growth of microalgae was investigated previously (11,12), but to our knowledge, no research has been conducted to investigate nutritional components as the variables in the wastewater treatment.

The present study was conducted for the first time with the aim of determining the optimal levels of two factors of the micronutrients and N/P ratio in the microalgal post-treatment of synthetic meat processing wastewater. For this purpose, two types of bed packing, which are commercially available, were used in this study. In addition, the effect of the beds was also compared in this study.

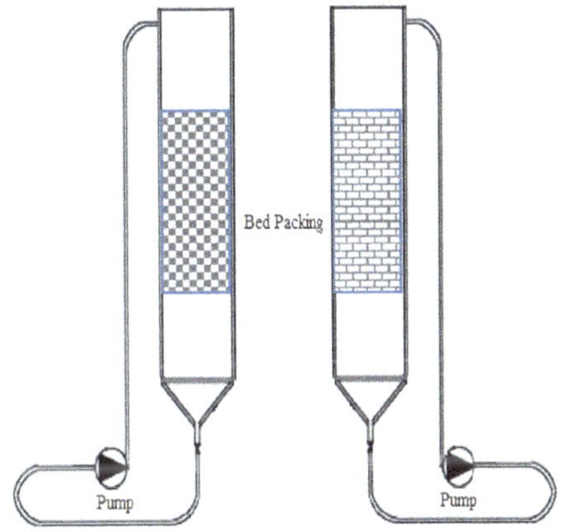

Figure 1. Diagram of photobioreactors used for algal attachment on the bed packings.

Methods

Microalgae, *Scenedesmus quadricauda*, was prepared (13) and cultured in Bold's Basal Medium (BBM) at temperature of 25°C, light of 3000 lux for approximately 10 days to prepare inoculum. It was then inoculated into two 6-L glass columns (with an internal radius of 9 cm and height of 100 cm, at 25°C, 3000 lux and 16:8 photoperiod) to reach an approximate density of 3×10^7 (3,14). Then, two types of bed packing (Yellow Sponge Filter and High Friction Grooved Rasching Ring Packing [HFGRRP]) were added to the middle section of each column so that the microalgal biofilm was formed on their specific surface area under a laminar flow (Figure 1) (2,3,14).

Bed packings characteristics are presented in Table 1.

The synthetic wastewater that was prepared equivalent to the major compounds of an industrial meat processing unit and used in the experiments, containing (analytical grade): magnesium sulfate (73.93 mg/L), sodium nitrate (21.934 mg/L), sodium carbonate (73.42 mg/L), ammonium chloride (68.3 mg/L), potassium dihydrogen phosphate (0.34 mg/L), dipotassium hydrogen phosphate (0.226 mg/L). Organic carbon was not used in the preparation of the synthetic wastewater in order to limit the bacterial growth (2).

Experiments were designed by the factorial design method with three factors: Micro-elements of BBM culture medium (13), N/P ratio, and changes in the type of bed packing (both with the same volume) with two central points. The levels of these factors are shown in Table 2.

The bed packings covered with microalgal biofilm were transferred to the flasks (3). Then, the performance of the biofilm of *Scenedesmus* was assessed in the synthetic meat processing wastewater at different levels of the factors to determine the maximum removal efficiency by Minitab software (version 14). During the removal process, the concentration of nutrient compounds (ammonium, nitrate and phosphate) was measured by standard water and wastewater analysis (4500-NH_3 C, 4500-NO_3^- B, 4500-P D, respectively) (15). Furthermore, parameters of pH, total dissolved solids (TDS), and electrical conductivity (EC) were measured in several days. Then, the mean of three replicates was used in the Factorial design.

Table 1. Bed packings characteristics

	Sponge Filter	HFGRRP
Material	Polyurethane	Polyethylene
Specific surface area (m^2/m^3)	800	899

Results

Variations of the three variables of ammonium, nitrate and phosphate concentrations during the growth of microalgal biofilm of *S. quadricauda* formed on the PB are presented in Figure 2. These results were obtained based on the combination of experimental design of the two factors of the N/P ratio and the volumetric percentage of nutrient compounds in the synthetic wastewater. As shown in Figure 2, ammonium was gradually declined during the removal process, however, nitrate was slightly decreased in the first few days. Phosphate removal also had a declining trend during the experiments (Figure 2).

The removal efficiency of nitrate and ammonium in PB varied between 20% and 55%, and its highest rate was measured at N/P ratio of 10.5 and BBM micronutrient

Table 2. Levels of factors in the Factorial design

Factor	High Level	Center Point	Low Level
BBM Nutritional components	100% (volumetric) of micronutrient solution	50% (volumetric) of micronutrient solution	-
N/P ratio	7	10.5	14
Bed	Sponge Filter (SF)	Packed Bed (PB)	
Total run	20		

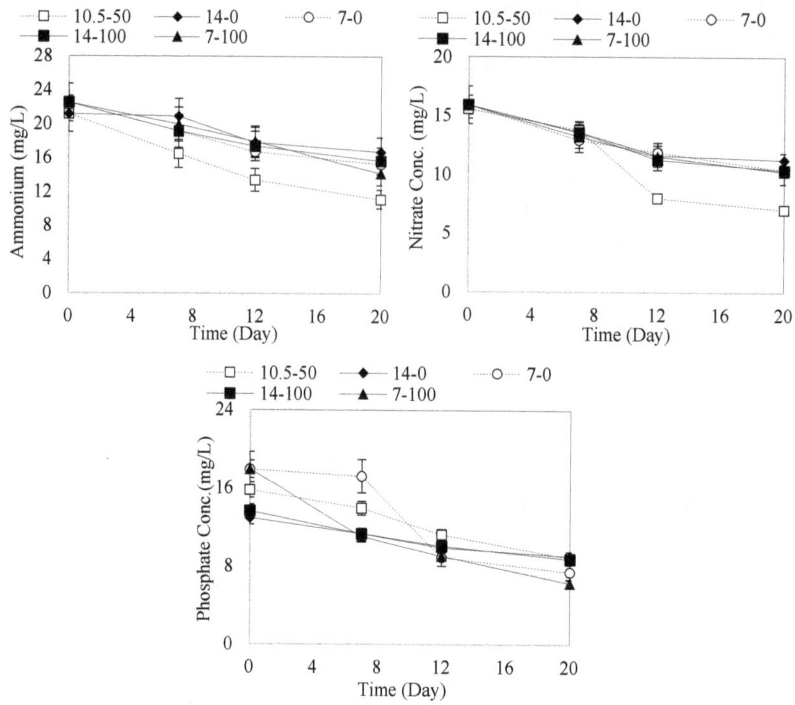

Figure 2. Ammonium, nitrate and phosphate concentrations of wastewater during the growth of *Scenedesmus quadricauda* on PB in different N/P ratios and volumes of the micronutrients.

solution in the volumetric percentage of 50%, which was obtained 55% and 47%, respectively. The maximum removal efficiency of phosphate was obtained at a lower N/P ratio, so that in the presence of BBM micronutrient solution and the N/P ratio of 7, about 62% of the initial phosphate was decreased after 20 days.

The effects of the variables on microalgal biofilm of *S. quadricauda* formed on the SF bed were investigated in a synthetic wastewater; the results of wastewater removal are illustrated in Figure 3. In comparison with the previous graph, it can be seen that the trends are similar, so that the highest removal efficiency of nitrate and ammonium obtained in the same condition, was 55.6% and 47%, respectively. At a lower N/P ratio, the phosphate removal process had a better condition, so that the maximum removal efficiency obtained in the presence of micronutrient solution and at N/P ratio of 7, was 66.6%, while at the center point of experimental design, phosphate removal efficiency was 48.4% (Figure 3).

Figure 4 shows the optimal points (presented in red color) and desirability (presented as "d") for the simultaneous removal of all three compounds. The values obtained for the removal efficiencies (presented in blue color as "y"), were adjusted in the software. The software reported the N/P ratio of 10.5, BBM micronutrient solution in the volumetric percentage of 50% and SF as the optimum condition (Figure 4).

Figure 5 shows the trends of EC and TDS in the treated wastewater by *Scenedesmus* grown on PB and SF. The experimental data showed a maximum reduction of 64% by PB and 77% by SF in TDS as a result of the treatment process (Figure 5).

Discussion

The biological removal trends of ammonium and nitrate were the same in both beds, so that the concentration of nitrate in the first 8 days had an approximate slight reduction, however, the microalgal removal rate then increased by more than 80%, while the removal rate of ammonium decreased during the first days. This can be due to the preferred form of nitrogen source in *S. quadricauda* (16). In a study carried out on this strain, unlike ammonium, variations of nitrate concentration were low in the first 7 days (17). In another study, no significant changes were observed in nitrate concentration during the first 5 days (18). In the present study, nitrate concentration had also little changes after 8 days.

In addition, the micronutrient solution had a positive effect on the biological treatment. This can be due to the higher production rate of chlorophyll in the presence of BBM micronutrient solution (12,19). In a study conducted on *S. quadricauda*, the maximum removal of nitrogen source was observed in the presence of a BBM micronutrient solution (20).

Variations of the N/P ratio were the same in both sources of nitrogen, which was also observed in other studies (21), because despite the priority of the use of ammonium source, both sources were finally eliminated with the approximate same value. In addition, in both beds, the optimum removal rate for ammonium and nitrate was at

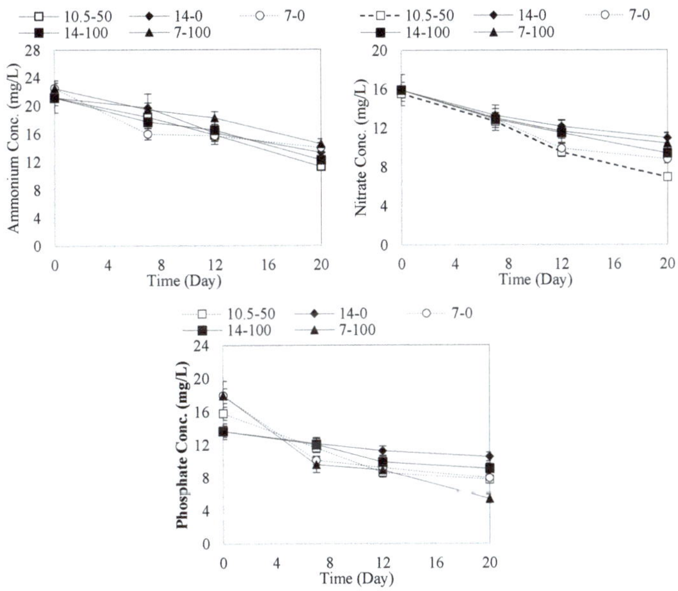

Figure 3. Ammonium, nitrate and phosphate concentrations of wastewater during the growth of *Scenedesmus quadricauda* on SF in different N/P ratios and volumes of the micronutrients.

the N/P ratio of 10.5 and in the presence of micronutrient solution. One of the reasons for the importance of this ratio and its effect on the growth and removal of nitrogen and phosphorus sources is the effect of this parameter on the production of chlorophyll. This was reported in a study that the maximum chlorophyll was produced at N/P ratios between 10 and 12 (22). In another report, the maximum removal of TN and biomass production was measured at N/P ratio of 10, in addition, TN removal was significantly reduced at N/P ratio of 15 (23). In a report by Zhang et al, *S quadricauda* which was immobilized on calcium alginate, was exploited to treat urban wastewater, with

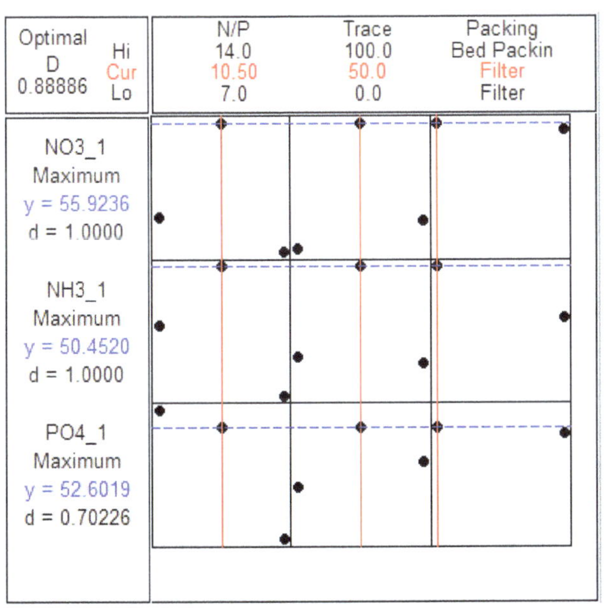

Figure 4. Results of optimum conditions based on the maximum removal efficiency of ammonium, nitrate and phosphate.

the optimal N/P ratio of 10 for ammonium removal (4). In another study on urban wastewater treatment using *S. quadricauda*, the rate of biomass production and biological removal of TN increased and then decreased by increasing the N/P ratio to 10, so that the rate of biomass production at the ratio of 10 was 2.9 g/L/d, which decreased to 1.9 g/L/d by doubling the N/P ratio. Like phosphate removal rate in the present study, the maximum removal of TP was obtained at N/P ratio of less than 10, and with increasing this ratio, the phosphate removal rate was decreased (24). In another study conducted on the biofilms, the removal percentage for phosphate reached from 45% to 100% with increasing its concentration from 0.9 to 1.55 mg/L (2).

In the present study, the maximum removal levels of ammonium and nitrate were obtained at N/P ratio of 10 with an initial concentration of 21.2 mg/L and 40 mg/L, respectively, in both beds. On the other hand, the removal rate increased by increasing the amount of phosphate in the growth medium of both beds, so that the maximum removal efficiency of phosphate was measured at N/P ratio of 7. This is due to the increased phosphate absorption with an increase in its concentration (2).

After 20 days of culture in the two beds under optimal conditions, the removal efficiency of nitrate in the PB and SF beds was 55 and 56%, respectively. In addition, the ammonium removal efficiency in the two beds was 47.3 and 47%, respectively. This indicates that changing the type of beds had less effect on the results compared to the other factors. Nevertheless, the removal efficiency of phosphate in the SF bed was better under optimal conditions, as the removal efficiency was 66.6 and 61.9% for the SF and PB, respectively. This is due to the high surface area of the SF compared to the PB. Hence, the nutrient uptake increases by increasing the specific

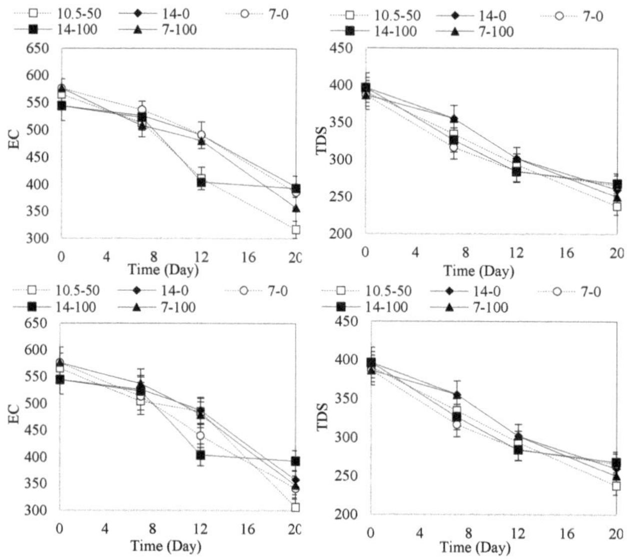

Figure 5. EC and TDS of treated wastewater by *Scenedesmus* grown on the PB (above) and SF (below).

surface area (9), while low diffusion rate of the nutrient compounds into the biofilm and incomplete ammonium removal can be effective in the reduction of assimilations rates through the biofilm (2,14).

In the present study, the maximum rate of removal for phosphate, ammonium, and nitrate in the PB was 2 mg/L/d, 0.7 mg/L/d, and 1.4 mg/L/day, respectively. This rate for the micronutrients in the SF bed was 2.4 mg/L/d, 0.6 mg/L/d, and 1.4 mg/L/d, respectively. In a report published in 2007, two strains of *Chlorella vulgaris* and *Scenedesmus rubescens* were immobilized on a bed in which the removal rate of phosphate, ammonium, and nitrate was 1 mg/L, 2.2 mg/L, and 0.25 mg/L at the initial concentrations of 3 mg/L, 21 mg/L, and 2.5 mg/L, respectively (21). These results are similar to the present study. Furthermore, Boelee et al, in their study on phosphate, ammonium, and nitrate assimilation using a biofilm microalgae consortium, reported the removal rates of 2 mg/L/d, 2.5 mg/L/d, and 0.6 mg/L/d, respectively (2). Similarly, Wei et al reported a removal rate of 4 mg/L/d for ammonium and 80% removal of ammonium (with initial concentration of 18 mg/L/d) using a biofilm containing five different types of microalgae. They also reported the removal rate of 2 mg/L/d for phosphate at concentration of 10 mg/L/d (9). The results indicate that the rate and final removal efficiency of compounds in the present study are consistent with those of previous studies.

It can be observed that the use of minerals by microalgae leads to a decrease in EC and TDS according to Figure 5. This could be an indication of ammonium, nitrate and phosphate removal by *S. quadricauda* microalgae.

Conclusion

The present study showed the ability of microalgal biofilms in the post-treatment of synthetic meat processing wastewater. A maximum removal efficiency of 47% for ammonium, 55% for nitrate and 66% for phosphate was achieved during the experiments. However, as the results reveal that inadequate phosphorus and nutritional components can significantly affect the removal efficiencies, therefore, an optimization is needed for increasing the effectiveness of microalgae-based wastewater treatment.

Acknowledgments

The authors would like to thank the staff of Biotechnology Department of Iranian Research Organization for Science and Technology with a special thanks to Mr. Sheikhi Nejhad.

Competing interests

The authors declared no competing interests.

Authors' Contributions

All authors contributed equally to the study, and all authors participated in data acquisition, analysis, and interpretation. All authors critically reviewed, refined, and approved the manuscript.

References

1. Aslan S, Kapdan IK. Batch kinetics of nitrogen and phosphorus removal from synthetic wastewater by algae. Ecol Eng 2006; 28(1): 64-70. doi: 10.1016/j.ecoleng.2006.04.003.
2. Boelee N, Temmink H, Janssen M, Buisman C, Wijffels R. Nitrogen and phosphorus removal from municipal wastewater effluent using microalgal biofilms. Water Res

2011; 45(18): 5925-33. doi: 10.1016/j.watres.2011.08.044.
3. Irving TE, Allen DG. Species and material considerations in the formation and development of microalgal biofilms. Appl Microbiol Biotechnol 2011; 92(2): 283-94. doi: 10.1007/s00253-011-3341-0.
4. Zhang E, Wang B, Wang Q, Zhang S, Zhao B. Ammonia–nitrogen and orthophosphate removal by immobilized *Scenedesmus sp.* isolated from municipal wastewater for potential use in tertiary treatment. Bioresour Technol 2008; 99(9): 3787-93. doi: 10.1016/j.biortech.2007.07.011.
5. Su Y. Settleable algal-bacterial culture for municipal wastewater treatment [dissertation]. Lüneburg: Leuphana Universität Lüneburg; 2012.
6. Babu M. Effect of Effect of Algal Biofilm and Operational Conditions on Nitrogen Removal in Waste Stabilization Ponds: UNESCO-IHE PhD Thesis. CRC Press; 2011.
7. Hoffmann JP. Wastewater treatment with suspended and nonsuspended algae. J Phycol 1998; 34(5): 757-63. doi: 10.1046/j.1529-8817.1998.340757.x.
8. Sabater S, Guasch H, Romaní A, Muñoz I. The effect of biological factors on the efficiency of river biofilms in improving water quality. Hydrobiologia 2002; 469(1-3): 149-56. doi: 10.1023/A:1015549404082.
9. Wei Q, Hu Z, Li G, Xiao B, Sun H, Tao M. Removing nitrogen and phosphorus from simulated wastewater using algal biofilm technique. Front Environ Sci Eng China 2008; 2(4): 446-51. doi: 10.1007/s11783-008-0064-2.
10. Guzzon A, Bohn A, Diociaiuti M, Albertano P. Cultured phototrophic biofilms for phosphorus removal in wastewater treatment. Water Res 2008; 42(16): 4357-67. doi: 10.1016/j.watres.2008.07.029.
11. Al-Shatri AHA, Ali E, Al-Shorgani NKN, Kalil MS. Growth of *Scenedesmus dimorphus* in different algal media and pH profile due to secreted metabolites. African Journal of Biotechnology 2014; 13(16):1714-20. doi: 10.5897/AJB2013.13455.
12. Chia MA, Musa RI. Effect of indigo dye effluent on the growth, biomass production and phenotypic plasticity of *Scenedesmus quadricauda* (Chlorococcales). An Acad Bras Cienc 2014; 86(1): 419-28.
13. Heidari S, Farhadian O, Soofiani NM. Biomass Production and ammonia and nitrite removal from fish farm effluent by *Scenedesmus quadricauda* culture. J Environ Stud 2011; 37(59):7.
14. Travieso L, Benitez F, Weiland P, Sánchez E, Dupeyrón R, Dominguez AR. Experiments on immobilization of microalgae for nutrient removal in wastewater treatments. Bioresour Technol 1996; 55(3): 181-6. doi: 10.1016/0960-8524(95)00196-4.
15. Greenberg Arnold E, Clesceri Lenore S. Standard methods for the examination of water and wastewater. USA: American Public Health Association; 1992.
16. Ruiz-Martinez A, Serralta J, Seco A, Ferrer J. Effect of temperature on ammonium removal in *Scenedesmus sp.* Bioresour Technol 2015; 191: 346-9. doi: 10.1016/j.biortech.2015.05.070.
17. He P, Mao B, Lü F, Shao L, Lee D, Chang J. The combined effect of bacteria and Chlorella vulgaris on the treatment of municipal wastewaters. Bioresour Technol 2013; 146: 562-8. doi: 10.1016/j.biortech.2013.07.111.
18. Roudsari FP, Mehrnia MR, Asadi A, Moayedi Z, Ranjbar R. Effect of microalgae/activated sludge ratio on cooperative treatment of anaerobic effluent of municipal wastewater. Appl Biochem Biotechnol 2014; 172(1): 131-40. doi: 10.1007/s12010-013-0480-z.
19. Sreesai S, Pakpain P. Nutrient recycling by *Chlorella vulgaris* from septage effluent of the Bangkok City, Thailand. ScienceAsia 2007;33(3):293-9.
20. Anand J, Arumugam M. Enhanced lipid accumulation and biomass yield of *Scenedesmus quadricauda* under nitrogen starved condition. Bioresour Technol 2015; 188: 190-4. doi: 10.1016/j.biortech.2014.12.097.
21. Shi J, Podola B, Melkonian M. Removal of nitrogen and phosphorus from wastewater using microalgae immobilized on twin layers: an experimental study. J Appl Phycol 2007; 19(5): 417-23. doi: 10.1007/s10811-006-9148-1.
22. Smith VH. The nitrogen and phosphorus dependence of algal biomass in lakes: an empirical and theoretical analysis. Limnol Oceanogr 1982; 27(6): 1101-11. doi: 10.4319/lo.1982.27.6.1101.
23. Xin L, Hong-ying H, Ke G, Ying-xue S. Effects of different nitrogen and phosphorus concentrations on the growth, nutrient uptake, and lipid accumulation of a freshwater microalga *Scenedesmus sp.* Bioresour Technol 2010; 101(14): 5494-500. doi: 10.1016/j.biortech.2010.02.016.
24. Choi HJ, Lee SM. Effect of the N/P ratio on biomass productivity and nutrient removal from municipal wastewater. Bioprocess Biosyst Eng 2015;38(4): 761-6. doi: 10.1007/s00449-014-1317-z.

Geohelminthic: human ascariasis and trichuriasis in Mazandaran province, northern Iran

Hajar Ziaei[1], Fatemeh Sayyahi[2], Mahboobeh Hoseiny[3], Mohammad Vahedi[4], Shirzad Gholami[5*]

[1]Associate Professor, Toxoplasmosis Research Center, Mazandaran University of Medical Sciences, Sari, Iran
[2]Medical Student, Research Committee, Faculty of Medicine, Mazandaran University of Medical Sciences, Sari, Iran
[3]MSC Statistic, GIS Research Center, Mazandaran University of Medical Sciences, Sari, Iran
[4]MSC Microbiology, Faculty Member, Department of Microbiology, Mazandaran University of Medical Sciences, Sari, Iran
[5]Associate Professor, Molecular and Cell Biology Research Center, Department of Parasitology and Mycology, Mazandaran University of Medical Sciences, Sari, Iran

Abstract
Background: Ascariasis and trichuriasis are the most common intestinal geohelminthic diseases, and as such they are significant in terms of clinical and public health. This study was done to determine prevalence, status and geographic distribution patterns for Ascariasis and Trichuriasis. The study was done in the period 1991-2014 in northern Iran using Aregis 9.2 software.
Methods: This was a review study, using description and analysis, of geographical distribution of *Ascaris* and *Trichuris* relating to townships in Mazandran province, northern Iran, covering a 23-year period. Data were collected from a review of the relevant literature, summarized and classified using Arc GIS, 9.2 to design maps and tables.
Results: Based on results presented in tables and maps, means for prevalence of *Ascaris* and *Trichuris* were divided into five groups. The maximum prevalence rate of *Ascaris* was 16.3% reported in rural areas of Tonkabon in 1981-1982. Prevalence means for *Ascaris* in the central and western areas of Mazandaran province were 2%-4% and 4%, respectively. The maximum prevalence of *Trichuris* in the rural area of Tonkabon was 22.5% and the lowest 0.06% was among the cattle breeder's in rural areas of the province at in 2002-2003.
Conclusion: Data presented in this study provides information useful to health care workers researchers and health administrators, especially for physicians, clinicians and for future research. Also, it is necessary to control and prevent geohelminthic parasitic infections, particularly in rural areas by public education for families, health authorities and health care systems.
Keywords: Geohelminths, Ascariasis, Trichuriasis, Geographical distribution

***Correspondence to:**
Shirzad Gholami
Email: sgholami200@gmail.com

Introduction

Intestinal helminths parasitic infections are closely related to personal, public health and environmental conditions, especially *Ascaris* and *Trichuris.* These soil transmitted helminths, transmitted by fecal contamination of soil, foodstuff and water supplies, are considered as a health concern in Iran and in developing countries (1-5). Among the parasitic infections, the soil-transmitted helminthes (STH) are a common group of intestinal parasites that cause human infections through contact with parasite eggs or larvae (in warm and moist soil), particularly attributed to the class of the nematoda, such as *Ascaris lumbricoides, Trichuris trichiura* and the hookworms (*Ancylostoma duodenale* and *Necator americanus*) (1,4,5). Approximately 1 billion people worldwide are affected by soil-transmitted infections and school children are the most heavily affected group. Estimates determine that over one billion people are infected with *Ascaris lumbricoides*, 800 million with hookworm and 770 million with *Trichuris trichiura* (2). Geohelminths is the second most common cause of mortality in children under 6 years of age in Africa (3). *Ascaris Lumbricoides* and *Trichuris trichiura* are among the most common STH helminthes and as such they constitute a major concern for public health, particularly in countries with low and middle-incomes population (1,6,7). They cause infection in about one quarter of world's population (1,5). Helminths worms are a major health concern, particularly for children worldwide; in both developing and under developed countries, tropical, semitropical and temperate regions, as well as areas affected by poverty with poor sanitary and hygiene conditions and waste disposal facilities. Parasitic infections are also prevalent among emigrants and asylum seekers in developed countries (8). Higher rates of these infections have been reported in people with compromised immunity and malnutrition (9). These soil-transmitted nematodes are a major health concern worldwide. Infected patients develop nutrition and intestinal problems, as well as other complications contributing to high rates of hospital admissions and cases of surgery as a result of *Ascaris lumbricoides,* annually (10-14).

According to the World Health Organization (WHO) survey, more than 60 000 people die of *Ascariasis* complications per year (7,15). In addition, annual hospital costs to treat re-infected patients are high because infections can remain untreated. This impaired chain could continue over a long period (15-17). *Ascariasis* has been one of the most common helminths infections in Iran (18). Various factors could have an effect on prevalence of the disease, such as use of human biofertilizer, unhygienic human waste disposal, suitable environmental conditions (moisture, temperature, and shade and clay soil), human exposure to environmental contamination from sources, such as water, vegetables or soil. The trichuriasis mode of transmission is similar to ascariasis and parasite infections are more prevalent in rainy and tropical regions and resistance of parasite eggs against cold is higher than that in *Ascaris* (5,18).

Reports show higher prevalence of geohelminths infection in rural areas compared to urban. Studies in Hamedan province (Iran) up to the year 1997, indicate a high rate of ascariasis prevalence. In some rural areas of Hamedan province, this infection has reached a rate of more than 70% (19). The rate of ascariasis in the Caspian Sea coastal area was reported 14%-60% (18,20). Rokni (18) in a review of the related literature showed prevalence of ascariasis 0.01%-0.03% in Iran (18). In past decades, the highest rate of trichuriasis was reported 78.3% in coastal areas of the Caspian Sea (20). However, in the Iranian cities of Rasht and Anzali, rates were reported 1.5% and 3.1%, respectively (21). Ashrafi (21) showed prevalence of this worm in rural areas of Anzali, northern Iran 0.3%.

Studying the geographic distribution of parasite infection based on area, year and prevalence of infection provides useful knowledge on the status of diseases in a particular environment, and provides indications for status among individuals, families and public hygiene. Furthermore, there is the possibility of contaminated soil and vegetables caused by the use of wastewater, feces and well water in irrigated food production, humans are exposed to contamination by a lack politeness around fields. Mazandran province in Iran has climatic conditions that support the transmission of parasites (21). Conditions such as heavy rainfall and high humidity provide suitable conditions for parasites (22). Thus, ascariasis and trichuriasis are the most common intestinal geohelminths diseases with clinical and public health significance. Mazandran province is a tourist destination. Therefore, study of ascariasis and trichuriasis distribution as the main soil-transmitted intestinal parasites in different regions of this province has been considered as a main health concern in recent decades. Hence, present study was done to determine prevalence, status and geographic distribution patterns of ascariasis and trichuriasis using the Aregis 9.2 Software, in Mazandaran province, northern Iran in 1991-2014.

Methods

In this review study, a description and analysis of the geographical distribution of *Ascaris* and *Trichuris* prevalence were made. The study was based on a township in Mazandran province, northern Iran covering a 23-year period (Figure 1). Data were collected from 1991 to 2014 through electronic databases, such as PubMed, Google Scholar, Science Direct, Scopus, Web of Science, Magiran, Irandoc and Iran Medex, Scientific Information Database (SID), Global Health, and LILACS. Abstracts from scientific congresses and dissertations were also included in the study. The data were summarized and classified. Using the software Arc GIS 9.2, a map was designed to show geographical distributions of disease prevalence (23-25).

Table arrangement

Two tables were prepared showing classification of data collected from the relevant literatures. The tables present data according places of distribution, year of the report, percentage of infection, title of papers, and authors. Collected information were reported from various studies related to Mazandran province from 1991 to 2014.

Map preparation

Referring to the tables and the map of Mazandaran province, Arc Gis 9.2 software was used to prepare a database. Maps showing data for the two helminths diseases were drawn from information based on means of prevalence and geographic distribution reported from different areas of the province (15 townships and 38 districts) over a 22-year period.

Results

According to the study design, the prevalence, status and pattern of geographic distribution of ascariasis and trichuriasis parasitic diseases in humans were determined and described in Mazandaran province in the study period. This research contributes to databases for etiologic agency, distribution area and year of publication; percentages of infections were prepared and these shown in Tables 1 and 2. Based on these tables, disease in Mazandaran province was shown in terms of prevalence of infection and distribution area for each parasite, using the Arc Gis 9.2 software. The highest rate of infection of ascariasis was 16.3%, shown in the rural area of Tonecabon township in 1991-1992 and the lowest, was 0.12% observed in patients with gastroenteritis referring to health centers in the townships of Babol and Babolsar in 2005-2006 (Table 1). Due to the prevalence of ascariasis and its means per year, a map of disease distribution was prepared, considering the township in Mazandaran province (Figure 2). Plotting distribution of ascariasis on a map was designed according to the following 5 groups:

First group: Ramsar (western region), Savadkooh (central

Table 1. Distribution status of ascariasis, based on the distribution area, year of publication, percentage of infection reported from various study in Mazandaran province, Iran, 1991 to 2013

Distribution place or report	Year of publication	Infection rate %	Title of the paper	Author
Inhabitants from 27 rural areas of Tonekabon from the hill stations and flat areas	1991-1992	16.3%	Prevalence of intestinal parasites in the rural areas of Tonekabon, Iran.	Rezaeian et al (20)
The rural areas of the Noor and Behshahr Township	1995	2.4% Behshahr; 2.3% Noor	Study on the frequency of intestinal helminth from the rural areas of behshahr and Noor Township of Mazandran province	Saboori et al (22)
The rural areas of Amol township	1998-1999	3.1%	Study on the frequency of intestinal parasite from the rural areas of Amol township	Hashemzadeh Omran and Mobedi (23)
The flat and forest areas of 28 rural areas of Sari township	1998-1999	Total 0.4%; Farmer 0.61%; Housewife 0.85%; Student 0.35%	Prevalence of intestinal parasites in the rural areas of Sari county	Rohani et al (26)
All of the elementary students at the East Bandpay areas of Babol township	1999	0.1%	Contamination of intestinal parasites in primary school in the Bandpey area, eastern region Babol 1999.	Ghahremanlo et al (27)
East health center of Babol Narivaran Shohada	2000	4%	Frequency of intestinal parasitic contamination in Babolian schools girls, 1998.	Sajjadi et al (37)
East Mazandran	2000	1.5%	The prevalence of parasite infection in eastern areas of Mazandaran.	Abedian and Parsaei (39)
Qaemshahr township	2002	The individual infected with parasite 0.3%	Relationship between serum IgE and intestinal parasites.	Jalalian et al (29)
Cattle breeders at rural areas of 9 township in Mazandran province	2002-2003	Total 0.4%; Cattle breeders infected with parasite 25%	Intestinal parasitic protozoan and helminthic infections in cattle breeders in rural regions of Mazandaran province in 2003.	Gholami et al (4)
The gastroenteritis patients referring to the 7 health care centers of the Babol and Babolsar township	2005-2006	0.12%	Frequency of cryptosporidiosis, isosporiasis and other enteropathogenic parasites in gastroenteritic patients (Babol and Babolsar; 2005-2006).	Ghorbannia Delavar et al (38)

Table 2. The trichuriasis distribution status based on the distribution place, the year of report, percentage of infection reported from various studies in Mazandaran province from 1991 to 2013

Distribution place or report	Year of publication	Infection rate %	Title of paper	Author
Inhabitants of 27 rural areas of Tonekabon from the hill stations and flat regions	1991-1992	22.5%	Prevalence of Intestinal Parasites in the rural areas of Tonekabon, Iran.	Rezaeian and Hoshiar (20)
The rural areas of Amol township	1998-1999	6.6%	Study on the frequency of intestinal parasite from the rural areas of Amol township	Hashemzadeh Omran and Mobedi (23)
The mental retarded primary school children of Babol city	1999	3.3%	Prevalence of intestinal parasitic infections in primary school children with the exceptional mental of Babel county.	Gholampour (28)
All of the primary school students in the east Bandpay	1999	0.5%	Contamination of intestinal parasites in primary school in the Bandpey area, eastern region Babol 1999.	Ghahremanlo et al (27)
East health center of Babol Narivaran Shohada	2000	1%	Frequency of intestinal parasitic contamination in Babolian schools girls, 1998.	Sajjadi et al., 2000 (37)
East Mazandaran	2000	0.5%	The prevalence of parasite infection in eastern areas of Mazandaran.	Abedian and Parsaei (39)
Qaemshahr township	2002	0.3%	Relationship between serum IgE and intestinal parasites.	Jalalian et al (29)
Cattle breeders at rural areas of 9 Township in Mazandaran province	2002-2003	0.06%	Intestinal parasitic protozoan and helminthic infections in cattle breeders in rural regions of Mazandaran province in 2003.	Gholami et al (4)
The cities and the rural areas of Qaemshahr city	2004	1%	Prevalence of Intestinal Parasites in Ghaemshahr in 2004.	Ranjbar Bahadori et al (31)

region) and Mayandorod (eastern region) with no reports of parasitic infection.

Second group: Sari and Qaemshahr (central region); with mean prevalence of ascaraisis less than 1%.

Third group: Galogah, Behshahr, and Nekah in the east and Babol in the central region of Mazandran province; townships with mean prevalence of ascariasis 1%-2% (26,27).

Fourth group: Townships of Amol and Noor (central region of the province) with mean prevalence of 2%-4%.

Fifth group: The township of Tonekabon with mean prevalence of 4% in the western area of the province (Figure 2).

Results of this study indicate prevalence of trichuriasis 22.5% in the rural areas of Tokenkabon in western of Mazandaran province in 1991-1992 and 0.06% in cattle breeders of the 9 townships in the province in 2002-2003 (Table 2). Gholampour (28) reported prevalence of *Trichuris trichiura* in infection the mentally retarded elementary school students of Babol township 3.3% (Table 2). That figure does not agree with the rate of 0.5% in the same year recorded in the primary school students in the Bandpey area, eastern region of Babol (Table 2). Data as described by the map, indicate the mean of infection prevalence rate of *Trichuris trichiura* more than 5% in Amol and Tonekabon (western areas), 1%-2% in Babol and lower than 1% in the townships of Noor, Qaemshahr (central area), Sari, Nekah, Behshahr and Galogah (eastern area) in Mazandaran province (Figure 3).

Discussion

Ascariasis and trichuriasis are two important intestinal geohelminthic diseases with clinical and public health significance (1,5,7). In recent years, medical sciences researchers, in order to extend knowledge on distribution of infectious diseases for planning and implementing effective treatment measures, have used geographic information system (GIS) and the occurrence of disease in a particular past period to monitor the disease and health conditions. Some researchers, using GIS and plotting distribution of a disease on a map have investigated *Fascioliasis* in northern Iran and presented a map of the *Visceral leishmaniasis* disease, indicating the high risk regions in Ardabil province in 1999 (24,25). This review study was done to determine prevalence, status and pattern of geographical distribution of *Ascaris* and *Trichuris* in Mazandaran province, northern Iran in1991-2014.

The areas have suitable environmental conditions in terms of climate, humidity and water temperature to facilitate the transfer of parasite eggs and larvae to humans. However, a review of the relevant literatures highlighted a reduced rate of infection from intestinal helminths in Mazandaran province with a few exceptions, in particular cases of ascariasis and trichuriasis, as STH. Though, ascariasis had an apparent decline in incidence rate from 16.3% to 0.1% from 1981 through 1999 (20,23,26-28), but the following years showed a decline of helminthic infection (24,29-31). The latest report on prevalence of gastroenteritis patients referring to the Babol health center indicated 0.12% in 2009-2010 (32).

In Mazandaran province, limited studies have been done in different social groups; the obtained data indicated a partial decline in the rate of *Ascaris* prevalence in cities after 2000. *Trichuris* occurrence showed a significant reduction in 2005 compared to 2001 (from 22.5% to 0.3% in Tonekabon and to 0.9% in Behshahr and Noor (22,28,30). Prevalence rates of 6.6% and 3.3% were reported in Amol and special students in Babol, respectively (27,28,32). However, in the same year, a prevalence rate of 0.5% was reported in elementary school students in the east Bandpay region of Babol and there was a similar rate in the following year (27). The latest infection rate in Qaemshahr was reported 0.8% in 2004 (31). The maps showed higher rates of helminthic infection in Tonekabon, Amol and Babol. One limited study on mentally retarded elementary school students showed trichuriasis 3.3% (28), but no study has been done to date on infection rates of *Ascaris* and *Trichuris* in different age groups, particularly among children in contact with soil.

In the current study, infection distribution maps of *Ascaris* and *Trichuris* have shown prevalence of 2 parasitic diseases reported sporadically in the rural population in different years and areas. However, an epidemiologic study using the system of geographical informatics needs to be done in order to evaluate the conditions of diseases and

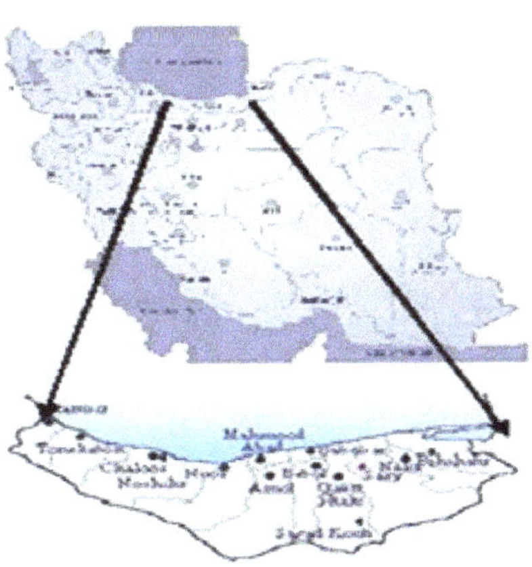

Figure 1. The Mazandran province map based on the townships and their location.

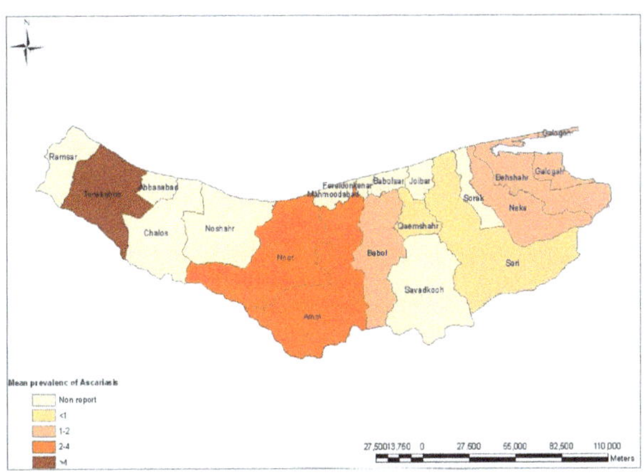

Figure 2. The geographical distribution of the mean prevalence rate of *Ascariasis* in the Mazandran province, from 1991 to 2013.

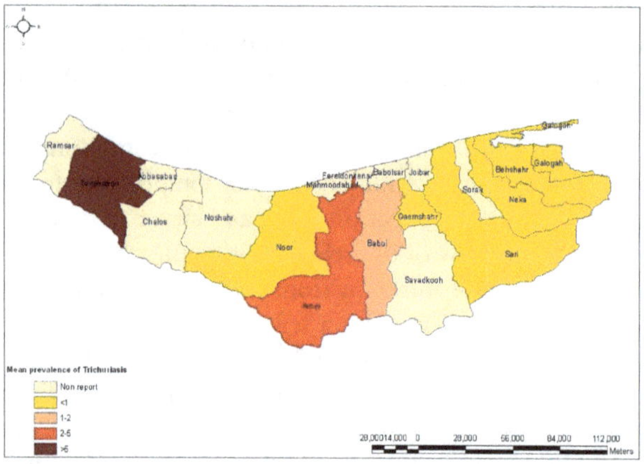

Figure 3. The geographical distribution of the mean prevalence rate of *Trichuriasis* in the Mazandran province, from 1991 to 2013.

distribution patterns in Mazandaran province (northern Iran). However, a study that to cover both urban and rural distribution status of *Ascaris* and *Trichuris* is needed, particularly in high risk areas. *Ascaris* and *Trichuris* are STH, so incidence increases with increased consumption of raw materials from the soil, such as root of vegetables. With the current increasing trend of organic fertilizer application, parasites should be identified in soil and drinking water, rather than among people with direct and indirect contact with water and/or soil. The severity of geohelminths disease particular attention because of its widespread geographic distribution (33); deleterious effects on nutrition and immune disorder (34,35). Therefore, authorities should be more attentive to the field of education and the control of food-borne helminthes and protozoan parasitic diseases in various parts of Iran (30,36).

The map in this study showed higher mean rate of *Ascaris* infection in Tonekabon (western Mazandaran), Amol and Noor (central area), Nekah and Behshahr (eastern area). A comparative study on distribution of rate of ascariasis and trichuriasis were conducted more in Amol (central area) and Tonekabon (western area) showed distribution of these 2 parasites in flat and forest areas of Mazandaran from 1991 to 2014. Although conditions of general hygiene have significantly improved in recent years particularly in relation to food stuff and the quality of drinking water, has also improved in the provincial townships. Townships of Ramsar, Chalos, Savadkoh, Joybar, Mahmood Abad, Babolsar and Galogah were not included in this investigation (Figures 1 and 2) (37-39). Thus, it is recommended for future studies in order to obtain correct patterns of disease distribution.

Conclusion

Data presented in this study provide useful information for health care workers, especially physicians and clinicians, researchers, health centers and health administrators. Results of this study demonstrate prevalence and geographical distribution of *Ascaris* and *Trichuris* and their relation to public health in northern Iran. This study emphasizes the need to establish rates of incidence, prevalence, disease burden and distribution of helminthes and protozoan infection in many parts of the Iran as well as in the wider Middle Eastern region, so that, regional, national and global estimates can be made. Therefore, it is necessary to control and to prevent the most common parasitic infections, particularly in the townships in the rural areas of the province with higher rates of prevalence. Hence, public education is necessary for families, and health care systems.

Competing interests

The authors declare that they have no competing interests.

Authors' contributions

The study designed by SG and all authors contributed equally to the data acquisition, analysis and interpretation. All authors also critically reviewed, refined and approved the manuscript.

References

1. Ojha SC, Jaide C, Jinawath N, Rotjanapan P, Baral P. Geohelminths: public health significance. J Infect Dev Ctries 2014; 8(1): 5-16.
2. Ogbe MG, Edet EE, Isichel MN. Intestinal helminth infection in primary School Children in areas of operation of shell petroleum development company of Nigeria (SPDC) western division in delta state. Nig J Parasitiol 2002; 23(1): 3-10.
3. Nock IH, Duniya D, Galadima M. Geohelminth eggs in the soil and stool of pupils of some primary schools in Samaru, Zaria Nigeria. Nig J Parasitiol 2003; 24(1): 115-122.
4. Gholami S, Sharif M, Mobedi E, Ziaei H, Ali Mohammadpour R, Kianiyan H. Intestinal parasitic protozoan and helminthic infections in cattle breeders in rural regions of Mazandaran province in 2003. Proceedings of the Fourth National Congress of Occupational Health; 2004; Hamadan Univ Med Sci; 519-529. [In Persian].
5. Yap P, Fürst T, Müller I, Kriemler S, Utzinger J, Steinmann P. Determining soil-transmitted helminth infection status and physical fitness of school-aged children. J Vis Exp 2012; (66): e3966.
6. Hotez PJ, Fenwick A, Savioli L, Molyneux DH. Rescuing the bottom billion through control of neglected tropical diseases. Lancet 2009; 373(9674): 1570-5.
7. WHO. Soil-transmitted helminth infections. http://www.who.int/mediacentre/factsheets/fs366/en/. Accessed May l, 2015.
8. Chan MS, Medley GF, Jamison D, Bundy DA. The evaluation of potential global morbidity attributable to intestinal nematode infections. Parasitology 1994; 109(Pt 3): 373-87.
9. Lili Z, Bingxiang Z, Hong T, Shuhua X, Hotez P, Bing Z, et al. Epidemiology of human geohelminth infections (ascariasis, trichuriasis and necatoriasis) in Lushui and Puer Counties, Yunnan Province, China. Southeast Asian J Trop Med Public Health 2000; 31(3): 448-53.
10. Ramareddy RS, Alladi A, Siddapa OS, Deepti V, Akthar T, Mamata B. Surgical complications of Ascaris lumbricoides in children. J Indian Assoc Pediatr Surg 2012; 17(3): 116-9.
11. Mukhopadhyay M. Biliary ascariasis in the Indian subcontinent: A study of 42 cases. Saudi J Gastroenterol 2009; 15(2): 121-4.

12. Esposito C, Settimi A, De Marco M, De Fazio C, Giurin I, Savanelli A, et al. Surgical complications of ascariasis in children. J Pediatric Surg Spec 2008; 2(3): 8-12.
13. Refeidi A. Live Ascaris lumbricoides in the peritoneal cavity. Ann Saudi Med 2007; 27(2): 118-21.
14. Sanai FM, Al-Karawi MA. Biliary ascariasis: report of a complicated case and literature review. Saudi J Gastroenterol 2007; 13(1): 25-32.
15. WHO. Prevention and control of schistosomiasis and soil-transmitted helminthiasis. Geneva: WHO; 2002.
16. Hwang SW. Mortality among men using homeless shelters in Toronto, Ontario. JAMA 2000; 283(16): 2152-7.
17. Cheung AM, Hwang SW. Risk of death among homeless women: a cohort study and review of the literature. CMAJ 2004; 170(8): 1243-7.
18. Rokni MB. The present status of human helminthic diseases in Iran. Ann Trop Med Parasitol 2008; 102(4): 283-95.
19. Fallah M, Azimian MH, Nabiee M, Hojati M. Epidemiological study of Ascariasis in Hamadan city, west of Iran, 2001. Journal of Hamadan University of Medical Sciences 2004; 11(1): 55-60. [In Persian].
20. Rezaiian M, Hooshyar H. Prevalence of Intestinal parasitic infection in rural areas of Tonekabon, Iran. Iranian Journal of Public Health 1996; 25(3-4): 47-58. [In Persian].
21. Ashrafi K. Medical Helminthology. Ist ed. Rasht: Payame Sobhan; 2009. [In Persian].
22. Saboori A, Masood J, Mobedi E, Rezaeian M. Prevalence of intestinal parasites in the Behshahr and Noor cities [Thesis]. Tehran: Tehran University and Medical Sciences; 1995. [In Persian].
23. Hashemzadeh Omran A, Mobedi E. Evaluation of intestinal parasites prevalence in the rural areas of Amol county [Thesis] Tehran: Tehran University and Medical Sciences; 1998-1999. [In Persian].
24. Salahi Moghaddam A. Epidemiology of human Fascioliasis in Iran. J Kerman Univ Med Sci 2009; 16(4): 385-98 [In Persian].
25. Sayahi F, Hossini M, Ziaei H, Gholami S. Geographical distribution of enteobiasis in Mazandaran province, 1981-2013. J Mazandaran Univ Med Sci 2015; 24 (121): 481-9 [In Persian].
26. Rohani S, Kianiyan H, Athari A. Prevalence of intestinal parasities in villages of Sari (1998-99). J Zanjan Univ Med Sci 2001; 9(34): 33-40. [In Persian].
27. Ghahramanloo M, Hassanjani Roshan MR , Haji Ahmadi M. Prevalence of intestinal parasites in primary school children, Eastern Bandpay, Babol, 1999. J Babol Univ Med Sci 2001; 3(2): 47-51. [In Persian].
28. Gholampour A. Prevalence of intestinal parasitic infections in primary school children with the exceptional mental of Babol county. Islamic Azad Univ Sci 1999; 3(5): 158. [In Persian].
29. Jalalian M, Rezaiian M, Kia EB, Massoud J, Mahdavi M, Rokni MB. Relationship between serum IgE and intestinal parasites. Iran J Publ Health 2004; 33(1): 18-21.
30. Gholami SH, Mohammadpour Tahamtan RA, Sharif M, Ziaei H, Euroji A,GohardehI SH, et al. Intestinal parasite infections in cattle breeders in rural regions of Babol town during 2003. J Babol Univ Med Sci 2005; 7(4): 83-7. [In Persian].
31. Ranjbar Bahadori S, Dastorian A, Heidari B. Prevalence of intestinal parasites in Ghaemshahr in 2004. J Medical Sciences 2005 2004; 15(3): 151-5. [In Persian].
32. Rahimi-Esboei B, Gholami S, Ghorbani Pasha Kolaei A, Pour Haji Baqer M, Hasannia H, Shaban R, et al. The Prevalence of Intestinal Parasitic Infections among the People Living in the Central Areas of Mazandaran Province (2009-2010). Medical Laboratory Journal 2013; 7(2): 43-8 [In Persian].
33. WHO. Prevention and control of intestinal parasitic infections. Geneva: WHO; 1987.
34. Allen LH. Nutritional influences on linear growth: a general review. Eur J Clin Nutr 1994; 48 Suppl 1: S75-89.
35. Maizels R, Yazdanbakhsh M. Immune regulation by helminth parasites: cellular and molecular mechanisms. Nat Rev Immunol 2003; 3(9): 733-44.
36. Avazpoor M, Yousefipoor M, Dusty M, Mehdipour M, Seifipour F, Gholami Z. Determination of the level of parasitic infection (Cryptosporidium and Giardia) of the vegetables marketed in Ilam city. Environmental Health Engineering and Management Journal 2015; 2(1): 37-40.
37. Sajjadi P, Seddighian F, Alaeddoleie H. The prevalence of intestinal parasitic infection among high school girls in Babol (1996). J Babol Univ Med Sci 2000; 2(1): 39-43. [In Persian].
38. Ghorbannia Delavar A, Nahrevanian H, Assmar M, Amirkhany A, Esfandiary B. Frequency of cryptosporidiosis, isosporiasis and other enteropathogenic parasites in gastroenteritic patients (Babol and Babolsar; 2005-2006). JBUMS 2008; 10(2): 56-61. [In Persian].
39. Abedian S, Parsaei MR. The prevalence of parasite infection in eastern areas of Mazandaran. Third congress of Parasitology and Parasite Infection; Iran, Sari; 1999.

Permissions

All chapters in this book were first published in EHERC, by Kerman University of Medical Sciences; hereby published with permission under the Creative Commons Attribution License or equivalent. Every chapter published in this book has been scrutinized by our experts. Their significance has been extensively debated. The topics covered herein carry significant findings which will fuel the growth of the discipline. They may even be implemented as practical applications or may be referred to as a beginning point for another development.

The contributors of this book come from diverse backgrounds, making this book a truly international effort. This book will bring forth new frontiers with its revolutionizing research information and detailed analysis of the nascent developments around the world.

We would like to thank all the contributing authors for lending their expertise to make the book truly unique. They have played a crucial role in the development of this book. Without their invaluable contributions this book wouldn't have been possible. They have made vital efforts to compile up to date information on the varied aspects of this subject to make this book a valuable addition to the collection of many professionals and students.

This book was conceptualized with the vision of imparting up-to-date information and advanced data in this field. To ensure the same, a matchless editorial board was set up. Every individual on the board went through rigorous rounds of assessment to prove their worth. After which they invested a large part of their time researching and compiling the most relevant data for our readers.

The editorial board has been involved in producing this book since its inception. They have spent rigorous hours researching and exploring the diverse topics which have resulted in the successful publishing of this book. They have passed on their knowledge of decades through this book. To expedite this challenging task, the publisher supported the team at every step. A small team of assistant editors was also appointed to further simplify the editing procedure and attain best results for the readers.

Apart from the editorial board, the designing team has also invested a significant amount of their time in understanding the subject and creating the most relevant covers. They scrutinized every image to scout for the most suitable representation of the subject and create an appropriate cover for the book.

The publishing team has been an ardent support to the editorial, designing and production team. Their endless efforts to recruit the best for this project, has resulted in the accomplishment of this book. They are a veteran in the field of academics and their pool of knowledge is as vast as their experience in printing. Their expertise and guidance has proved useful at every step. Their uncompromising quality standards have made this book an exceptional effort. Their encouragement from time to time has been an inspiration for everyone.

The publisher and the editorial board hope that this book will prove to be a valuable piece of knowledge for researchers, students, practitioners and scholars across the globe.

List of Contributors

Ali Asghar Najafpoor
Associate Professor, Health Sciences Research Center, Department of Environmental Health Engineering, School of Health, Mashhad University of Medical Sciences, Mashhad, Iran

Ghorban Soleimani
MS of Environmental Health Engineering, Shahid Sadoughi University of Medical Sciences, Yazd, Iran

Mohammad Hasan Ehrampoush
Professor, Environmental Health Engineering Department, Shahid Sadoughi University of Medical Sciences, Yazd, Iran

Mohammad Taghi Ghaneian
Associate Professor, Environmental Health Engineering Department, Shahid Sadoughi University of Medical Sciences, Yazd, Iran

Elham Rahmanpour Salmani and Maryam Dolatabadi Takabi
MSc student of Environmental Health Engineering, Student Research Committee, Department of Environmental Health, School of Health, Mashhad University of Medical Sciences, Mashhad, Iran

Naseh Babakhani
MSc Student, Department of Environment, Faculty of the Basic Knowledge, Hamedan Branch, Islamic Azad University, Hamedan, Iran

Mahdi Reyahi-Khoram
Assistant Professor, Department of Environment, Faculty of the Basic Knowledge, Hamedan Branch, Islamic Azad University, Hamedan, Iran

Soheil Sobhanardakani
Associate Professor, Department of Environment, Faculty of the Basic Knowledge, Hamedan Branch, Islamic Azad University, Hamedan, Iran

Ahmad Reza Yazdanbakhsh
Associate Professor, Department of Environmental Health Engineering, Shahid Beheshti University of Medical Sciences, Tehran, Iran

Majid Kermani
Assistant Professor, Department of Environmental Health Engineering, Iran University of Medical Sciences, Tehran, Iran

Sanaz Komasi
MSc of Environmental Engineering, Department of Environmental Engineering, Islamic Azad University, West Tehran Branch, Tehran, Iran

Ehsan Aghayani
PhD Student, Department of Environmental Health Engineering, Tarbiat Modares University of Medical Sciences, Tehran, Iran

Amir Sheikhmohammadi
PhD Student, Department of Environmental Health Engineering, Shahid Beheshti University of Medical Sciences, Tehran, Iran

Mohammad Malakootian
Professor, Environmental Health Engineering Research Center, Department of Environmental Health, Kerman University of Medical Sciences, Kerman, Iran

Behnam Hatami
MSc, Department of Environmental Health, Kerman University of Medical Sciences, Kerman, Iran

Shidwash Dowlatshahi and Ahmad Rajabizadeh
Instructure, Environmental Health Engineering Research Center, Department of Environmental Health, Kerman University of Medical Sciences, Kerman, Iran

Zabihollah Yousefi
Professor, Department of Environmental Health Engineering, School of Public Health, Mazandaran University of Medical Sciences, Sari, Iran

Reza Ali Mohammadpour Tahmtan
Associate Professor, Department of Biostatistics, School of Public Health, Mazandaran University of Medical Sciences, Sari, Iran

Farzad Kazemi
MSc Student, Department of Environmental Health Engineering, School of Public Health, Mazandaran University of Medical Sciences, Sari, Iran

Zabihollah Yousefi
Professor, Department of Environmental Health, Mazandaran University of Medical Sciences, Sari, Iran

Ramazan Motallebi
MSc student, Department of Water Engineering, Islamic Azad University, Bandar Abbas Branch, Hormozgan, Iran

Afshin Takdastan
Associate Professor, Environmental Technologies Research Center, Ahvaz Jundishapur University of Medical Sciences, Ahvaz, Iran

Dariush Naghipour
Assistant Professor of Environmental Health Engineering, Department of Environmental Health Engineering, School of Health, Guilan University of Medical Sciences, Rasht, Iran

Zeinab Shaabaninezhad
MSc of Environmental Health Engineering, Student Research Committee, School of Health, Guilan University of Medical Sciences, Rasht, Iran

Abdoliman Amouei
Associate Professor of Environmental Health Engineering, Department of Environmental Health Engineering, School of Health, Babol University of Medical Sciences, Babol, Iran

Neamat Jaafarzadeh Haghighi Fard
Professor, Environmental Technologies Research Center, Ahvaz Jundishapur University of Medical Sciences, Ahvaz, Iran

Neamat Jaafarzadeh Haghighi Fard
Professor, Department of Environmental Health Engineering, School of Public Health, Ahvaz Jundishapur University of Medical Sciences, Ahvaz, Iran

Behnam Moradi, Mokhtar Abbasi, Rahman Alivar Babadi, Hossein Bahrani and Azadeh Mirzaie
BSc, Department of Environmental Health Engineering, School of Public Health, Ahvaz Jundishapur University of Medical Sciences, Ahvaz, Iran

Behnam Moradi and Maryam Ravanbakhsh
Student Research Committee, Ahvaz Jundishapur University of Medical Sciences, Ahvaz, Iran

Ahmad Zare Javid
PhD, Nutrition and Metabolic Diseases Research Center, Ahvaz Jundishapur University of Medical Sciences, Ahvaz, Iran

Maryam Ravanbakhsh
MSc, Department of Environmental Health Engineering, School of Public Health, Ahvaz Jundishapur University of Medical Sciences, Ahvaz, Iran

Mohammad Malakootian
Professor, Environmental Health Engineering Research Center, Department of Environmental Health, School of Public Health, Kerman University of Medical Sciences, Kerman, Iran

Mohammad Seddiq Mortazavi
Associate Professor, Persian Gulf and Oman Sea Ecological Research Institute, IFSRI, AREEO, Bandar Abbas, Hormozgan, Iran

Abdolkarim Ahmadi
MSc, Department of Environmental Health, School of Public Health, Kerman University of Medical Sciences, Kerman, Iran

Monireh Majlessi
Associate Professor, Department of Environmental Health Engineering, School of Public Health, Shahid Beheshti University of Medical Science, Tehran, Iran

Aida Vaezi
MA of Environmental Science, Islamic Azad University, Science and Research Branch, Tehran, Iran

Mohsen Mehdipour Rabori
PhD Student of Environmental Health in SBMU, Member of Environmental Health Engineering Research Center, Kerman University of Medical Sciences, Kerman, Iran

Mahdi Farzadkia
Associate Professor, Department of Environmental Health Engineering, Iran University of Medical Sciences, Tehran, Iran

Mohammad Hassan Ehrampoush
Professor, Department of Environmental Health Engineering, Shahid Sadoughi University of Medical Sciences, Yazd, Iran

Shahram Sadeghi
MSc Student, Environmental Health Engineering, Student Research Committee, Kurdistan University of Medical Sciences, Sanandaj, Iran

Majid Kermani
Assistant Professor, Department of Environmental Health Engineering, Iran University of Medical Sciences, Tehran, Iran

Mohammad Taghi Ghaneian
Associate Professor, Department of Environmental Health Engineering, Shahid Sadoughi University of Medical Sciences, Yazd, Iran

Vahid Ghelmani
Bachelor of Civil Engineering, Manager of Wastewater Treatment Plant in Yazd, Iran

Ehsan Abouee Mehrizi
PhD Student, Environment Research Center, Iran University of Medical Sciences, Department of Environmental Health Engineering, Faculty of Public Health, North Khorasan University of Medical Sciences- Bojnurd, Bojnurd, Iran

Ali Ahmad Aghapour
Assistant Professor of Environmental Health Engineering, Department of Environmental Health Engineering, School of Health, Urmia University of Medical Sciences, Urmia, Iran

Sepideh Nemati, Amir Mohammadi and Sima Karimzadeh
MSc of Environmental Health Engineering, Student Research Committee, School of Health, Urmia University of Medical Sciences, Urmia, Iran

Heshmatollah Nourmoradi
Assistant Professor of Environmental Health Engineering, Department of Environmental Health Engineering, School of Health, Ilam University of Medical Sciences, Ilam, Iran

Kavoos Dindarloo and Parvin Lakbala
Assistant Professor, Environmental Health Engineering Research Center, Instructor Department of Health, Hormozgan University of Medical Sciences, Bandar Abbas, Iran

Hamza Ali Jamali
Assistant Professor, Department of Environmental Health Engineering, Qazvin University of Medical Sciences, Qazvin, Iran

Hamid Mahmoodi
MSc, Environmental Health Engineering Research Center, Instructor Department of Health, Hormozgan University of Medical Sciences, Bandar Abbas, Iran

Fatemeh Kazemi
MSc, Environmental Health Engineering Research Center, Hormozgan University of Medical Sciences, Bandar Abbas, Iran

Shidvash Dowlatshahi and Mahshid Loloei
Instructor, Environmental Health Engineering Research Center and Department of Environmental Health, School of Public Health, Kerman University of Medical Sciences, Kerman, Iran

Ahmad Reza Haratinezhad Torbati
M.Sc Student of Environmental Health Engineering, Department of Environmental Health, School of Public Health, Kerman University of Medical Sciences, Kerman, Iran

Mohammad Malakootian
Professor, Environmental Health Engineering Research Center and Department of Environmental Health, Kerman University of Medical Sciences, Kerman, Iran

Behnam Hatami
MSc, Department of Environmental Health, Kerman University of Medical Sciences, Kerman, Iran

Shidwash Dowlatshahi and Ahmad Rajabizadeh
Instructure, Environmental Health Engineering Research Center, Department of Environmental Health, Kerman University of Medical Sciences, Kerman, Iran

Naser Samadi
MSc of Cellular and Mulecular Research Center, Faculty of medicine, Urmia University of Medical Sciences, Urmia, Iran

Rahim Aali
Assistant Professor of Department of Environmental Health Engineering, Faculty of Nursing and Health, Urmia University of Medical Sciences, Urmia, Iran

Esrafil Asgari
Instructor of Department of Environmental Health Engineering, Faculty of Nursing and Health, Urmia University of Medical Sciences, Urmia, Iran

Hamed Mirhosaeini
Instructor of Medical Microobiology, Department of Environmental Health Engineering, Faculty of Health, Arak University of Medical Sciences, Arak, Iran

Ali Shahriari
Assistant Professor of Department of Environmental Health Engineering, Environmental Health Research Center, Golestan University of Medical Sciences, Gorgan, Iran

Farhad Mahmoodi
PhD student of Department of Environmental Engineering, Ahvaz Branch, Islamic Azad University, Ahvaz, Iran

Fardin Nouri, Asad Hamdi, Farzad Fanaie, Shahla Yosefi and Saied Dadashi
BSc of Environmental Health, Student Research Committee, Urmia University of Medical Sciences, Urmia, Iran

Maryam Khodadadi
Ph.D. Student of Environmental Health, International Campus, Shahid Sadoughi University of Medical Sciences, Yazd, Iran

Alireza Mesdaghinia and Simin Nasseri
Professor, Center for Water Quality Research (CWQR), Institute for Environmental Research (IER), Tehran University of Medical Sciences, Tehran, Iran

Mohammad Taghi Ghaneian
Associate professor, Department of Environmental Health, School of Public Health, Shahid Sadoughi University of Medical Sciences, Yazd, Iran

Mohammad Hassan Ehrampoush
Professor, Department of Environmental Health, School of Public Health, Shahid Sadoughi University of Medical Sciences, Yazd, Iran

Mahdi Hadi
Ph.D. student, Center for Water Quality Research (CWQR), Institute for Environmental Research (IER), Tehran University of Medical Sciences, Tehran, Iran

Seyed Hashem Hosseini and Mohammad Reza Khaleghi
Department of Range and Watershed Management, Torbat-e-Jam Branch, Islamic Azad University, Torbat-e-Jam, Iran

Hasan Jami and Sajjad Baygi
Department of Civil Engineering, Torbat-e-Jam Branch, Islamic Azad University, Torbat-e-Jam, Iran

Yusef Omidi Khaniabadi
MSc of Environmental Health, Health Care System of Karoon, Ahvaz Jundishapur University of Medical Sciences, Ahvaz, Iran

Mohammad Javad Mohammadi
PhD, Abadan School of Medical Sciences, Abadan, Iran

Mohammad Javad Mohammadi
PhD, Student Research Committee, Department of Environmental Health Engineering, School of Public Health and Environmental Technologies Research Center, Ahvaz Jundishapur University of Medical Sciences, Ahvaz, Iran

Mojtaba Shegerd
BSc of Occupational Health, Health Center of East, Ahvaz Jundishapur University of Medical Sciences, Ahvaz, Iran

Shahram Sadeghi
MSc of Environmental Health, Environmental Health Research Center, Kurdistan University of Medical Sciences, Sanandaj, Iran

Sedigheh Saeedi
MSc of Environmental Health, Sama Technical and Vocatinal College, Islamic Azad University, Izeh Branch, Izah, Iran

Hassan Basiri
MSc of Environmental Health, Rural Water and Wastewater Company, Semnan, Iran

MSc of Environmental Health, Department of Environmental Health Engineering, School of Health, Lorestan University of Medical Sciences, Lorestan, Iran

Hossein Sasani
PhD Student of Environmental Engineering, Department of Environmental Engineering, School of Environment, University of Tehran, Tehran, Iran

Naser Mehrdadi
Assistant Professor of Environmental Engineering, Department of Environmental Engineering, Graduate Faculty of Environment, University of Tehran, Iran

Behnoush Aminzadeh
Assistant Professor of Environmental Engineering, Department of Environmental Engineering, School of Environment, University of Tehran, Tehran, Iran

Afshin Takdastan
Associate Professor of Environmental Health Engineering, Department of Environmental Health Engineering, School of Public Health, Ahvaz Jundishapur University of Medical Sciences, Ahvaz, Iran

Akbar Eslami
Environmental and Occupational Hazards Control Research Center, Shahid Beheshti University of Medical Sciences, Tehran, Iran

Mohtasham Ghaffari
Department of Education and Health, School of Public Health, Shahid Beheshti University of Medical Sciences, Tehran, Iran

Behnam Barikbin
Department of Environmental Health Engineering, School of Public Health, Birjand University of Medical Sciences, Birjand, Iran

Farzaneh Fanaei
Department of Environmental Health Engineering, School of Public Health, Shahid Beheshti University of Medical Sciences, Tehran, Iran

Maryam Moradnia and Masoud Panahifard
MSc Student in Environmental Health Engineering, Health Faculty, Qazvin University of Medical Science, Qazvin, Iran

Kavoos Dindarlo
Assistant Professor of Environmental Health Engineering, Health Faculty, Hormozgan University of Medical Science, Hormozgan, Iran

Hamzeh Ali Jamali
Assistant Professor of Environmental Health Engineering, Health Faculty, Qazvin University of Medical Science, Qazvin, Iran

Zabihollah Yousefi and Mohammad Behbodi
Department of Environmental Health Engineering, School of Public Health, Mazandaran University of Medical Sciences, Sari, Iran

Reza Ali Mohammadpour
Department of Biostatistics, School of Public Health, Mazandaran University of Medical Sciences, Sari, Iran

Mohammad Malakootian and Farnaz Iranmanesh
Environmental Health Engineering Research Center, Kerman University of Medical Sciences, Kerman, Iran

Mohammad Malakootian and Marzie Javdan
Department of Environmental Health, School of Public Health, Kerman University of Medical Sciences, Kerman, Iran

Soheila Chavoshan and Aliyeh Hosseinnejad
Student Research Committee, Birjand University of Medical Sciences, Birjand, Iran

Maryam Khodadadi
Social Determinants of Health Research Center, Birjand University of Medical Science, Birjand, Iran

Negin Nasseh
Yong Research and Elites Club, Science and Research Branch Islamic Azad University, Tehran, Iran

Ayat Hossein Panahi
Office of Research and Technology, Birjand University of Medical Sciences, South Khorasan, Iran

Mohammad Hossein Tarmahi
MSc in Water and Wastewater Engineering, Department of Chemistry, College of Science, Bandar Abbas Branch, Islamic Azad University, Bandar Abbas, Iran

Farid Moeinpour
Assistant Professor of Chemistry, Department of Chemistry, College of Science, Bandar Abbas Branch, Islamic Azad University, Bandar Abbas, Iran

Farzad Kazemi
MSc, Department of Environmental Health Engineering, Student Research Committee, Faculty of Health, Mazandaran University of Medical Sciences, Sari, Iran

Zabihollah Yousefi
Professor, Department of Environmental Health Engineering, Faculty of Health, Mazandaran University of Medical Sciences, Sari, Iran

Reza Ali Mohammadpour
Associate Professor, Department of Biostatistics, Faculty of Health, Mazandaran University of Medical Sciences, Sari, Iran

Sisay Derso Mengesha, Abel Weldetinsae and Kirubel Tesfaye
Environmental Health Research Team, Ethiopian Public Health Institute, Addis Ababa, Ethiopia

Girum Taye
Health System Research Directorate, Ethiopian Public Health Institute, Addis Ababa, Ethiopia

Mohammad Rafiee
Assistant Professor of Environmental Health Engineering, Department of Environmental Health, School of Public Health, Shahid Beheshti University of Medical Sciences, Tehran, Iran

Mahsa Jahangiri Rad
Assistant Professor of Environmental Health Engineering, Department of Environmental Health Engineering, Tehran Medical Sciences Branch, Islamic Azad University, Tehran, Iran

Arash Afshari
MSc Student of Environmental Engineering, Department of Environmental Engineering, Tehran Shomal Branch, Islamic Azad University, Tehran, Iran

Ahmadreza Yari
Assistant Professor, Research Center for Environmental Pollutants, Qom University of Medical Sciences, Qom, Iran

Hossein Jafari Mansoorian
Lecturer, Environmental Health Engineering Research Center, Department of Environmental Health, Kerman University of Medical Sciences, Kerman, Iran

Gharib Majidi and Ali Sayfouri
MSc in Environmental Health Engineering, School of Public Health, Qom University of Medical Sciences, Qom, Iran

Ghazal Yazdanpanah
MSc of Microbiology, Environmental Health Engineering Research Center, Kerman University of Medical Sciences, Kerman, Iran

Soroosh Danaee, Hamideh Ofoghi, Seyed Mohammad Heydarian and Abbas Farazmand
Biotechnology Department, Iranian Research Organization for Science and Technology, Tehran, Iran

Hajar Ziaei
Associate Professor, Toxoplasmosis Research Center, Mazandaran University of Medical Sciences, Sari, Iran

Fatemeh Sayyahi
Medical Student, Research Committee, Faculty of Medicine, Mazandaran University of Medical Sciences, Sari, Iran

Mahboobeh Hoseiny
MSC Statistic, GIS Research Center, Mazandaran University of Medical Sciences, Sari, Iran

Mohammad Vahedi
MSC Microbiology, Faculty Member, Department of Microbiology, Mazandaran University of Medical Sciences, Sari, Iran

Shirzad Gholami
Associate Professor, Molecular and Cell Biology Research Center, Department of Parasitology and Mycology, Mazandaran University of Medical Sciences, Sari, Iran

Index

www.ingramcontent.com/pod-product-compliance
Lightning Source LLC
LaVergne TN
LVHW060529210926
848485LV00005B/1173